Dedicated to

Monika and Albert

and

Tina, Carla, David, and Carrie

with

$$\left\{ \begin{pmatrix} \sin(2t) \\ \sin(3t) \end{pmatrix} : 0 \le t \le \pi \right\}$$

MARTIN BOHNER
ALLAN PETERSON

DYNAMIC EQUATIONS ON TIME SCALES

AN INTRODUCTION WITH APPLICATIONS

SPRINGER SCIENCE+BUSINESS MEDIA, LLC

Martin Bohner
Department of Mathematics
Univeristy of Missouri-Rolla
Rolla, Missouri 65409-0020
U.S.A.

Allan Peterson
Department of Mathematics
University of Nebraska
Lincoln, Nebraska 68588-0323
U.S.A.

Library of Congress Cataloging-in-Publication Data

A CIP catalogue record for this book is available from the Library of Congress,
Washington D.C., USA.

AMS Subject Classifications: 39Axx, 39A10, 39A12, 39A13, 65Lxx, 65L05, 65L10, 65L15

Printed on acid-free paper.
©2001 Springer Science+Business Media New York
Originally published by Birkhäuser Boston in 2001
Softcover reprint of the hardcover 1st edition 2001

ISBN 978-1-4612-6659-4   ISBN 978-1-4612-0201-1 (eBook)   SPIN 10796629
DOI 10.1007/978-1-4612-0201-1

Typeset by the authors in LaTeX

9 8 7 6 5 4 3 2 1

Contents

Preface

On becoming familiar with difference equations and their close relation to differential equations, I was in hopes that the theory of difference equations could be brought completely abreast with that for ordinary differential equations.

[HUGH L. TURRITTIN, *My Mathematical Expectations*,
Springer Lecture Notes 312 (page 10), 1973]

A major task of mathematics today is to harmonize the continuous and the discrete, to include them in one comprehensive mathematics, and to eliminate obscurity from both.

[E. T. BELL, *Men of Mathematics*,
Simon and Schuster, New York (page 13/14), 1937]

The theory of time scales, which has recently received a lot of attention, was introduced by Stefan Hilger in his PhD thesis [159] in 1988 (supervised by Bernd Aulbach) in order to unify continuous and discrete analysis. This book is an introduction to the study of dynamic equations on time scales. Many results concerning differential equations carry over quite easily to corresponding results for difference equations, while other results seem to be completely different in nature from their continuous counterparts. The study of dynamic equations on time scales reveals such discrepancies, and helps avoid proving results twice, once for differential equations and once for difference equations. The general idea is to prove a result for a dynamic equation where the domain of the unknown function is a so-called time scale, which is an arbitrary nonempty closed subset of the reals. By choosing the time scale to be the set of real numbers, the general result yields a result concerning an ordinary differential equation as studied in a first course in differential equations. On the other hand, by choosing the time scale to be the set of integers, the same general result yields a result for difference equations. However, since there are many other time scales than just the set of real numbers or the set of integers, one has a much more general result. We may summarize the above and state that

Unification and Extension

are the two *main features* of the time scales calculus.

The time scales calculus has a tremendous potential for *applications*. For example, it can model insect populations that are continuous while in season (and may follow a difference scheme with variable step-size), die out in (say) winter, while their eggs are incubating or dormant, and then hatch in a new season, giving rise to a nonoverlapping population.

The *audience* for this book is as follows:

1. Most parts of this book are appropriate for students who have had a first course in calculus and linear algebra. Usually, a first course in differential equations does not even consider the discrete case, which students encounter in numerous applications, for example, in biology, engineering, economics, physics, neural networks, social sciences, and so on. A course taught out of this book would simultaneously teach the continuous and the discrete theory, which would better prepare students for these applications. The first four chapters can be used for an introductory course on time scales. They contain plenty of exercises, and we included many of the solutions at the end of the book. Altogether, there are 212 exercises, many of them consisting of several separate parts.

2. The last four chapters can be used for an advanced course on time scales at the beginning graduate level. Also, a special topics course is appropriate. These chapters also contain many exercises, however, most of their answers are not included in the solutions section at the end of the book.

3. A third audience might be graduate students who are interested in the subject for a thesis project at the masters or doctoral level. Some of the exercises describe open problems that can be used as a starting point for such a project. The "Notes and References" sections at the end of each chapter also point out directions of further possible research.

4. Finally, researchers with a knowledge of differential or difference equations, who want a rather complete introduction into the time scales calculus without going through all the current literature, may also find this book very useful.

Most of the results given in this book have recently been investigated by Stefan Hilger and by the authors of this book (together with their research collaborators R. P. Agarwal, C. Ahlbrandt, E. Akın, S. Clark, O. Došlý, P. Eloe, L. Erbe, B. Kaymakçalan, D. Lutz, R. Mathsen, and J. Ridenhour). Other results presented or results related to the presented ones have been obtained by D. Anderson, F. Atıcı, B. Aulbach, J. Davis, G. Guseinov, J. Henderson, R. Hilscher, S. Keller, V. Lakshmikantham, C. Pötzsche, Z. Pospíšil, S. Siegmund, and S. Sivasundaram. Many of these results are presented here at a level that will be easy for an undergraduate mathematics student to understand.

In Chapter 1 *the time scales calculus* as developed in [160] by Stefan Hilger is introduced. A time scale $\mathbb{T}$ is an arbitrary nonempty closed subset of the reals. For functions $f : \mathbb{T} \to \mathbb{R}$ we introduce a derivative and an integral. Fundamental results, e.g., the product rule and the quotient rule, are presented. Further results concerning differentiability and integrability, which have been previously unpublished but are easy to derive, are given as they are needed in the remaining parts of the book. Important examples of time scales, which we will consider frequently throughout the book, are given in this chapter. Such examples contain of course $\mathbb{R}$ (the set of all real numbers, which gives rise to differential equations) and $\mathbb{Z}$ (the set of all integers, which gives rise to difference equations), but also the set of all integer multiples of a number $h > 0$ and the set of all integer powers of a number $q > 1$, including 0 (this time scale gives rise to so-called q-difference equations, see, e.g., [58, 247, 253]). Other examples are sets of disjoint closed intervals (which have applications, e.g., in population dynamics) or even "exotic" time scales such as the Cantor set. After discussing these examples, we also derive analogues of the chain rule. Taylor's formula is presented, which is helpful in the study of boundary value problems.

In Chapter 2 we introduce the Hilger complex plane, following closely Hilger's paper [**164**]. We use the so-called cylinder transformation to introduce the exponential function on time scales. This exponential function is then shown to satisfy an initial value problem involving a first order linear dynamic equation. We derive many properties of the exponential function and use it to solve all initial value problems involving *first order linear dynamic equations*. For the nonhomogeneous cases we utilize a variation of constants technique.

Next, we consider *second order linear dynamic equations* in Chapter 3. Again, there are several kinds of second order linear homogeneous equations, which we solve (in the constant coefficient case) using hyperbolic and trigonometric functions. Wronskian determinants are introduced and Abel's theorem is used to develop a reduction of order technique to find a second solution in case one solution is already known. Certain dynamic equations of second order with nonconstant coefficients (e.g., the Euler–Cauchy equation) are also considered. We also present a variation of constants formula that helps in solving nonhomogeneous second order linear dynamic equations. The Laplace transformation on a general time scale is introduced and many of its properties are derived.

In Chapter 4, we study *self-adjoint dynamic equations* on time scales. Such equations have been well studied in the continuous case (where they are also called Sturm–Liouville equations) and in the discrete case (where they are called Sturm–Liouville difference equations). In this chapter we only consider such equations of second order. We investigate disconjugacy of self-adjoint equations and use corresponding Green's functions to study boundary value problems. Also the theory of Riccati equations is developed in the general setting of time scales, and we present a characterization of disconjugacy in terms of a certain quadratic functional. An analogue of the classical Prüfer transformation, which has proved to be a useful tool in oscillation theory of Sturm–Liouville equations, is given as well. In the last section, we examine eigenvalue problems on time scales. For the case of separated boundary conditions we present an oscillation result on the number of zeros of the kth eigenfunction. Such a result goes back to Sturm in the continuous case, and its discrete counterpart is contained in the book by Kelley and Peterson [**191**, Theorem 7.6]. Further results on eigenvalue problems contain a comparison theorem and Rayleigh's principle.

Chapter 5 is concerned with *linear systems of dynamic equations* on a time scale. Uniqueness and existence theorems are presented, and the matrix exponential on a time scale is introduced. We also examine fundamental systems and their Wronskian determinants and give a variation of constants formula. The case of constant coefficient matrices is also investigated, and a Putzer algorithm from [**31**] is presented. This chapter contains a section on self-adjoint vector equations. Such equations are a special case of symplectic systems as discussed in Chapter 7. They are closely connected to certain matrix Riccati equations, and in this section we also discuss oscillation results for those systems. Further topics contain a discussion on the asymptotic behavior of solutions of linear systems of dynamic equations. Related results are time scales versions of Levinson's perturbation lemma and the Hartman–Wintner theorem. Finally we study *higher order linear dynamic equations* on a time scale. We give conditions that imply that corresponding initial value problems have unique solutions. Abel's formula is given, and the notion of a generalized zero of a solution is introduced.

Chapter 6 is concerned with *dynamic inequalities* on time scales. Analogues of Gronwall's inequality, Hölder's inequality, and Jensen's inequality are presented. We also derive Opial's inequality and point out its applications in the study of initial or boundary value problems. Opial inequalities have proved to be a useful tool in differential equations and in difference equations, and in fact there is an entire book [19] devoted to them. Next, we prove Lyapunov's inequality for Sturm–Liouville equations of second order with positive coefficients. It can be used to give sufficient conditions for disconjugacy. We also offer an extension of Lyapunov's inequality to the case of linear Hamiltonian dynamic systems. Further results in this section concern upper and lower solutions of boundary value problems and are contained in the article [32] by Akın.

In Chapter 7 we consider *linear symplectic dynamic systems* on time scales. This is a very general class of systems that contains, for example, linear Hamiltonian dynamic systems which in turn contain Sturm–Liouville dynamic equations of higher order (and hence of course also of order two) and self-adjoint vector dynamic equations. We derive a Wronskian identity for such systems as well as a close connection to certain matrix Riccati dynamic equations. Disconjugacy of symplectic systems is introduced as well. Some of the results in this chapter are due to Roman Hilscher, who considered Hamiltonian systems on time scales in [166, 167, 168]. Other results are contained in a paper by Ondřej Došlý and Roman Hilscher [121].

Chapter 8 contains several possible *extensions* of the time scales calculus. In the first section we present an introduction to the concept of measure chains as introduced by Hilger in [159]. The second section contains the proofs of the main local and global existence theorems, that are needed throughout the book. Another extension, which is considered in this last chapter, concerns alpha derivatives. The time scales calculus as presented in this book is a special case of this concept.

Parts or earlier versions of this book have been proofread by D. Anderson, R. Avery, R. Chiquet, C. J. Chyan, L. Cole, J. Davis, L. Erbe, K. Fick, J. Henderson, J. Hoffacker, K. Howard, N. Hummel, B. Karna, R. Mathsen, K. Messer, A. Rosidian, P. Singh, and W. Yin. In particular, we would like to thank Elvan Akın, Roman Hilscher, Billûr Kaymakçalan, and Stefan Siegmund for proofreading the entire manuscript. Finally, we wish to express our appreciation to "Birkhäuser Boston", in particular to Ann Kostant, for accomplished handling of this manuscript.

Martin Bohner and Allan Peterson
May 2001

CHAPTER 1

The Time Scales Calculus

1.1. Basic Definitions

A *time scale* (which is a special case of a *measure chain*, see Chapter 8) is an arbitrary nonempty closed subset of the real numbers. Thus

$$\mathbb{R}, \quad \mathbb{Z}, \quad \mathbb{N}, \quad \mathbb{N}_0,$$

i.e., the real numbers, the integers, the natural numbers, and the nonnegative integers are examples of time scales, as are

$$[0,1] \cup [2,3], \quad [0,1] \cup \mathbb{N}, \quad \text{and the Cantor set,}$$

while

$$\mathbb{Q}, \quad \mathbb{R} \setminus \mathbb{Q}, \quad \mathbb{C}, \quad (0,1),$$

i.e., the rational numbers, the irrational numbers, the complex numbers, and the open interval between 0 and 1, are *not* time scales. Throughout this book we will denote a time scale by the symbol $\mathbb{T}$. We assume throughout that a time scale $\mathbb{T}$ has the topology that it inherits from the real numbers with the standard topology.

The calculus of time scales was initiated by Stefan Hilger in [**159**] in order to create a theory that can unify discrete and continuous analysis. Indeed, we will introduce the delta derivative f^Δ for a function f defined on $\mathbb{T}$, and it turns out that

(i) $f^\Delta = f'$ is the usual derivative if $\mathbb{T} = \mathbb{R}$ and
(ii) $f^\Delta = \Delta f$ is the usual forward difference operator if $\mathbb{T} = \mathbb{Z}$.

In this section we introduce the basic notions connected to time scales and differentiability of functions on them, and we offer the above two cases as examples. However, the general theory is of course applicable to many more time scales $\mathbb{T}$, and we will give some examples of such time scales in Section 1.3 and many more examples throughout the rest of this book. Let us start by defining the forward and backward jump operators.

Definition 1.1. Let $\mathbb{T}$ be a time scale. For $t \in \mathbb{T}$ we define the *forward jump operator* $\sigma : \mathbb{T} \to \mathbb{T}$ by

$$\sigma(t) := \inf\{s \in \mathbb{T} : s > t\},$$

while the *backward jump operator* $\rho : \mathbb{T} \to \mathbb{T}$ is defined by

$$\rho(t) := \sup\{s \in \mathbb{T} : s < t\}.$$

In this definition we put $\inf \emptyset = \sup \mathbb{T}$ (i.e., $\sigma(t) = t$ if $\mathbb{T}$ has a maximum t) and $\sup \emptyset = \inf \mathbb{T}$ (i.e., $\rho(t) = t$ if $\mathbb{T}$ has a minimum t), where $\emptyset$ denotes the empty set. If $\sigma(t) > t$, we say that t is *right-scattered*, while if $\rho(t) < t$ we say that t is *left-scattered*. Points that are right-scattered and left-scattered at the same time

Table 1.1. Classification of Points

t right-scattered	$t < \sigma(t)$
t right-dense	$t = \sigma(t)$
t left-scattered	$\rho(t) < t$
t left-dense	$\rho(t) = t$
t isolated	$\rho(t) < t < \sigma(t)$
t dense	$\rho(t) = t = \sigma(t)$

Figure 1.1. Classifications of Points

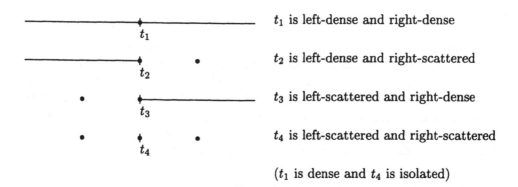

t_1 is left-dense and right-dense

t_2 is left-dense and right-scattered

t_3 is left-scattered and right-dense

t_4 is left-scattered and right-scattered

(t_1 is dense and t_4 is isolated)

are called *isolated*. Also, if $t < \sup \mathbb{T}$ and $\sigma(t) = t$, then t is called *right-dense*, and if $t > \inf \mathbb{T}$ and $\rho(t) = t$, then t is called *left-dense*. Points that are right-dense and left-dense at the same time are called *dense*. Finally, the *graininess function* $\mu : \mathbb{T} \to [0, \infty)$ is defined by

$$\mu(t) := \sigma(t) - t.$$

See Table 1.1 for a classification and Figure 1.1 for a schematic classification of points in $\mathbb{T}$. Note that in the definition above both $\sigma(t)$ and $\rho(t)$ are in $\mathbb{T}$ when $t \in \mathbb{T}$. This is because of our assumption that $\mathbb{T}$ is a closed subset of $\mathbb{R}$. We also need below the set $\mathbb{T}^\kappa$ which is derived from the time scale $\mathbb{T}$ as follows: If $\mathbb{T}$ has a left-scattered maximum m, then $\mathbb{T}^\kappa = \mathbb{T} - \{m\}$. Otherwise, $\mathbb{T}^\kappa = \mathbb{T}$. In summary,

$$\mathbb{T}^\kappa = \begin{cases} \mathbb{T} \setminus (\rho(\sup \mathbb{T}), \sup \mathbb{T}] & \text{if} \quad \sup \mathbb{T} < \infty \\ \mathbb{T} & \text{if} \quad \sup \mathbb{T} = \infty. \end{cases}$$

Finally, if $f : \mathbb{T} \to \mathbb{R}$ is a function, then we define the function $f^\sigma : \mathbb{T} \to \mathbb{R}$ by

$$f^\sigma(t) = f(\sigma(t)) \quad \text{for all} \quad t \in \mathbb{T},$$

i.e., $f^\sigma = f \circ \sigma$.

Example 1.2. Let us briefly consider the two examples $\mathbb{T} = \mathbb{R}$ and $\mathbb{T} = \mathbb{Z}$.

(i) If $\mathbb{T} = \mathbb{R}$, then we have for any $t \in \mathbb{R}$

$$\sigma(t) = \inf\{s \in \mathbb{R} : s > t\} = \inf(t, \infty) = t$$

and similarly $\rho(t) = t$. Hence every point $t \in \mathbb{R}$ is dense. The graininess function μ turns out to be

$$\mu(t) \equiv 0 \quad \text{for all} \quad t \in \mathbb{T}.$$

(ii) If $\mathbb{T} = \mathbb{Z}$, then we have for any $t \in \mathbb{Z}$

$$\sigma(t) = \inf\{s \in \mathbb{Z} : s > t\} = \inf\{t+1, t+2, t+3, \dots\} = t+1$$

and similarly $\rho(t) = t-1$. Hence every point $t \in \mathbb{Z}$ is isolated. The graininess function μ in this case is

$$\mu(t) \equiv 1 \quad \text{for all} \quad t \in \mathbb{T}.$$

For the two cases discussed above, the graininess function is a constant function. We will see below that the graininess function plays a central role in the analysis on time scales. For the general case, many formulae will have some term containing the factor $\mu(t)$. This term is there in case $\mathbb{T} = \mathbb{Z}$ since $\mu(t) \equiv 1$. However, for the case $\mathbb{T} = \mathbb{R}$ this term disappears since $\mu(t) \equiv 0$ in this case. In various cases this fact is the reason for certain differences between the continuous and the discrete case. One of the many examples of this that we will see later is the so-called scalar Riccati equation on a general time scale $\mathbb{T}$ (see formula (4.18))

$$z^{\Delta} + q(t) + \frac{z^2}{p(t) + \mu(t)z} = 0.$$

Note that if $\mathbb{T} = \mathbb{R}$, then we get the well-known Riccati differential equation

$$z' + q(t) + \frac{1}{p(t)}z^2 = 0,$$

and if $\mathbb{T} = \mathbb{Z}$, then we get the Riccati difference equation (see [**191**, Chapter 6])

$$\Delta z + q(t) + \frac{z^2}{p(t) + z} = 0.$$

Of course, for the case of a general time scale, the graininess function might very well be a function of $t \in \mathbb{T}$, as the reader can verify in the next exercise. For more such examples we refer to Section 1.3.

Exercise 1.3. For each of the following time scales $\mathbb{T}$, find σ, ρ, and μ, and classify each point $t \in \mathbb{T}$ as left-dense, left-scattered, right-dense, or right-scattered:

(i) $\mathbb{T} = \{2^n : n \in \mathbb{Z}\} \cup \{0\}$;
(ii) $\mathbb{T} = \{\frac{1}{n} : n \in \mathbb{N}\} \cup \{0\}$;
(iii) $\mathbb{T} = \{\frac{n}{2} : n \in \mathbb{N}_0\}$;
(iv) $\mathbb{T} = \{\sqrt{n} : n \in \mathbb{N}_0\}$;
(v) $\mathbb{T} = \{\sqrt[3]{n} : n \in \mathbb{N}_0\}$.

Exercise 1.4. Give examples of time scales $\mathbb{T}$ and points $t \in \mathbb{T}$ such that the following equations are not true. Also determine the conditions on t under which those equations are true:

(i) $\sigma(\rho(t)) = t$;
(ii) $\rho(\sigma(t)) = t$.

Exercise 1.5. Is $\sigma : \mathbb{T} \to \mathbb{T}$ one-to-one? Is it onto? If it is not onto, determine the range $\sigma(\mathbb{T})$ of σ. How about $\rho : \mathbb{T} \to \mathbb{T}$? This exercise was suggested by Roman Hilscher.

Exercise 1.6. If $\mathbb{T}$ consists of finitely many points, calculate $\sum_{t \in \mathbb{T}} \mu(t)$.

Throughout this book we make the blanket assumption that a and b are points in $\mathbb{T}$. Often we assume $a \leq b$. We then define the interval $[a, b]$ in $\mathbb{T}$ by

$$[a, b] := \{t \in \mathbb{T} : \ a \leq t \leq b\} .$$

Open intervals and half-open intervals etc. are defined accordingly. Note that $[a, b]^\kappa = [a, b]$ if b is left-dense and $[a, b]^\kappa = [a, b) = [a, \rho(b)]$ if b is left-scattered.

Sometimes the following *induction principle* is a useful tool.

Theorem 1.7 (Induction Principle). *Let $t_0 \in \mathbb{T}$ and assume that*

$$\{S(t) : \ t \in [t_0, \infty)\}$$

is a family of statements satisfying:

I. The statement $S(t_0)$ is true.
II. If $t \in [t_0, \infty)$ is right-scattered and $S(t)$ is true, then $S(\sigma(t))$ is also true.
III. If $t \in [t_0, \infty)$ is right-dense and $S(t)$ is true, then there is a neighborhood U of t such that $S(s)$ is true for all $s \in U \cap (t, \infty)$.
IV. If $t \in (t_0, \infty)$ is left-dense and $S(s)$ is true for all $s \in [t_0, t)$, then $S(t)$ is true.

Then $S(t)$ is true for all $t \in [t_0, \infty)$.

Proof. Let

$$S^* := \{t \in [t_0, \infty) : \ S(t) \text{ is not true}\}.$$

We want to show $S^* = \emptyset$. To achieve a contradiction we assume $S^* \neq \emptyset$. But since S^* is closed and nonempty, we have

$$\inf S^* =: t^* \in \mathbb{T}.$$

We claim that $S(t^*)$ is true. If $t^* = t_0$, then $S(t^*)$ is true from (i). If $t^* \neq t_0$ and $\rho(t^*) = t^*$, then $S(t^*)$ is true from (iv). Finally if $\rho(t^*) < t^*$, then $S(t^*)$ is true from (ii). Hence, in any case,

$$t^* \notin S^*.$$

Thus, t^* cannot be right-scattered, and $t^* \neq \max \mathbb{T}$ either. Hence t^* is right-dense. But now (iii) leads to a contradiction. $\square$

Remark 1.8. A dual version of the induction principle also holds for a family of statements $S(t)$ for t in an interval of the form $(-\infty, t_0]$: To show that $S(t)$ is true for all $t \in (-\infty, t_0]$ we have to show that $S(t_0)$ is true, that $S(t)$ is true at a left-scattered t implies $S(\rho(t))$ is true, that $S(t)$ is true at a left-dense t implies $S(r)$ is true for all r in a left neighborhood of t, and that $S(r)$ is true for all $r \in (t, t_0]$ where t is right-dense implies $S(t)$ is true.

Exercise 1.9. Prove Remark 1.8.

1.2. Differentiation

Now we consider a function $f : \mathbb{T} \to \mathbb{R}$ and define the so-called *delta* (or *Hilger*) *derivative* of f at a point $t \in \mathbb{T}^\kappa$.

Definition 1.10. Assume $f : \mathbb{T} \to \mathbb{R}$ is a function and let $t \in \mathbb{T}^\kappa$. Then we define $f^\Delta(t)$ to be the number (provided it exists) with the property that given any $\varepsilon > 0$, there is a neighborhood U of t (i.e., $U = (t - \delta, t + \delta) \cap \mathbb{T}$ for some $\delta > 0$) such that

$$\left| [f(\sigma(t)) - f(s)] - f^\Delta(t)[\sigma(t) - s] \right| \le \varepsilon \left| \sigma(t) - s \right| \quad \text{for all} \quad s \in U.$$

We call $f^\Delta(t)$ the *delta* (or *Hilger*) *derivative* of f at t.

Moreover, we say that f is *delta* (or *Hilger*) *differentiable* (or in short: *differentiable*) on $\mathbb{T}^\kappa$ provided $f^\Delta(t)$ exists for all $t \in \mathbb{T}^\kappa$. The function $f^\Delta : \mathbb{T}^\kappa \to \mathbb{R}$ is then called the (delta) derivative of f on $\mathbb{T}^\kappa$.

Exercise 1.11. Prove that the delta derivative is well defined.

Exercise 1.12. Sometimes it is convenient to have $f^\Delta(t)$ also defined at a point $t \in \mathbb{T} \setminus \mathbb{T}^\kappa$. At such a point we use the same definition as given in Definition 1.10. Prove that an $f : \mathbb{T} \to \mathbb{R}$ has any $\alpha \in \mathbb{R}$ as its derivative at points $t \in \mathbb{T} \setminus \mathbb{T}^\kappa$.

Example 1.13.   (i) If $f : \mathbb{T} \to \mathbb{R}$ is defined by $f(t) = \alpha$ for all $t \in \mathbb{T}$, where $\alpha \in \mathbb{R}$ is constant, then $f^\Delta(t) \equiv 0$. This is clear because for any $\varepsilon > 0$,

$$|f(\sigma(t)) - f(s) - 0 \cdot [\sigma(t) - s]| = |\alpha - \alpha| = 0 \le \varepsilon |\sigma(t) - s|$$

holds for all $s \in \mathbb{T}$.

(ii) If $f : \mathbb{T} \to \mathbb{R}$ is defined by $f(t) = t$ for all $t \in \mathbb{T}$, then $f^\Delta(t) \equiv 1$. This follows since for any $\varepsilon > 0$,

$$|f(\sigma(t)) - f(s) - 1 \cdot [\sigma(t) - s]| = |\sigma(t) - s - (\sigma(t) - s)| = 0 \le \varepsilon |\sigma(t) - s|$$

holds for all $s \in \mathbb{T}$.

Exercise 1.14.   (i) Define $f : \mathbb{T} \to \mathbb{R}$ by $f(t) = t^2$ for all $t \in \mathbb{T}$. Find f^Δ.
(ii) Define g by $g(t) = \sqrt{t}$ for all $t \in \mathbb{T}$ with $t > 0$. Find g^Δ.

Exercise 1.15. Using Definition 1.10 show that if $t \in \mathbb{T}^\kappa$ ($t \neq \min \mathbb{T}$) satisfies $\rho(t) = t < \sigma(t)$, then the jump operator σ is not delta differentiable at t.

Some easy and useful relationships concerning the delta derivative are given next.

Theorem 1.16. *Assume $f : \mathbb{T} \to \mathbb{R}$ is a function and let $t \in \mathbb{T}^\kappa$. Then we have the following:*

(i) *If f is differentiable at t, then f is continuous at t.*
(ii) *If f is continuous at t and t is right-scattered, then f is differentiable at t with*

$$f^\Delta(t) = \frac{f(\sigma(t)) - f(t)}{\mu(t)}.$$

(iii) *If t is right-dense, then f is differentiable at t iff the limit*

$$\lim_{s \to t} \frac{f(t) - f(s)}{t - s}$$

exists as a finite number. In this case

$$f^\Delta(t) = \lim_{s \to t} \frac{f(t) - f(s)}{t - s}.$$

(iv) *If f is differentiable at t, then*

$$f(\sigma(t)) = f(t) + \mu(t)f^\Delta(t).$$

Proof. Part (i). Assume that f is differentiable at t. Let $\varepsilon \in (0,1)$. Define

$$\varepsilon^* = \varepsilon[1 + |f^\Delta(t)| + 2\mu(t)]^{-1}.$$

Then $\varepsilon^* \in (0,1)$. By Definition 1.10 there exists a neighborhood U of t such that

$$|f(\sigma(t)) - f(s) - [\sigma(t) - s]f^\Delta(t)| \le \varepsilon^*|\sigma(t) - s| \quad \text{for all} \quad s \in U.$$

Therefore we have for all $s \in U \cap (t - \varepsilon^*, t + \varepsilon^*)$

$$
\begin{aligned}
|f(t) - f(s)| &= \left|\{f(\sigma(t)) - f(s) - f^\Delta(t)[\sigma(t) - s]\}\right. \\
&\quad \left. -\{f(\sigma(t)) - f(t) - \mu(t)f^\Delta(t)\} + (t - s)f^\Delta(t)\right| \\
&\le \varepsilon^*|\sigma(t) - s| + \varepsilon^*\mu(t) + |t - s||f^\Delta(t)| \\
&\le \varepsilon^*[\mu(t) + |t - s| + \mu(t) + |f^\Delta(t)|] \\
&< \varepsilon^*[1 + |f^\Delta(t)| + 2\mu(t)] \\
&= \varepsilon.
\end{aligned}
$$

It follows that f is continuous at t.

Part (ii). Assume f is continuous at t and t is right-scattered. By continuity

$$\lim_{s \to t} \frac{f(\sigma(t)) - f(s)}{\sigma(t) - s} = \frac{f(\sigma(t)) - f(t)}{\sigma(t) - t} = \frac{f(\sigma(t)) - f(t)}{\mu(t)}.$$

Hence, given $\varepsilon > 0$, there is a neighborhood U of t such that

$$\left|\frac{f(\sigma(t)) - f(s)}{\sigma(t) - s} - \frac{f(\sigma(t)) - f(t)}{\mu(t)}\right| \le \varepsilon$$

for all $s \in U$. It follows that

$$\left|[f(\sigma(t)) - f(s)] - \frac{f(\sigma(t)) - f(t)}{\mu(t)}[\sigma(t) - s]\right| \le \varepsilon|\sigma(t) - s|$$

for all $s \in U$. Hence we get the desired result

$$f^\Delta(t) = \frac{f(\sigma(t)) - f(t)}{\mu(t)}.$$

Part (iii). Assume f is differentiable at t and t is right-dense. Let $\varepsilon > 0$ be given. Since f is differentiable at t, there is a neighborhood U of t such that

$$\left|[f(\sigma(t)) - f(s)] - f^\Delta(t)[\sigma(t) - s]\right| \le \varepsilon|\sigma(t) - s|$$

for all $s \in U$. Since $\sigma(t) = t$ we have that

$$\left|[f(t) - f(s)] - f^\Delta(t)(t - s)\right| \le \varepsilon|t - s|$$

for all $s \in U$. It follows that

$$\left|\frac{f(t) - f(s)}{t - s} - f^\Delta(t)\right| \le \varepsilon$$

for all $s \in U$, $s \neq t$. Therefore we get the desired result

$$f^{\Delta}(t) = \lim_{s \to t} \frac{f(t) - f(s)}{t - s}.$$

The remaining part of the proof of part (iii) is Exercise 1.17.

Part (iv). If $\sigma(t) = t$, then $\mu(t) = 0$ and we have that

$$f(\sigma(t)) = f(t) = f(t) + \mu(t)f^{\Delta}(t).$$

On the other hand if $\sigma(t) > t$, then by (ii)

$$\begin{aligned} f(\sigma(t)) &= f(t) + \mu(t) \cdot \frac{f(\sigma(t)) - f(t)}{\mu(t)} \\ &= f(t) + \mu(t)f^{\Delta}(t), \end{aligned}$$

and the proof of part (iv) is complete. □

Exercise 1.17. Prove the converse part of the statement in Theorem 1.16 (iii): Let $t \in \mathbb{T}^{\kappa}$ be right-dense. If

$$\lim_{s \to t} \frac{f(t) - f(s)}{t - s}$$

exists as a finite number, then f is differentiable at t and

$$f^{\Delta}(t) = \lim_{s \to t} \frac{f(t) - f(s)}{t - s}.$$

Example 1.18. Again we consider the two cases $\mathbb{T} = \mathbb{R}$ and $\mathbb{T} = \mathbb{Z}$.

(i) If $\mathbb{T} = \mathbb{R}$, then Theorem 1.16 (iii) yields that $f : \mathbb{R} \to \mathbb{R}$ is delta differentiable at $t \in \mathbb{R}$ iff

$$f'(t) = \lim_{s \to t} \frac{f(t) - f(s)}{t - s} \quad \text{exists},$$

i.e., iff f is differentiable (in the ordinary sense) at t. In this case we then have

$$f^{\Delta}(t) = \lim_{s \to t} \frac{f(t) - f(s)}{t - s} = f'(t)$$

by Theorem 1.16 (iii).

(ii) If $\mathbb{T} = \mathbb{Z}$, then Theorem 1.16 (ii) yields that $f : \mathbb{Z} \to \mathbb{R}$ is delta differentiable at $t \in \mathbb{Z}$ with

$$f^{\Delta}(t) = \frac{f(\sigma(t)) - f(t)}{\mu(t)} = \frac{f(t+1) - f(t)}{1} = f(t+1) - f(t) = \Delta f(t),$$

where Δ is the usual *forward difference operator* defined by the last equation above.

Exercise 1.19. For each of the following functions $f : \mathbb{T} \to \mathbb{R}$, use Theorem 1.16 to find f^{Δ}. Write your final answer in terms of $t \in \mathbb{T}$:

(i) $f(t) = \sigma(t)$ for $t \in \mathbb{T} := \{\frac{1}{n} : n \in \mathbb{N}\} \cup \{0\}$;

(ii) $f(t) = t^2$ for $t \in \mathbb{T} := \mathbb{N}_0^{\frac{1}{2}} := \{\sqrt{n} : n \in \mathbb{N}_0\}$;

(iii) $f(t) = t^2$ for $t \in \mathbb{T} := \{\frac{n}{2} : n \in \mathbb{N}_0\}$;

(iv) $f(t) = t^3$ for $t \in \mathbb{T} := \mathbb{N}_0^{\frac{1}{3}} := \{\sqrt[3]{n} : n \in \mathbb{N}_0\}$.

Next, we would like to be able to find the derivatives of sums, products, and quotients of differentiable functions. This is possible according to the following theorem.

Theorem 1.20. *Assume* $f, g : \mathbb{T} \to \mathbb{R}$ *are differentiable at* $t \in \mathbb{T}^\kappa$. *Then:*

(i) *The sum* $f + g : \mathbb{T} \to \mathbb{R}$ *is differentiable at* t *with*

$$(f + g)^\Delta(t) = f^\Delta(t) + g^\Delta(t).$$

(ii) *For any constant* α, $\alpha f : \mathbb{T} \to \mathbb{R}$ *is differentiable at* t *with*

$$(\alpha f)^\Delta(t) = \alpha f^\Delta(t).$$

(iii) *The product* $fg : \mathbb{T} \to \mathbb{R}$ *is differentiable at* t *with*

$$(fg)^\Delta(t) = f^\Delta(t)g(t) + f(\sigma(t))g^\Delta(t) = f(t)g^\Delta(t) + f^\Delta(t)g(\sigma(t)).$$

(iv) *If* $f(t)f(\sigma(t)) \neq 0$, *then* $\frac{1}{f}$ *is differentiable at* t *with*

$$\left(\frac{1}{f}\right)^\Delta(t) = -\frac{f^\Delta(t)}{f(t)f(\sigma(t))}.$$

(v) *If* $g(t)g(\sigma(t)) \neq 0$, *then* $\frac{f}{g}$ *is differentiable at* t *and*

$$\left(\frac{f}{g}\right)^\Delta(t) = \frac{f^\Delta(t)g(t) - f(t)g^\Delta(t)}{g(t)g(\sigma(t))}.$$

Proof. Assume that f and g are delta differentiable at $t \in \mathbb{T}^\kappa$.

Part (i). Let $\varepsilon > 0$. Then there exist neighborhoods U_1 and U_2 of t with

$$|f(\sigma(t)) - f(s) - f^\Delta(t)(\sigma(t) - s)| \leq \frac{\varepsilon}{2}|\sigma(t) - s| \quad \text{for all} \quad s \in U_1$$

and

$$|g(\sigma(t)) - g(s) - g^\Delta(t)(\sigma(t) - s)| \leq \frac{\varepsilon}{2}|\sigma(t) - s| \quad \text{for all} \quad s \in U_2.$$

Let $U = U_1 \cap U_2$. Then we have for all $s \in U$

$$
\begin{aligned}
&|(f + g)(\sigma(t)) - (f + g)(s) - [f^\Delta(t) + g^\Delta(t)](\sigma(t) - s)| \\
&= \quad |f(\sigma(t)) - f(s) - f^\Delta(t)(\sigma(t) - s) + g(\sigma(t)) - g(s) - g^\Delta(t)(\sigma(t) - s)| \\
&\leq \quad |f(\sigma(t)) - f(s) - f^\Delta(t)(\sigma(t) - s)| + |g(\sigma(t)) - g(s) - g^\Delta(t)(\sigma(t) - s)| \\
&\leq \quad \frac{\varepsilon}{2}|\sigma(t) - s| + \frac{\varepsilon}{2}|\sigma(t) - s| \\
&= \quad \varepsilon|\sigma(t) - s|.
\end{aligned}
$$

Therefore $f + g$ is differentiable at t and $(f + g)^\Delta = f^\Delta + g^\Delta$ holds at t.

Part (iii). Let $\varepsilon \in (0, 1)$. Define $\varepsilon^* = \varepsilon[1 + |f(t)| + |g(\sigma(t))| + |g^\Delta(t)|]^{-1}$. Then $\varepsilon^* \in (0, 1)$ and hence there exist neighborhoods U_1, U_2, and U_3 of t such that

$$|f(\sigma(t)) - f(s) - f^\Delta(t)(\sigma(t) - s)| \leq \varepsilon^*|\sigma(t) - s| \quad \text{for all} \quad s \in U_1,$$

$$|g(\sigma(t)) - g(s) - g^\Delta(t)(\sigma(t) - s)| \leq \varepsilon^*|\sigma(t) - s| \quad \text{for all} \quad s \in U_2,$$

and (from Theorem 1.16 (i))

$$|f(t) - f(s)| \leq \varepsilon^* \quad \text{for all} \quad s \in U_3.$$

Put $U = U_1 \cap U_2 \cap U_3$ and let $s \in U$. Then

$$
\begin{aligned}
&|(fg)(\sigma(t)) - (fg)(s) - [f^\Delta(t)g(\sigma(t)) + f(t)g^\Delta(t)](\sigma(t) - s)| \\
&= |[f(\sigma(t)) - f(s) - f^\Delta(t)(\sigma(t) - s)]g(\sigma(t)) \\
&\quad + [g(\sigma(t)) - g(s) - g^\Delta(t)(\sigma(t) - s)]f(t) \\
&\quad + [g(\sigma(t)) - g(s) - g^\Delta(t)(\sigma(t) - s)][f(s) - f(t)] \\
&\quad + (\sigma(t) - s)g^\Delta(t)[f(s) - f(t)]| \\
&\leq \varepsilon^*|\sigma(t) - s||g(\sigma(t))| + \varepsilon^*|\sigma(t) - s||f(t)| \\
&\quad + \varepsilon^*\varepsilon^*|\sigma(t) - s| + \varepsilon^*|\sigma(t) - s||g^\Delta(t)| \\
&= \varepsilon^*|\sigma(t) - s|[|g(\sigma(t))| + |f(t)| + \varepsilon^* + |g^\Delta(t)|] \\
&\leq \varepsilon^*|\sigma(t) - s|[1 + |f(t)| + |g(\sigma(t))| + |g^\Delta(t)|] \\
&= \varepsilon|\sigma(t) - s|.
\end{aligned}
$$

Thus $(fg)^\Delta = f^\Delta g^\sigma + fg^\Delta$ holds at t. The other product rule in part (iii) of this theorem follows from this last equation by interchanging the functions f and g.

For the quotient formula (v), we use (ii) and (iv) to calculate

$$
\begin{aligned}
\left(\frac{f}{g}\right)^\Delta(t) &= \left(f \cdot \frac{1}{g}\right)^\Delta(t) \\
&= f(t)\left(\frac{1}{g}\right)^\Delta(t) + f^\Delta(t)\frac{1}{g(\sigma(t))} \\
&= -f(t)\frac{g^\Delta(t)}{g(t)g(\sigma(t))} + f^\Delta(t)\frac{1}{g(\sigma(t))} \\
&= \frac{f^\Delta(t)g(t) - f(t)g^\Delta(t)}{g(t)g(\sigma(t))}.
\end{aligned}
$$

The reader is asked in Exercise 1.21 to prove (ii) and (iv). □

Exercise 1.21. Prove parts (ii) and (iv) of Theorem 1.20.

Exercise 1.22. Prove that if x, y, and z are delta differentiable at t, then

$$(xyz)^\Delta = x^\Delta yz + x^\sigma y^\Delta z + x^\sigma y^\sigma z^\Delta$$

holds at t. Write down the generalization of this formula for n functions.

Exercise 1.23. We have by Theorem 1.20 (iii)

(1.1) $$(f^2)^\Delta = (f \cdot f)^\Delta = f^\Delta f + f^\sigma f^\Delta = (f + f^\sigma)f^\Delta.$$

Give the generalization of this formula for the derivative of the $(n+1)$st power of f, $n \in \mathbb{N}$, i.e., for $(f^{n+1})^\Delta$.

Theorem 1.24. *Let α be constant and $m \in \mathbb{N}$.*

(i) *For f defined by $f(t) = (t - \alpha)^m$ we have*

$$f^\Delta(t) = \sum_{\nu=0}^{m-1}(\sigma(t) - \alpha)^\nu(t - \alpha)^{m-1-\nu}.$$

(ii) *For g defined by $g(t) = \frac{1}{(t-\alpha)^m}$ we have*

$$g^\Delta(t) = -\sum_{\nu=0}^{m-1} \frac{1}{(\sigma(t)-\alpha)^{m-\nu}(t-\alpha)^{\nu+1}},$$

provided $(t-\alpha)(\sigma(t)-\alpha) \neq 0$.

Proof. We will prove the first formula by induction. If $m = 1$, then $f(t) = t - \alpha$, and clearly $f^\Delta(t) = 1$ holds by Example 1.13 (i), (ii), and Theorem 1.20 (i). Now we assume that

$$f^\Delta(t) = \sum_{\nu=0}^{m-1} (\sigma(t)-\alpha)^\nu (t-\alpha)^{m-1-\nu}$$

holds for $f(t) = (t-\alpha)^m$ and let $F(t) = (t-\alpha)^{m+1} = (t-\alpha)f(t)$. We use the product rule (Theorem 1.20 (iii)) to obtain

$$
\begin{aligned}
F^\Delta(t) &= f(\sigma(t)) + (t-\alpha)f^\Delta(t) \\
&= (\sigma(t)-\alpha)^m + (t-\alpha)\sum_{\nu=0}^{m-1}(\sigma(t)-\alpha)^\nu(t-\alpha)^{m-1-\nu} \\
&= (\sigma(t)-\alpha)^m + \sum_{\nu=0}^{m-1}(\sigma(t)-\alpha)^\nu(t-\alpha)^{m-\nu} \\
&= \sum_{\nu=0}^{m}(\sigma(t)-\alpha)^\nu(t-\alpha)^{m-\nu}.
\end{aligned}
$$

Hence, by mathematical induction, part (i) holds.

Next, for $g(t) = \frac{1}{(t-\alpha)^m} = \frac{1}{f(t)}$ we apply Theorem 1.20 (iv) to obtain

$$
\begin{aligned}
g^\Delta(t) &= -\frac{f^\Delta(t)}{f(t)f^\sigma(t)} \\
&= -\frac{\sum_{\nu=0}^{m-1}(\sigma(t)-\alpha)^\nu(t-\alpha)^{m-1-\nu}}{(t-\alpha)^m(\sigma(t)-\alpha)^m} \\
&= -\sum_{\nu=0}^{m-1}\frac{1}{(t-\alpha)^{\nu+1}(\sigma(t)-\alpha)^{m-\nu}},
\end{aligned}
$$

provided $(t-\alpha)(\sigma(t)-\alpha) \neq 0$. $\qquad\square$

Example 1.25. The derivative of t^2 is

$$t + \sigma(t).$$

The derivative of $1/t$ is

$$-\frac{1}{t\sigma(t)}.$$

Exercise 1.26. Use Theorem 1.24 to find the derivatives in Exercise 1.19.

We define higher order derivatives of functions on time scales in the usual way.

Definition 1.27. For a function $f : \mathbb{T} \to \mathbb{R}$ we shall talk about the second derivative $f^{\Delta\Delta}$ provided f^{Δ} is differentiable on $\mathbb{T}^{\kappa^2} = (\mathbb{T}^{\kappa})^{\kappa}$ with derivative $f^{\Delta\Delta} = (f^{\Delta})^{\Delta} : \mathbb{T}^{\kappa^2} \to \mathbb{R}$. Similarly we define higher order derivatives $f^{\Delta^n} : \mathbb{T}^{\kappa^n} \to \mathbb{R}$. Finally, for $t \in \mathbb{T}$, we denote $\sigma^2(t) = \sigma(\sigma(t))$ and $\rho^2(t) = \rho(\rho(t))$, and $\sigma^n(t)$ and $\rho^n(t)$ for $n \in \mathbb{N}$ are defined accordingly. For convenience we also put

$$\rho^0(t) = \sigma^0(t) = t, \quad f^{\Delta^0} = f, \quad \text{and} \quad \mathbb{T}^{\kappa^0} = \mathbb{T}.$$

Exercise 1.28. Find the second derivative of each of the functions given in Exercise 1.19.

Exercise 1.29. Find the second derivative of f on an arbitrary time scale:

    (i) $f(t) \equiv 1$;
    (ii) $f(t) = t$;
    (iii) $f(t) = t^2$.

Exercise 1.30. For an arbitrary time scale, try to find a function f such that $f^{\Delta\Delta} = 1$.

Example 1.31. In general, fg is not twice differentiable even if both f and g are twice differentiable. We have

$$(fg)^{\Delta} = f^{\Delta}g + f^{\sigma}g^{\Delta}.$$

If f and g are twice differentiable and if f^{σ} is differentiable, then

$$\begin{aligned} (fg)^{\Delta} &= (f^{\Delta}g + f^{\sigma}g^{\Delta})^{\Delta} \\ &= f^{\Delta\Delta}g + f^{\Delta^{\sigma}}g^{\Delta} + f^{\sigma^{\Delta}}g^{\Delta} + f^{\sigma^{\sigma}}g^{\Delta\Delta} \\ &= f^{\Delta\Delta}g + (f^{\Delta\sigma} + f^{\sigma\Delta})g^{\Delta} + f^{\sigma\sigma}g^{\Delta\Delta}, \end{aligned}$$

where we wrote $f^{\Delta\sigma}$ for $f^{\Delta^{\sigma}}$ etc.; we shall also use such notation in the sequel for combinations of more than two "exponents" of the form Δ or σ. The formula for the nth derivative under certain conditions is given in the following result.

Theorem 1.32 (Leibniz Formula). *Let $S_k^{(n)}$ be the set consisting of all possible strings of length n, containing exactly k times σ and $n - k$ times Δ. If*

$$f^{\Lambda} \quad \text{exists for all} \quad \Lambda \in S_k^{(n)},$$

then

(1.2) $$(fg)^{\Delta^n} = \sum_{k=0}^{n} \left(\sum_{\Lambda \in S_k^{(n)}} f^{\Lambda} \right) g^{\Delta^k}$$

holds for all $n \in \mathbb{N}$.

Proof. We will show (1.2) by induction. First, if $n = 1$, then (1.2) is true by Example 1.31. (With the convention that $\sum_{\Lambda \in \emptyset} f^{\Lambda} = f$, (1.2) also holds for $n = 0$.) Next, we assume that (1.2) is true for $n = m \in \mathbb{N}$. Then, using Theorem 1.20 (i) and (iii), we obtain

$$(fg)^{\Delta^{m+1}} = \left\{ \sum_{k=0}^{m} \left(\sum_{\Lambda \in S_k^{(m)}} f^{\Lambda} \right) g^{\Delta^k} \right\}^{\Delta}$$

$$= \sum_{k=0}^{m} \left\{ \left(\sum_{\Lambda \in S_k^{(m)}} f^{\Lambda} \right)^{\sigma} g^{\Delta^{k+1}} + \left(\sum_{\Lambda \in S_k^{(m)}} f^{\Lambda} \right)^{\Delta} g^{\Delta^k} \right\}$$

$$= \sum_{k=1}^{m+1} \left(\sum_{\Lambda \in S_{k-1}^{(m)}} f^{\Lambda\sigma} \right) g^{\Delta^k} + \sum_{k=0}^{m} \left(\sum_{\Lambda \in S_k^{(m)}} f^{\Lambda\Delta} \right) g^{\Delta^k}$$

$$= \left(\sum_{\Lambda \in S_m^{(m)}} f^{\Lambda\sigma} \right) g^{\Delta^{m+1}} + \left(\sum_{\Lambda \in S_0^{(m)}} f^{\Lambda\Delta} \right) g$$

$$+ \sum_{k=1}^{m} \left(\sum_{\Lambda \in S_{k-1}^{(m)}} f^{\Lambda\sigma} + \sum_{\Lambda \in S_k^{(m)}} f^{\Lambda\Delta} \right) g^{\Delta^k}$$

$$= \left(\sum_{\Lambda \in S_{m+1}^{(m+1)}} f^{\Lambda} \right) g^{\Delta^{m+1}} + \left(\sum_{\Lambda \in S_0^{(m+1)}} f^{\Lambda} \right) g + \sum_{k=1}^{m} \left(\sum_{\Lambda \in S_k^{(m+1)}} f^{\Lambda} \right) g^{\Delta^k}$$

$$= \sum_{k=0}^{m+1} \left(\sum_{\Lambda \in S_k^{(m+1)}} f^{\Lambda} \right) g^{\Delta^k}$$

so that (1.2) holds for $n = m+1$. By the principle of mathematical induction, (1.2) holds for all $n \in \mathbb{N}_0$. □

Example 1.33. If $\mathbb{T} = \mathbb{R}$, then

$$f^{\Lambda} = f^{(n-k)} \quad \text{for all} \quad \Lambda \in S_k^{(n)},$$

where $f^{(n)}$ denotes the nth (usual) derivative of f, if it exists, and since

$$\left| S_k^{(n)} \right| = \binom{n}{k},$$

where $|M|$ denotes the cardinality of the set M, we have

$$\sum_{\Lambda \in S_k^{(n)}} f^{\Lambda} = \sum_{\Lambda \in S_k^{(n)}} f^{(n-k)} = f^{(n-k)} \sum_{\Lambda \in S_k^{(n)}} 1 = \binom{n}{k} f^{(n-k)}$$

and therefore

$$(fg)^{\Delta^n} = \sum_{k=0}^{n} \left(\sum_{\Lambda \in S_k^{(n)}} f^{\Lambda} \right) g^{\Delta^k} = \sum_{k=0}^{n} \binom{n}{k} f^{(n-k)} g^{(k)}.$$

This is the usual Leibniz formula from calculus.

Exercise 1.34. Use Theorem 1.32 to find $\Delta^n(fg)$, i.e., $(fg)^{\Delta^n}$ if $\mathbb{T} = \mathbb{Z}$.

Exercise 1.35. Show that in general, even if both f^{Δ^σ} and f^{σ^Δ} exist,

(1.3) $$f^{\Delta^\sigma} = f^{\sigma^\Delta}$$

does not hold. Is (1.3) true for $\mathbb{T} = \mathbb{R}$ and $\mathbb{T} = \mathbb{Z}$? Give a sufficient condition that guarantees that (1.3) holds.

Exercise 1.36. Suppose μ is differentiable.

(i) If f^{Δ^σ} and f^{σ^Δ} both exist, give a formula that actually relates these two functions.

(ii) Give a similar formula that relates the functions (if they exist) $f^{\sigma\sigma\Delta}$, $f^{\sigma\Delta\sigma}$, and $f^{\Delta\sigma\sigma}$.

(iii) Give the corresponding formula that relates the functions (if they exist) $f^{\sigma^n\Delta}$ and $f^{\Delta\sigma^n}$, $n \in \mathbb{N}$.

1.3. Examples and Applications

In this section we will discuss some examples of time scales that are considered throughout this book.

Example 1.37. Let $h > 0$ and
$$\mathbb{T} = h\mathbb{Z} = \{hk : k \in \mathbb{Z}\}.$$
Then we have for $t \in \mathbb{T}$
$$\sigma(t) = \inf\{s \in \mathbb{T} : s > t\} = \inf\{t + nh : n \in \mathbb{N}\} = t + h$$
and similarly $\rho(t) = t - h$. Hence every point $t \in \mathbb{T}$ is isolated and
$$\mu(t) = \sigma(t) - t = t + h - t \equiv h \quad \text{for all} \quad t \in \mathbb{T}$$
so that μ in this example is constant. For a function $f : \mathbb{T} \to \mathbb{R}$ we have
$$f^\Delta(t) = \frac{f(\sigma(t)) - f(t)}{\mu(t)} = \frac{f(t + h) - f(t)}{h} \quad \text{for all} \quad t \in \mathbb{T}.$$

Next,

$$
\begin{aligned}
f^{\Delta\Delta}(t) &= \frac{f^\Delta(\sigma(t)) - f^\Delta(t)}{\mu(t)} \\
&= \frac{f^\Delta(t + h) - f^\Delta(t)}{h} \\
&= \frac{\frac{f(t+2h)-f(t+h)}{h} - \frac{f(t+h)-f(t)}{h}}{h} \\
&= \frac{f(t + 2h) - f(t + h) - f(t + h) + f(t)}{h^2} \\
&= \frac{f(t + 2h) - 2f(t + h) + f(t)}{h^2}.
\end{aligned}
$$

It would be too tedious to calculate $f^{\Delta^n}(t)$ in a similar way, so we consider another approach. First, note that
$$\sigma^n(t) = t + nh \quad \text{and} \quad \rho^n(t) = t - nh \quad \text{for all} \quad n \in \mathbb{N}_0.$$
We now introduce an operator Δ_h by
$$\Delta_h = \frac{1}{h}(\sigma - I), \quad I \text{ being the identity operator.}$$

Figure 1.2. Some Time Scales

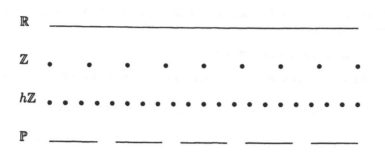

Recall that the binomial theorem says that

$$\sum_{k=0}^{n} \binom{n}{k} a^k b^{n-k} = (a+b)^n.$$

We use an operator version of this binomial theorem to obtain the nth power of Δ_h as

$$\Delta_h^n = \frac{1}{h^n} (\sigma - I)^n = \frac{1}{h^n} \sum_{k=0}^{n} \binom{n}{k} \sigma^k (-I)^{n-k}.$$

Applying the obtained operator to the function f, we find

$$f^{\Delta^n}(t) = \frac{1}{h^n} \sum_{k=0}^{n} \binom{n}{k} (-1)^{n-k} f(t + kh).$$

This time scale is of particular interest: It is in some cases possible to obtain "continuous" results, i.e., results for $\mathbb{T} = \mathbb{R}$, via letting h tend to zero from above in the corresponding "discrete" results, i.e., results for $\mathbb{T} = h\mathbb{Z}$.

We now give some examples of time scales with nonconstant graininess.

Example 1.38. Let $a, b > 0$ and consider the time scale

$$\mathbb{P}_{a,b} = \bigcup_{k=0}^{\infty} [k(a+b), k(a+b) + a].$$

Then

$$\sigma(t) = \begin{cases} t & \text{if} \quad t \in \bigcup_{k=0}^{\infty} [k(a+b), k(a+b) + a) \\ t+b & \text{if} \quad t \in \bigcup_{k=0}^{\infty} \{k(a+b) + a\} \end{cases}$$

and

$$\mu(t) = \begin{cases} 0 & \text{if} \quad t \in \bigcup_{k=0}^{\infty} [k(a+b), k(a+b) + a) \\ b & \text{if} \quad t \in \bigcup_{k=0}^{\infty} \{k(a+b) + a\}. \end{cases}$$

See Figure 1.3 for a graph of the forward jump operator for this time scale.

Figure 1.3. Forward Jump Operator for $\mathbb{P}_{a,b}$

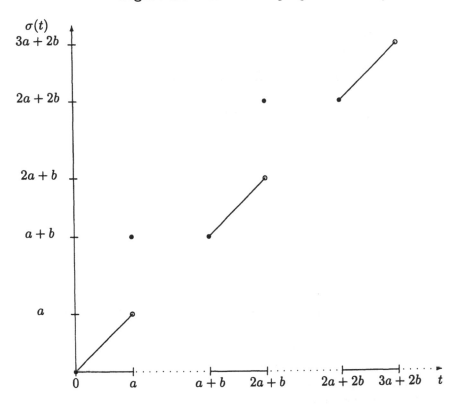

Example 1.39. Assume that the life span of a certain species is one unit of time. Suppose that just before the species dies out, eggs are laid which are hatched one unit of time later. Hence we are interested in the time scale

$$\mathbb{P}_{1,1} = \bigcup_{k=0}^{\infty} [2k, 2k+1].$$

For this time scale,

$$\mu(t) = \begin{cases} 0 & \text{for} \quad t \in \bigcup_{k=0}^{\infty} [2k, 2k+1) \\ 1 & \text{for} \quad t \in \bigcup_{k=0}^{\infty} \{2k+1\}. \end{cases}$$

For a specific example of this type see Christiansen and Fenchel [**96**, page 7ff]. A couple of examples of this type, where the time scale consists of a sequence of disjoint closed intervals, are the 17 year cicada *magicicada septendecim* which lives as a larva for 17 years and as an adult for perhaps a week, and the common mayfly *stenonema canadense* which lives as a larva for a year and as an adult for less than a day.

Example 1.40 (S. Keller [**190**, Beispiel 2.1.13]). Consider a simple electric circuit with resistance R, inductance L, and capacitance C (see Figure 1.4). Suppose we discharge the capacitor periodically every time unit and assume that the discharging

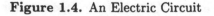

Figure 1.4. An Electric Circuit

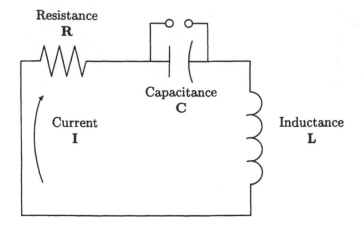

takes $\delta > 0$ (but small) time units. Then this simulation can be modeled using the time scale

$$\mathbb{P}_{1-\delta,\delta} = \bigcup_{k \in \mathbb{N}_0} [k, k + 1 - \delta].$$

If $Q(t)$ is the total charge on the capacitor at time t and $I(t)$ is the current as a function of time t, then we have

$$Q^\Delta(t) = \begin{cases} bQ(t) & \text{if} \quad t \in \bigcup_{k \in \mathbb{N}} \{k - \delta\} \\ I & \text{otherwise} \end{cases}$$

and

$$I^\Delta(t) = \begin{cases} 0 & \text{if} \quad t \in \bigcup_{k \in \mathbb{N}} \{k - \delta\} \\ -\frac{1}{LC}Q(t) - \frac{R}{L}I(t) & \text{otherwise,} \end{cases}$$

where b is a constant satisfying $-1 < b\delta < 0$.

Example 1.41. Let $q > 1$ and

$$q^{\mathbb{Z}} := \{q^k : k \in \mathbb{Z}\} \quad \text{and} \quad \overline{q^{\mathbb{Z}}} := q^{\mathbb{Z}} \cup \{0\}.$$

Here we consider the time scale $\mathbb{T} = \overline{q^{\mathbb{Z}}}$. We have

$$\sigma(t) = \inf\{q^n : n \in [m + 1, \infty)\} = q^{m+1} = qq^m = qt$$

if $t = q^m \in \mathbb{T}$ and obviously $\sigma(0) = 0$. So we obtain

$$\sigma(t) = qt \quad \text{and} \quad \rho(t) = \frac{t}{q} \quad \text{for all} \quad t \in \mathbb{T}$$

and consequently

$$\mu(t) = \sigma(t) - t = (q - 1)t \quad \text{for all} \quad t \in \mathbb{T}.$$

Hence 0 is a right-dense minimum and every other point in $\mathbb{T}$ is isolated. For a function $f : \mathbb{T} \to \mathbb{R}$ we have

$$f^{\Delta}(t) = \frac{f(\sigma(t)) - f(t)}{\mu(t)} = \frac{f(qt) - f(t)}{(q-1)t} \quad \text{for all} \quad t \in \mathbb{T} \setminus \{0\}$$

and

$$f^{\Delta}(0) = \lim_{s \to 0} \frac{f(0) - f(s)}{0 - s} = \lim_{s \to 0} \frac{f(s) - f(0)}{s}$$

provided this limit exists. Now we calculate the second derivative of f at $t \neq 0$ (refer to Definition 1.27 for how $f^{\Delta\Delta}$ is defined) as

$$
\begin{aligned}
f^{\Delta\Delta}(t) &= \frac{f^{\Delta}(\sigma(t)) - f^{\Delta}(t)}{\mu(t)} \\
&= \frac{f^{\Delta}(qt) - f^{\Delta}(t)}{(q-1)t} \\
&= \frac{\frac{f(q^2 t) - f(qt)}{q(q-1)t} - \frac{f(qt) - f(t)}{(q-1)t}}{(q-1)t} \\
&= \frac{f(q^2 t) - f(qt) - qf(qt) + qf(t)}{q(q-1)^2 t^2} \\
&= \frac{f(q^2 t) - (q+1)f(qt) + qf(t)}{q(q-1)^2 t^2}.
\end{aligned}
$$

Notice that $\mu(t) = t$ above in the particular case $q = 2$.

Exercise 1.42. Let $q > 1$. For the time scale $\mathbb{T} = \overline{q^{\mathbb{Z}}}$, evaluate

(i) σ^{Δ};
(ii) μ^{Δ}.

Exercise 1.43. Find f^{Δ^3} for the time scale $\mathbb{T} = \overline{q^{\mathbb{Z}}}$. Find f^{Δ^4} and finally find a formula for f^{Δ^n} for any natural number n.

Example 1.44. Consider the time scale

$$\mathbb{T} = \mathbb{N}_0^2 = \{n^2 : n \in \mathbb{N}_0\}.$$

We have $\sigma(n^2) = (n+1)^2$ for $n \in \mathbb{N}_0$ and

$$\mu(n^2) = \sigma(n^2) - n^2 = (n+1)^2 - n^2 = 2n + 1.$$

Hence

$$\sigma(t) = (\sqrt{t} + 1)^2 \quad \text{and} \quad \mu(t) = 1 + 2\sqrt{t} \quad \text{for} \quad t \in \mathbb{T}.$$

Example 1.45. Let H_n be the so-called *harmonic numbers*

$$H_0 = 0 \quad \text{and} \quad H_n = \sum_{k=1}^{n} \frac{1}{k} \quad \text{for } n \in \mathbb{N}.$$

Consider the time scale

$$\mathbb{T} = \{H_n : n \in \mathbb{N}_0\}.$$

Then $\sigma(H_n) = H_{n+1}$ for all $n \in \mathbb{N}_0$, $\rho(H_n) = H_{n-1}$ when $n \in \mathbb{N}$, and $\rho(H_0) = H_0$. The graininess is given by

$$\mu(H_n) = \sigma(H_n) - H_n = H_{n+1} - H_n = \frac{1}{n+1}$$

Table 1.2. Examples of Time Scales

$\mathbb{T}$	$\mu(t)$	$\sigma(t)$	$\rho(t)$
$\mathbb{R}$	0	t	t
$\mathbb{Z}$	1	$t+1$	$t-1$
$h\mathbb{Z}$	h	$t+h$	$t-h$
$q^{\mathbb{N}}$	$(q-1)t$	qt	$\frac{t}{q}$
$2^{\mathbb{N}}$	t	$2t$	$\frac{t}{2}$
$\mathbb{N}_0^2$	$2\sqrt{t}+1$	$(\sqrt{t}+1)^2$	$(\sqrt{t}-1)^2$

for all $n \in \mathbb{N}_0$. If $f : \mathbb{T} \to \mathbb{R}$ is a function, then

$$f^{\Delta}(H_n) = \frac{f(H_{n+1}) - f(H_n)}{\mu(H_n)} = (n+1)\Delta f(H_n).$$

Example 1.46. We let $\{\alpha_n\}_{n \in \mathbb{N}_0}$ be a sequence of real numbers with $\alpha_n > 0$ for all $n \in \mathbb{N}$ and put

$$t_n = \sum_{k=0}^{n-1} \alpha_k.$$

Consider the time scale

$$\mathbb{T} = \{t_n : n \in \mathbb{N}\}$$

if $\sum_{k=0}^{\infty} \alpha_k = \infty$ or

$$\mathbb{T} = \{t_n : n \in \mathbb{N}\} \cup \{L\}$$

if $\sum_{k=0}^{\infty} \alpha_k = L$ converges. We have

$$\sigma(t_n) = t_{n+1} \quad \text{and} \quad \mu(t_n) = \alpha_n$$

for all $n \in \mathbb{N}$. For a function $y : \mathbb{T} \to \mathbb{R}$ we find

$$y^{\Delta}(t_n) = \frac{y(t_{n+1}) - y(t_n)}{\alpha_n} = \frac{\Delta y(t_n)}{\alpha_n}$$

for all $n \in \mathbb{N}$.

We remark that using the harmonic series above corresponds to Example 1.45, while using the geometric series corresponds to Example 1.41.

Example 1.47 (The Cantor Set). Consider $K_0 = [0,1]$. We obtain a subset K_1 of K_0 by removing the open "middle third" of K_0, i.e., the open interval $(1/3, 2/3)$, from K_0. K_2 is obtained by removing the two open middle thirds of K_1, i.e., the two open intervals $(1/9, 2/9)$ and $(7/9, 8/9)$ from K_1. Proceeding in this manner, we obtain a sequence $\{K_n\}_{n \in \mathbb{N}_0}$ of subsets of $[0,1]$. See Figure 1.5 for K_0, K_1, K_2, and K_3. The Cantor set C is now defined as

$$C = \bigcap_{n=0}^{\infty} K_n$$

and hence is closed. Therefore $\mathbb{T} = C$ is a time scale. Each $x \in [0,1]$ can be represented in its *ternary* expansion as

$$x = \sum_{k=1}^{\infty} \frac{a_k}{3^k}, \quad \text{where} \quad a_k \in \{0,1,2\} \text{ for each } k \in \mathbb{N}.$$

It is known that a number x is an element of C if and only if it can be represented by a ternary expansion, where the a_k are either 0 or 2 (see, e.g., [**142**, page 38]). Let L denote the set of all the left-hand end points of the open intervals removed, i.e.,

$$L = \left\{ \sum_{k=1}^{m} \frac{a_k}{3^k} + \frac{1}{3^{m+1}} : m \in \mathbb{N} \text{ and } a_k \in \{0,2\} \text{ for all } 1 \leq k \leq m \right\}.$$

Then $L \subset \mathbb{T}$. The set of all right-hand end points of the open intervals removed is given by

$$R = \left\{ \sum_{k=1}^{m} \frac{a_k}{3^k} + \frac{2}{3^{m+1}} : m \in \mathbb{N} \text{ and } a_k \in \{0,2\} \text{ for all } 1 \leq k \leq m \right\},$$

and we also have $R \subset \mathbb{T}$. It follows that

$$\sigma(t) = t + \frac{1}{3^{m+1}} \quad \text{whenever} \quad t = \sum_{k=1}^{m} \frac{a_k}{3^k} + \frac{1}{3^{m+1}} \in L.$$

Each point $t \in \mathbb{T} \setminus L$ has other points of $\mathbb{T}$ in any neighborhood of t, and therefore satisfies $\sigma(t) = t$. Altogether,

$$\sigma(t) = \begin{cases} t + \frac{1}{3^{m+1}} & \text{if } t = \sum_{k=1}^{m} \frac{a_k}{3^k} + \frac{1}{3^{m+1}} \in L \\ t & \text{if } t \in \mathbb{T} \setminus L \end{cases}$$

and similarly

$$\rho(t) = \begin{cases} t - \frac{1}{3^{m+1}} & \text{if } t = \sum_{k=1}^{m} \frac{a_k}{3^k} + \frac{2}{3^{m+1}} \in R \\ t & \text{if } t \in \mathbb{T} \setminus R. \end{cases}$$

Now we obtain the graininess function μ of the Cantor set as

$$\mu(t) = \begin{cases} \frac{1}{3^{m+1}} & \text{if } t = \sum_{k=1}^{m} \frac{a_k}{3^k} + \frac{1}{3^{m+1}} \in L \\ 0 & \text{if } t \in \mathbb{T} \setminus L. \end{cases}$$

Hence L consists of the right-scattered elements of $\mathbb{T}$, and R consists of the left-scattered elements of $\mathbb{T}$. Thus, $\mathbb{T}$ does not contain any isolated points.

We now discuss some examples for the time scale $\mathbb{T} = \mathbb{Z}$.

Definition 1.48. Let $t \in \mathbb{C}$ (i.e., t is a complex number) and $k \in \mathbb{Z}$. The *factorial function* $t^{(k)}$ is defined as follows:

(i) If $k \in \mathbb{N}$, then

$$t^{(k)} = t(t-1)\cdots(t-k+1).$$

(ii) If $k = 0$, then

$$t^{(0)} = 1.$$

Figure 1.5. The Cantor Set

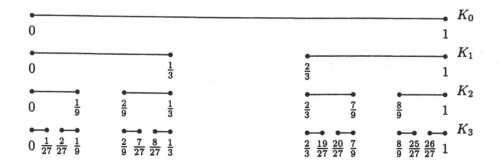

(iii) If $-k \in \mathbb{N}$, then

$$t^{(k)} = \frac{1}{(t+1)(t+2)\cdots(t-k)}$$

for $t \neq -1, -2, \cdots, k$.

In general

(1.4)
$$t^{(k)} := \frac{\Gamma(t+1)}{\Gamma(t-k+1)}$$

for all $t, k \in \mathbb{C}$ such that the right-hand side of (1.4) makes sense, where Γ is the gamma function. See [191] for some results concerning the gamma and factorial functions.

Exercise 1.49. Show that the general definition (1.4) of the factorial function $t^{(k)}$ gives parts (i), (ii), and (iii) in Definition 1.48 as special cases.

Exercise 1.50. Show that for any constant c

$$u(t) = ca^t \frac{\Gamma(t-t_1)\Gamma(t-t_2)\cdots\Gamma(t-t_n)}{\Gamma(t-s_1)\Gamma(t-s_2)\cdots\Gamma(t-s_m)}$$

is a solution of the recurrence relation

$$u(t+1) = a \frac{(t-t_1)(t-t_2)\cdots(t-t_n)}{(t-s_1)(t-s_2)\cdots(t-s_m)} u(t),$$

where $a, t_1, \ldots, t_n, s_1, \ldots, s_m$ are real constants and $n, m \in \mathbb{N}$. Use this to solve the difference equations

  (i) $\Delta u = \frac{4t+6}{t^2+5t+6} u$, where $t \in \mathbb{N}$;
  (ii) $\Delta u = \frac{2t+5}{2t+1} u$, where $t \in \mathbb{N}_0$.

Next we define a general binomial coefficient $\binom{\alpha}{\beta}$.

Definition 1.51. We define the *binomial coefficient* $\binom{\alpha}{\beta}$ by

$$\binom{\alpha}{\beta} = \frac{\alpha^{(\beta)}}{\Gamma(\beta+1)}$$

for all $\alpha, \beta \in \mathbb{C}$ such that the right-hand side of this equation makes sense.

Exercise 1.52. Assume $\alpha, k \in \mathbb{C}$ and $^\Delta$ is differentiation with respect to t on the time scale $\mathbb{T} = \mathbb{Z}$. Show that

(i) $\left[(t + \alpha)^{(k)}\right]^\Delta = k(t + \alpha)^{(k-1)}$;

(ii) $(\alpha^t)^\Delta = (\alpha - 1)\alpha^t$;

(iii) $\binom{t}{\alpha}^\Delta = \binom{t}{\alpha-1}$.

Exercise 1.53. Prove the following well-known formula concerning binomial coefficients:

$$\binom{\alpha}{\beta} + \binom{\alpha}{\beta+1} = \binom{\alpha+1}{\beta+1}.$$

Exercise 1.54. Introduce some of the above concepts for the time scale $\mathbb{T} = h\mathbb{Z}$ and prove some of the above results for this time scale.

We conclude this section by giving some examples concerning the jump operator.

Example 1.55 (σ is in general not continuous). In Figure 1.3 we already have seen an example of a time scale $\mathbb{T}$ whose jump function $\sigma : \mathbb{T} \to \mathbb{T}$ is not continuous at points $t \in \mathbb{T}$ which are left-dense and right-scattered at the same time. Here we present another such example. This example is due to Douglas Anderson. Let

$$\mathbb{T} = \{t_n = -1/n : n \in \mathbb{N}\} \cup \mathbb{N}_0.$$

Then

$$\sigma(t_n) = t_{n+1} = -\frac{1}{n+1} \to 0 \neq 1 = \sigma(0), \quad n \to \infty,$$

and hence $\lim_{s \to 0} \sigma(s) \neq \sigma(0)$ so σ is not continuous at 0. According to Theorem 1.16 (i), σ is not differentiable at 0 either, and this can also be shown directly in this case using the definition of differentiability (Definition 1.10). However, note that σ is continuous at right-dense points and that $\lim_{s \to t^-} \sigma(s)$ exists at left-dense points $t \in \mathbb{T}$.

Example 1.56 (σ is in general not differentiable). Here we present an example of a time scale $\mathbb{T}$ whose jump function $\sigma : \mathbb{T} \to \mathbb{T}$ is continuous but not differentiable at a right-dense point $t \in \mathbb{T}$. Let

$$\mathbb{T} = \left\{t_n = (1/2)^{2^n} : n \in \mathbb{N}_0\right\} \cup \{0, -1\}.$$

Then

$$\sigma(t_n) = t_{n-1} \to 0 = \sigma(0), \quad n \to \infty,$$

and hence $\lim_{s \to 0} \sigma(s) = \sigma(0)$ so σ is continuous at 0. But

$$\begin{aligned}
\lim_{s \to 0} \frac{\sigma(\sigma(0)) - \sigma(s)}{\sigma(0) - s} &= \lim_{s \to 0} \frac{\sigma(s)}{s} \\
&= \lim_{s \to 0} \frac{\sqrt{s}}{s} \\
&= \lim_{s \to 0} \frac{1}{\sqrt{s}} \\
&= \infty
\end{aligned}$$

so that σ is not differentiable at 0 (regardless, in fact, whether 0 is left-dense or left-scattered). Note that t is twice differentiable at 0 (see Example 1.13) while $t \cdot t = t^2$ is not.

1.4. Integration

In order to describe classes of functions that are "integrable", we introduce the following two concepts.

Definition 1.57. A function $f : \mathbb{T} \to \mathbb{R}$ is called *regulated* provided its right-sided limits exist (finite) at all right-dense points in $\mathbb{T}$ and its left-sided limits exist (finite) at all left-dense points in $\mathbb{T}$.

Definition 1.58. A function $f : \mathbb{T} \to \mathbb{R}$ is called *rd-continuous* provided it is continuous at right-dense points in $\mathbb{T}$ and its left-sided limits exist (finite) at left-dense points in $\mathbb{T}$. The set of rd-continuous functions $f : \mathbb{T} \to \mathbb{R}$ will be denoted in this book by

$$C_{rd} = C_{rd}(\mathbb{T}) = C_{rd}(\mathbb{T}, \mathbb{R}).$$

The set of functions $f : \mathbb{T} \to \mathbb{R}$ that are differentiable and whose derivative is rd-continuous is denoted by

$$C_{rd}^1 = C_{rd}^1(\mathbb{T}) = C_{rd}^1(\mathbb{T}, \mathbb{R}).$$

Exercise 1.59. Are the operators σ, ρ, and μ

  (i) continuous;
 (ii) rd-continuous;
(iii) regulated?

Some results concerning rd-continuous and regulated functions are contained in the following theorem.

Theorem 1.60. *Assume* $f : \mathbb{T} \to \mathbb{R}$.

  (i) *If f is continuous, then f is rd-continuous.*
 (ii) *If f is rd-continuous, then f is regulated.*
(iii) *The jump operator σ is rd-continuous.*
(iv) *If f is regulated or rd-continuous, then so is f^σ.*
 (v) *Assume f is continuous. If $g : \mathbb{T} \to \mathbb{R}$ is regulated or rd-continuous, then $f \circ g$ has that property too.*

Exercise 1.61. Prove Theorem 1.60.

Definition 1.62. A continuous function $f : \mathbb{T} \to \mathbb{R}$ is called *pre-differentiable* with (region of differentiation) D, provided $D \subset \mathbb{T}^\kappa$, $\mathbb{T}^\kappa \setminus D$ is countable and contains no right-scattered elements of $\mathbb{T}$, and f is differentiable at each $t \in D$.

Example 1.63. Let $\mathbb{T} := \mathbb{P}_{2,1}$ and let $f : \mathbb{T} \to \mathbb{R}$ be defined by

$$f(t) = \begin{cases} 0 & \text{if} \quad t \in \bigcup_{k=0}^{\infty} [3k, 3k+1] \\ t - 3k - 1 & \text{if} \quad t \in [3k+1, 3k+2], \ k \in \mathbb{N}_0. \end{cases}$$

Then f is pre-differentiable with

$$D := \mathbb{T} \setminus \bigcup_{k=0}^{\infty} \{3k+1\}.$$

Exercise 1.64. For each of the following determine if f is regulated on $\mathbb{T}$, if f is rd-continuous on $\mathbb{T}$, and if f is pre-differentiable. If f is pre-differentiable, find its region of differentiability D.

(i) The function f is defined on a time scale $\mathbb{T}$ and every point $t \in \mathbb{T}$ is isolated.

(ii) Assume $\mathbb{T} = \mathbb{R}$ and
$$f(t) = \begin{cases} 0 & \text{if} \quad t = 0 \\ \frac{1}{t} & \text{if} \quad t \in \mathbb{R} \setminus \{0\}. \end{cases}$$

(iii) Assume $\mathbb{T} = \mathbb{N}_0 \cup \{1 - 1/n : n \in \mathbb{N}\}$ and
$$f(t) = \begin{cases} 0 & \text{if} \quad t \in \mathbb{N} \\ t & \text{otherwise.} \end{cases}$$

(iv) Assume $\mathbb{T} = \mathbb{R}$ and $f(t) = |t|$, $t \in \mathbb{R}$.

(v) Assume $\mathbb{T} = \mathbb{P}_{1,1}$ and
$$f(t) = \begin{cases} 0 & \text{if} \quad t = 2k + 1, \ k \in \mathbb{N}_0 \\ t - 2k & \text{if} \quad t \in [2k, 2k+1), \ k \in \mathbb{N}_0. \end{cases}$$

(vi) Assume $\mathbb{T} = \mathbb{P}_{1,1}$ and
$$f(t) = k, \quad t \in [2k, 2k+1], \quad k \in \mathbb{N}_0.$$

Theorem 1.65. *Every regulated function on a compact interval is bounded.*

Proof. Assume $f : [a,b] \to \mathbb{R}$ is unbounded, i.e., for each $n \in \mathbb{N}$ there exists $t_n \in [a,b]$ with $|f(t_n)| > n$. Since
$$\{t_n : n \in \mathbb{N}\} \subset [a,b],$$
there exists a convergent subsequence $\{t_{n_k}\}_{k \in \mathbb{N}}$, i.e.,

(1.5) $$\lim_{k \to \infty} t_{n_k} = t_0 \quad \text{for some} \quad t_0 \in [a,b].$$

Note that $t_0 \in \mathbb{T}$ since $\{t_{n_k} : k \in \mathbb{N}\} \subset \mathbb{T}$ and $\mathbb{T}$ is closed. By (1.5), t_0 cannot be isolated, and there exists either a subsequence that tends to t_0 from above or a subsequence that tends to t_0 from below, and in any case the limit of $f(t)$ as $t \to t_0$ has to be finite according to regularity, a contradiction. $\qquad\square$

Remark 1.66. If f is regulated or even if $f \in C_{rd}$, $\max_{a \le t \le b} f(t)$ and $\min_{a \le t \le b} f(t)$ need not exist. See Exercise 1.64 (iii) for an example of a function which is rd-continuous but does not attain its supremum on $[0,1]$.

The following mean value theorem holds for pre-differentiable functions and will be used to prove the main existence theorems for pre-antiderivatives and antiderivatives later on in this section. Its proof is an application of the induction principle.

Theorem 1.67 (Mean Value Theorem). *Let f and g be real-valued functions defined on $\mathbb{T}$, both pre-differentiable with D. Then*
$$|f^{\Delta}(t)| \le g^{\Delta}(t) \quad \text{for all} \quad t \in D$$
implies
$$|f(s) - f(r)| \le g(s) - g(r) \quad \text{for all} \quad r, s \in \mathbb{T}, \ r \le s.$$

Proof. Let $r, s \in \mathbb{T}$ with $r \le s$ and denote $[r,s) \setminus D = \{t_n : n \in \mathbb{N}\}$. Let $\varepsilon > 0$. We now show by induction that
$$S(t): \quad |f(t) - f(r)| \le g(t) - g(r) + \varepsilon\left(t - r + \sum_{t_n < t} 2^{-n}\right)$$

holds for all $t \in [r, s]$. Note that once we have shown this, the claim of the mean value theorem follows. We now check the four conditions given in Theorem 1.7.

I. The statement $S(r)$ is trivially satisfied.

II. Let t be right-scattered and assume that $S(t)$ holds. Then $t \in D$ and

$$
\begin{aligned}
|f(\sigma(t)) - f(r)| &= |f(t) + \mu(t)f^\Delta(t) - f(r)| \\
&\leq \mu(t)|f^\Delta(t)| + |f(t) - f(r)| \\
&\leq \mu(t)g^\Delta(t) + g(t) - g(r) + \varepsilon\left(t - r + \sum_{t_n < t} 2^{-n}\right) \\
&= g(\sigma(t)) - g(r) + \varepsilon\left(t - r + \sum_{t_n < \sigma(t)} 2^{-n}\right) \\
&< g(\sigma(t)) - g(r) + \varepsilon\left(\sigma(t) - r + \sum_{t_n < \sigma(t)} 2^{-n}\right).
\end{aligned}
$$

Therefore $S(\sigma(t))$ holds.

III. Suppose $S(t)$ holds and $t \neq s$ is right-dense, i.e., $\sigma(t) = t$. We consider two cases, namely $t \in D$ and $t \notin D$. First of all, suppose $t \in D$. Then f and g are differentiable at t and hence there exists a neighborhood U of t with

$$
|f(t) - f(\tau) - f^\Delta(t)(t - \tau)| \leq \frac{\varepsilon}{2}|t - \tau| \quad \text{for all} \quad \tau \in U
$$

and

$$
|g(t) - g(\tau) - g^\Delta(t)(t - \tau)| \leq \frac{\varepsilon}{2}|t - \tau| \quad \text{for all} \quad \tau \in U.
$$

Thus

$$
|f(t) - f(\tau)| \leq \left[|f^\Delta(t)| + \frac{\varepsilon}{2}\right]|t - \tau| \quad \text{for all} \quad \tau \in U
$$

and

$$
g(\tau) - g(t) - g^\Delta(t)(\tau - t) \geq -\frac{\varepsilon}{2}|t - \tau| \quad \text{for all} \quad \tau \in U.
$$

Hence we have for all $\tau \in U \cap (t, \infty)$

$$
\begin{aligned}
|f(\tau) - f(r)| &\leq |f(\tau) - f(t)| + |f(t) - f(r)| \\
&\leq \left[|f^\Delta(t)| + \frac{\varepsilon}{2}\right]|t - \tau| + |f(t) - f(r)| \\
&\leq \left[g^\Delta(t) + \frac{\varepsilon}{2}\right]|t - \tau| + g(t) - g(r) + \varepsilon\left(t - r + \sum_{t_n < t} 2^{-n}\right) \\
&= g^\Delta(t)(\tau - t) + \frac{\varepsilon}{2}(\tau - t) + g(t) - g(r) + \varepsilon(t - r) + \varepsilon\sum_{t_n < t} 2^{-n} \\
&\leq g(\tau) - g(t) + \frac{\varepsilon}{2}|t - \tau| + \frac{\varepsilon}{2}(\tau - t) + g(t) - g(r) \\
&\quad + \varepsilon(t - r) + \varepsilon\sum_{t_n < t} 2^{-n} \\
&= g(\tau) - g(r) + \varepsilon\left(\tau - r + \sum_{t_n < \tau} 2^{-n}\right)
\end{aligned}
$$

so that $S(\tau)$ follows for all $\tau \in U \cap (t, \infty)$.

For the second case, suppose $t \notin D$. Then $t = t_m$ for some $m \in \mathbb{N}$. Since f and g are pre-differentiable, they both are continuous and hence there exists a neighborhood U of t with

$$|f(\tau) - f(t)| \leq \frac{\varepsilon}{2} 2^{-m} \quad \text{for all} \quad \tau \in U$$

and

$$|g(\tau) - g(t)| \leq \frac{\varepsilon}{2} 2^{-m} \quad \text{for all} \quad \tau \in U.$$

Therefore

$$g(\tau) - g(t) \geq -\frac{\varepsilon}{2} 2^{-m} \quad \text{for all} \quad \tau \in U$$

and hence

$$
\begin{aligned}
|f(\tau) - f(r)| &\leq |f(\tau) - f(t)| + |f(t) - f(r)| \\
&\leq \frac{\varepsilon}{2} 2^{-m} + g(t) - g(r) + \varepsilon \left(t - r + \sum_{t_n < t} 2^{-n} \right) \\
&\leq \frac{\varepsilon}{2} 2^{-m} + g(\tau) + \frac{\varepsilon}{2} 2^{-m} - g(r) + \varepsilon \left(\tau - r + \sum_{t_n < t} 2^{-n} \right) \\
&= \varepsilon 2^{-m} + g(\tau) - g(r) + \varepsilon \left(\tau - r + \sum_{t_n < t} 2^{-n} \right) \\
&\leq g(\tau) - g(r) + \varepsilon \left(\tau - r + \sum_{t_n < \tau} 2^{-n} \right)
\end{aligned}
$$

so that again $S(\tau)$ follows for all $\tau \in U \cap (t, \infty)$.

IV. Now let t be left-dense and suppose $S(\tau)$ is true for all $\tau < t$. Then

$$
\begin{aligned}
\lim_{\tau \to t^-} |f(\tau) - f(r)| &\leq \lim_{\tau \to t^-} \left\{ g(\tau) - g(r) + \varepsilon \left(\tau - r + \sum_{t_n < \tau} 2^{-n} \right) \right\} \\
&\leq \lim_{\tau \to t^-} \left\{ g(\tau) - g(r) + \varepsilon \left(\tau - r + \sum_{t_n < t} 2^{-n} \right) \right\}
\end{aligned}
$$

implies $S(t)$ as both f and g are continuous at t.

An application of Theorem 1.7 finishes the proof. $\square$

Corollary 1.68. *Suppose f and g are pre-differentiable with D.*

(i) *If U is a compact interval with endpoints $r, s \in \mathbb{T}$, then*

$$|f(s) - f(r)| \leq \left\{ \sup_{t \in U^\kappa \cap D} |f^\Delta(t)| \right\} |s - r|.$$

(ii) *If $f^\Delta(t) = 0$ for all $t \in D$, then f is a constant function.*

(iii) *If $f^\Delta(t) = g^\Delta(t)$ for all $t \in D$, then*

$$g(t) = f(t) + C \quad \text{for all} \quad t \in \mathbb{T},$$

where C is a constant.

Proof. Suppose f is pre-differentiable with D and let $r, s \in \mathbb{T}$ with $r \leq s$. If we define

$$g(t) := \left\{ \sup_{\tau \in [r,s]^\kappa \cap D} |f^\Delta(\tau)| \right\} (t - r) \quad \text{for} \quad t \in \mathbb{T},$$

then

$$g^\Delta(t) = \sup_{\tau \in [r,s]^\kappa \cap D} |f^\Delta(\tau)| \geq |f^\Delta(t)| \quad \text{for all} \quad t \in D \cap [r, s]^\kappa.$$

By Theorem 1.67,

$$g(t) - g(r) \geq |f(t) - f(r)| \quad \text{for all} \quad t \in [r, s]$$

so that

$$|f(s) - f(r)| \leq g(s) - g(r) = g(s) = \left\{ \sup_{\tau \in [r,s]^\kappa \cap D} |f^\Delta(\tau)| \right\} (s - r).$$

This completes the proof of part (i). Part (ii) follows immediately from (i), and (iii) follows from (ii). □

Exercise 1.69. Prove Theorem 1.68 (ii) and (iii).

The main existence theorem for pre-antiderivatives now reads as follows. We will prove this theorem in a more general form in Chapter 8.

Theorem 1.70 (Existence of Pre-Antiderivatives). *Let f be regulated. Then there exists a function F which is pre-differentiable with region of differentiation D such that*

$$F^\Delta(t) = f(t) \quad \text{holds for all} \quad t \in D.$$

Proof. See the proof of Theorem 8.13. □

Definition 1.71. Assume $f : \mathbb{T} \to \mathbb{R}$ is a regulated function. Any function F as in Theorem 1.70 is called a *pre-antiderivative* of f. We define the *indefinite integral* of a regulated function f by

$$\int f(t)\Delta t = F(t) + C,$$

where C is an arbitrary constant and F is a pre-antiderivative of f. We define the *Cauchy integral* by

$$\int_r^s f(t)\Delta t = F(s) - F(r) \quad \text{for all} \quad r, s \in \mathbb{T}.$$

A function $F : \mathbb{T} \to \mathbb{R}$ is called an *antiderivative* of $f : \mathbb{T} \to \mathbb{R}$ provided

$$F^\Delta(t) = f(t) \quad \text{holds for all} \quad t \in \mathbb{T}^\kappa.$$

Example 1.72. If $\mathbb{T} = \mathbb{Z}$, evaluate the indefinite integral

$$\int a^t \Delta t,$$

where $a \neq 1$ is a constant. Since

$$\left(\frac{a^t}{a-1} \right)^\Delta = \Delta \left(\frac{a^t}{a-1} \right) = \frac{a^{t+1} - a^t}{a-1} = a^t,$$

we get that

$$\int a^t \Delta t = \frac{a^t}{a-1} + C,$$

where C is an arbitrary constant.

Exercise 1.73. Show that if $\mathbb{T} = \mathbb{Z}$, $k \neq -1$, and $\alpha \in \mathbb{R}$, then

(i) $\int (t+\alpha)^{(k)} \Delta t = \frac{(t+\alpha)^{(k+1)}}{k+1} + C$;

(ii) $\int \binom{t}{\alpha} \Delta t = \binom{t}{\alpha+1} + C$.

Theorem 1.74 (Existence of Antiderivatives). *Every rd-continuous function has an antiderivative. In particular if $t_0 \in \mathbb{T}$, then F defined by*

$$F(t) := \int_{t_0}^{t} f(\tau) \Delta \tau \quad for \quad t \in \mathbb{T}$$

is an antiderivative of f.

Proof. Suppose f is an rd-continuous function. By Theorem 1.60 (ii), f is regulated. Let F be a function guaranteed to exist by Theorem 1.70, together with D, satisfying

$$F^{\Delta}(t) = f(t) \quad \text{for all} \quad t \in D.$$

This F is pre-differentiable with D. We have to show that $F^{\Delta}(t) = f(t)$ holds for all $t \in \mathbb{T}^{\kappa}$ (this, of course, includes all points in $\mathbb{T}^{\kappa} \setminus D$). So let $t \in \mathbb{T}^{\kappa} \setminus D$. Then t is right-dense because $\mathbb{T}^{\kappa} \setminus D$ cannot contain any right-scattered points according to Definition 1.62. Since f is rd-continuous, it is continuous at t. Let $\varepsilon > 0$. Then there exists a neighborhood U of t with

$$|f(s) - f(t)| \leq \varepsilon \quad \text{for all} \quad s \in U.$$

Define

$$h(\tau) := F(\tau) - f(t)(\tau - t_0) \quad \text{for} \quad \tau \in \mathbb{T}.$$

Then h is pre-differentiable with D and we have

$$h^{\Delta}(\tau) = F^{\Delta}(\tau) - f(t) = f(\tau) - f(t) \quad \text{for all} \quad \tau \in D.$$

Hence

$$|h^{\Delta}(s)| = |f(s) - f(t)| \leq \varepsilon \quad \text{for all} \quad s \in D \cap U.$$

Therefore

$$\sup_{s \in D \cap U} |h^{\Delta}(s)| \leq \varepsilon.$$

Thus, by Corollary 1.68, we have for $r \in U$

$$
\begin{aligned}
|F(t) - F(r) - f(t)(t-r)| &= |h(t) + f(t)(t-t_0) - [h(r) + f(t)(r-t_0)] \\
&\qquad -f(t)(t-r)| \\
&= |h(t) - h(r)| \\
&\leq \left\{ \sup_{s \in D \cap U} |h^{\Delta}(s)| \right\} |t - r| \\
&\leq \varepsilon |t - r|.
\end{aligned}
$$

But this shows that F is differentiable at t with $F^{\Delta}(t) = f(t)$. $\square$

Table 1.3. The two most important examples

Time scale $\mathbb{T}$	$\mathbb{R}$	$\mathbb{Z}$
Backward jump operator $\rho(t)$	t	$t-1$
Forward jump operator $\sigma(t)$	t	$t+1$
Graininess $\mu(t)$	0	1
Derivative $f^{\Delta}(t)$	$f'(t)$	$\Delta f(t)$
Integral $\int_a^b f(t)\Delta t$	$\int_a^b f(t)dt$	$\sum_{t=a}^{b-1} f(t)$ (if $a<b$)
Rd-continuous f	continuous f	any f

Theorem 1.75. *If* $f \in C_{\mathrm{rd}}$ *and* $t \in \mathbb{T}^{\kappa}$, *then*

$$\int_t^{\sigma(t)} f(\tau)\Delta\tau = \mu(t)f(t).$$

Proof. By Theorem 1.74, there exists an antiderivative F of f, and

$$\begin{aligned}
\int_t^{\sigma(t)} f(\tau)\Delta\tau &= F(\sigma(t)) - F(t) \\
&= \mu(t)F^{\Delta}(t) \\
&= \mu(t)f(t),
\end{aligned}$$

where the second equation holds because of Theorem 1.16 (iv). □

Theorem 1.76. *If* $f^{\Delta} \geq 0$, *then* f *is nondecreasing.*

Proof. Let $f^{\Delta} \geq 0$ on $[a,b]$ and let $s,t \in \mathbb{T}$ with $a \leq s \leq t \leq b$. Then

$$f(t) = f(s) + \int_s^t f^{\Delta}(\tau)\Delta\tau \geq f(s)$$

so that the conclusion follows. □

Theorem 1.77. *If* $a,b,c \in \mathbb{T}$, $\alpha \in \mathbb{R}$, *and* $f,g \in C_{\mathrm{rd}}$, *then*

(i) $\int_a^b [f(t) + g(t)]\Delta t = \int_a^b f(t)\Delta t + \int_a^b g(t)\Delta t$;

(ii) $\int_a^b (\alpha f)(t)\Delta t = \alpha \int_a^b f(t)\Delta t$;

(iii) $\int_a^b f(t)\Delta t = -\int_b^a f(t)\Delta t$;

(iv) $\int_a^b f(t)\Delta t = \int_a^c f(t)\Delta t + \int_c^b f(t)\Delta t$;

(v) $\int_a^b f(\sigma(t))g^{\Delta}(t)\Delta t = (fg)(b) - (fg)(a) - \int_a^b f^{\Delta}(t)g(t)\Delta t$;

(vi) $\int_a^b f(t)g^{\Delta}(t)\Delta t = (fg)(b) - (fg)(a) - \int_a^b f^{\Delta}(t)g(\sigma(t))\Delta t$;

(vii) $\int_a^a f(t)\Delta t = 0$;

(viii) *if* $|f(t)| \le g(t)$ *on* $[a,b)$, *then*

$$\left| \int_a^b f(t)\Delta t \right| \le \int_a^b g(t)\Delta t;$$

(ix) *if* $f(t) \ge 0$ *for all* $a \le t < b$, *then* $\int_a^b f(t)\Delta t \ge 0$.

Proof. These results follow easily from Definition 1.71, Theorem 1.20, and Theorem 1.67. We only prove (i), (iv), and (v), and leave the rest of the proof as an exercise (see Exercise 1.78). Since f and g are rd-continuous, they possess antiderivatives F and G by Theorem 1.74. By Theorem 1.20 (i), $F+G$ is an antiderivative of $f+g$ so that

$$\begin{aligned}
\int_a^b (f+g)(t)\Delta t &= (F+G)(b) - (F+G)(a) \\
&= F(b) - F(a) + G(b) - G(a) \\
&= \int_a^b f(t)\Delta t + \int_a^b g(t)\Delta t.
\end{aligned}$$

Also

$$\begin{aligned}
\int_a^b f(t)\Delta t &= F(b) - F(a) \\
&= F(c) - F(a) + F(b) - F(c) \\
&= \int_a^c f(t)\Delta t + \int_c^b f(t)\Delta t.
\end{aligned}$$

Finally, since fg is an antiderivative of $f^\sigma g^\Delta + f^\Delta g$,

$$\int_a^b \left(f^\sigma g^\Delta + f^\Delta g \right)(t)\Delta t = (fg)(b) - (fg)(a),$$

so that (v) follows by using (i). □

Note that the formulas in Theorem 1.77 (v) and (vi) are called *integration by parts* formulas. Also note that all of the formulas given in Theorem 1.77 also hold for the case that f and g are only regulated functions.

Exercise 1.78. Finish the proof of Theorem 1.77. Also prove each item of Theorem 1.77 assuming that the functions f and g are merely regulated rather than rd-continuous.

Theorem 1.79. *Let* $a, b \in \mathbb{T}$ *and* $f \in C_{\mathrm{rd}}$.

(i) *If* $\mathbb{T} = \mathbb{R}$, *then*

$$\int_a^b f(t)\Delta t = \int_a^b f(t)dt,$$

where the integral on the right is the usual Riemann integral from calculus.
(ii) *If* $[a,b]$ *consists of only isolated points, then*

$$\int_a^b f(t)\Delta t = \begin{cases} \sum_{t\in[a,b)} \mu(t)f(t) & \text{if} \quad a < b \\ 0 & \text{if} \quad a = b \\ -\sum_{t\in[b,a)} \mu(t)f(t) & \text{if} \quad a > b. \end{cases}$$

(iii) *If* $\mathbb{T} = h\mathbb{Z} = \{hk : k \in \mathbb{Z}\}$, *where* $h > 0$, *then*

$$\int_a^b f(t)\Delta t = \begin{cases} \sum_{k=\frac{a}{h}}^{\frac{b}{h}-1} f(kh)h & \text{if} \quad a < b \\ 0 & \text{if} \quad a = b \\ -\sum_{k=\frac{b}{h}}^{\frac{a}{h}-1} f(kh)h & \text{if} \quad a > b. \end{cases}$$

(iv) *If* $\mathbb{T} = \mathbb{Z}$, *then*

$$\int_a^b f(t)\Delta t = \begin{cases} \sum_{t=a}^{b-1} f(t) & \text{if} \quad a < b \\ 0 & \text{if} \quad a = b \\ -\sum_{t=b}^{a-1} f(t) & \text{if} \quad a > b. \end{cases}$$

Proof. Part (i) follows from Example 1.18 (i) and the standard fundamental theorem of calculus. We now prove (ii). First note that $[a, b]$ contains only finitely many points since each point in $[a, b]$ is isolated. Assume that $a < b$ and let $[a, b] = \{t_0, t_1, \ldots, t_n\}$, where

$$a = t_0 < t_1 < t_2 < \cdots < t_n = b.$$

By Theorem 1.77 (iv),

$$\begin{aligned} \int_a^b f(t)\Delta t &= \sum_{i=0}^{n-1} \int_{t_i}^{t_{i+1}} f(t)\Delta t \\ &= \sum_{i=0}^{n-1} \int_{t_i}^{\sigma(t_i)} f(t)\Delta t \\ &= \sum_{i=0}^{n-1} \mu(t_i)f(t_i) \\ &= \sum_{t\in[a,b)} \mu(t)f(t), \end{aligned}$$

where the third equation above follows from Theorem 1.75. If $b < a$, then the result follows from what we just proved and Theorem 1.77 (iii). If $a = b$, then $\int_a^b f(t)\Delta t = 0$ by Theorem 1.77 (vii). Parts (iii) and (iv) are special cases of (ii) (see Exercise 1.80). $\square$

Exercise 1.80. Prove that Theorem 1.79 (iii) and (iv) follow from Theorem 1.79 (ii).

Exercise 1.81. Let $a \in \mathbb{T}$, where $\mathbb{T}$ is an arbitrary time scale and evaluate $\int_a^t 1\Delta s$. Also evaluate $\int_0^t s\Delta s$ for $t \in \mathbb{T}$, for $\mathbb{T} = \mathbb{R}$, for $\mathbb{T} = \mathbb{Z}$, for $\mathbb{T} = h\mathbb{Z}$, and for $\mathbb{T} = [0, 1] \cup [2, 3]$.

We next define the improper integral $\int_a^\infty f(t)\Delta t$ as one would expect.

Definition 1.82. If $a \in \mathbb{T}$, $\sup \mathbb{T} = \infty$, and f is rd-continuous on $[a, \infty)$, then we define the *improper integral* by

$$\int_a^\infty f(t)\Delta t := \lim_{b\to\infty} \int_a^b f(t)\Delta t$$

provided this limit exists, and we say that the improper integral converges in this case. If this limit does not exist, then we say that the improper integral diverges.

We now give two exercises concerning improper integrals.

Exercise 1.83. Evaluate the integral

$$\int_1^\infty \frac{1}{t^2} \Delta t$$

if $\mathbb{T} = q^{\mathbb{N}_0}$, where $q > 1$.

Exercise 1.84. Assume $a \in \mathbb{T}$, $a > 0$ and $\sup \mathbb{T} = \infty$. Evaluate

$$\int_a^\infty \frac{1}{t\sigma(t)} \Delta t.$$

1.5. Chain Rules

If $f, g : \mathbb{R} \to \mathbb{R}$, then the chain rule from calculus is that if g is differentiable at t and if f is differentiable at $g(t)$, then

$$(f \circ g)'(t) = f'(g(t)) g'(t).$$

The next example shows that the chain rule as we know it in calculus does not hold for all time scales.

Example 1.85. Assume $f, g : \mathbb{Z} \to \mathbb{Z}$ are defined by $f(t) = t^2$, $g(t) = 2t$ and our time scale is $\mathbb{T} = \mathbb{Z}$. It is easy to see that

$$(f \circ g)^\Delta(t) = 8t + 4 \neq 8t + 2 = f^\Delta(g(t)) g^\Delta(t) \quad \text{for all} \quad t \in \mathbb{Z}.$$

Hence the chain rule as we know it in calculus does not hold in this setting.

Exercise 1.86. Assume that $\mathbb{T} = \mathbb{Z}$ and $f(t) = g(t) = t^2$. Show that for all $t \neq 0$,

$$(f \circ g)^\Delta(t) \neq f^\Delta(g(t)) g^\Delta(t).$$

Sometimes the following substitute of the "continuous" chain rule is useful.

Theorem 1.87 (Chain Rule). *Assume $g : \mathbb{R} \to \mathbb{R}$ is continuous, $g : \mathbb{T} \to \mathbb{R}$ is delta differentiable on $\mathbb{T}^\kappa$, and $f : \mathbb{R} \to \mathbb{R}$ is continuously differentiable. Then there exists c in the real interval $[t, \sigma(t)]$ with*

$$(1.6) \qquad (f \circ g)^\Delta(t) = f'(g(c)) g^\Delta(t).$$

Proof. Fix $t \in \mathbb{T}^\kappa$. First we consider the case where t is right-scattered. In this case

$$(f \circ g)^\Delta(t) = \frac{f(g(\sigma(t))) - f(g(t))}{\mu(t)}.$$

If $g(\sigma(t)) = g(t)$, then we get $(f \circ g)^\Delta(t) = 0$ and $g^\Delta(t) = 0$ and so (1.6) holds for any c in the real interval $[t, \sigma(t)]$. Hence we can assume $g(\sigma(t)) \neq g(t)$. Then

$$\begin{aligned}
(f \circ g)^\Delta(t) &= \frac{f(g(\sigma(t))) - f(g(t))}{g(\sigma(t)) - g(t)} \cdot \frac{g(\sigma(t)) - g(t)}{\mu(t)} \\
&= f'(\xi) g^\Delta(t)
\end{aligned}$$

by the mean value theorem, where ξ is between $g(t)$ and $g(\sigma(t))$. Since $g : \mathbb{R} \to \mathbb{R}$ is continuous, there is $c \in [t, \sigma(t)]$ such that $g(c) = \xi$, which gives us the desired result.

It remains to consider the case when t is right-dense. In this case

$$(f \circ g)^{\Delta}(t) = \lim_{s \to t} \frac{f(g(t)) - f(g(s))}{t - s}$$

$$= \lim_{s \to t} \left\{ f'(\xi_s) \cdot \frac{g(t) - g(s)}{t - s} \right\}$$

by the mean value theorem in calculus, where ξ_s is between $g(s)$ and $g(t)$. By the continuity of g we get that $\lim_{s \to t} \xi_s = g(t)$ which gives us the desired result.   □

Example 1.88. Given $\mathbb{T} = \mathbb{Z}$, $f(t) = t^2$, $g(t) = 2t$, find directly the value c guaranteed by Theorem 1.87 so that

$$(f \circ g)^{\Delta}(3) = f'(g(c)) g^{\Delta}(3)$$

and show that c is in the interval guaranteed by Theorem 1.87. Using the calculations we made in Example 1.85 we get that this last equation becomes

$$28 = (4c)2.$$

Solving for c we get that $c = \frac{7}{2}$ which is in the real interval $[3, \sigma(3)] = [3, 4]$ as we are guaranteed by Theorem 1.87.

Exercise 1.89. Assume that $\mathbb{T} = \mathbb{Z}$ and $f(t) = g(t) = t^2$. Find directly the c guaranteed by Theorem 1.87 so that $(f \circ g)^{\Delta}(2) = f'(g(c)) g^{\Delta}(2)$ and be sure to note that c is in the interval guaranteed by Theorem 1.87.

Now we present a chain rule which calculates $(f \circ g)^{\Delta}$, where

$$g : \mathbb{T} \to \mathbb{R} \quad \text{and} \quad f : \mathbb{R} \to \mathbb{R}.$$

This chain rule is due to Christian Pötzsche, who derived it first in 1998 (see also Stefan Keller's PhD thesis [190] and [234]).

Theorem 1.90 (Chain Rule). *Let $f : \mathbb{R} \to \mathbb{R}$ be continuously differentiable and suppose $g : \mathbb{T} \to \mathbb{R}$ is delta differentiable. Then $f \circ g : \mathbb{T} \to \mathbb{R}$ is delta differentiable and the formula*

$$(f \circ g)^{\Delta}(t) = \left\{ \int_0^1 f'(g(t) + h\mu(t)g^{\Delta}(t)) \, dh \right\} g^{\Delta}(t)$$

holds.

Proof. First of all we apply the ordinary substitution rule from calculus to find

$$f(g(\sigma(t))) - f(g(s)) = \int_{g(s)}^{g(\sigma(t))} f'(\tau) d\tau$$

$$= [g(\sigma(t)) - g(s)] \int_0^1 f'(hg(\sigma(t)) + (1 - h)g(s)) dh.$$

Let $t \in \mathbb{T}^{\kappa}$ and $\varepsilon > 0$ be given. Since g is differentiable at t, there exists a neighborhood U_1 of t such that

$$|g(\sigma(t)) - g(s) - g^{\Delta}(t)(\sigma(t) - s)| \le \varepsilon^* |\sigma(t) - s| \quad \text{for all} \quad s \in U_1,$$

where

$$\varepsilon^* = \frac{\varepsilon}{1 + 2 \int_0^1 |f'(hg(\sigma(t)) + (1 - h)g(t))| dh}.$$

Moreover, f' is continuous on $\mathbb{R}$, and therefore it is uniformly continuous on closed subsets of $\mathbb{R}$, and (observe also that g is continuous as it is differentiable, see Theorem 1.16 (i)) hence there exists a neighborhood U_2 of t such that

$$|f'(hg(\sigma(t)) + (1-h)g(s)) - f'(hg(\sigma(t)) + (1-h)g(t))| \le \frac{\varepsilon}{2(\varepsilon^* + |g^\Delta(t)|)}$$

for all $s \in U_2$. To see this, note also that

$$
\begin{aligned}
|hg(\sigma(t)) + (1-h)g(s) - (hg(\sigma(t)) + (1-h)g(t))| &= (1-h)|g(s) - g(t)| \\
&\le |g(s) - g(t)|
\end{aligned}
$$

holds for all $0 \le h \le 1$. We then define $U = U_1 \cap U_2$ and let $s \in U$. For convenience we put

$$\alpha = hg(\sigma(t)) + (1-h)g(s) \quad \text{and} \quad \beta = hg(\sigma(t)) + (1-h)g(t).$$

Then we have

$$\left| (f \circ g)(\sigma(t)) - (f \circ g)(s) - (\sigma(t) - s)g^\Delta(t) \int_0^1 f'(\beta)dh \right|$$

$$= \left| [g(\sigma(t)) - g(s)] \int_0^1 f'(\alpha)dh - (\sigma(t) - s)g^\Delta(t) \int_0^1 f'(\beta)dh \right|$$

$$= \left| [g(\sigma(t)) - g(s) - (\sigma(t) - s)g^\Delta(t)] \int_0^1 f'(\alpha)dh \right.$$

$$\left. + (\sigma(t) - s)g^\Delta(t) \int_0^1 (f'(\alpha) - f'(\beta))dh \right|$$

$$\le |g(\sigma(t)) - g(s) - (\sigma(t) - s)g^\Delta(t)| \int_0^1 |f'(\alpha)|dh$$

$$+ |\sigma(t) - s||g^\Delta(t)| \int_0^1 |f'(\alpha) - f'(\beta)|dh$$

$$\le \varepsilon^*|\sigma(t) - s| \int_0^1 |f'(\alpha)|dh + |\sigma(t) - s||g^\Delta(t)| \int_0^1 |f'(\alpha) - f'(\beta)|dh$$

$$\le \varepsilon^*|\sigma(t) - s| \int_0^1 |f'(\beta)|dh + [\varepsilon^* + |g^\Delta(t)|]|\sigma(t) - s| \int_0^1 |f'(\alpha) - f'(\beta)|dh$$

$$\le \frac{\varepsilon}{2}|\sigma(t) - s| + \frac{\varepsilon}{2}|\sigma(t) - s|$$

$$= \varepsilon|\sigma(t) - s|.$$

Therefore $f \circ g$ is differentiable at t and the derivative is as claimed above. $\qquad \square$

Example 1.91. We define $g : \mathbb{Z} \to \mathbb{R}$ and $f : \mathbb{R} \to \mathbb{R}$ by

$$g(t) = t^2 \quad \text{and} \quad f(x) = \exp(x).$$

Then

$$g^\Delta(t) = (t+1)^2 - t^2 = 2t + 1 \quad \text{and} \quad f'(x) = \exp(x).$$

Hence we have by Theorem 1.90

$$
\begin{aligned}
(f \circ g)^{\Delta}(t) &= \left\{ \int_0^1 f'\left(g(t) + h\mu(t)g^{\Delta}(t)\right) dh \right\} g^{\Delta}(t) \\
&= (2t+1) \int_0^1 \exp(t^2 + h(2t+1))dh \\
&= (2t+1)\exp(t^2) \int_0^1 \exp(h(2t+1))dh \\
&= (2t+1)\exp(t^2)\frac{1}{2t+1} \left[\exp(h(2t+1))\right]_{h=0}^{h=1} \\
&= (2t+1)\exp(t^2)\frac{1}{2t+1}(\exp(2t+1) - 1) \\
&= \exp(t^2)(\exp(2t+1) - 1).
\end{aligned}
$$

On the other hand, it is easy to check that we have indeed

$$
\begin{aligned}
\Delta f(g(t)) &= f(g(t+1)) - f(g(t)) \\
&= \exp((t+1)^2) - \exp(t^2) \\
&= \exp(t^2 + 2t + 1) - \exp(t^2) \\
&= \exp(t^2)(\exp(2t+1) - 1).
\end{aligned}
$$

In the remainder of this section we present some results related to results in the paper by C. D. Ahlbrandt, M. Bohner, and J. Ridenhour [24]. Let $\mathbb{T}$ be a time scale and $\nu : \mathbb{T} \to \mathbb{R}$ be a strictly increasing function such that $\tilde{\mathbb{T}} = \nu(\mathbb{T})$ is also a time scale. By $\tilde{\sigma}$ we denote the jump function on $\tilde{\mathbb{T}}$ and by $\tilde{\Delta}$ we denote the derivative on $\tilde{\mathbb{T}}$. Then $\nu \circ \sigma = \tilde{\sigma} \circ \nu$.

Exercise 1.92. Prove that $\nu \circ \sigma = \tilde{\sigma} \circ \nu$ under the hypotheses of the above paragraph.

Theorem 1.93 (Chain Rule). *Assume that $\nu : \mathbb{T} \to \mathbb{R}$ is strictly increasing and $\tilde{\mathbb{T}} := \nu(\mathbb{T})$ is a time scale. Let $w : \tilde{\mathbb{T}} \to \mathbb{R}$. If $\nu^{\Delta}(t)$ and $w^{\tilde{\Delta}}(\nu(t))$ exist for $t \in \mathbb{T}^{\kappa}$, then*

$$
(w \circ \nu)^{\Delta} = (w^{\tilde{\Delta}} \circ \nu)\nu^{\Delta}.
$$

Proof. Let $0 < \varepsilon < 1$ be given and define $\varepsilon^* = \varepsilon \left[1 + |\nu^{\Delta}(t)| + |w^{\tilde{\Delta}}(\nu(t))|\right]^{-1}$. Note that $0 < \varepsilon^* < 1$. According to the assumptions, there exist neighborhoods $\mathcal{N}_1$ of t and $\mathcal{N}_2$ of $\nu(t)$ such that

$$
|\nu(\sigma(t)) - \nu(s) - (\sigma(t) - s)\nu^{\Delta}(t)| \leq \varepsilon^*|\sigma(t) - s| \quad \text{for all} \quad s \in \mathcal{N}_1
$$

and

$$
|w(\tilde{\sigma}(\nu(t))) - w(r) - (\tilde{\sigma}(\nu(t)) - r)w^{\tilde{\Delta}}(\nu(t))| \leq \varepsilon^*|\tilde{\sigma}(\nu(t)) - r|, \ r \in \mathcal{N}_2.
$$

Put $\mathcal{N} = \mathcal{N}_1 \cap \nu^{-1}(\mathcal{N}_2)$ and let $s \in \mathcal{N}$. Then $s \in \mathcal{N}_1$ and $\nu(s) \in \mathcal{N}_2$ and

$$
\begin{aligned}
&|w(\nu(\sigma(t))) - w(\nu(s)) - (\sigma(t) - s)[w^{\tilde{\Delta}}(\nu(t))\nu^{\Delta}(t)]| \\
&= \quad |w(\nu(\sigma(t))) - w(\nu(s)) - (\tilde{\sigma}(\nu(t)) - \nu(s))w^{\tilde{\Delta}}(\nu(t)) \\
&\qquad + [\tilde{\sigma}(\nu(t)) - \nu(s) - (\sigma(t) - s)\nu^{\Delta}(t)]w^{\tilde{\Delta}}(\nu(t))| \\
&\leq \quad \varepsilon^*|\tilde{\sigma}(\nu(t)) - \nu(s)| + \varepsilon^*|\sigma(t) - s||w^{\tilde{\Delta}}(\nu(t))| \\
&\leq \quad \varepsilon^* \left\{ |\tilde{\sigma}(\nu(t)) - \nu(s) - (\sigma(t) - s)\nu^{\Delta}(t)| + |\sigma(t) - s||\nu^{\Delta}(t)| \right. \\
&\qquad \left. + |\sigma(t) - s||w^{\tilde{\Delta}}(\nu(t))| \right\} \\
&\leq \quad \varepsilon^* \left\{ \varepsilon^*|\sigma(t) - s| + |\sigma(t) - s||\nu^{\Delta}(t)| + |\sigma(t) - s||w^{\tilde{\Delta}}(\nu(t))| \right\} \\
&= \quad \varepsilon^*|\sigma(t) - s| \left\{ \varepsilon^* + |\nu^{\Delta}(t)| + |w^{\tilde{\Delta}}(\nu(t))| \right\} \\
&\leq \quad \varepsilon^* \left\{ 1 + |\nu^{\Delta}(t)| + |w^{\tilde{\Delta}}(\nu(t))| \right\} |\sigma(t) - s| \\
&= \quad \varepsilon|\sigma(t) - s|.
\end{aligned}
$$

This proves the claim. $\square$

Example 1.94. Let $\mathbb{T} = \mathbb{N}_0$ and $\nu(t) = 4t + 1$. Hence

$$
\tilde{\mathbb{T}} = \nu(\mathbb{T}) = \{4n + 1 : n \in \mathbb{N}_0\} = \{1, 5, 9, 13, \ldots\}.
$$

Moreover, let $w : \tilde{\mathbb{T}} \to \mathbb{R}$ be defined by $w(t) = t^2$. Then

$$
(w \circ \nu)(t) = w(\nu(t)) = w(4t + 1) = (4t + 1)^2
$$

and hence

$$
\begin{aligned}
(w \circ \nu)^{\Delta}(t) &= [4(t + 1) + 1]^2 - (4t + 1)^2 \\
&= (4t + 5)^2 - (4t + 1)^2 \\
&= 16t^2 + 40t + 25 - 16t^2 - 8t - 1 \\
&= 32t + 24.
\end{aligned}
$$

Now we apply Theorem 1.93 to obtain the derivative of this composite function. We first calculate $\nu^{\Delta}(t) \equiv 4$ and then

$$
w^{\tilde{\Delta}}(t) = \frac{w(\tilde{\sigma}(t)) - w(t)}{\tilde{\sigma}(t) - t} = \frac{(t + 4)^2 - t^2}{t + 4 - t} = \frac{8t + 16}{4} = 2t + 4
$$

and therefore

$$
(w^{\tilde{\Delta}} \circ \nu)(t) = w^{\tilde{\Delta}}(\nu(t)) = w^{\tilde{\Delta}}(4t + 1) = 2(4t + 1) + 4 = 8t + 6.
$$

Thus we obtain

$$
[(w^{\tilde{\Delta}} \circ \nu)\nu^{\Delta}](t) = (8t + 6)4 = 32t + 24 = (w \circ \nu)^{\Delta}(t).
$$

Exercise 1.95. Let $\mathbb{T} = \mathbb{N}_0$, $\nu(t) = t^2$, $\tilde{\mathbb{T}} = \nu(\mathbb{T})$, and $w(t) = 2t^2 + 3$. Show directly as in Example 1.94 that

$$
(w \circ \nu)^{\Delta} = (w^{\tilde{\Delta}} \circ \nu)\nu^{\Delta}.
$$

Exercise 1.96. Find a time scale $\mathbb{T}$ and a strictly increasing function $\nu : \mathbb{T} \to \mathbb{R}$ such that $\nu(\mathbb{T})$ is not a time scale.

As a consequence of Theorem 1.93 we can now write down a formula for the derivative of the inverse function.

Theorem 1.97 (Derivative of the Inverse). *Assume $\nu : \mathbb{T} \to \mathbb{R}$ is strictly increasing and $\tilde{\mathbb{T}} := \nu(\mathbb{T})$ is a time scale. Then*

$$\frac{1}{\nu^{\Delta}} = (\nu^{-1})^{\tilde{\Delta}} \circ \nu$$

at points where ν^{Δ} is different from zero.

Proof. Let $w = \nu^{-1} : \tilde{\mathbb{T}} \to \mathbb{T}$ in the previous theorem.   □

Another consequence of Theorem 1.93 is the substitution rule for integrals.

Theorem 1.98 (Substitution). *Assume $\nu : \mathbb{T} \to \mathbb{R}$ is strictly increasing and $\tilde{\mathbb{T}} := \nu(\mathbb{T})$ is a time scale. If $f : \mathbb{T} \to \mathbb{R}$ is an rd-continuous function and ν is differentiable with rd-continuous derivative, then for $a, b \in \mathbb{T}$,*

$$\int_a^b f(t)\nu^{\Delta}(t)\Delta t = \int_{\nu(a)}^{\nu(b)} (f \circ \nu^{-1})(s)\tilde{\Delta}s.$$

Proof. Since $f\nu^{\Delta}$ is an rd-continuous function, it possesses an antiderivative F by Theorem 1.74, i.e., $F^{\Delta} = f\nu^{\Delta}$, and

$$
\begin{aligned}
\int_a^b f(t)\nu^{\Delta}(t)\Delta t &= \int_a^b F^{\Delta}(t)\Delta t \\
&= F(b) - F(a) \\
&= (F \circ \nu^{-1})(\nu(b)) - (F \circ \nu^{-1})(\nu(a)) \\
&= \int_{\nu(a)}^{\nu(b)} (F \circ \nu^{-1})^{\tilde{\Delta}}(s)\tilde{\Delta}s \\
&= \int_{\nu(a)}^{\nu(b)} (F^{\Delta} \circ \nu^{-1})(s)(\nu^{-1})^{\tilde{\Delta}}(s)\tilde{\Delta}s \\
&= \int_{\nu(a)}^{\nu(b)} ((f\nu^{\Delta}) \circ \nu^{-1})(s)(\nu^{-1})^{\tilde{\Delta}}(s)\tilde{\Delta}s \\
&= \int_{\nu(a)}^{\nu(b)} (f \circ \nu^{-1})(s)[(\nu^{\Delta} \circ \nu^{-1})(\nu^{-1})^{\tilde{\Delta}}](s)\tilde{\Delta}s \\
&= \int_{\nu(a)}^{\nu(b)} (f \circ \nu^{-1})(s)\tilde{\Delta}s,
\end{aligned}
$$

where for the fifth equal sign we have used Theorem 1.93 and in the last step we have used Theorem 1.97.   □

Example 1.99. In this example we use the method of substitution (Theorem 1.98) to evaluate the integral

$$\int_0^t \left(\sqrt{\tau^2 + 1} + \tau\right) 3^{\tau^2} \Delta\tau$$

for $t \in \mathbb{T} := \mathbb{N}_0^{\frac{1}{2}} = \{\sqrt{n} : n \in \mathbb{N}_0\}$. We take

$$\nu(t) = t^2$$

for $t \in \mathbb{N}_0^{\frac{1}{2}}$. Then $\nu : \mathbb{N}_0^{\frac{1}{2}} \to \mathbb{R}$ is strictly increasing and $\nu\left(\mathbb{N}_0^{\frac{1}{2}}\right) = \mathbb{N}_0$ is a time scale. From Exercise 1.19 we get that

$$\nu^\Delta(t) = \sqrt{t^2 + 1} + t.$$

Hence if $f(t) := 3^{t^2}$, we get from Theorem 1.98 that

$$
\begin{aligned}
\int_0^t \left(\sqrt{\tau^2 + 1} + \tau\right) 3^{\tau^2} \Delta\tau &= \int_0^t f(\tau)\nu^\Delta(\tau)\Delta\tau \\
&= \int_0^{t^2} f(\sqrt{s})\tilde{\Delta}s \\
&= \int_0^{t^2} 3^s \tilde{\Delta}s \\
&= \left[\frac{1}{2}3^s\right]_{s=0}^{s=t^2} \\
&= \frac{1}{2}(3^{t^2} - 1).
\end{aligned}
$$

Exercise 1.100. Evaluate the integral

$$\int_0^t 2\tau(2\tau - 1)\Delta\tau$$

for $t \in \mathbb{T} := \{\frac{n}{2} : n \in \mathbb{N}_0\}$ by applying Theorem 1.98 with $\nu(t) = 2t$.

Exercise 1.101. Evaluate the integral

$$\int_0^t \left[(\tau^3 + 1)^{\frac{2}{3}} + \tau(\tau^3 + 1)^{\frac{1}{3}} + \tau^2\right] 2^{\tau^3} \Delta\tau$$

for $t \in \mathbb{T} := \{\sqrt[3]{n} : n \in \mathbb{N}_0\}$.

1.6. Polynomials

An antiderivative of 0 is 1, an antiderivative of 1 is t, but it is not possible to find a closed formula (for an arbitrary time scale) of an antiderivative of t. Certainly $t^2/2$ is not the solution, as the derivative of $t^2/2$ is

$$\frac{t + \sigma(t)}{2} = t + \frac{\mu(t)}{2}$$

which is, as we know, e.g., by Example 1.56, not even necessarily a differentiable function (although it is the product of two differentiable functions). Similarly, none of the "classical" polynomials are necessarily more than once differentiable, see Theorem 1.24. So the question arises which function plays the role of, e.g., $t^2/2$, in the time scales calculus. It could be either

$$\int_0^t \sigma(\tau)\Delta\tau \quad \text{or} \quad \int_0^t \tau\Delta\tau.$$

In fact, if we define

$$g_2(t, s) = \int_s^t (\sigma(\tau) - s)\Delta\tau \quad \text{and} \quad h_2(t, s) = \int_s^t (\tau - s)\Delta\tau,$$

then we find the following relation between g_2 and h_2:

$$
\begin{aligned}
g_2(t,s) &= \int_s^t (\sigma(\tau) - s)\Delta\tau \\
&= \int_s^t (\sigma(\tau) + \tau)\Delta\tau - \int_s^t \tau\Delta\tau - \int_s^t s\Delta\tau \\
&= \int_s^t (\tau^2)^\Delta \Delta\tau + \int_t^s \tau\Delta\tau - s(t-s) \\
&= \int_t^s \tau\Delta\tau + t^2 - s^2 - s(t-s) \\
&= \int_t^s (\tau - t)\Delta\tau \\
&= h_2(s,t).
\end{aligned}
$$

In this section we give a Taylor's formula for functions on a general time scale. Many of the results in this section can be found in R. P. Agarwal and M. Bohner [9]. The generalized polynomials, that also occur in Taylor's formula, are the functions $g_k, h_k : \mathbb{T}^2 \to \mathbb{R}$, $k \in \mathbb{N}_0$, defined recursively as follows: The functions g_0 and h_0 are

(1.7) $g_0(t,s) = h_0(t,s) \equiv 1$ for all $s, t \in \mathbb{T}$,

and, given g_k and h_k for $k \in \mathbb{N}_0$, the functions g_{k+1} and h_{k+1} are

(1.8) $g_{k+1}(t,s) = \int_s^t g_k(\sigma(\tau), s)\Delta\tau$ for all $s, t \in \mathbb{T}$

and

(1.9) $h_{k+1}(t,s) = \int_s^t h_k(\tau, s)\Delta\tau$ for all $s, t \in \mathbb{T}.$

Note that the functions g_k and h_k are all well defined according to Theorem 1.60 and Theorem 1.74. If we let $h_k^\Delta(t,s)$ denote for each fixed s the derivative of $h_k(t,s)$ with respect to t, then

$$h_k^\Delta(t,s) = h_{k-1}(t,s) \quad \text{for} \quad k \in \mathbb{N}, \ t \in \mathbb{T}^\kappa.$$

Similarly

$$g_k^\Delta(t,s) = g_{k-1}(\sigma(t), s) \quad \text{for} \quad k \in \mathbb{N}, \ t \in \mathbb{T}^\kappa.$$

The above definitions obviously imply

$$g_1(t,s) = h_1(t,s) = t - s \quad \text{for all} \quad s, t \in \mathbb{T}.$$

However, finding g_k and h_k for $k > 1$ is not easy in general. But for a particular given time scale it might be easy to find these functions. We will consider several examples first before we present Taylor's formula in general.

Example 1.102. For the cases $\mathbb{T} = \mathbb{R}$ and $\mathbb{T} = \mathbb{Z}$ it is easy to find the functions g_k and h_k:

(i) First, consider $\mathbb{T} = \mathbb{R}$. Then $\sigma(t) = t$ for $t \in \mathbb{R}$ so that $g_k = h_k$ for $k \in \mathbb{N}_0$. We have

$$
\begin{aligned}
g_2(t,s) &= h_2(t,s) \\
&= \int_s^t (\tau - s) d\tau \\
&= \left. \frac{(\tau - s)^2}{2} \right|_{\tau=s}^{\tau=t} \\
&= \frac{(t - s)^2}{2}.
\end{aligned}
$$

We claim that for $k \in \mathbb{N}_0$

$$
(1.10) \qquad g_k(t,s) = h_k(t,s) = \frac{(t-s)^k}{k!} \qquad \text{for all} \quad s,t \in \mathbb{R}
$$

as we will now show using the principle of mathematical induction: Obviously (1.10) holds for $k = 0$. Assume that (1.10) holds with k replaced by some $m \in \mathbb{N}_0$. Then

$$
\begin{aligned}
g_{m+1}(t,s) &= h_{m+1}(t,s) \\
&= \int_s^t \frac{(\tau - s)^m}{m!} d\tau \\
&= \left. \frac{(\tau - s)^{m+1}}{(m+1)!} \right|_{\tau=s}^{\tau=t} \\
&= \frac{(t - s)^{m+1}}{(m+1)!},
\end{aligned}
$$

i.e., (1.10) holds with k replaced by $m + 1$. We note that, for an n times differentiable function $f : \mathbb{R} \to \mathbb{R}$, the following well-known Taylor's formula holds: Let $\alpha \in \mathbb{R}$ be arbitrary. Then, for all $t \in \mathbb{R}$, the representations

$$
\begin{aligned}
f(t) &= \sum_{k=0}^{n-1} \frac{(t-\alpha)^k}{k!} f^{(k)}(\alpha) + \frac{1}{(n-1)!} \int_\alpha^t (t-\tau)^{n-1} f^{(n)}(\tau) d\tau \\
(1.11) \qquad &= \sum_{k=0}^{n-1} h_k(t,\alpha) f^{(k)}(\alpha) + \int_\alpha^t h_{n-1}(t,\sigma(\tau)) f^{(n)}(\tau) d\tau \\
(1.12) \qquad &= \sum_{k=0}^{n-1} (-1)^k g_k(\alpha,t) f^{(k)}(\alpha) + \int_\alpha^t (-1)^{n-1} g_{n-1}(\sigma(\tau),t) f^{(n)}(\tau) d\tau
\end{aligned}
$$

are valid, where $f^{(k)}$ denotes as usual the kth derivative of f. Above we have used the relationship

$$
(1.13) \qquad (-1)^k g_k(s,t) = (-1)^k \frac{(s-t)^k}{k!} = \frac{(t-s)^k}{k!} = h_k(t,s),
$$

which holds for all $k \in \mathbb{N}_0$.

(ii) Next, consider $\mathbb{T} = \mathbb{Z}$. Then $\sigma(t) = t + 1$ for $t \in \mathbb{Z}$. We have for $s, t \in \mathbb{Z}$

$$
\begin{aligned}
h_2(t, s) &= \int_s^t h_1(\tau, s) \Delta s \\
&= \left[\frac{(\tau - s)^{(2)}}{2} \right]_{\tau=s}^{\tau=t} \\
&= \binom{t - s}{2}.
\end{aligned}
$$

We claim that for $k \in \mathbb{N}_0$ we have

$$
(1.14) \qquad h_k(t, s) = \frac{(t - s)^{(k)}}{k!} = \binom{t - s}{k} \qquad \text{for all} \quad s, t \in \mathbb{Z}.
$$

Assume (1.14) holds for k replaced by m. Then

$$
\begin{aligned}
h_{m+1}(t, s) &= \int_s^t h_m(\tau, s) \Delta \tau \\
&= \int_s^t \frac{(\tau - s)^{(m)}}{m!} \Delta \tau \\
&= \frac{(t - s)^{(m+1)}}{(m + 1)!},
\end{aligned}
$$

which is (1.14) with k replaced by $m + 1$. Hence by mathematical induction we get that (1.14) holds for all $k \in \mathbb{N}_0$. Similarly it is possible to show that

$$
(1.15) \qquad g_k(t, s) = \frac{(t - s + k - 1)^{(k)}}{k!} \qquad \text{for all} \quad s, t \in \mathbb{Z}
$$

holds for all $k \in \mathbb{N}_0$. As before we observe that the relationship

$$
\begin{aligned}
(1.16) \quad (-1)^k g_k(s, t) &= (-1)^k \frac{(s - t + k - 1)^{(k)}}{k!} \\
&= (-1)^k \frac{(s - t + k - 1)(s - t + k - 2) \cdots (s - t)}{k!} \\
&= \frac{(t - s) \cdots (t - s + 2 - k)(t - s + 1 - k)}{k!} \\
&= \frac{(t - s)^{(k)}}{k!} \\
&= h_k(t, s)
\end{aligned}
$$

holds for all $k \in \mathbb{N}_0$. The well-known discrete version of Taylor's formula (see, e.g., [5]) reads as follows: Let $f : \mathbb{Z} \to \mathbb{R}$ be a function, and let $\alpha \in \mathbb{Z}$. Then, for all $t \in \mathbb{Z}$ with $t > \alpha + n$, the representations

$$
\begin{aligned}
f(t) &= \sum_{k=0}^{n-1} \frac{(t - \alpha)^{(k)}}{k!} \Delta^k f(\alpha) + \frac{1}{(n - 1)!} \sum_{\tau=\alpha}^{t-n} (t - \tau - 1)^{(n-1)} \Delta^n f(\tau) \\
(1.17) \qquad &= \sum_{k=0}^{n-1} h_k(t, \alpha) \Delta^k f(\alpha) + \sum_{\tau=\alpha}^{t-n} h_{n-1}(t, \sigma(\tau)) \Delta^n f(\tau) \\
(1.18) \qquad &= \sum_{k=0}^{n-1} (-1)^k g_k(\alpha, t) \Delta^k f(\alpha) + \sum_{\tau=\alpha}^{t-n} (-1)^{n-1} g_{n-1}(\sigma(\tau), t) \Delta^n f(\tau)
\end{aligned}
$$

hold, where Δ^k is the usual k times iterated forward difference operator.

Exercise 1.103. Verify that formula (1.15) holds for all $k \in \mathbb{N}_0$.

Example 1.104. We consider the time scale

$$\mathbb{T} = \overline{q^{\mathbb{Z}}} \quad \text{for some} \quad q > 1$$

from Example 1.41. The claim is that

$$(1.19) \qquad h_k(t,s) = \prod_{\nu=0}^{k-1} \frac{t - q^\nu s}{\sum_{\mu=0}^{\nu} q^\mu} \quad \text{for all} \quad s,t \in \mathbb{T}$$

holds for all $k \in \mathbb{N}_0$. Obviously, for $k = 0$ (observe that the empty product is considered to be 1, as usual), the claim (1.19) holds. Now we assume (1.19) holds with k replaced by some $m \in \mathbb{N}_0$. Then

$$\left\{ \prod_{\nu=0}^{m} \frac{t - q^\nu s}{\sum_{\mu=0}^{\nu} q^\mu} \right\}^\Delta = \frac{\prod_{\nu=0}^{m} \frac{\sigma(t) - q^\nu s}{\sum_{\mu=0}^{\nu} q^\mu} - \prod_{\nu=0}^{m} \frac{t - q^\nu s}{\sum_{\mu=0}^{\nu} q^\mu}}{\mu(t)}$$

$$= \frac{\frac{qt - q^m s}{\sum_{\mu=0}^{m} q^\mu} h_m(\sigma(t),s) - \frac{t - q^m s}{\sum_{\mu=0}^{m} q^\mu} h_m(t,s)}{\mu(t)}$$

$$= \frac{qt - q^m s}{\mu(t) \sum_{\mu=0}^{m} q^\mu} \left\{ h_m(t,s) + \mu(t) h_m^\Delta(t,s) \right\} - \frac{t - q^m s}{\mu(t) \sum_{\mu=0}^{m} q^\mu} h_m(t,s)$$

$$= \frac{qt - t}{\mu(t) \sum_{\mu=0}^{m} q^\mu} h_m(t,s) + \frac{qt - q^m s}{\sum_{\mu=0}^{m} q^\mu} h_m^\Delta(t,s)$$

$$= \frac{1}{\sum_{\mu=0}^{m} q^\mu} h_m(t,s) + \frac{qt - q^m s}{\sum_{\mu=0}^{m} q^\mu} h_{m-1}(t,s)$$

$$= \frac{1}{\sum_{\mu=0}^{m} q^\mu} \left\{ h_m(t,s) + (qt - q^m s) \prod_{\nu=0}^{m-2} \frac{t - q^\nu s}{\sum_{\mu=0}^{\nu} q^\mu} \right\}$$

$$= \frac{1}{\sum_{\mu=0}^{m} q^\mu} \left\{ h_m(t,s) + q \left(\sum_{\mu=0}^{m-1} q^\mu \right) \prod_{\nu=0}^{m-1} \frac{t - q^\nu s}{\sum_{\mu=0}^{\nu} q^\mu} \right\}$$

$$= \frac{h_m(t,s)}{\sum_{\mu=0}^{m} q^\mu} \left\{ 1 + q \sum_{\mu=0}^{m-1} q^\mu \right\}$$

$$= h_m(t,s) \frac{1 + \sum_{\mu=1}^{m} q^\mu}{\sum_{\mu=0}^{m} q^\mu}$$

$$= h_m(t,s)$$

so that (1.19) follows with k replaced by $m + 1$. Hence, by the principle of mathematical induction, (1.19) holds for all $k \in \mathbb{N}_0$.

As a special case, we consider the choice $q = 2$. This yields

$$h_k(t,s) = \prod_{\nu=0}^{k-1} \frac{t - 2^\nu s}{2^{\nu+1} - 1} \quad \text{for all} \quad s,t \in \mathbb{T}.$$

E.g., we have

$$h_2(t,s) = \frac{(t - s)(t - 2s)}{3}, \quad h_3(t,s) = \frac{(t - s)(t - 2s)(t - 4s)}{21},$$

and

$$h_4(t,s) = \frac{(t-s)(t-2s)(t-4s)(t-8s)}{315}.$$

Exercise 1.105. Find the functions g_k for $k \in \mathbb{N}_0$, where $\mathbb{T}$ is the time scale considered in Example 1.104.

Exercise 1.106. Let $\mathbb{T} = \overline{q^{\mathbb{Z}}}$ for some $q > 1$. For $n \in \mathbb{N}$, evaluate

$$\int_0^t s^n \Delta s.$$

Exercise 1.107. Find the functions $h_k(\cdot, 0)$, for $k = 0, 1, 2, 3$ if $\mathbb{T} := [0,1] \cup [3,4]$.

Now we will present and prove Taylor's formula for the case of a general time scale $\mathbb{T}$. First we need three preliminary results.

Lemma 1.108. *Let $n \in \mathbb{N}$. If f is n times differentiable and p_k, $0 \le k \le n-1$, are differentiable at some $t \in \mathbb{T}$ with*

$$(1.20) \qquad p_{k+1}^\Delta(t) = p_k^\sigma(t) \quad \text{for all} \quad 0 \le k \le n-2,$$

then we have at t

$$\left[\sum_{k=0}^{n-1}(-1)^k f^{\Delta^k} p_k\right]^\Delta = (-1)^{n-1} f^{\Delta^n} p_{n-1}^\sigma + f p_0^\Delta.$$

Proof. Using Theorem 1.20 (i), (ii), (iii), and (1.20) we find that

$$\left[\sum_{k=0}^{n-1}(-1)^k f^{\Delta^k} p_k\right]^\Delta = \sum_{k=0}^{n-1}(-1)^k \left[f^{\Delta^k} p_k\right]^\Delta$$

$$= \sum_{k=0}^{n-1}(-1)^k \left[f^{\Delta^{k+1}} p_k^\sigma + f^{\Delta^k} p_k^\Delta\right]$$

$$= \sum_{k=0}^{n-2}(-1)^k f^{\Delta^{k+1}} p_k^\sigma + (-1)^{n-1} f^{\Delta^n} p_{n-1}^\sigma + \sum_{k=1}^{n-1}(-1)^k f^{\Delta^k} p_k^\Delta + f p_0^\Delta$$

$$= \sum_{k=0}^{n-2}(-1)^k f^{\Delta^{k+1}} p_k^\sigma + (-1)^{n-1} f^{\Delta^n} p_{n-1}^\sigma - \sum_{k=0}^{n-2}(-1)^k f^{\Delta^{k+1}} p_{k+1}^\Delta + f p_0^\Delta$$

$$= (-1)^{n-1} f^{\Delta^n} p_{n-1}^\sigma + f p_0^\Delta$$

holds at t. This proves the lemma. $\square$

Lemma 1.109. *The functions g_k defined in (1.7) and (1.8) satisfy for all $t \in \mathbb{T}$*

$$g_n(\rho^k(t), t) = 0 \quad \text{for all} \quad n \in \mathbb{N} \quad \text{and all} \quad 0 \le k \le n-1.$$

Proof. We prove this result by induction. For $k = 0$, we have

$$g_n(\rho^0(t), t) = g_n(t, t) = 0.$$

To complete the induction it suffices to show that

$$g_{n-1}(\rho^k(t), t) = g_n(\rho^k(t), t) = 0 \quad \text{with} \quad 0 \le k < n$$

implies that

$$g_n(\rho^{k+1}(t), t) = 0.$$

If $\rho^k(t)$ is left-dense, then $\rho^{k+1}(t) = \rho^k(t)$ so that

$$g_n(\rho^{k+1}(t), t) = g_n(\rho^k(t), t) = 0.$$

If $\rho^k(t)$ is not left-dense, then it is left-scattered, and $\sigma(\rho^{k+1}(t)) = \rho^k(t)$ so that by Theorem 1.16 (iv) and (1.8)

$$
\begin{aligned}
g_n(\rho^{k+1}(t), t) &= g_n(\sigma(\rho^{k+1}(t)), t) - \mu(\rho^{k+1}(t))g_n^\Delta(\rho^{k+1}(t), t) \\
&= g_n(\rho^k(t), t) - \mu(\rho^{k+1}(t))g_{n-1}(\sigma(\rho^{k+1}(t)), t) \\
&= g_n(\rho^k(t), t) - \mu(\rho^{k+1}(t))g_{n-1}(\rho^k(t), t) \\
&= 0
\end{aligned}
$$

(observe $n \neq 1$). This proves our claim. $\square$

Lemma 1.110. *Let $n \in \mathbb{N}$, $t \in \mathbb{T}$, and suppose that f is $(n-1)$ times differentiable at $\rho^{n-1}(t)$. Then we have*

(1.21)
$$\sum_{k=0}^{n-1} (-1)^k f^{\Delta^k}(\rho^{n-1}(t)) g_k(\rho^{n-1}(t), t) = f(t),$$

where the functions g_k are defined by (1.7) and (1.8).

Proof. First we have

$$(-1)^0 f^{\Delta^0}(\rho^0(t)) g_0(\rho^0(t), t) = f(t) g_0(t, t) = f(t)$$

so that (1.21) is true for $n = 1$. Suppose now that (1.21) holds with n replaced by some $m \in \mathbb{N}$. Then we consider two cases. First, if $\rho^{m-1}(t)$ is left-dense, then $\rho^m(t) = \rho^{m-1}(t)$ and hence

$$
\begin{aligned}
\sum_{k=0}^{m} (-1)^k f^{\Delta^k}(\rho^m(t)) g_k(\rho^m(t), t) &= \sum_{k=0}^{m-1} (-1)^k f^{\Delta^k}(\rho^m(t)) g_k(\rho^m(t), t) \\
&\quad + (-1)^m f^{\Delta^m}(\rho^m(t)) g_m(\rho^m(t), t) \\
&= \sum_{k=0}^{m-1} (-1)^k f^{\Delta^k}(\rho^{m-1}(t)) g_k(\rho^{m-1}(t), t) + (-1)^m f^{\Delta^m}(\rho^{m-1}(t)) g_m(\rho^{m-1}(t), t) \\
&= f(t) + (-1)^m f^{\Delta^m}(\rho^{m-1}(t)) g_m(\rho^{m-1}(t), t) \\
&= f(t),
\end{aligned}
$$

where we used Lemma 1.109 to obtain the last equation. Hence (1.21) holds with n replaced by $m + 1$. We have to draw the same conclusion for the case that $\rho^{m-1}(t)$ is left-scattered. In this case we have $\sigma(\rho^m(t)) = \rho^{m-1}(t)$ and hence by Theorem 1.16 (iv) and (1.8) for $k \in \mathbb{N}$

$$
\begin{aligned}
g_k(\rho^{m-1}(t), t) &= g_k(\sigma(\rho^m(t)), t) \\
&= g_k(\rho^m(t), t) + \mu(\rho^m(t)) g_k^\Delta(\rho^m(t), t) \\
&= g_k(\rho^m(t), t) + \mu(\rho^m(t)) g_{k-1}(\sigma(\rho^m(t)), t) \\
&= g_k(\rho^m(t), t) + \mu(\rho^m(t)) g_{k-1}(\rho^{m-1}(t), t).
\end{aligned}
$$

Therefore we conclude (apply Lemma 1.109 for the third equation)

$$\sum_{k=0}^{m}(-1)^k f^{\Delta^k}(\rho^m(t))g_k(\rho^m(t),t)$$

$$= f(\rho^m(t)) + \sum_{k=1}^{m}(-1)^k f^{\Delta^k}(\rho^m(t))g_k(\rho^m(t),t)$$

$$= f(\rho^m(t))$$
$$+ \sum_{k=1}^{m}(-1)^k f^{\Delta^k}(\rho^m(t))\left[g_k(\rho^{m-1}(t),t) - \mu(\rho^m(t))g_{k-1}(\rho^{m-1}(t),t)\right]$$

$$= f(\rho^m(t)) + \sum_{k=1}^{m-1}(-1)^k f^{\Delta^k}(\rho^m(t))g_k(\rho^{m-1}(t),t)$$
$$+ \sum_{k=1}^{m}(-1)^{k-1}\mu(\rho^m(t))f^{\Delta^k}(\rho^m(t))g_{k-1}(\rho^{m-1}(t),t)$$

$$= \sum_{k=0}^{m-1}(-1)^k f^{\Delta^k}(\rho^m(t))g_k(\rho^{m-1}(t),t)$$
$$+ \sum_{k=0}^{m-1}(-1)^k \mu(\rho^m(t))f^{\Delta^{k+1}}(\rho^m(t))g_k(\rho^{m-1}(t),t)$$

$$= \sum_{k=0}^{m-1}(-1)^k \left[f^{\Delta^k}(\rho^m(t)) + \mu(\rho^m(t))(f^{\Delta^k})^{\Delta}(\rho^m(t))\right] g_k(\rho^{m-1}(t),t)$$

$$= \sum_{k=0}^{m-1}(-1)^k (f^{\Delta^k})^{\sigma}(\rho^m(t))g_k(\rho^{m-1}(t),t)$$

$$= \sum_{k=0}^{m-1}(-1)^k (f^{\Delta^k})(\sigma(\rho^m(t)))g_k(\rho^{m-1}(t),t)$$

$$= \sum_{k=0}^{m-1}(-1)^k (f^{\Delta^k})(\rho^{m-1}(t))g_k(\rho^{m-1}(t),t)$$

$$= f(t).$$

As before, (1.21) holds with n replaced by $m+1$, and an application of the principle of mathematical induction finishes the proof. $\qquad\square$

Theorem 1.111 (Taylor's Formula). *Let $n \in \mathbb{N}$. Suppose f is n times differentiable on $\mathbb{T}^{\kappa^n}$. Let $\alpha \in \mathbb{T}^{\kappa^{n-1}}$, $t \in \mathbb{T}$, and define the functions g_k by (1.7) and (1.8), i.e.,*

$$g_0(r,s) \equiv 1 \quad and \quad g_{k+1}(r,s) = \int_s^r g_k(\sigma(\tau),s)\Delta\tau \text{ for } k \in \mathbb{N}_0.$$

Then we have

$$f(t) = \sum_{k=0}^{n-1}(-1)^k g_k(\alpha,t)f^{\Delta^k}(\alpha) + \int_\alpha^{\rho^{n-1}(t)}(-1)^{n-1}g_{n-1}(\sigma(\tau),t)f^{\Delta^n}(\tau)\Delta\tau.$$

Proof. By Lemma 1.108 we have

$$\left[\sum_{k=0}^{n-1}(-1)^k g_k(\cdot,t)f^{\Delta^k}\right]^{\Delta}(\tau) = (-1)^{n-1}g_{n-1}(\sigma(\tau),t)f^{\Delta^n}(\tau)$$

for all $\tau \in \mathbb{T}^{\kappa^n}$. Since $\alpha, \rho^{n-1}(t) \in \mathbb{T}^{\kappa^{n-1}}$, we may integrate the above equation from α to $\rho^{n-1}(t)$ to obtain

$$\int_{\alpha}^{\rho^{n-1}(t)} (-1)^{n-1}g_{n-1}(\sigma(\tau),t)f^{\Delta^n}(\tau)\Delta\tau = \int_{\alpha}^{\rho^{n-1}(t)} \left[\sum_{k=0}^{n-1}(-1)^k g_k(\cdot,t)f^{\Delta^k}\right]^{\Delta}(\tau)\Delta\tau$$

$$= \sum_{k=0}^{n-1}(-1)^k g_k(\rho^{n-1}(t),t)f^{\Delta^k}(\rho^{n-1}(t)) - \sum_{k=0}^{n-1}(-1)^k g_k(\alpha,t)f^{\Delta^k}(\alpha)$$

$$= f(t) - \sum_{k=0}^{n-1}(-1)^k g_k(\alpha,t)f^{\Delta^k}(\alpha),$$

where we used formula (1.21) from Lemma 1.110. $\square$

Our first application of Theorem 1.111 yields an alternative form of Taylor's formula in terms of the functions h_k rather than g_k.

Theorem 1.112. *The functions g_k and h_k defined in (1.7), (1.8), and (1.9) satisfy*

$$h_n(t,s) = (-1)^n g_n(s,t) \quad \text{for all} \quad t \in \mathbb{T} \quad \text{and all} \quad s \in \mathbb{T}^{\kappa^n}.$$

Proof. We let $t \in \mathbb{T}$, $s \in \mathbb{T}^{\kappa^n}$, and apply Theorem 1.111 with

$$\alpha = s \quad \text{and} \quad f = h_n(\cdot,s).$$

This yields $f^{\Delta} = h_{n-1}(\cdot,s)$ and (apply (1.9) successively)

$$f^{\Delta^k} = h_{n-k}(\cdot,s) \quad \text{for all} \quad 0 \le k \le n.$$

Therefore

$$f^{\Delta^k}(s) = h_{n-k}(s,s) = 0 \quad \text{for all} \quad 0 \le k \le n-1,$$

$f^{\Delta^n}(s) = h_0(s,s) = 1$, and $f^{\Delta^{n+1}}(\tau) \equiv 0$. An application of Theorem 1.111 now shows

$$\begin{aligned}
h_n(t,s) &= f(t) \\
&= \sum_{k=0}^{n}(-1)^k g_k(\alpha,t)f^{\Delta^k}(\alpha) + \int_{\alpha}^{\rho^n(t)}(-1)^n g_n(\sigma(\tau),t)f^{\Delta^{n+1}}(\tau)\Delta\tau \\
&= \sum_{k=0}^{n}(-1)^k g_k(s,t)f^{\Delta^k}(s) + \int_{s}^{\rho^n(t)}(-1)^n g_n(\sigma(\tau),t)f^{\Delta^{n+1}}(\tau)\Delta\tau \\
&= (-1)^n g_n(s,t)f^{\Delta^n}(s) \\
&= (-1)^n g_n(s,t),
\end{aligned}$$

and this completes the proof. $\square$

Theorem 1.113 (Taylor's Formula). *Let $n \in \mathbb{N}$. Suppose f is n times differentiable on $\mathbb{T}^{\kappa^n}$. Let $\alpha \in \mathbb{T}^{\kappa^{n-1}}$, $t \in \mathbb{T}$, and define the functions h_k by (1.7) and (1.9), i.e.,*

$$h_0(r,s) \equiv 1 \quad \text{and} \quad h_{k+1}(r,s) = \int_s^r h_k(\tau,s)\Delta\tau \text{ for } k \in \mathbb{N}_0.$$

Then we have

$$f(t) = \sum_{k=0}^{n-1} h_k(t, \alpha) f^{\Delta^k}(\alpha) + \int_\alpha^{\rho^{n-1}(t)} h_{n-1}(t, \sigma(\tau)) f^{\Delta^n}(\tau) \Delta\tau.$$

Proof. This is a direct consequence of Theorem 1.111 and Theorem 1.112. □

Remark 1.114. The reader may compare Example 1.102 (i.e., the cases $\mathbb{T} = \mathbb{R}$ and $\mathbb{T} = \mathbb{Z}$) to the above presented theory. Theorem 1.112 is reflected in formulas (1.10) and (1.16). While the first version of Taylor's formula (Theorem 1.111) corresponds to formulas (1.12) and (1.18), the second version, Theorem 1.113, corresponds to formulas (1.11) and (1.17).

1.7. Further Basic Results

We now state the intermediate value theorem for a continuous function on a time scale.

Theorem 1.115 (Intermediate Value Theorem). *Assume $x : \mathbb{T} \to \mathbb{R}$ is continuous, $a < b$ are points in $\mathbb{T}$, and*

$$x(a)x(b) < 0.$$

Then there exists $c \in [a, b)$ such that either $x(c) = 0$ or

$$x(c)x^\sigma(c) < 0.$$

Exercise 1.116. Prove the above intermediate value theorem.

We will use the following result later. By $f^\Delta(t, \tau)$ in the following theorem we mean for each fixed τ the derivative of $f(t, \tau)$ with respect to t.

Theorem 1.117. *Let $a \in \mathbb{T}^\kappa$, $b \in \mathbb{T}$ and assume $f : \mathbb{T} \times \mathbb{T}^\kappa \to \mathbb{R}$ is continuous at (t, t), where $t \in \mathbb{T}^\kappa$ with $t > a$. Also assume that $f^\Delta(t, \cdot)$ is rd-continuous on $[a, \sigma(t)]$. Suppose that for each $\varepsilon > 0$ there exists a neighborhood U of t, independent of $\tau \in [a, \sigma(t)]$, such that*

$$|f(\sigma(t), \tau) - f(s, \tau) - f^\Delta(t, \tau)(\sigma(t) - s)| \le \varepsilon|\sigma(t) - s| \quad \text{for all} \quad s \in U,$$

where f^Δ denotes the derivative of f with respect to the first variable. Then

   (i) $g(t) := \int_a^t f(t, \tau)\Delta\tau$ *implies* $g^\Delta(t) = \int_a^t f^\Delta(t, \tau)\Delta\tau + f(\sigma(t), t)$;
   (ii) $h(t) := \int_t^b f(t, \tau)\Delta\tau$ *implies* $h^\Delta(t) = \int_t^b f^\Delta(t, \tau)\Delta\tau - f(\sigma(t), t)$.

Proof. We only prove (i) while the proof of (ii) is similar and is left to the reader in Exercise 1.118. Let $\varepsilon > 0$. By assumption there exists a neighborhood U_1 of t such that

$$|f(\sigma(t), \tau) - f(s, \tau) - f^\Delta(t, \tau)(\sigma(t) - s)| \le \frac{\varepsilon}{2(\sigma(t) - a)}|\sigma(t) - s| \quad \text{for all} \quad s \in U_1.$$

Since f is continuous at (t, t), there exists a neighborhood U_2 of t such that

$$|f(s, \tau) - f(t, t)| \le \frac{\varepsilon}{2} \quad \text{whenever} \quad s, \tau \in U_2.$$

Now define $U = U_1 \cap U_2$ and let $s \in U$. Then

$$\left| g(\sigma(t)) - g(s) - \left[f(\sigma(t), t) + \int_a^t f^\Delta(t, \tau) \Delta\tau \right] (\sigma(t) - s) \right|$$

$$= \left| \int_a^{\sigma(t)} f(\sigma(t), \tau) \Delta\tau - \int_a^s f(s, \tau) \Delta\tau - (\sigma(t) - s) f(\sigma(t), t) \right.$$

$$\left. - (\sigma(t) - s) \int_a^t f^\Delta(t, \tau) \Delta\tau \right|$$

$$= \left| \int_a^{\sigma(t)} \left[f(\sigma(t), \tau) - f(s, \tau) - f^\Delta(t, \tau)(\sigma(t) - s) \right] \Delta\tau \right.$$

$$\left. - \int_{\sigma(t)}^s f(s, \tau) \Delta\tau - (\sigma(t) - s) f(\sigma(t), t) - (\sigma(t) - s) \int_{\sigma(t)}^t f^\Delta(t, \tau) \Delta\tau \right|$$

$$= \left| \int_a^{\sigma(t)} \left[f(\sigma(t), \tau) - f(s, \tau) - f^\Delta(t, \tau)(\sigma(t) - s) \right] \Delta\tau \right.$$

$$\left. - \int_{\sigma(t)}^s f(s, \tau) \Delta\tau - (\sigma(t) - s) f(\sigma(t), t) + (\sigma(t) - s) \mu(t) f^\Delta(t, t) \right|$$

$$= \left| \int_a^{\sigma(t)} \left[f(\sigma(t), \tau) - f(s, \tau) - f^\Delta(t, \tau)(\sigma(t) - s) \right] \Delta\tau \right.$$

$$\left. + \int_s^{\sigma(t)} f(s, \tau) \Delta\tau - (\sigma(t) - s) f(t, t) \right|$$

$$= \left| \int_a^{\sigma(t)} \left[f(\sigma(t), \tau) - f(s, \tau) - f^\Delta(t, \tau)(\sigma(t) - s) \right] \Delta\tau \right.$$

$$\left. + \int_s^{\sigma(t)} [f(s, \tau) - f(t, t)] \Delta\tau \right|$$

$$\leq \int_a^{\sigma(t)} \left| f(\sigma(t), \tau) - f(s, \tau) - f^\Delta(t, \tau)(\sigma(t) - s) \right| \Delta\tau$$

$$+ \left| \int_s^{\sigma(t)} |f(s, \tau) - f(t, t)| \, \Delta\tau \right|$$

$$\leq \int_a^{\sigma(t)} \frac{\varepsilon}{2(\sigma(t) - a)} |\sigma(t) - s| \Delta\tau + \left| \int_s^{\sigma(t)} \frac{\varepsilon}{2} \Delta\tau \right|$$

$$= \frac{\varepsilon}{2} |\sigma(t) - s| + \frac{\varepsilon}{2} |\sigma(t) - s|$$

$$= \varepsilon |\sigma(t) - s|,$$

where we also have used Theorem 1.75 and Theorem 1.16 (iv). □

Exercise 1.118. Prove Theorem 1.117 (ii).

Finally, we present several versions of L'Hôpital's rule. We let

$$\overline{\mathbb{T}} = \mathbb{T} \cup \{\sup \mathbb{T}\} \cup \{\inf \mathbb{T}\}.$$

If $\infty \in \overline{\mathbb{T}}$, we call ∞ left-dense, and $-\infty$ is called right-dense provided $-\infty \in \overline{\mathbb{T}}$. For any left-dense $t_0 \in \mathbb{T}$ and any $\varepsilon > 0$, the set

$$L_\varepsilon(t_0) = \{t \in \mathbb{T} : 0 < t_0 - t < \varepsilon\}$$

is nonempty, and so is $L_\varepsilon(\infty) = \{t \in \mathbb{T} : t > \frac{1}{\varepsilon}\}$ if $\infty \in \overline{\mathbb{T}}$. The sets $R_\varepsilon(t_0)$ for right-dense $t_0 \in \overline{\mathbb{T}}$ and $\varepsilon > 0$ are defined accordingly. For a function $h : \mathbb{T} \to \mathbb{R}$ we define

$$\liminf_{t \to t_0^-} h(t) = \lim_{\varepsilon \to 0^+} \inf_{t \in L_\varepsilon(t_0)} h(t) \quad \text{for left-dense} \quad t_0 \in \overline{\mathbb{T}},$$

and $\liminf_{t \to t_0^+} h(t)$, $\limsup_{t \to t_0^-} h(t)$, $\limsup_{t \to t_0^+} h(t)$ are defined analogously.

Theorem 1.119 (L'Hôpital's Rule). *Assume f and g are differentiable on $\mathbb{T}$ with*

$$(1.22) \qquad \lim_{t \to t_0^-} f(t) = \lim_{t \to t_0^-} g(t) = 0 \quad \text{for some left-dense} \quad t_0 \in \overline{\mathbb{T}}.$$

Suppose there exists $\varepsilon > 0$ with

$$(1.23) \qquad g(t) > 0, \ g^\Delta(t) < 0 \quad \text{for all} \quad t \in L_\varepsilon(t_0).$$

Then we have

$$\liminf_{t \to t_0^-} \frac{f^\Delta(t)}{g^\Delta(t)} \le \liminf_{t \to t_0^-} \frac{f(t)}{g(t)} \le \limsup_{t \to t_0^-} \frac{f(t)}{g(t)} \le \limsup_{t \to t_0^-} \frac{f^\Delta(t)}{g^\Delta(t)}.$$

Proof. Let $\delta \in (0, \varepsilon]$ and put $\alpha = \inf_{\tau \in L_\delta(t_0)} \frac{f^\Delta(\tau)}{g^\Delta(\tau)}$, $\beta = \sup_{\tau \in L_\delta(t_0)} \frac{f^\Delta(\tau)}{g^\Delta(\tau)}$. Then

$$\alpha g^\Delta(\tau) \ge f^\Delta(\tau) \ge \beta g^\Delta(\tau) \quad \text{for all} \quad \tau \in L_\delta(t_0)$$

by (1.23) and hence by Theorem 1.77 (vii)

$$\int_s^t \alpha g^\Delta(\tau) \Delta\tau \ge \int_s^t f^\Delta(\tau) \Delta\tau \ge \int_s^t \beta g^\Delta(\tau) \Delta\tau \text{ for all } s, t \in L_\delta(t_0), \ s < t$$

so that

$$\alpha g(t) - \alpha g(s) \ge f(t) - f(s) \ge \beta g(t) - \beta g(s) \text{ for all } s, t \in L_\delta(t_0), \ s < t.$$

Now, letting $t \to t_0^-$, we find from (1.22)

$$-\alpha g(s) \ge -f(s) \ge -\beta g(s) \quad \text{for all} \quad s \in L_\delta(t_0)$$

and hence by (1.23)

$$\inf_{\tau \in L_\delta(t_0)} \frac{f^\Delta(\tau)}{g^\Delta(\tau)} = \alpha \le \inf_{s \in L_\delta(t_0)} \frac{f(s)}{g(s)} \le \sup_{s \in L_\delta(t_0)} \frac{f(s)}{g(s)} \le \beta = \sup_{\tau \in L_\delta(t_0)} \frac{f^\Delta(\tau)}{g^\Delta(\tau)}.$$

Letting $\delta \to 0^+$ yields our desired result. $\qquad \square$

Theorem 1.120 (L'Hôpital's Rule). *Assume f and g are differentiable on $\mathbb{T}$ with*

$$(1.24) \qquad \lim_{t \to t_0^-} g(t) = \infty \quad \text{for some left-dense} \quad t_0 \in \overline{\mathbb{T}}.$$

Suppose there exists $\varepsilon > 0$ with

$$(1.25) \qquad g(t) > 0, \ g^\Delta(t) > 0 \quad \text{for all} \quad t \in L_\varepsilon(t_0).$$

Then $\lim_{t \to t_0^-} \frac{f^\Delta(t)}{g^\Delta(t)} = r \in \overline{\mathbb{R}}$ implies $\lim_{t \to t_0^-} \frac{f(t)}{g(t)} = r$.

Proof. First suppose $r \in \mathbb{R}$. Let $c > 0$. Then there exists $\delta \in (0, \varepsilon]$ such that

$$\left| \frac{f^{\Delta}(\tau)}{g^{\Delta}(\tau)} - r \right| \le c \quad \text{for all} \quad \tau \in L_{\delta}(t_0)$$

and hence by (1.25)

$$-cg^{\Delta}(\tau) \le f^{\Delta}(\tau) - rg^{\Delta}(\tau) \le cg^{\Delta}(\tau) \quad \text{for all} \quad \tau \in L_{\delta}(t_0).$$

We integrate as in the proof of Theorem 1.119 and use (1.25) to obtain

$$(r - c)\left(1 - \frac{g(s)}{g(t)}\right) \le \frac{f(t)}{g(t)} - \frac{f(s)}{g(t)} \le (r + c)\left(1 - \frac{g(s)}{g(t)}\right) \quad \text{for all } s, t \in L_{\delta}(t_0); s < t.$$

Letting $t \to t_0^-$ and applying (1.24) yields

$$r - c \le \liminf_{t \to t_0^-} \frac{f(t)}{g(t)} \le \limsup_{t \to t_0^-} \frac{f(t)}{g(t)} \le r + c.$$

Now we let $c \to 0^+$ to see that $\lim_{t \to t_0^-} \frac{f(t)}{g(t)}$ exists and equals r.

Next, if $r = \infty$ (and similarly if $r = -\infty$), let $c > 0$. Then there exists $\delta \in (0, \varepsilon]$ with

$$\frac{f^{\Delta}(\tau)}{g^{\Delta}(\tau)} \ge \frac{1}{c} \quad \text{for all} \quad \tau \in L_{\delta}(t_0)$$

and hence by (1.25)

$$f^{\Delta}(\tau) \ge \frac{1}{c} g^{\Delta}(\tau) \quad \text{for all} \quad \tau \in L_{\delta}(t_0).$$

We integrate again to get

$$\frac{f(t)}{g(t)} - \frac{f(s)}{g(t)} \ge \frac{1}{c}\left(1 - \frac{g(s)}{g(t)}\right) \quad \text{for all} \quad s, t \in L_{\delta}(t_0); s < t.$$

Thus, letting $t \to t_0^-$ and applying (1.24), we find $\liminf_{t \to t_0^-} \frac{f(t)}{g(t)} \ge \frac{1}{c}$, and then, letting $c \to 0^+$, we obtain $\lim_{t \to t_0^-} \frac{f(t)}{g(t)} = \infty = r$. $\square$

1.8. Notes and References

The calculus of measure chains originated in 1988, when Stefan Hilger completed his PhD thesis [159] "Ein Maßkettenkalkül mit Anwendung auf Zentrumsmannigfaltigkeiten" at Universität Würzburg, Germany, under the supervision of Bernd Aulbach. The first publications on the subject of measure chains are Hilger [160] and Aulbach and Hilger [49, 50].

The basic definitions of jump operators and delta differentiation are due to Hilger. We have included some previously unpublished but easy examples in the section on differentiation, e.g., the derivatives of t^k for $k \in \mathbb{Z}$ in Theorem 1.24 and the Leibniz formula for the nth derivative of a product of two functions in Theorem 1.32. This first section also contains the induction principle on time scales. It is essentially contained in Dieudonné [110]. For the proofs of some of the main existence results, this induction principle is utilized, while the remaining results presented in this book can be derived without using this induction principle. The induction principle requires the distinction of several cases, depending on whether the point under consideration is left-dense, left-scattered, right-dense, or right-scattered. However, it is one of the key features of the time scales calculus to unify proofs for the continuous and the discrete cases, and hence it should, wherever possible, be avoided to start discussing those several cases independently. For that purpose it is helpful to have formulas available that are valid at each element of the time scale, as, e.g., in Theorem 1.16 (iv) or the product and quotient rules in Theorem 1.20.

Section 1.3 contains many examples that are considered throughout this book. Concerning the two main examples of the time scales calculus, we refer to the classical books [103, 152, 237] for differential equations and [5, 125, 191, 200, 216] for difference equations. Other examples discussed in this section contain the integer multiples of a number $h > 0$, denoted by $h\mathbb{Z}$; the union of closed intervals, denoted by $\mathbb{P}$; the integer powers of a number $q > 1$, denoted by $\overline{q^{\mathbb{Z}}}$; the integer squares, denoted by $\mathbb{Z}^2$; the harmonic numbers, denoted by H_n; and the Cantor set. Many of these examples are considered here for the first time or are contained in [90]. The example on the electric circuit is taken from Stefan Keller's PhD thesis [190], 1999, at Universität Augsburg, Germany, under the supervision of Bernd Aulbach.

In the section on integration we define the two crucial notions of rd-continuity and regularity. These are classes of functions that possess an antiderivative and a pre-antiderivative, respectively. We present the corresponding existence theorems; however, delay most of their proofs to the last chapter in a more general setting. A rather weak form of an integral, the Cauchy integral, is defined in terms of antiderivatives. It would be of interest to derive an integral for a more general class of functions than the ones presented here, and this might be one direction of future research. Theorem 1.65 has been discussed with Roman Hilscher and Ronald Mathsen.

Section 1.5 contains several versions of the chain rule, which, in the time scales setting, does not appear in its usual form. Theorem 1.90 is due to Christian Pötzsche [234], and it also appears in Keller [190]. The chain rule given in Theorem 1.93 originated in the study of so-called alpha derivatives. These alpha derivatives are introduced in Calvin Ahlbrandt, Martin Bohner, and Jerry Ridenhour [24], and some related results are presented in the last chapter. This topic on alpha derivatives could become a future area of research.

The time scales versions of L'Hôpital's rules given in the last section are taken from Ravi Agarwal and Martin Bohner [9]. Most of the results presented in the section on polynomials are contained in [9] as well. The functions g_k and h_k are the time scales "substitutes" for the usual polynomials. There are two of them, and that is why there are also two versions of Taylor's theorem, one using the functions g_k and the other one using the functions h_k as coefficients. The functions g_k and h_k could also be used when expressing functions defined on time scales in terms of "power" series, but this question is not addressed in this book. It also remains an open area of future research, and one could investigate how solutions of dynamic equations (see the next chapter) may be expressed in terms of such "power" series.

CHAPTER 2

First Order Linear Equations

2.1. Hilger's Complex Plane

Definition 2.1. Suppose $f : \mathbb{T} \times \mathbb{R}^2 \to \mathbb{R}$. Then the equation

$$(2.1) \qquad\qquad y^\Delta = f(t, y, y^\sigma)$$

is called a first order *dynamic equation*, sometimes also a *differential equation*. If

$$f(t, y, y^\sigma) = f_1(t)y + f_2(t) \quad \text{or} \quad f(t, y, y^\sigma) = f_1(t)y^\sigma + f_2(t)$$

for functions f_1 and f_2, then (2.1) is called a *linear equation*. A function $y : \mathbb{T} \to \mathbb{R}$ is called a *solution* of (2.1) if

$$y^\Delta(t) = f(t, y(t), y(\sigma(t))) \quad \text{is satisfied for all} \quad t \in \mathbb{T}^\kappa.$$

The *general solution* of (2.1) is defined to be the set of all solutions of (2.1). Given $t_0 \in \mathbb{T}$ and $y_0 \in \mathbb{R}$, the problem

$$y^\Delta = f(t, y, y^\sigma), \quad y(t_0) = y_0$$

is called an initial value problem (abbreviated by IVP), and a solution y of (2.1) with $y(t_0) = y_0$ is called a solution of this IVP.

In this section we will study linear first order dynamic equations and initial value problems for them. In fact, we will construct the solution of the initial value problem

$$y^\Delta = p(t)y, \quad y(t_0) = 1$$

explicitly, and we will call this solution the *exponential function* associated with the given time scale. Many of the results in this section can be found in Hilger [**164**]. We now introduce what we call the *Hilger complex plane*.

Definition 2.2. For $h > 0$ we define the *Hilger complex numbers*, the *Hilger real axis*, the *Hilger alternating axis*, and the *Hilger imaginary circle* as

$$\mathbb{C}_h := \left\{ z \in \mathbb{C} : z \neq -\frac{1}{h} \right\},$$

$$\mathbb{R}_h := \left\{ z \in \mathbb{C}_h : z \in \mathbb{R} \text{ and } z > -\frac{1}{h} \right\},$$

$$\mathbb{A}_h := \left\{ z \in \mathbb{C}_h : z \in \mathbb{R} \text{ and } z < -\frac{1}{h} \right\},$$

$$\mathbb{I}_h := \left\{ z \in \mathbb{C}_h : \left| z + \frac{1}{h} \right| = \frac{1}{h} \right\},$$

respectively. For $h = 0$, let $\mathbb{C}_0 := \mathbb{C}$, $\mathbb{R}_0 := \mathbb{R}$, $\mathbb{I}_0 = i\mathbb{R}$, and $\mathbb{A}_0 := \emptyset$.

The sets introduced in Definition 2.2 for $h > 0$ are drawn in Figure 2.1.

Figure 2.1. Hilger's Complex Plane

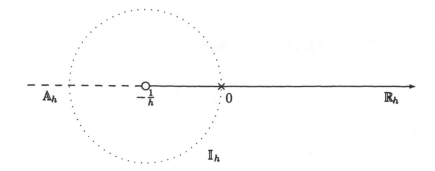

Definition 2.3. Let $h > 0$ and $z \in \mathbb{C}_h$. We define the Hilger real part of z by

$$\text{Re}_h(z) := \frac{|zh + 1| - 1}{h}$$

and the Hilger imaginary part of z by

$$\text{Im}_h(z) := \frac{\text{Arg}(zh + 1)}{h},$$

where $\text{Arg}(z)$ denotes the principal argument of z (i.e., $-\pi < \text{Arg}(z) \leq \pi$).

Note that $\text{Re}_h(z)$ and $\text{Im}_h(z)$ satisfy

$$-\frac{1}{h} < \text{Re}_h(z) < \infty \quad \text{and} \quad -\frac{\pi}{h} < \text{Im}_h(z) \leq \frac{\pi}{h},$$

respectively. In particular, $\text{Re}_h(z) \in \mathbb{R}_h$ (see Figure 2.2).

Definition 2.4. Let $-\frac{\pi}{h} < \omega \leq \frac{\pi}{h}$. We define the Hilger purely imaginary number $\overset{\circ}{i}\omega$ by

(2.2)
$$\overset{\circ}{i}\omega = \frac{e^{i\omega h} - 1}{h}.$$

For $z \in \mathbb{C}_h$, we have that $\overset{\circ}{i}\,\text{Im}_h(z) \in \mathbb{I}_h$ (see Figure 2.2).

Exercise 2.5. Prove that

$$\lim_{h \to 0} [\text{Re}_h z + \overset{\circ}{i}\,\text{Im}_h z] = \text{Re}(z) + i\,\text{Im}(z).$$

Also try to visualize this by looking at what happens to the Hilger complex plane as $h \to 0$.

Hilger calls formula (2.3) in the following theorem the arc-chord formula because it gives the length of the chord of the Hilger imaginary circle $\mathbb{I}_h$ from the origin to $\overset{\circ}{i}\omega$.

Figure 2.2. Hilger's Complex Numbers

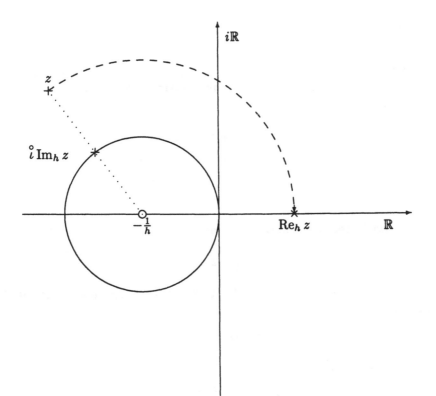

Theorem 2.6. *If* $-\frac{\pi}{h} < \omega \leq \frac{\pi}{h}$*, then*

$$(2.3) \qquad |\overset{\circ}{\iota}\omega|^2 = \frac{4}{h^2} \sin^2 \frac{\omega h}{2}.$$

Proof. Using (2.2) we get that

$$
\begin{aligned}
|\overset{\circ}{\iota}\omega|^2 &= (\overset{\circ}{\iota}\omega)\overline{(\overset{\circ}{\iota}\omega)} \\
&= \left(\frac{e^{i\omega h} - 1}{h}\right)\left(\frac{e^{-i\omega h} - 1}{h}\right) \\
&= \frac{2 - e^{i\omega h} - e^{-i\omega h}}{h^2} \\
&= \frac{2}{h^2}\left(1 - \cos(\omega h)\right) \\
&= \frac{4}{h^2}\sin^2\frac{\omega h}{2},
\end{aligned}
$$

which is our desired result. $\qquad\square$

Theorem 2.7. *If we define the "circle plus" addition $\oplus$ on $\mathbb{C}_h$ by*

$$z \oplus w := z + w + zwh,$$

then $(\mathbb{C}_h, \oplus)$ is an Abelian group.

Proof. To prove that we have closure under the addition $\oplus$, note that clearly, for $z, w \in \mathbb{C}_h$, $z \oplus w$ is a complex number. It only remains to show that $z \oplus w \neq -\frac{1}{h}$, but this follows from

$$
\begin{aligned}
1 + h(z \oplus w) &= 1 + h(z + w + wzh) \\
&= 1 + hz + hw + hzwh \\
&= (1 + hz)(1 + hw) \\
&\neq 0.
\end{aligned}
$$

Hence $\mathbb{C}_h$ is closed under the addition $\oplus$. Since

$$z \oplus 0 = 0 \oplus z = z,$$

0 is the additive identity for $\oplus$. For $z \in \mathbb{C}_h$, to find the additive inverse of z under $\oplus$, we must solve

$$z \oplus w = 0$$

for w. Hence we must solve

$$z + w + zwh = 0$$

for w. Thus

$$w = -\frac{z}{1 + zh}$$

is the additive inverse of z under the addition $\oplus$. The fact that the associative law holds is the next exercise. Hence $(\mathbb{C}_h, \oplus)$ is a group. Since

$$
\begin{aligned}
z \oplus w &= z + w + zwh \\
&= w + z + wzh \\
&= w \oplus z,
\end{aligned}
$$

the commutative law holds, and hence $(\mathbb{C}_h, \oplus)$ is an Abelian group. □

Exercise 2.8. Show that the addition $\oplus$ on $\mathbb{C}_h$ satisfies the associative law.

Exercise 2.9. Assuming that $z \in \mathbb{C}_h$ and $w \in \mathbb{C}$, simplify the expression $z \oplus \frac{w}{1+hz}$.

Theorem 2.10. *For $z \in \mathbb{C}_h$ we have*

$$z = \operatorname{Re}_h z \oplus i \operatorname{Im}_h z.$$

Proof. Let $z \in \mathbb{C}_h$. Then

$$\operatorname{Re}_h z \oplus i \operatorname{Im}_h z = \frac{|zh+1|-1}{h} \oplus i \frac{\operatorname{Arg}(zh+1)}{h}$$

$$= \frac{|zh+1|-1}{h} \oplus \frac{\exp(i \operatorname{Arg}(zh+1))-1}{h}$$

$$= \frac{|zh+1|-1}{h} + \frac{\exp(i \operatorname{Arg}(zh+1))-1}{h}$$

$$\quad + \frac{|zh+1|-1}{h} \frac{\exp(i \operatorname{Arg}(zh+1))-1}{h} h$$

$$= \frac{1}{h} \left\{ |zh+1| - 1 + \exp(i \operatorname{Arg}(zh+1)) - 1 \right.$$

$$\left. + [|zh+1|-1][\exp(i \operatorname{Arg}(zh+1))-1] \right\}$$

$$= \frac{1}{h} \left\{ |zh+1| \exp(i \operatorname{Arg}(zh+1)) - 1 \right\}$$

$$= \frac{(zh+1)-1}{h} = z,$$

which proves the claim. $\qquad \square$

Exercise 2.11. If $n \in \mathbb{N}$ and $z \in \mathbb{C}_h$, then we define the "circle dot" multiplication $\odot$ by

$$n \odot z := z \oplus z \oplus z \oplus \cdots \oplus z,$$

where we have n terms on the right-hand side of this last equation. Show that

$$n \odot z = \frac{(zh+1)^n - 1}{h}.$$

In the proof of Theorem 2.7 we saw that if $z \in \mathbb{C}_h$, then the additive inverse of z under the operation $\oplus$ is

$$(2.4) \qquad \ominus z := \frac{-z}{1+zh}.$$

Exercise 2.12. Show directly from (2.4) that if $z \in \mathbb{C}_h$, then $\ominus(\ominus z) = z$.

Definition 2.13. We define the "circle minus" subtraction $\ominus$ on $\mathbb{C}_h$ by

$$(2.5) \qquad z \ominus w := z \oplus (\ominus w).$$

Exercise 2.14. If $z, w \in \mathbb{C}_h$ with $h \geq 0$, show directly from (2.5) that

(i) $z \ominus z = 0$;
(ii) $z \ominus w = \frac{z-w}{1+wh}$;
(iii) $z \ominus w = z - w$ if $h = 0$.

Exercise 2.15. Show that if $z \in \mathbb{C}_h$, then $\bar{z} = \ominus z$ iff $z \in \mathbb{I}_h$.

Exercise 2.16. Show that $\ominus(\overset{\circ}{i}\omega) = \overline{\overset{\circ}{i}\omega}$.

In Table 2.1 we show various complex numbers and their additive inverses.

Exercise 2.17. Graph the complex numbers z and their additive inverses $\ominus z$ given in Table 2.1 in the Hilger complex plane. Do you notice any geometric symmetries between z and $\ominus z$?

Table 2.1. Additive Inverses

z	0	1	i	$\frac{1}{h}$	$-\frac{2}{h}$	$\frac{i-1}{h}$
$\ominus z$	0	$-\frac{1}{1+h}$	$\frac{1}{i-h}$	$-\frac{1}{2h}$	$-\frac{2}{h}$	$-\frac{i+1}{h}$

Definition 2.18. If $z \in \mathbb{C}_h$, then the generalized square of z is defined by

$$z^{\circledtwo} := (-z)(\ominus z) = \frac{z^2}{1+zh}.$$

The next theorem gives various properties of the generalized square.

Theorem 2.19. *For* $z \in \mathbb{C}_h$,

(2.6)
$$(\ominus z)^{\circledtwo} = z^{\circledtwo},$$

(2.7)
$$1 + zh = \frac{z^2}{z^{\circledtwo}},$$

(2.8)
$$z + (\ominus z) = z^{\circledtwo} \cdot h,$$

(2.9)
$$z \oplus z^{\circledtwo} = z + z^2,$$

(2.10)
$$z^{\circledtwo} \in \mathbb{R} \iff z \in \mathbb{R}_h \cup \mathbb{A}_h \cup \mathbb{I}_h,$$

and for $-\frac{\pi}{h} < \omega \leq \frac{\pi}{h}$,

(2.11)
$$-(\overset{\circ}{i}\omega)^{\circledtwo} = \frac{4}{h^2} \sin^2\left(\frac{\omega h}{2}\right).$$

Proof. Throughout this proof assume $z \in \mathbb{C}_h$. To prove (2.6), consider

$$\begin{aligned}
(\ominus z)^{\circledtwo} &= \frac{(\ominus z)^2}{1 + (\ominus z)h} \\
&= \frac{\frac{z^2}{(1+zh)^2}}{1 + \frac{-zh}{1+zh}} \\
&= \frac{z^2}{(1+zh)^2} \cdot \frac{1+zh}{1+zh-zh} \\
&= \frac{z^2}{1+zh} \\
&= z^{\circledtwo}.
\end{aligned}$$

Formula (2.7) follows immediately from the definition of $z^{\circledtwo}$. To prove (2.8), note that

$$\begin{aligned}
z + (\ominus z) &= z + \frac{-z}{1+zh} \\
&= \frac{z + z^2 h - z}{1+zh} \\
&= \frac{z^2}{1+zh} \cdot h \\
&= z^{\circledtwo} \cdot h.
\end{aligned}$$

The proof of (2.9) is left to the reader in Exercise 2.20. To prove (2.10), let $z = u + iv$ and note that

$$z^{\circledR} \in \mathbb{R} \iff \frac{(u+iv)^2}{(1+uh)+ivh} \in \mathbb{R}$$
$$\iff (u^2 + 2iuv - v^2)(1 + uh - ivh) \in \mathbb{R}$$
$$\iff 2uv(1+uh) - vh(u^2 - v^2) = 0$$
$$\iff u^2vh + 2uv + v^3h = 0$$
$$\iff v = 0 \text{ or } u^2h + 2u + v^2h = 0$$
$$\iff v = 0 \text{ or } \left(u + \frac{1}{h}\right)^2 + v^2 = \frac{1}{h^2}$$
$$\iff z \in \mathbb{R}_h \cup \mathbb{A}_h \text{ or } z \in \mathbb{I}_h$$
$$\iff z \in \mathbb{R}_h \cup \mathbb{A}_h \cup \mathbb{I}_h.$$

To prove (2.11), observe that we have

$$-(\overset{\circ}{i}\omega)^{\circledR} = (\overset{\circ}{i}\omega) \cdot [\ominus(\overset{\circ}{i}\omega)]$$
$$= (\overset{\circ}{i}\omega) \cdot \overline{(\overset{\circ}{i}\omega)}$$
$$= |\overset{\circ}{i}\omega|^2$$
$$= \frac{4}{h^2} \sin^2\left(\frac{\omega h}{2}\right),$$

where we have used Exercise 2.16 for the second equality and Theorem 2.6 for the last equality. $\square$

Note that if $z \in \mathbb{C}$, then

$$z^2 \in \mathbb{R} \iff z \in \mathbb{R} \cup \mathbb{I}_0,$$

where $\mathbb{I}_0$ is the set of purely imaginary numbers. Hence (2.10) is just the corresponding statement for Hilger complex numbers.

Exercise 2.20. Prove formula (2.9).

For $h > 0$, let $\mathbb{Z}_h$ be the strip

$$\mathbb{Z}_h := \left\{ z \in \mathbb{C} : -\frac{\pi}{h} < \mathrm{Im}(z) \leq \frac{\pi}{h} \right\},$$

and for $h = 0$, let $\mathbb{Z}_0 := \mathbb{C}$.

Definition 2.21. For $h > 0$, we define the cylinder transformation $\xi_h : \mathbb{C}_h \to \mathbb{Z}_h$ by

$$\xi_h(z) = \frac{1}{h} \mathrm{Log}(1 + zh),$$

where Log is the principal logarithm function. For $h = 0$, we define $\xi_0(z) = z$ for all $z \in \mathbb{C}$.

We call ξ_h the cylinder transformation because when $h > 0$ we can view $\mathbb{Z}_h$ as a cylinder if we glue the bordering lines $\mathrm{Im}(z) = -\frac{\pi}{h}$ and $\mathrm{Im}(z) = \frac{\pi}{h}$ of $\mathbb{Z}_h$ together to form a cylinder. We define addition on $\mathbb{Z}_h$ by

$$(2.12) \qquad z + w := z + w \quad \left(\mathrm{mod}\frac{2\pi i}{h}\right) \quad \text{for} \quad z, w \in \mathbb{Z}_h.$$

Exercise 2.22. Show that the cylinder transformation ξ_h when $h > 0$ maps open rays emanating from the point $-\frac{1}{h}$ in $\mathbb{C}$ onto horizontal lines on the cylinder $\mathbb{Z}_h$. Also show that circles with center at $-\frac{1}{h}$ are mapped onto vertical lines on the strip $\mathbb{Z}_h$ (circles on the cylinder $\mathbb{Z}_h$).

Exercise 2.23. Show that the inverse transformation of the cylinder transformation ξ_h when $h > 0$ is given by

$$\xi_h^{-1}(z) = \frac{1}{h}\left(e^{zh} - 1\right)$$

for $z \in \mathbb{Z}_h$. Also find $\xi_0^{-1}(z)$.

Theorem 2.24. *The cylinder transformation ξ_h is a group homomorphism from $(\mathbb{C}_h, \oplus)$ onto $(\mathbb{Z}_h, +)$, where the addition $+$ on $\mathbb{Z}_h$ is defined by (2.12).*

Proof. Let $h > 0$ and $z, w \in \mathbb{C}_h$ and consider

$$
\begin{aligned}
\xi_h(z \oplus w) &= \frac{1}{h}\,\mathrm{Log}(1 + (z \oplus w)h) \\
&= \frac{1}{h}\,\mathrm{Log}(1 + zh + wh + zwh^2) \\
&= \frac{1}{h}\,\mathrm{Log}[(1 + zh)(1 + wh)] \\
&= \frac{1}{h}\,\mathrm{Log}(1 + zh) + \frac{1}{h}\,\mathrm{Log}(1 + wh) \\
&= \xi_h(z) + \xi_h(w).
\end{aligned}
$$

This proves our result for $h > 0$. The case $h = 0$ is trivial. $\qquad\square$

2.2. The Exponential Function

In this section we use the cylinder transformation introduced in Section 2.1 to define a generalized exponential function for an arbitrary time scale $\mathbb{T}$. First we make some preliminary definitions.

Definition 2.25. We say that a function $p : \mathbb{T} \to \mathbb{R}$ is *regressive* provided

$$(2.13) \qquad\qquad 1 + \mu(t)p(t) \neq 0 \quad \text{for all} \quad t \in \mathbb{T}^\kappa$$

holds. The set of all regressive and rd-continuous functions $f : \mathbb{T} \to \mathbb{R}$ will be denoted in this book by

$$\mathcal{R} = \mathcal{R}(\mathbb{T}) = \mathcal{R}(\mathbb{T}, \mathbb{R}).$$

Exercise 2.26. Show that $\mathcal{R}$ is an Abelian group under the "circle plus" addition $\oplus$ defined by

$$(p \oplus q)(t) := p(t) + q(t) + \mu(t)p(t)q(t) \quad \text{for all} \quad t \in \mathbb{T}^\kappa,$$

$p, q \in \mathcal{R}$. This group is called the *regressive group*.

Exercise 2.27. Show that if $p, q \in \mathcal{R}$, then the function $p \oplus q$ and the function $\ominus p$ defined by

$$(\ominus p)(t) := -\frac{p(t)}{1 + \mu(t)p(t)} \quad \text{for all} \quad t \in \mathbb{T}^\kappa$$

are also elements of $\mathcal{R}$.

Exercise 2.28. We define the "circle minus" subtraction $\ominus$ on $\mathcal{R}$ by

$$(2.14) \qquad (p \ominus q)(t) := (p \oplus (\ominus q))(t) \quad \text{for all} \quad t \in \mathbb{T}^\kappa.$$

Suppose $p, q \in \mathcal{R}$. Show directly from the definition that

    (i) $p \ominus p = 0$;
    (ii) $\ominus(\ominus p) = p$;
    (iii) $p \ominus q \in \mathcal{R}$;
    (iv) $p \ominus q = \frac{p-q}{1+\mu q}$;
    (v) $\ominus(p \ominus q) = q \ominus p$;
    (vi) $\ominus(p \oplus q) = (\ominus p) \oplus (\ominus q)$.

Exercise 2.29. If $p \in \mathcal{R}$ and $q : \mathbb{T}^\kappa \to \mathbb{R}$, simplify $p \oplus \frac{q}{1+\mu p}$.

The *generalized exponential function* $e_p(t, s)$ is defined next.

Definition 2.30. If $p \in \mathcal{R}$, then we define the exponential function by

$$(2.15) \qquad e_p(t, s) = \exp\left(\int_s^t \xi_{\mu(\tau)}(p(\tau))\Delta\tau \right) \quad \text{for} \quad s, t \in \mathbb{T},$$

where the cylinder transformation $\xi_h(z)$ is introduced in Definition 2.21.

Lemma 2.31. *If $p \in \mathcal{R}$, then the semigroup property*

$$e_p(t, r)e_p(r, s) = e_p(t, s) \quad \text{for all} \quad r, s, t \in \mathbb{T}$$

is satisfied.

Proof. Suppose $p \in \mathcal{R}$. Let $r, s, t \in \mathbb{T}$. Then we have by Definition 2.30

$$\begin{aligned}
e_p(t, r)e_p(r, s) &= \exp\left(\int_r^t \xi_{\mu(\tau)}(p(\tau))\Delta\tau \right) \exp\left(\int_s^r \xi_{\mu(\tau)}(p(\tau))\Delta\tau \right) \\
&= \exp\left(\int_r^t \xi_{\mu(\tau)}(p(\tau))\Delta\tau + \int_s^r \xi_{\mu(\tau)}(p(\tau))\Delta\tau \right) \\
&= \exp\left(\int_s^t \xi_{\mu(\tau)}(p(\tau))\Delta\tau \right) \\
&= e_p(t, s),
\end{aligned}$$

where we have used Theorem 1.77 (iv). $\qquad\square$

Definition 2.32. If $p \in \mathcal{R}$, then the first order linear dynamic equation

$$(2.16) \qquad\qquad\qquad y^\Delta = p(t)y$$

is called *regressive*.

Theorem 2.33. *Suppose (2.16) is regressive and fix $t_0 \in \mathbb{T}$. Then $e_p(\cdot, t_0)$ is a solution of the initial value problem*

$$(2.17) \qquad\qquad y^\Delta = p(t)y, \quad y(t_0) = 1$$

on $\mathbb{T}$.

Proof. Fix $t_0 \in \mathbb{T}^\kappa$ and assume (2.16) is regressive. First note that

$$e_p(t_0, t_0) = 1.$$

It remains to show that $e_p(t, t_0)$ satisfies the dynamic equation $y^\Delta = p(t)y$. Fix $t \in \mathbb{T}^\kappa$. There are two cases.

Case 1. Assume $\sigma(t) > t$. By Exercise 2.23 and Lemma 2.31,

$$
\begin{aligned}
e_p^\Delta(t, t_0) &= \frac{\exp\left(\int_{t_0}^{\sigma(t)} \xi_{\mu(r)}(p(r))\Delta r\right) - \exp\left(\int_{t_0}^{t} \xi_{\mu(r)}(p(r))\Delta r\right)}{\mu(t)} \\
&= \frac{\exp\left(\int_{t}^{\sigma(t)} \xi_{\mu(r)}(p(r))\Delta r\right) - 1}{\mu(t)} e_p(t, t_0) \\
&= \frac{e^{\xi_{\mu(t)}(p(t))\mu(t)} - 1}{\mu(t)} e_p(t, t_0) \\
&= \xi_{\mu(t)}^{-1}(\xi_{\mu(t)}(p(t))) \cdot e_p(t, t_0) \\
&= p(t) \cdot e_p(t, t_0).
\end{aligned}
$$

Case 2. Assume $\sigma(t) = t$. If $y(t) := e_p(t, t_0)$, then we want to show that $y^\Delta(t) = p(t)y(t)$. Using Lemma 2.31 we obtain

$$
\begin{aligned}
|y(t) - y(s) - p(t)y(t)(t - s)| &= |e_p(t, t_0) - e_p(s, t_0) - p(t)e_p(t, t_0)(t - s)| \\
&= |e_p(t, t_0)| \cdot |1 - e_p(s, t) - p(t)(t - s)| \\
&= |e_p(t, t_0)| \left|1 - \int_s^t \xi_{\mu(\tau)}(p(\tau))\Delta\tau - e_p(s, t) + \int_s^t \xi_{\mu(\tau)}(p(\tau))\Delta\tau - p(t)(t - s)\right| \\
&\leq |e_p(t, t_0)| \cdot \left|1 - \int_s^t \xi_{\mu(\tau)}(p(\tau))\Delta\tau - e_p(s, t)\right| \\
&\quad + |e_p(t, t_0)| \cdot \left|\int_s^t \xi_{\mu(\tau)}(p(\tau))\Delta\tau - p(t)(t - s)\right| \\
&\leq |e_p(t, t_0)| \cdot \left|1 - \int_s^t \xi_{\mu(\tau)}(p(\tau))\Delta\tau - e_p(s, t)\right| \\
&\quad + |e_p(t, t_0)| \cdot \left|\int_s^t [\xi_{\mu(\tau)}(p(\tau)) - \xi_0(p(t))]\Delta\tau\right|.
\end{aligned}
$$

Let $\varepsilon > 0$ be given. We now show that there is a neighborhood U of t so that the right-hand side of the last inequality is less than $\varepsilon|t - s|$, and the proof will be complete. Since $\sigma(t) = t$ and $p \in C_{rd}$, it follows that (see Exercise 2.34)

$$(2.18) \qquad \lim_{r \to t} \xi_{\mu(r)}(p(r)) = \xi_0(p(t)).$$

This implies that there is a neighborhood U_1 of t such that

$$\left|\xi_{\mu(\tau)}(p(\tau)) - \xi_0(p(t))\right| < \frac{\varepsilon}{3|e_p(t, t_0)|} \qquad \text{for all} \quad \tau \in U_1.$$

Let $s \in U_1$. Then

$$(2.19) \qquad |e_p(t, t_0)| \cdot \left|\int_s^t [\xi_{\mu(\tau)}(p(\tau)) - \xi_0(p(t))]\Delta\tau\right| < \frac{\varepsilon}{3}|t - s|.$$

Next, by L'Hôpital's rule

$$\lim_{z \to 0} \frac{1 - z - e^{-z}}{z} = 0,$$

so there is a neighborhood U_2 of t so that if $s \in U_2$, then

$$\left| \frac{1 - \int_s^t \xi_{\mu(\tau)}(p(\tau))\Delta\tau - e_p(s,t)}{\int_s^t \xi_{\mu(\tau)}(p(\tau))\Delta\tau} \right| < \varepsilon^*,$$

where

$$\varepsilon^* = \min\left\{1, \frac{\varepsilon}{1 + 3|p(t)e_p(t,t_0)|}\right\}.$$

Let $s \in U := U_1 \cap U_2$. Then

$$|e_p(t,t_0)| \cdot \left|1 - \int_s^t \xi_{\mu(\tau)}(p(\tau))\Delta\tau - e_p(s,t)\right| < |e_p(t,t_0)|\varepsilon^* \left|\int_s^t \xi_{\mu(\tau)}(p(\tau))\Delta\tau\right|$$

$$\leq |e_p(t,t_0)| \cdot \varepsilon^* \left\{\left|\int_s^t [\xi_{\mu(\tau)}(p(\tau)) - \xi_0(p(t))]\Delta\tau\right| + |p(t)||t-s|\right\}$$

$$\leq |e_p(t,t_0)| \cdot \left|\int_s^t [\xi_{\mu(\tau)}(p(\tau)) - \xi_0(p(t))]\Delta\tau\right| + |e_p(t,t_0)|\varepsilon^*|p(t)||t-s|$$

$$\leq \frac{\varepsilon}{3}|t-s| + |e_p(t,t_0)|\varepsilon^*|p(t)||t-s|$$

$$\leq \frac{\varepsilon}{3}|t-s| + \frac{\varepsilon}{3}|t-s|$$

$$= \frac{2\varepsilon}{3}|t-s|,$$

using (2.19). $\qquad\square$

Exercise 2.34. Verify (2.18). This exercise is due to Ronald Mathsen.

We now have confirmed the existence of a solution to the initial value problem (2.17). Now we address the uniqueness of such a solution.

Theorem 2.35. *If (2.16) is regressive, then the only solution of (2.17) is given by* $e_p(\cdot, t_0)$.

Proof. Assume y is a solution of (2.17) and consider the quotient $y/e_p(\cdot, t_0)$ (note that by Definition 2.30 we have that $e_p(t,s) \neq 0$ for all $t, s \in \mathbb{T}$). By Theorem 1.20 (v) we have

$$\left(\frac{y}{e_p(\cdot, t_0)}\right)^\Delta (t) = \frac{y^\Delta(t)e_p(t,t_0) - y(t)e_p^\Delta(t,t_0)}{e_p(t,t_0)e_p(\sigma(t),t_0)}$$

$$= \frac{p(t)y(t)e_p(t,t_0) - y(t)p(t)e_p(t,t_0)}{e_p(t,t_0)e_p(\sigma(t),t_0)}$$

$$= 0$$

so that $y/e_p(\cdot, t_0)$ is constant according to Corollary 1.68 (ii). Hence

$$\frac{y(t)}{e_p(t,t_0)} \equiv \frac{y(t_0)}{e_p(t_0,t_0)} = \frac{1}{1} = 1$$

and therefore $y = e_p(\cdot, t_0)$. $\qquad\square$

We proceed by collecting some important properties of the exponential function.

Theorem 2.36. *If* $p, q \in \mathcal{R}$*, then*

(i) $e_0(t, s) \equiv 1$ *and* $e_p(t, t) \equiv 1$;

(ii) $e_p(\sigma(t), s) = (1 + \mu(t)p(t))e_p(t, s)$;

(iii) $\frac{1}{e_p(t,s)} = e_{\ominus p}(t, s)$;

(iv) $e_p(t, s) = \frac{1}{e_p(s,t)} = e_{\ominus p}(s, t)$;

(v) $e_p(t, s)e_p(s, r) = e_p(t, r)$;

(vi) $e_p(t, s)e_q(t, s) = e_{p \oplus q}(t, s)$;

(vii) $\frac{e_p(t,s)}{e_q(t,s)} = e_{p \ominus q}(t, s)$;

(viii) $\left(\frac{1}{e_p(\cdot,s)} \right)^{\Delta} = -\frac{p(t)}{e_p^{\sigma}(\cdot,s)}$.

Proof. Part (i). The function $y(t) \equiv 1$ is obviously a solution of the initial value problem $y^{\Delta} = 0y$, $y(s) = 1$, and since this problem has only one solution according to Theorem 2.35, namely $e_0(t, s)$ by Theorem 2.33, we have that $e_0(t, s) \equiv y(t) \equiv 1$.

Part (ii). By Theorem 1.16 (iv) we have

$$
\begin{aligned}
e_p(\sigma(t), s) &= e_p^{\sigma}(t, s) \\
&= e_p(t, s) + \mu(t)e_p^{\Delta}(t, s) \\
&= e_p(t, s) + \mu(t)p(t)e_p(t, s) \\
&= (1 + \mu(t)p(t))e_p(t, s),
\end{aligned}
$$

where we have used Theorem 2.33.

Part (iii). Consider the initial value problem

$$(2.20) \qquad\qquad y^{\Delta} = (\ominus p)(t)y, \quad y(s) = 1.$$

From Exercise 2.27 the dynamic equation in (2.20) is regressive. We show that for each fixed s, $y(t) = 1/e_p(t, s)$ satisfies (2.20), and then the claim follows with Theorem 2.33 and Theorem 2.35. Indeed, $y(s) = 1$ is obvious. We use the quotient rule (Theorem 1.20 (iv)) to obtain

$$
\begin{aligned}
y^{\Delta}(t) &= \left(\frac{1}{e_p(\cdot, s)} \right)^{\Delta}(t) \\
&= -\frac{e_p^{\Delta}(t, s)}{e_p(t, s)e_p(\sigma(t), s)} \\
&= -\frac{p(t)e_p(t, s)}{e_p(t, s)e_p(\sigma(t), s)} \\
&= -\frac{p(t)}{e_p(\sigma(t), s)} \\
&= -\frac{p(t)}{(1 + \mu(t)p(t))e_p(t, s)} \\
&= (\ominus p)(t)y(t),
\end{aligned}
$$

where we have also used part (ii) in the second to the last step.

Part (iv). By equation (2.15), it follows that

$$e_p(t, s) = \frac{1}{e_p(s, t)},$$

and this is equal to $e_{\ominus p}(s, t)$ according to part (iii).

Part (v). This semigroup property was already shown in Lemma 2.31, based on the definition of the exponential function in terms of the cylinder transformation. However, let us here deduce the semigroup property using only the fact that exponential functions are unique solutions of initial value problems (2.17). Consider the initial value problem

$$(2.21) \qquad y^\Delta = p(t)y, \quad y(r) = 1.$$

We show that $y(t) := e_p(t,s)e_p(s,r)$ satisfies (2.21), and then the claim follows from Theorem 2.33 and Theorem 2.35. It is obvious that $y^\Delta(t) = p(t)y(t)$, and

$$y(r) = e_p(r,s)e_p(s,r) = 1$$

follows from part (iv).

Part (vi). Consider the initial value problem

$$(2.22) \qquad y^\Delta = (p \oplus q)(t)y, \quad y(s) = 1.$$

From Exercise 2.27 the dynamic equation in (2.22) is regressive. We show that $y(t) = e_p(t,s)e_q(t,s)$ satisfies (2.22), and then the claim follows using Theorem 2.33 and Theorem 2.35. We have $y(s) = e_p(s,s)e_q(s,s) = 1$, and we use the product rule (Theorem 1.20 (iii)) to calculate

$$
\begin{aligned}
y^\Delta(t) &= (e_p(\cdot,s)e_q(\cdot,s))^\Delta(t) \\
&= p(t)e_p(t,s)e_q(\sigma(t),s) + e_p(t,s)q(t)e_q(t,s) \\
&= p(t)e_p(t,s)[1 + \mu(t)q(t)]e_q(t,s) + e_p(t,s)q(t)e_q(t,s) \\
&= \{p(t)[1 + \mu(t)q(t)] + q(t)\}\,e_p(t,s)e_q(t,s) \\
&= (p \oplus q)(t)e_p(t,s)e_q(t,s) \\
&= (p \oplus q)(t)y(t),
\end{aligned}
$$

where we have used part (ii) of this theorem.

Part (vii). This follows easily using parts (iii) and (vi) of this theorem.

Part (viii). We calculate

$$
\begin{aligned}
\left(\frac{1}{e_p(\cdot,s)}\right)^\Delta &= (e_{\ominus p}(\cdot,s))^\Delta \\
&= (\ominus p)e_{\ominus p}(\cdot,s) \\
&= \frac{-p}{1 + \mu p}\frac{1}{e_p(\cdot,s)} \\
&= \frac{-p}{e_p^\sigma(\cdot,s)},
\end{aligned}
$$

where we have used parts (ii) and (iii) of this theorem. $\qquad\square$

Exercise 2.37. Let $p \in \mathcal{C}_{\mathrm{rd}}$ and $r, s, t \in \mathbb{T}$. Simplify $e_p(t, \sigma(s))e_p(s,r)$.

Two useful results, which are needed later on, are presented next.

Theorem 2.38. *If $p, q \in \mathcal{R}$, then*

$$e_{p \ominus q}^\Delta(\cdot, t_0) = (p - q)\frac{e_p(\cdot, t_0)}{e_q^\sigma(\cdot, t_0)}.$$

Proof. We have

$$
\begin{aligned}
e^{\Delta}_{p\ominus q}(t,t_0) &= (p\ominus q)(t)e_{p\ominus q}(t,t_0) \\
&= \frac{p(t)-q(t)}{1+\mu(t)q(t)}\frac{e_p(t,t_0)}{e_q(t,t_0)} \\
&= \frac{(p(t)-q(t))e_p(t,t_0)}{e_q(\sigma(t),t_0)}
\end{aligned}
$$

where we have used Theorem 2.36 (ii) and (vii). □

Theorem 2.39. *If $p \in \mathcal{R}$ and $a,b,c \in \mathbb{T}$, then*

$$[e_p(c,\cdot)]^{\Delta} = -p[e_p(c,\cdot)]^{\sigma}$$

and

$$\int_a^b p(t)e_p(c,\sigma(t))\Delta t = e_p(c,a) - e_p(c,b).$$

Proof. We use many of the properties from Theorem 2.36 to find

$$
\begin{aligned}
p(t)e_p(c,\sigma(t)) &= p(t)e_{\ominus p}(\sigma(t),c) \\
&= p(t)[1+\mu(t)(\ominus p)(t)]e_{\ominus p}(t,c) \\
&= p(t)\left[1-\frac{\mu(t)p(t)}{1+\mu(t)p(t)}\right]e_{\ominus p}(t,c) \\
&= p(t)\frac{1}{1+\mu(t)p(t)}e_{\ominus p}(t,c) \\
&= -(\ominus p)(t)e_{\ominus p}(t,c) \\
&= -e^{\Delta}_{\ominus p}(t,c) \\
&= -[e_p(c,t)]^{\Delta},
\end{aligned}
$$

where Δ denotes differentiation with respect to t. Thus

$$
\begin{aligned}
\int_a^b p(t)e_p(c,\sigma(t))\Delta t &= -\int_a^b [e_p(c,\cdot)]^{\Delta}(t)\Delta t \\
&= e_p(c,a) - e_p(c,b),
\end{aligned}
$$

which proves our desired identity. □

Exercise 2.40. Assume $\frac{2}{t}$, $\frac{5}{t}$ are regressive on $\mathbb{T}\cap(0,\infty)$ and let $t_0 \in \mathbb{T}\cap(0,\infty)$. Evaluate the integral

$$\int_{t_0}^t \frac{e_{\frac{5}{s}}(s,t_0)}{se_{\frac{2}{s}}^{\sigma}(s,t_0)}\Delta s.$$

Later when we study the Euler–Cauchy dynamic equation we will use the result in the following exercise.

Exercise 2.41. Show that if $\alpha \in \mathbb{R}$ is constant and the exponentials below exist, then

$$\frac{e_{\frac{\alpha^2}{t}-\frac{(\alpha-1)^2}{\sigma(t)}}(t,t_0)}{e_{\frac{\alpha}{t}}^2(t,t_0)} = \frac{t_0}{t}$$

by first showing that

$$\frac{e_{\frac{\alpha^2}{t} - \frac{(\alpha-1)^2}{\sigma(t)}}(t,t_0)}{e_{\frac{\alpha}{t}}(t,t_0)} = e_{\frac{\alpha-1}{\sigma(t)}}(t,t_0)$$

and then showing that

$$\frac{e_{\frac{\alpha-1}{\sigma(t)}}(t,t_0)}{e_{\frac{\alpha}{t}}(t,t_0)} = e_{-\frac{1}{\sigma(t)}}(t,t_0) = \frac{t_0}{t}$$

for $t, t_0 \in \mathbb{T} \cap (0, \infty)$.

Exercise 2.42. Let $s, t \in \mathbb{T}$. Assuming we can interchange the order of differentiation and integration, show that

$$\frac{d}{dz}[e_z(t,s)] = \left(\int_s^t \frac{1}{1+\mu(\tau)z} \Delta\tau \right) e_z(t,s)$$

for those $z \in \mathbb{C}$ satisfying $1 + \mu(\tau)z \neq 0$ for τ between s and t.

In the remainder of this section we study the sign of the exponential function. Lemma 2.43 and Theorem 2.48 below appear in Akın, Erbe, Kaymakçalan, and Peterson [35].

Lemma 2.43. *Let $p \in \mathcal{R}$. Suppose there exists a sequence of distinct points $\{t_n\}_{n \in \mathbb{N}} \subset \mathbb{T}^\kappa$ such that*

$$1 + \mu(t_n)p(t_n) < 0 \quad \text{for all} \quad n \in \mathbb{N}.$$

Then $\lim_{n \to \infty} |t_n| = \infty$. In particular, if there exists a bounded interval $J \subset \mathbb{T}^\kappa$ such that $1 + \mu(t)p(t) < 0$ for all $t \in J$, then $|J| < \infty$.

Proof. Let $J \subset \mathbb{T}^\kappa$ be a bounded interval and assume there exists a sequence of distinct points $\{t_n\}_{n \in \mathbb{N}} \subset J$ such that

$$1 + \mu(t_n)p(t_n) < 0 \quad \text{for all} \quad n \in \mathbb{N}.$$

Since the sequence $\{t_n\}_{n \in \mathbb{N}}$ is bounded, it has a convergent subsequence. We assume without loss of generality that the sequence $\{t_n\}_{n \in \mathbb{N}}$ is itself convergent, i.e.,

$$\lim_{n \to \infty} t_n = t_0.$$

Since $\mathbb{T}$ is closed, $t_0 \in \mathbb{T}$. Since $1 + \mu(t_n)p(t_n) < 0$, we have $\mu(t_n) > 0$ and

$$(2.23) \qquad p(t_n) < -\frac{1}{\mu(t_n)} \quad \text{for all} \quad n \in \mathbb{N}.$$

There is a subsequence $\{t_{n_k}\}_{k \in \mathbb{N}}$ of $\{t_n\}_{n \in \mathbb{N}}$ with $\lim_{k \to \infty} t_{n_k} = t_0$ such that $\{t_{n_k}\}_{k \in \mathbb{N}}$ is either strictly decreasing or strictly increasing. If $\{t_{n_k}\}_{k \in \mathbb{N}}$ is strictly decreasing, then $\lim_{k \to \infty} \mu(t_{n_k}) = 0$ because of

$$0 < \mu(t_{n_k}) = \sigma(t_{n_k}) - t_{n_k} \le t_{n_{k-1}} - t_{n_k}.$$

If $\{t_{n_k}\}_{k \in \mathbb{N}}$ is strictly increasing, then $\lim_{k \to \infty} \mu(t_{n_k}) = 0$ because of

$$0 < \mu(t_{n_k}) = \sigma(t_{n_k}) - t_{n_k} \le t_{n_{k+1}} - t_{n_k}.$$

So in either case $\lim_{k \to \infty} \mu(t_{n_k}) = 0$ and hence using (2.23),

$$\lim_{k \to \infty} p(t_k) = -\infty.$$

But this contradicts the fact that $p \in C_{\text{rd}}$. $\qquad\qquad\qquad\qquad\qquad\qquad \square$

Theorem 2.44. *Assume $p \in \mathcal{R}$ and $t_0 \in \mathbb{T}$.*

(i) *If $1 + \mu p > 0$ on $\mathbb{T}^\kappa$, then $e_p(t, t_0) > 0$ for all $t \in \mathbb{T}$.*

(ii) *If $1 + \mu p < 0$ on $\mathbb{T}^\kappa$, then $e_p(t, t_0) = \alpha(t, t_0)(-1)^{n_t}$ for all $t \in \mathbb{T}$, where*

$$\alpha(t, t_0) := \exp\left(\int_{t_0}^t \frac{\log|1 + \mu(\tau)p(\tau)|}{\mu(\tau)} \Delta\tau\right) > 0$$

and

$$n_t = \begin{cases} |[t_0, t)| & \text{if} \quad t \geq t_0 \\ |[t, t_0)| & \text{if} \quad t < t_0. \end{cases}$$

Proof. Part (i) can be shown directly using Definition 2.30: Since $1 + \mu(t)p(t) > 0$, we have $\text{Log}[1 + \mu(t)p(t)] \in \mathbb{R}$ for all $t \in \mathbb{T}^\kappa$ and therefore

$$\xi_{\mu(t)}(p(t)) \in \mathbb{R} \quad \text{for all} \quad t \in \mathbb{T}^\kappa.$$

Hence $e_p(t, t_0) > 0$ for all $t \in \mathbb{T}$, by formula (2.15).

Part (ii) follows similarly: Since $1 + \mu(t)p(t) < 0$, we have

$$\text{Log}[1 + \mu(t)p(t)] = \log|1 + \mu(t)p(t)| + i\pi \quad \text{for all} \quad t \in \mathbb{T}^\kappa.$$

Note also that $\mu(t)$ never vanishes in this case and that

$$n_t < \infty$$

because of Lemma 2.43, and then we can calculate $e_p(t, t_0)$ as

$$
\begin{aligned}
e_p(t, t_0) &= \exp\left(\int_{t_0}^t \xi_{\mu(\tau)}(p(\tau))\Delta\tau\right) \\
&= \exp\left(\int_{t_0}^t \frac{\text{Log}(1 + \mu(\tau)p(\tau))}{\mu(\tau)} \Delta\tau\right) \\
&= \exp\left(\int_{t_0}^t \frac{\log|1 + \mu(\tau)p(\tau)| + i\pi}{\mu(\tau)} \Delta\tau\right) \\
&= \exp\left(\int_{t_0}^t \left\{\frac{\log|1 + \mu(\tau)p(\tau)|}{\mu(\tau)} + \frac{i\pi}{\mu(\tau)}\right\} \Delta\tau\right) \\
&= \alpha(t, t_0) \exp\left(i\pi \int_{t_0}^t \frac{\Delta\tau}{\mu(\tau)}\right).
\end{aligned}
$$

By *Euler's formula,*

$$
\begin{aligned}
\exp\left(i\pi \int_{t_0}^t \frac{\Delta\tau}{\mu(\tau)}\right) &= \cos\left(\pi \int_{t_0}^t \frac{\Delta\tau}{\mu(\tau)}\right) \\
&= \cos(n_t\pi) \\
&= (-1)^{n_t},
\end{aligned}
$$

where the second equation is true because of Theorem 1.79 and because of $n_t < \infty$. This leads to the desired result. $\square$

In view of Theorem 2.44 (i), we make the following definition.

Definition 2.45. We define the set $\mathcal{R}^+$ of all *positively regressive* elements of $\mathcal{R}$ by

$$\mathcal{R}^+ = \mathcal{R}^+(\mathbb{T}, \mathbb{R}) = \{p \in \mathcal{R} : 1 + \mu(t)p(t) > 0 \text{ for all } t \in \mathbb{T}\}.$$

Exercise 2.46. Assume $p \in C_{rd}$ and $p(t) \geq 0$ for all $t \in \mathbb{T}$. Show that $p \in \mathcal{R}^+$.

Lemma 2.47. $\mathcal{R}^+$ *is a subgroup of* $\mathcal{R}$.

Proof. Obviously we have that $\mathcal{R}^+ \subset \mathcal{R}$ and that $0 \in \mathcal{R}^+$, where 0 is the function f defined by $f(t) \equiv 0$ for all $t \in \mathbb{T}$. Now let $p, q \in \mathcal{R}^+$. Then

$$1 + \mu p > 0 \quad \text{and} \quad 1 + \mu q > 0 \quad \text{on} \quad \mathbb{T}.$$

Therefore

$$1 + \mu(p \oplus q) = (1 + \mu p)(1 + \mu q) > 0 \quad \text{on} \quad \mathbb{T}.$$

Hence we have

$$p \oplus q \in \mathcal{R}^+.$$

Next, let $p \in \mathcal{R}^+$. Then

$$1 + \mu p > 0 \quad \text{on} \quad \mathbb{T}.$$

This implies that

$$1 + \mu(\ominus p) = 1 - \frac{\mu p}{1 + \mu p} = \frac{1}{1 + \mu p} > 0 \quad \text{on} \quad \mathbb{T}.$$

Hence

$$\ominus p \in \mathcal{R}^+.$$

These calculations establish that $\mathcal{R}^+$ is a subgroup of $\mathcal{R}$. $\qquad\square$

Theorem 2.48 (Sign of the Exponential Function). *Let* $p \in \mathcal{R}$ *and* $t_0 \in \mathbb{T}$.

(i) *If* $p \in \mathcal{R}^+$, *then* $e_p(t, t_0) > 0$ *for all* $t \in \mathbb{T}$.

(ii) *If* $1 + \mu(t)p(t) < 0$ *for some* $t \in \mathbb{T}^\kappa$, *then*

$$e_p(t, t_0)e_p(\sigma(t), t_0) < 0.$$

(iii) *If* $1 + \mu(t)p(t) < 0$ *for all* $t \in \mathbb{T}^\kappa$, *then* $e_p(t, t_0)$ *changes sign at every point* $t \in \mathbb{T}$.

(iv) *Assume there exist sets* $T = \{t_i : i \in \mathbb{N}\} \subset \mathbb{T}^\kappa$ *and* $S = \{s_i : i \in \mathbb{N}\} \subset \mathbb{T}^\kappa$ *with*

$$\cdots < s_2 < s_1 < t_0 \leq t_1 < t_2 < \cdots$$

such that $1 + \mu(t)p(t) < 0$ *for all* $t \in S \cup T$ *and* $1 + \mu(t)p(t) > 0$ *for all* $t \in \mathbb{T}^\kappa \setminus (S \cup T)$. *Furthermore if* $|T| = \infty$, *then* $\lim_{n \to \infty} t_n = \infty$, *and if* $|S| = \infty$, *then* $\lim_{n \to \infty} s_n = -\infty$. *If* $T \neq \emptyset$ *and* $S \neq \emptyset$, *then*

$$e_p(\cdot, t_0) > 0 \quad \text{on} \quad [\sigma(s_1), t_1].$$

If $|T| = \infty$, *then*

$$(-1)^i e_p(\cdot, t_0) > 0 \quad \text{on} \quad [\sigma(t_i), t_{i+1}] \quad \text{for all} \quad i \in \mathbb{N}.$$

If $|T| = N \in \mathbb{N}$, *then*

$$(-1)^i e_p(\cdot, t_0) > 0 \quad \text{on} \quad [\sigma(t_i), t_{i+1}] \quad \text{for all} \quad 1 \leq i \leq N - 1$$

and

$$(-1)^N e_p(\cdot, t_0) > 0 \quad \text{on} \quad [\sigma(t_N), \infty).$$

If $T = \emptyset$ *and* $S \neq \emptyset$, *then*

$$e_p(\cdot, t_0) > 0 \quad \text{on} \quad [\sigma(s_1), \infty).$$

If $|S| = \infty$, *then*

$$(-1)^i e_p(\cdot, t_0) > 0 \quad \text{on} \quad [\sigma(s_{i+1}), s_i] \quad \text{for all} \quad i \in \mathbb{N}.$$

If $|S| = M \in \mathbb{N}$, then

$$(-1)^i e_p(\cdot, t_0) > 0 \quad on \quad [\sigma(s_{i+1}), s_i] \quad for\ all \quad 1 \le i \le M - 1$$

and

$$(-1)^M e_p(t, t_0) > 0 \quad on \quad (-\infty, s_M].$$

If $S = \emptyset$ and $T \ne \emptyset$, then

$$e_p(\cdot, t_0) > 0 \quad on \quad (-\infty, t_1].$$

In particular, the exponential function $e_p(\cdot, t_0)$ is a real-valued function that is never equal to zero but can be negative.

Proof. While (i) is just Theorem 2.44 (i), (ii) follows immediately from Theorem 2.36 (ii). Part (iii) is clear by Theorem 2.44 (ii). Now we prove (iv). By Lemma 2.43, the set of points in $\mathbb{T}$ where $1 + \mu(t)p(t) < 0$ is countable. If $|T| = \infty$, then $\lim_{n \to \infty} t_n = \infty$ by Lemma 2.43. We can assume $t_0 \le t_1 < t_2 < \cdots$. Consider the case where $|T| = \infty$ with $t_0 \le t_1 < t_2 < \cdots$ such that $1 + \mu(t)p(t) < 0$ for all $t \in T$ and $1 + \mu(t)p(t) > 0$ for all $t \in [t_0, \infty) \setminus T$. We prove the conclusion of (iv) for this case by mathematical induction with respect to the intervals $[t_0, t_1], [\sigma(t_1), t_2], [\sigma(t_2), t_3], \ldots$. First we prove that

$$e_p(\cdot, t_0) > 0 \quad on \quad [t_0, t_1].$$

Since $e_p(t_0, t_0) = 1$ and $1 + \mu(t)p(t) > 0$ on $[t_0, t_1)$, $e_p(\cdot, t_0) > 0$ on $[t_0, t_1)$. We now show that $e_p(t_1, t_0) > 0$. If $t_1 = t_0$, then $e_p(t_1, t_0) = 1 > 0$. Hence we can assume $t_1 > t_0$. There are two cases to consider: First, if t_1 is left-scattered, then

$$
\begin{aligned}
e_p(t_1, t_0) &= e_p(t_1, \rho(t_1))e_p(\rho(t_1), t_0) \\
&= [1 + \mu(\rho(t_1))p(\rho(t_1))]e_p(\rho(t_1), \rho(t_1))e_p(\rho(t_1), t_0) \\
&= [1 + \mu(\rho(t_1))p(\rho(t_1))]e_p(\rho(t_1), t_0) \\
&> 0.
\end{aligned}
$$

Next, if t_1 is left-dense, then the value of $\int_{t_0}^{t_1} \xi_{\mu(s)}(p(s))\Delta s$ does not depend on the value of the integrand at t_1 and so $e_p(t_1, t_0) > 0$. Therefore $e_p(\cdot, t_0) > 0$ on $[t_0, t_1]$. Assume $i \ge 0$ and $(-1)^i e_p(\cdot, t_0) > 0$ on $[\sigma(t_i), t_{i+1}]$ ($[t_0, t_1]$ if $i = 0$). It remains to show that

$$(-1)^{i+1} e_p(\cdot, t_0) > 0 \quad on \quad [\sigma(t_{i+1}), t_{i+2}].$$

Since $1 + \mu(t_{i+1})p(t_{i+1}) < 0$, we have by (ii) that $\sigma(t_{i+1}) > t_{i+1}$ and

$$e_p(t_{i+1}, t_0)e_p(\sigma(t_{i+1}), t_0) < 0.$$

Then

(2.24) $$(-1)^{i+1} e_p(\sigma(t_{i+1}), t_0) > 0.$$

Next we show that

$$(-1)^{i+1} e_p(\cdot, t_0) > 0 \quad on \quad [\sigma(t_{i+1}), t_{i+2}].$$

Let $t \in (\sigma(t_{i+1}), t_{i+2}]$. Then

$$
\begin{aligned}
(-1)^{i+1} e_p(t, t_0) &= (-1)^{i+1} e_p(t, \sigma(t_{i+1}))e_p(\sigma(t_{i+1}), t_0) \\
&= (-1)^{i+1} e^{\int_{\sigma(t_{i+1})}^{t} \xi_{\mu(s)}(p(s))\Delta s} e_p(\sigma(t_{i+1}), t_0) \\
&> 0
\end{aligned}
$$

by using equation (2.24) and the fact that $1 + \mu p > 0$ on $[\sigma(t_{i+1}), t_{i+2})$. The remaining cases are similar and hence are omitted.   □

Exercise 2.49. Show that if the constant $a < -1$ and $\mathbb{T} = \mathbb{Z}$, then the exponential function $e_a(\cdot, 0)$ changes sign at every point in $\mathbb{Z}$. In this case we say the exponential function $e_a(\cdot, 0)$ is *strongly oscillatory* on $\mathbb{Z}$.

2.3. Examples of Exponential Functions

Example 2.50. Let $\mathbb{T} = h\mathbb{Z}$ for $h > 0$. Let $\alpha \in \mathcal{R}$ be constant, i.e.,

$$\alpha \in \mathbb{C} \setminus \left\{ -\frac{1}{h} \right\}.$$

Then

(2.25) $e_\alpha(t, 0) = (1 + \alpha h)^{\frac{t}{h}}$ for all $t \in \mathbb{T}$.

To show this we note that y defined by the right-hand side of (2.25) satisfies

$$y(0) = (1 + \alpha h)^0 = 1$$

and

$$\begin{aligned}
y^\Delta(t) &= \frac{y(t + h) - y(t)}{h} \\
&= \frac{(1 + \alpha h)^{\frac{t+h}{h}} - (1 + \alpha h)^{\frac{t}{h}}}{h} \\
&= \frac{(1 + \alpha h)^{\frac{t}{h}}(1 + \alpha h - 1)}{h} \\
&= \alpha(1 + \alpha h)^{\frac{t}{h}} \\
&= \alpha y(t)
\end{aligned}$$

for all $t \in \mathbb{T}$.

Exercise 2.51. Show that if $\mathbb{T}$ has constant graininess $h \geq 0$ and if α is constant with $1 + \alpha h \neq 0$, then $e_\alpha(t + s, 0) = e_\alpha(t, 0)e_\alpha(s, 0)$ for all $s, t \in \mathbb{T}$.

Example 2.52. Consider the time scale

$$\mathbb{T} = \mathbb{N}_0^2 = \{n^2 : n \in \mathbb{N}_0\}$$

from Example 1.44. We claim that

(2.26) $e_1(t, 0) = 2^{\sqrt{t}}(\sqrt{t})!$ for $t \in \mathbb{T}$.

Let y be defined by the right-hand side of (2.26). Clearly, $y(0) = 1$, and for $t \in \mathbb{T}$ we have

$$\begin{aligned}
y(\sigma(t)) &= 2^{\sqrt{\sigma(t)}}(\sqrt{\sigma(t)})! \\
&= 2^{1+\sqrt{t}}(1 + \sqrt{t})! \\
&= 2 \cdot 2^{\sqrt{t}}(1 + \sqrt{t})(\sqrt{t})! \\
&= 2(1 + \sqrt{t})y(t) \\
&= (1 + \mu(t))y(t) \\
&= y(t) + \mu(t)y(t)
\end{aligned}$$

so that $y^\Delta(t) = y(t)$.

Figure 2.3. Plant Population

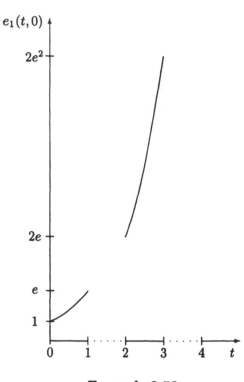

Example 2.58.

Thus, if $N(0) = 1$ is given, N is exactly $e_1(\cdot, 0)$ on the time scale $\mathbb{T}$. See Figure 2.3 for a graph of this exponential function on $\mathbb{T}$. We can calculate N as follows: If $k \in \mathbb{N}_0$ and $t \in [2k, 2k+1]$, then N satisfies $N' = N$ so that

$$N(t) = \alpha_k e^t \quad \text{for some} \quad \alpha_k \in \mathbb{R}.$$

Since $N(0) = 1$, we have

$$1 = N(0) = \alpha_0 e^0 = \alpha_0 \quad \text{and} \quad N(t) = \alpha_0 e^t = e^t \text{ for } 0 \le t \le 1.$$

Thus $N(1) = e$ and $N(2) = 2N(1) = 2e$. Now

$$2e = N(2) = \alpha_1 e^2 \quad \text{and} \quad N(t) = \alpha_1 e^t = \frac{2}{e} e^t = 2e^{t-1} \text{ for } 2 \le t \le 3.$$

Hence $N(3) = 2e^2$ and $N(4) = 2N(3) = 4e^2$. Next

$$4e^2 = N(4) = \alpha_2 e^4 \quad \text{and} \quad N(t) = \alpha_2 e^t = \frac{4}{e^2} e^t = 4e^{t-2} \text{ for } 4 \le t \le 5.$$

We now use mathematical induction to show that

$$N(t) = \left(\frac{2}{e}\right)^k e^t \quad \text{for} \quad t \in [2k, 2k+1].$$

Table 2.2. Plant Population

t	0	1	2	3	4	5	6	7
$e_1(t,0)$	1	e	$2e$	$2e^2$	$4e^2$	$4e^3$	$8e^3$	$8e^4$

The statement is already shown for $k = 0$. Assume it is true for $k = m \in \mathbb{N}_0$. Then $N(t) = (2/e)^m e^t$ for $t \in [2m, 2m+1]$ so that

$$N(2m+1) = \left(\frac{2}{e}\right)^m e^{2m+1} = 2^m e^{m+1}$$

and

$$N(2m+2) = 2N(2m+1) = 2 \cdot 2^m e^{m+1} = (2e)^{m+1}.$$

Therefore

$$(2e)^{m+1} = N(2m+2) = \alpha_{m+1} e^{2m+2}$$

and

$$N(t) = \alpha_{m+1} e^t = \left(\frac{2}{e}\right)^{m+1} e^t \quad \text{for} \quad 2m+2 \le t \le 2m+3.$$

In Table 2.2 we collected a few values of this exponential function.

Exercise 2.59. For $\alpha \ne -1$, find $e_\alpha(\cdot,0)$, where $\mathbb{T} = \mathbb{P}_{1,1}$ as in Example 2.58.

Exercise 2.60. Find the exponential function $e_{\frac{\lambda}{t}}(\cdot,1)$, where λ is a real constant for $\mathbb{T} = [1,\infty)$ and $\mathbb{T} = \mathbb{N}$. If $\mathbb{T} = \mathbb{N}$ and $\lambda \in \mathbb{N}$, simplify your answer.

Example 2.61. Let $\sum_{k=0}^\infty \alpha_k = L$ be a convergent series with $\alpha_k > 0$ for $k \ge 1$ and put

$$\mathbb{T} = \left\{ \sum_{k=0}^{n-1} \alpha_k : n \in \mathbb{N} \right\} \cup \{L\}.$$

Then by Exercise 2.54,

$$e_1(t,\alpha_0) = \prod_{k=1}^{n-1}(1+\alpha_k) \quad \text{for} \quad t = \sum_{k=0}^{n-1}\alpha_k,\ n \in \mathbb{N},$$

where we use the convention that $\prod_{k=1}^0(\cdots) = 1$. Therefore

$$\lim_{t\to L} e_1(t,\alpha_0) = e_1(L,\alpha_0) = \prod_{k=1}^\infty (1+\alpha_k).$$

In particular, if

$$\alpha_k = \frac{1}{4k^2-1} \quad \text{for all} \quad k \in \mathbb{N}_0,$$

then, using that the *Wallis product*

$$\prod_{k=1}^\infty (1+\alpha_k) = \prod_{k=1}^\infty \left(1+\frac{1}{4k^2-1}\right) = \prod_{k=1}^\infty \frac{4k^2}{4k^2-1} = \frac{\pi}{2}$$

converges, we find that

$$\lim_{t\to\infty} e_1(t,\alpha_0) = e_1(L,\alpha_0) = \frac{\pi}{2}.$$

Table 2.3. Exponential Functions

$\mathbb{T}$	$e_\alpha(t, t_0)$
$\mathbb{R}$	$e^{\alpha(t-t_0)}$
$\mathbb{Z}$	$(1+\alpha)^{t-t_0}$
$h\mathbb{Z}$	$(1+\alpha h)^{(t-t_0)/h}$
$\frac{1}{n}\mathbb{Z}$	$(1+\frac{\alpha}{n})^{n(t-t_0)}$
$q^{\mathbb{N}_0}$	$\prod_{s\in[t_0,t)}[1+(q-1)\alpha s]$ if $t > t_0$
$2^{\mathbb{N}_0}$	$\prod_{s\in[t_0,t)}(1+\alpha s)$ if $t > t_0$
$\{\sum_{k=1}^{n}\frac{1}{k} : n \in \mathbb{N}\}$	$\binom{n+\alpha-t_0}{n-t_0}$ if $t = \sum_{k=1}^{n}\frac{1}{k}$

Table 2.4. Exponential Functions

$\mathbb{T}$	t_0	$p(t)$	$e_p(t, t_0)$
$\mathbb{R}$	0	1	e^t
$\mathbb{Z}$	0	1	2^t
$h\mathbb{Z}$	0	1	$(1+h)^{t/h}$
$\frac{1}{n}\mathbb{Z}$	0	1	$\left[(1+\frac{1}{n})^n\right]^t$
$q^{\mathbb{N}_0}$	1	1	$\prod_{s\in(0,t)}[1+(q-1)s]$
$2^{\mathbb{N}_0}$	1	1	$\prod_{s\in(0,t)}(1+s)$
$q^{\mathbb{N}_0}$	1	$\frac{1-t}{(q-1)t^2}$	$\sqrt{t}e^{-\frac{\ln^2(t)}{2\ln(q)}}$
$2^{\mathbb{N}_0}$	1	$\frac{1-t}{t^2}$	$\sqrt{t}e^{-\frac{\ln^2(t)}{\ln(4)}}$
$\mathbb{N}_0^2$	0	1	$2^{\sqrt{t}}(\sqrt{t})!$
$\{\sum_{k=1}^{n}\frac{1}{k} : n \in \mathbb{N}\}$	0	1	$n+1$ if $t = \sum_{k=1}^{n}\frac{1}{k}$

2.4. Initial Value Problems

In this section we study the first order nonhomogeneous linear equation

(2.30) $$y^{\Delta} = p(t)y + f(t)$$

and the corresponding homogeneous equation

(2.31) $$y^{\Delta} = p(t)y$$

on a time scale $\mathbb{T}$. The results from Section 2.2 immediately yield the following theorem.

Theorem 2.62. *Suppose* (2.31) *is regressive. Let* $t_0 \in \mathbb{T}$ *and* $y_0 \in \mathbb{R}$. *The unique solution of the initial value problem*

(2.32) $$y^{\Delta} = p(t)y, \quad y(t_0) = y_0$$

is given by

$$y(t) = e_p(t, t_0)y_0.$$

Exercise 2.63. Verify directly that the function y given in Theorem 2.62 solves the initial value problem (2.32).

Definition 2.64. For $p \in \mathcal{R}$ we define an operator $L_1 : \mathrm{C}^1_{\mathrm{rd}} \to \mathrm{C}_{\mathrm{rd}}$ by

$$L_1 y(t) = y^{\Delta}(t) - p(t)y(t), \quad t \in \mathbb{T}^{\kappa}.$$

Then (2.31) can be written in the form $L_1 y(t) = 0$ and (2.30) can be written in the form $L_1 y(t) = f(t)$. Since L_1 is a linear operator we say that (2.30) is a linear equation. We say y is a *solution* of (2.30) on $\mathbb{T}$ provided $y \in \mathrm{C}^1_{\mathrm{rd}}$ and $L_1 y(t) = f(t)$ for $t \in \mathbb{T}^{\kappa}$.

Definition 2.65. The *adjoint operator* $L_1^* : \mathrm{C}^1_{\mathrm{rd}} \to \mathrm{C}_{\mathrm{rd}}$ is defined by

$$L_1^* x(t) = x^{\Delta}(t) + p(t)x^{\sigma}(t), \quad t \in \mathbb{T}^{\kappa}.$$

Example 2.66. The function

$$x(t) = (1 + \alpha h)^{\frac{-t}{h}}, \quad t \in h\mathbb{Z}$$

is a solution of the adjoint equation

$$x^{\Delta} + \alpha x^{\sigma} = 0, \quad t \in h\mathbb{Z},$$

where α is a regressive constant.

Exercise 2.67. Verify the above example by substituting x into the adjoint equation.

Theorem 2.68 (Lagrange Identity). *If* $x, y \in \mathrm{C}^1_{\mathrm{rd}}$, *then*

$$x^{\sigma} L_1 y + y L_1^* x = (xy)^{\Delta} \quad on \quad \mathbb{T}^{\kappa}.$$

Proof. Assume $x, y \in \mathrm{C}^1_{\mathrm{rd}}$ and consider

$$\begin{aligned}
(xy)^{\Delta} &= x^{\sigma} y^{\Delta} + x^{\Delta} y \\
&= x^{\sigma}(y^{\Delta} - py) + y(x^{\Delta} + px^{\sigma}) \\
&= x^{\sigma} L_1 y + y L_1^* x
\end{aligned}$$

on $\mathbb{T}^{\kappa}$. $\square$

The next result follows immediately from the Lagrange identity.

Corollary 2.69. *If x and y are solutions of $L_1 y = 0$ and $L_1^* x = 0$, respectively, then*

$$x(t)y(t) = C \quad for \quad t \in \mathbb{T},$$

where C is a constant.

It follows from this corollary that if a nontrivial y satisfies $L_1 y = 0$, then $x := \frac{1}{y}$ satisfies the adjoint equation $L_1^* x = 0$.

Exercise 2.70. Show directly by substitution into $L_1^* x = 0$ that if y is a nontrivial solution of $L_1 y = 0$, then $x := \frac{1}{y}$ is a nontrivial solution of the adjoint equation $L_1^* x = 0$.

For later reference we also cite the following existence and uniqueness result for the adjoint initial value problem.

Theorem 2.71. *Suppose $p \in \mathcal{R}$. Let $t_0 \in \mathbb{T}$ and $x_0 \in \mathbb{R}$. The unique solution of the initial value problem*

$$(2.33) \qquad x^\Delta = -p(t)x^\sigma, \quad x(t_0) = x_0$$

is given by

$$x(t) = e_{\ominus p}(t, t_0)x_0.$$

Exercise 2.72. Show that Theorem 2.71 follows directly from Corollary 2.69. Also verify Theorem 2.71 by showing directly that the function y given in Theorem 2.71 solves the initial value problem (2.33).

We now turn our attention to the nonhomogeneous problem

$$(2.34) \qquad x^\Delta = -p(t)x^\sigma + f(t), \quad x(t_0) = x_0.$$

Let us assume that x is a solution of (2.34). We multiply both sides of the dynamic equation in (2.34) by the so-called *integrating factor* $e_p(t, t_0)$ and obtain

$$\begin{aligned}
[e_p(\cdot, t_0)x]^\Delta(t) &= e_p(t, t_0)x^\Delta(t) + p(t)e_p(t, t_0)x^\sigma(t) \\
&= e_p(t, t_0)\left[x^\Delta(t) + p(t)x^\sigma(t)\right] \\
&= e_p(t, t_0)f(t),
\end{aligned}$$

and now we integrate both sides from t_0 to t to conclude

$$(2.35) \qquad e_p(t, t_0)x(t) - e_p(t_0, t_0)x(t_0) = \int_{t_0}^t e_p(\tau, t_0)f(\tau)\Delta\tau.$$

This integration is possible according to Theorem 1.74 provided $f \in C_{rd}$. Hence the following definition is useful.

Definition 2.73. The equation (2.30) is called *regressive* provided (2.31) is regressive and $f : \mathbb{T} \to \mathbb{R}$ is rd-continuous.

We next give the variation of constants formula for the adjoint equation $L_1^* x = f$.

Theorem 2.74 (Variation of Constants). *Suppose* (2.30) *is regressive. Let* $t_0 \in \mathbb{T}$ *and* $x_0 \in \mathbb{R}$. *The unique solution of the initial value problem*

(2.36) $$x^\Delta = -p(t)x^\sigma + f(t), \quad x(t_0) = x_0$$

is given by

$$x(t) = e_{\ominus p}(t, t_0)x_0 + \int_{t_0}^t e_{\ominus p}(t, \tau)f(\tau)\Delta\tau.$$

Proof. First, it is easily verified that x given above solves the initial value problem (2.36) (see Exercise 2.76 below). Next, if x is a solution of (2.36), then we have seen above that (2.35) holds. Hence we obtain

$$e_p(t, t_0)x(t) = x_0 + \int_{t_0}^t e_p(\tau, t_0)f(\tau)\Delta\tau.$$

We solve for x and apply Theorem 2.36 (iii) to arrive at

$$x(t) = e_{\ominus p}(t, t_0)x_0 + \int_{t_0}^t \frac{e_p(\tau, t_0)}{e_p(t, t_0)}f(\tau)\Delta\tau.$$

But since $e_p(t, \tau)e_p(\tau, t_0) = e_p(t, t_0)$ according to Theorem 2.36 (v), the final formula given in the theorem above follows with another application of Theorem 2.36 (iii). $\qquad\square$

Remark 2.75. Because of Theorem 2.36 (v), an alternative form of the solution of the initial value problem (2.36) is given by

$$x(t) = e_{\ominus p}(t, t_0)\left[x_0 + \int_{t_0}^t e_{\ominus p}(t_0, \tau)f(\tau)\Delta\tau\right].$$

Exercise 2.76. Verify directly that the function x given in Theorem 2.74 solves the IVP (2.36).

Next we give the variation of constants formula for $L_1 y = f$.

Theorem 2.77 (Variation of Constants). *Suppose* (2.30) *is regressive. Let* $t_0 \in \mathbb{T}$ *and* $y_0 \in \mathbb{R}$. *The unique solution of the initial value problem*

(2.37) $$y^\Delta = p(t)y + f(t), \quad y(t_0) = y_0$$

is given by

$$y(t) = e_p(t, t_0)y_0 + \int_{t_0}^t e_p(t, \sigma(\tau))f(\tau)\Delta\tau.$$

Proof. We equivalently rewrite $y^\Delta = p(t)y + f(t)$ as

$$y^\Delta = p(t)\left[y^\sigma - \mu(t)y^\Delta\right] + f(t),$$

i.e.,

$$(1 + \mu(t)p(t))y^\Delta = p(t)y^\sigma + f(t),$$

i.e., (use $p \in \mathcal{R}$)

$$y^\Delta = -(\ominus p)(t)y^\sigma + \frac{f(t)}{1 + \mu(t)p(t)}$$

and apply Theorem 2.74 to find the solution of (2.37) as (use $(\ominus(\ominus p))(t) = p(t)$)

$$y(t) = y_0 e_p(t, t_0) + \int_{t_0}^t e_p(t, \tau)\frac{f(\tau)}{1 + \mu(\tau)p(\tau)}\Delta\tau.$$

For the final calculation

$$\frac{e_p(t,\tau)}{1+\mu(\tau)p(\tau)} = \frac{e_p(t,\tau)}{e_p(\sigma(\tau),\tau)} = e_p(t,\sigma(\tau)),$$

we use Theorem 2.36 (ii) and (v). $\qquad\qquad\qquad\qquad\qquad\qquad\qquad\qquad\qquad\quad$ □

Remark 2.78. Because of Theorem 2.36 (v), an alternative form of the solution of the initial value problem (2.37) is given by

$$y(t) = e_p(t,t_0)\left[y_0 + \int_{t_0}^{t} e_p(t_0,\sigma(\tau))f(\tau)\Delta\tau\right].$$

Exercise 2.79. Use the variation of constants formula from Theorem 2.77 to solve the following initial value problems on the indicated time scales:

(i) $y^\Delta = 2y + t$, $y(0) = 0$, where $\mathbb{T} = \mathbb{R}$;
(ii) $y^\Delta = 2y + 3^t$, $y(0) = 0$, where $\mathbb{T} = \mathbb{Z}$;
(iii) $y^\Delta = p(t)y + e_p(t,t_0)$, $y(t_0) = 0$, where $\mathbb{T}$ is an arbitrary time scale and $p \in \mathcal{R}$.

2.5. Notes and References

Most of the results presented in Section 2.1 are taken from Hilger [164]. This section is mainly concerned with the time scale $h\mathbb{Z}$, but the cylinder transformation defined in Definition 2.21 can be used to find solutions of first order linear dynamic equations on an arbitrary time scale. We define the exponential function on an arbitrary time scale explicitly in terms of the cylinder transformation, and then show that it solves the initial value problem (2.17). Alternatively, we could have used the main existence theorem presented in the last chapter in order to establish the existence of the exponential function. In fact, the explicit representation of the exponential function in terms of the cylinder transformation is of no further use in this book except for proofs of Theorem 2.44 and Theorem 2.48, but it might be useful for finding estimates that are needed for stability results.

Besides rd-continuity, we utilize the concept of regressivity when defining the exponential function. All of these concepts are given in Hilger [160]. Many properties of the exponential function are derived, and some of them are contained in Bohner and Peterson [89, 90]. Theorem 2.39 has been found as Hilfssatz 2.5.1 in Keller [190]. The results on the sign of the exponential function are from Elvan Akın, Billûr Kaymakçalan, Lynn Erbe, and Allan Peterson [35]. As in [160], the set of positively regressive and rd-continuous functions is defined in this section.

Section 2.3 contains several examples of exponential functions on various time scales. Dynamic equations on the time scale $q^{\mathbb{N}}$ are called q-difference equations, and they are discussed independently of the time scales setting in many places throughout the literature. Here we have included the references [58, 109, 247, 253]. A plant population model is studied in Example 2.58, and we find the corresponding exponential function. Such types of examples can also be studied for animal population models. We refer to the book [96] for some explicit cases of models where our example applies. Of course one can also study the existence of two or several kinds of different species living on the time scale $\mathbb{P}$, which gives rise to predator-prey models. However, this is not pursued in this book and leaves room for future research. Many of the examples of exponential functions are from [89, 90]. For several time scales the form of the corresponding exponential function is rather unexpected and involves factorials, binomial coefficients, or the Wallis product. It certainly would be of interest to discuss exponential functions on other time scales, and other unexpected results may arise.

In the last section we solve initial value problems for first order linear equations. In contrast to the differential equations case, there are two forms of such equations. They both can be solved using exponential functions. The nonhomogeneous case is also solved, using a variation of parameters technique.

CHAPTER 3

Second Order Linear Equations

3.1. Wronskians

In this section we consider the second order linear dynamic equation

(3.1) $$y^{\Delta\Delta} + p(t)y^{\Delta} + q(t)y = f(t),$$

where we assume that $p, q, f \in C_{rd}$. If we introduce the operator $L_2 : C^2_{rd} \to C_{rd}$ by

$$L_2 y(t) = y^{\Delta\Delta}(t) + p(t)y^{\Delta}(t) + q(t)y(t)$$

for $t \in \mathbb{T}^{\kappa^2}$, then (3.1) can be rewritten as $L_2 y = f$. If $y \in C^2_{rd}$ and $L_2 y(t) = f(t)$ for all $t \in \mathbb{T}^{\kappa^2}$, then we say y is a *solution* of $L_2 y = f$ on $\mathbb{T}$. The fact that L_2 is a linear operator (see Theorem 3.1) is why we call equation (3.1) a linear equation. If $f(t) = 0$ for all $t \in \mathbb{T}^{\kappa^2}$, then we get the homogeneous dynamic equation $L_2 y = 0$. Otherwise we say the equation $L_2 y = f$ is nonhomogeneous. The following *principle of superposition* is easy to prove and is left as an exercise.

Theorem 3.1. *The operator $L_2 : C^2_{rd} \to C_{rd}$ is a linear operator, i.e.,*

$$L_2(\alpha y_1 + \beta y_2) = \alpha L_2(y_1) + \beta L_2(y_2) \quad \text{for all} \quad \alpha, \beta \in \mathbb{R} \text{ and } y_1, y_2 \in C^2_{rd}.$$

If y_1 and y_2 solve the homogeneous equation $L_2 y = 0$, then so does $y = \alpha y_1 + \beta y_2$, where α and β are any real constants.

Exercise 3.2. Prove Theorem 3.1. Show also that the sum of a solution of the homogeneous equation $L_2 y = 0$ and the nonhomogeneous equation (3.1) is a solution of (3.1).

Definition 3.3. Equation (3.1) is called *regressive* provided $p, q, f \in C_{rd}$ such that the regressivity condition

(3.2) $$1 - \mu(t)p(t) + \mu^2(t)q(t) \neq 0 \quad \text{for all} \quad t \in \mathbb{T}^{\kappa}$$

holds.

The following existence-uniqueness theorem is a special case of Corollary 5.90.

Theorem 3.4. *Assume that the dynamic equation (3.1) is regressive. If $t_0 \in \mathbb{T}^{\kappa}$, then the initial value problem*

$$L_2 y = f(t), \quad y(t_0) = y_0, \quad y^{\Delta}(t_0) = y_0^{\Delta},$$

where y_0 and y_0^{Δ} are given constants, has a unique solution, and this solution is defined on the whole time scale $\mathbb{T}$.

To motivate the next definition we now try to solve the initial value problem

$$(3.3) \qquad L_2 y = 0, \quad y(t_0) = y_0, \quad y^{\Delta}(t_0) = y_0^{\Delta},$$

where $t_0 \in \mathbb{T}^{\kappa}$ and $y_0, y_0^{\Delta} \in \mathbb{R}$. If y_1 and y_2 are two solutions of $L_2 y = 0$, then by Theorem 3.1

$$y(t) := \alpha y_1(t) + \beta y_2(t)$$

is a solution of $L_2 y = 0$ for all $\alpha, \beta \in \mathbb{R}$. Then we want to see if we can pick α and β so that

$$y_0 = y(t_0) = \alpha y_1(t_0) + \beta y_2(t_0)$$

and

$$y_0^{\Delta} = y^{\Delta}(t_0) = \alpha y_1^{\Delta}(t_0) + \beta y_2^{\Delta}(t_0),$$

i.e.,

$$(3.4) \qquad \begin{pmatrix} y_1(t_0) & y_2(t_0) \\ y_1^{\Delta}(t_0) & y_2^{\Delta}(t_0) \end{pmatrix} \begin{pmatrix} \alpha \\ \beta \end{pmatrix} = \begin{pmatrix} y_0 \\ y_0^{\Delta} \end{pmatrix}.$$

System (3.4) has a unique solution provided the occurring 2×2-matrix is invertible. This leads to the following definition.

Definition 3.5. For two differentiable functions y_1 and y_2 we define the *Wronskian* $W = W(y_1, y_2)$ by

$$W(t) = \det \begin{pmatrix} y_1(t) & y_2(t) \\ y_1^{\Delta}(t) & y_2^{\Delta}(t) \end{pmatrix}.$$

We say that two solutions y_1 and y_2 of $L_2 y = 0$ form a *fundamental set of solutions* (or a *fundamental system*) for $L_2 y = 0$ provided $W(y_1, y_2)(t) \neq 0$ for all $t \in \mathbb{T}^{\kappa}$.

Example 3.6. Suppose that $\mu(t) \neq 0$ for some $t \in \mathbb{T}^{\kappa}$. Then we can use Theorem 1.16 (ii) to compute an alternative form for the Wronskian of two functions y_1 and y_2 (at the argument t, which we omit below) as follows:

$$
\begin{aligned}
W(y_1, y_2) &= \det \begin{pmatrix} y_1 & y_2 \\ y_1^{\Delta} & y_2^{\Delta} \end{pmatrix} \\[2mm]
&= \det \begin{pmatrix} y_1 & y_2 \\ \dfrac{y_1^{\sigma} - y_1}{\mu} & \dfrac{y_2^{\sigma} - y_2}{\mu} \end{pmatrix} \\[2mm]
&= \det \begin{pmatrix} y_1 & y_2 \\ \dfrac{y_1^{\sigma}}{\mu} & \dfrac{y_2^{\sigma}}{\mu} \end{pmatrix} \\[2mm]
&= \frac{1}{\mu} \det \begin{pmatrix} y_1 & y_2 \\ y_1^{\sigma} & y_2^{\sigma} \end{pmatrix}.
\end{aligned}
$$

Theorem 3.7. *If the pair of functions y_1, y_2 forms a fundamental system of solutions for $L_2 y = 0$, then*

$$y(t) = \alpha y_1(t) + \beta y_2(t),$$

where α and β are constants, is a general solution of $L_2 y = 0$. By general solution we mean every function of this form is a solution and every solution is in this form. In particular the solution of the initial value problem (3.3) is given by

$$y(t) = \frac{y_2^\Delta(t_0)y_0 - y_2(t_0)y_0^\Delta}{W(y_1,y_2)(t_0)} y_1(t) + \frac{y_1(t_0)y_0^\Delta - y_1^\Delta(t_0)y_0}{W(y_1,y_2)(t_0)} y_2(t).$$

Proof. Assume that the pair of functions y_1, y_2 is a fundamental system of solutions for $L_2 y = 0$. By Theorem 3.1, any function of the form

$$y(t) = \alpha y_1(t) + \beta y_2(t),$$

where α and β are constants, is a solution of $L_2 y = 0$. Now we show that any solution of $L_2 y = 0$ is of this form. Let $y_0(t)$ be a fixed but arbitrary solution of $L_2 y = 0$. Fix $t_0 \in \mathbb{T}^\kappa$ and let

$$y_0 := y_0(t_0), \quad y_0^\Delta := y_0^\Delta(t_0).$$

Let

$$u(t) = \alpha y_1(t) + \beta y_2(t).$$

We want to pick α and β so that

$$u(t_0) = y_0 = \alpha y_1(t_0) + \beta y_2(t_0)$$

and

$$u^\Delta(t_0) = y_0^\Delta = \alpha y_1^\Delta(t_0) + \beta y_2^\Delta(t_0).$$

Hence we want to pick α and β so that (3.4) is satisfied. Since the determinant of the coefficient matrix in (3.4) is nonzero, there is a unique set of constants α_0, β_0 so that (3.4) is satisfied. It follows that

$$u(t) = \alpha_0 y_1(t) + \beta_0 y_2(t)$$

solves the initial value problem (3.3). By the uniqueness theorem (Theorem 3.4),

$$y_0(t) = u(t) = \alpha_0 y_1(t) + \beta_0 y_2(t).$$

To complete the proof of this theorem it remains to prove the last statement in this theorem which is the next exercise. $\square$

Exercise 3.8. Prove the last statement in Theorem 3.7.

It is an important fact that the Wronskian of two solutions of (3.1) is nonzero at a single point t_0 if and only if it is nonzero for all t. This is a consequence of the subsequent Abel's formula. Before we prove Abel's theorem we prove the following lemma concerning Wronskians.

Lemma 3.9. *Let y_1 and y_2 be twice differentiable. Then*

(i) $W(y_1,y_2) = \det \begin{pmatrix} y_1^\sigma & y_2^\sigma \\ y_1^\Delta & y_2^\Delta \end{pmatrix};$

(ii) $W^\Delta(y_1, y_2) = \det \begin{pmatrix} y_1^\sigma & y_2^\sigma \\ y_1^{\Delta\Delta} & y_2^{\Delta\Delta} \end{pmatrix}$;

(iii) $W^\Delta(y_1, y_2) = \det \begin{pmatrix} y_1^\sigma & y_2^\sigma \\ Ly_1 & Ly_2 \end{pmatrix} + (-p + \mu q)W(y_1, y_2)$.

Proof. By Definition 3.5 and Theorem 1.16 (iv), we have

$$W(y_1, y_2) = \det \begin{pmatrix} y_1 & y_2 \\ y_1^\Delta & y_2^\Delta \end{pmatrix}$$

$$= \det \begin{pmatrix} y_1^\sigma - \mu y_1^\Delta & y_2^\sigma - \mu y_2^\Delta \\ y_1^\Delta & y_2^\Delta \end{pmatrix}$$

$$= \det \begin{pmatrix} y_1^\sigma & y_2^\sigma \\ y_1^\Delta & y_2^\Delta \end{pmatrix}.$$

This proves part (i). For part (ii), we use the product formula in Theorem 1.20 (iii) to calculate

$$[W(y_1, y_2)]^\Delta = (y_1 y_2^\Delta - y_2 y_1^\Delta)^\Delta$$
$$= y_1^\sigma y_2^{\Delta\Delta} + y_1^\Delta y_2^\Delta - y_2^\sigma y_1^{\Delta\Delta} - y_2^\Delta y_1^\Delta$$
$$= y_1^\sigma y_2^{\Delta\Delta} - y_2^\sigma y_1^{\Delta\Delta}$$
$$= \det \begin{pmatrix} y_1^\sigma & y_2^\sigma \\ y_1^{\Delta\Delta} & y_2^{\Delta\Delta} \end{pmatrix}.$$

Finally we employ part (ii) to obtain

$$[W(y_1, y_2)]^\Delta = \det \begin{pmatrix} y_1^\sigma & y_2^\sigma \\ y_1^{\Delta\Delta} & y_2^{\Delta\Delta} \end{pmatrix}$$

$$= \det \begin{pmatrix} y_1^\sigma & y_2^\sigma \\ Ly_1 - py_1^\Delta - qy_1 & Ly_2 - py_2^\Delta - qy_2 \end{pmatrix}$$

$$= \det \begin{pmatrix} y_1^\sigma & y_2^\sigma \\ Ly_1 & Ly_2 \end{pmatrix} + \det \begin{pmatrix} y_1^\sigma & y_2^\sigma \\ -py_1^\Delta - qy_1 & -py_2^\Delta - qy_2 \end{pmatrix}$$

$$= \det \begin{pmatrix} y_1^\sigma & y_2^\sigma \\ Ly_1 & Ly_2 \end{pmatrix} + \det \begin{pmatrix} y_1^\sigma & y_2^\sigma \\ -py_1^\Delta - qy_1^\sigma + q\mu y_1^\Delta & -py_2^\Delta - qy_2^\sigma + q\mu y_2^\Delta \end{pmatrix}$$

$$= \det \begin{pmatrix} y_1^\sigma & y_2^\sigma \\ Ly_1 & Ly_2 \end{pmatrix} + \det \begin{pmatrix} y_1^\sigma & y_2^\sigma \\ -py_1^\Delta + q\mu y_1^\Delta & -py_2^\Delta + q\mu y_2^\Delta \end{pmatrix}$$

$$= \det \begin{pmatrix} y_1^\sigma & y_2^\sigma \\ Ly_1 & Ly_2 \end{pmatrix} + (-p + \mu q) \det \begin{pmatrix} y_1^\sigma & y_2^\sigma \\ y_1^\Delta & y_2^\Delta \end{pmatrix}.$$

From here, part (iii) follows by applying part (i). □

Theorem 3.10 (Abel's Theorem). *Let $t_0 \in \mathbb{T}^\kappa$ and assume that $L_2 y = 0$ is regressive. Suppose that y_1 and y_2 are two solutions of $L_2 y = 0$. Then their Wronskian $W = W(y_1, y_2)$ satisfies*

$$W(t) = e_{-p+\mu q}(t, t_0) W(t_0) \quad \text{for} \quad t \in \mathbb{T}^\kappa.$$

Proof. Using the fact that y_1 and y_2 are solutions of $L_2 y = 0$ and Lemma 3.9 (iii), we see that W is a solution of the initial value problem

$$(3.5) \qquad W^\Delta = [-p(t) + \mu(t)q(t)]W, \quad W(t_0) = W_0,$$

where we put $W_0 = W(t_0)$. Using condition (3.2) we get that the coefficient function $-p + \mu q$ is regressive. Since $p, q, \mu \in C_{rd}$, we have that $-p + \mu q \in C_{rd}$. Altogether $-p + \mu q \in \mathcal{R}$. By Theorem 2.62, the unique solution of (3.5) is $W(t) = e_{-p+\mu q}(t, t_0) W_0$ for $t \in \mathbb{T}^\kappa$. □

Alternatively, we may consider a linear dynamic equation of the form

$$(3.6) \qquad x^{\Delta\Delta} + p(t)x^{\Delta^\sigma} + q(t)x^\sigma = 0.$$

Definition 3.11. We say that (3.6) is *regressive* provided $p \in \mathcal{R}$ and $q \in C_{rd}$.

Theorem 3.12. *If (3.6) is regressive, then it is equivalent to a regressive equation of the form $L_2 y = 0$. Conversely, if $L_2 y = 0$ is regressive, then it is equivalent to a regressive equation of the form (3.6).*

Proof. Assume (3.6) is regressive. Then $p, q \in C_{rd}$ and $1 + \mu(t)p(t) \neq 0$ for $t \in \mathbb{T}^\kappa$. We use Theorem 1.16 (iv) twice (once for x and then for x^Δ) to write

$$\begin{aligned} x^{\Delta\Delta} + px^{\Delta^\sigma} + qx^\sigma &= x^{\Delta\Delta} + p(x^\Delta + \mu x^{\Delta\Delta}) + q(x + \mu x^\Delta) \\ &= x^{\Delta\Delta} + \mu p x^{\Delta\Delta} + (p + \mu q)x^\Delta + qx \\ &= (1 + \mu p)x^{\Delta\Delta} + (p + \mu q)x^\Delta + qx \\ &= [1 + \mu p]\left[x^{\Delta\Delta} + \frac{p + \mu q}{1 + \mu p}x^\Delta + \frac{q}{1 + \mu p}x \right]. \end{aligned}$$

Hence equation (3.6) is equivalent to

$$y^{\Delta\Delta} + p_1(t)y^\Delta + q_1(t)y = 0,$$

where

$$p_1 := \frac{p + \mu q}{1 + \mu p} \quad \text{and} \quad q_1 := \frac{q}{1 + \mu p}.$$

Note that $p_1, q_1 \in C_{rd}$, and since

$$
\begin{aligned}
1 - \mu p_1 + \mu^2 q_1 &= 1 - \mu \frac{p + \mu q}{1 + \mu p} + \mu^2 \frac{q}{1 + \mu p} \\
&= \frac{1 + \mu p - \mu p - \mu^2 q + \mu^2 q}{1 + \mu p} \\
&= \frac{1}{1 + \mu p} \\
&\neq 0,
\end{aligned}
$$

we get that (3.6) is equivalent to a regressive equation of the form (3.1).

Next assume that (3.1) is regressive. Then $1 - \mu(t)p(t) + \mu^2(t)q(t) \neq 0$ for $t \in \mathbb{T}^\kappa$ and $p, q \in C_{rd}$. Consider

$$
\begin{aligned}
y^{\Delta\Delta} + p y^{\Delta} + q y &= y^{\Delta\Delta} + p(y^{\Delta^\sigma} - \mu y^{\Delta\Delta}) + q(y^\sigma - \mu y^\Delta) \\
&= (1 - \mu p) y^{\Delta\Delta} + p y^{\Delta^\sigma} + q y^\sigma - \mu q(y^{\Delta^\sigma} - \mu y^{\Delta\Delta}) \\
&= (1 - \mu p + \mu^2 q) y^{\Delta\Delta} + (p - \mu q) y^{\Delta^\sigma} + q y^\sigma \\
&= (1 - \mu p + \mu^2 q) \left(y^{\Delta\Delta} + \frac{p - \mu q}{1 - \mu p + \mu^2 q} y^{\Delta^\sigma} + \frac{q}{1 - \mu p + \mu^2 q} y^\sigma \right).
\end{aligned}
$$

Hence $L_2 y = 0$ is equivalent to the equation

(3.7) $x^{\Delta\Delta} + p_2(t) x^{\Delta^\sigma} + q_2(t) x^\sigma = 0,$

where

$$p_2 := \frac{p - \mu q}{1 - \mu p + \mu^2 q} \quad \text{and} \quad q_2 := \frac{q}{1 - \mu p + \mu^2 q}.$$

Note that $p_2, q_2 \in C_{rd}$, and since

$$
\begin{aligned}
1 + \mu p_2 &= 1 + \mu \frac{p - \mu q}{1 - \mu p + \mu^2 q} \\
&= \frac{1}{1 - \mu p + \mu^2 q} \\
&\neq 0,
\end{aligned}
$$

we have $p_2 \in \mathcal{R}$ so that (3.7) is regressive.   □

Theorem 3.13 (Abel's Theorem for (3.6)). *Assume that (3.6) is regressive and let $t_0 \in \mathbb{T}^\kappa$. Suppose that x_1 and x_2 are two solutions of equation (3.6). Then their Wronskian $W = W(x_1, x_2)$ satisfies*

$$W(t) = W(t_0) e_{\ominus p}(t, t_0) \quad \text{for} \quad t \in \mathbb{T}^\kappa.$$

Proof. Let x_1 and x_2 be two solutions of (3.6). We use Theorem 3.9 (ii) to find

$$
\begin{aligned}
W^\Delta &= \det \begin{pmatrix} x_1^\sigma & x_2^\sigma \\ x_1^{\Delta\Delta} & x_2^{\Delta\Delta} \end{pmatrix} \\
&= \det \begin{pmatrix} x_1^\sigma & x_2^\sigma \\ -px_1^{\Delta^\sigma} - qx_1^\sigma & -px_2^{\Delta^\sigma} - qx_2^\sigma \end{pmatrix} \\
&= \det \begin{pmatrix} x_1^\sigma & x_2^\sigma \\ -px_1^{\Delta^\sigma} & -px_2^{\Delta^\sigma} \end{pmatrix} \\
&= -p \det \begin{pmatrix} x_1^\sigma & x_2^\sigma \\ x_1^{\Delta^\sigma} & x_2^{\Delta^\sigma} \end{pmatrix} \\
&= -p \left(\det \begin{pmatrix} x_1 & x_2 \\ x_1^\Delta & x_2^\Delta \end{pmatrix} \right)^\sigma \\
&= -pW^\sigma.
\end{aligned}
$$

Hence W is a solution of the initial value problem

$$W^\Delta = -p(t)W^\sigma, \quad W(t_0) = W_0,$$

where we put $W_0 = W(t_0)$. The unique solution of such an initial value problem of the form (2.33) is given in Theorem 2.71 (observe that the above first order linear equation is regressive due to Theorem 3.12 and in particular condition (2.13)), i.e., $W(t) = W_0 e_{\ominus p}(t, t_0)$. $\square$

Corollary 3.14. *Assume $q \in C_{\mathrm{rd}}$. Then the Wronskian of any two solutions of*

$$x^{\Delta\Delta} + q(t)x^\sigma = 0$$

is independent of t.

Proof. Let $t_0 \in \mathbb{T}$. The above assumptions ensure that (3.6) with $p \equiv 0$ is regressive. Since $\ominus p \equiv 0$, we have $e_{\ominus p}(t, t_0) \equiv 1$, and by Theorem 3.13,

$$W(x_1, x_2)(t) \equiv W(x_1, x_2)(t_0)$$

follows, where x_1 and x_2 can be any two solutions of (3.6). $\square$

3.2. Hyperbolic and Trigonometric Functions

Here and in the following section we consider the second order linear dynamic homogeneous equation with constant coefficients

(3.8) $$y^{\Delta\Delta} + \alpha y^\Delta + \beta y = 0 \quad \text{with} \quad \alpha, \beta \in \mathbb{R}$$

on a time scale $\mathbb{T}$. We assume throughout that the dynamic equation (3.8) is regressive, i.e.,

$$1 - \alpha\mu(t) + \beta\mu^2(t) \neq 0 \quad \text{for} \quad t \in \mathbb{T}^\kappa, \quad \text{i.e.,} \quad \beta\mu - \alpha \in \mathcal{R}.$$

We try to find numbers $\lambda \in \mathbb{C}$ with $1 + \lambda\mu(t) \neq 0$ for $t \in \mathbb{T}^\kappa$ such that

$$y(t) = e_\lambda(t, t_0)$$

is a solution of (3.8). Note that if $y(t) = e_\lambda(t, t_0)$, then

$$
\begin{aligned}
y^{\Delta\Delta}(t) + \alpha y^\Delta(t) + \beta y(t) &= \lambda^2 e_\lambda(t, t_0) + \alpha\lambda e_\lambda(t, t_0) + \beta e_\lambda(t, t_0) \\
&= \left(\lambda^2 + \alpha\lambda + \beta\right) e_\lambda(t, t_0).
\end{aligned}
$$

Since $e_\lambda(t, t_0)$ does not vanish, $y(t) = e_\lambda(t, t_0)$ is a solution iff λ satisfies the so-called *characteristic equation*

$$(3.9) \qquad\qquad\qquad \lambda^2 + \alpha\lambda + \beta = 0.$$

The solutions λ_1 and λ_2 of the characteristic equation (3.9) are given by

$$(3.10) \qquad \lambda_1 = \frac{-\alpha - \sqrt{\alpha^2 - 4\beta}}{2} \quad \text{and} \quad \lambda_2 = \frac{-\alpha + \sqrt{\alpha^2 - 4\beta}}{2}.$$

Exercise 3.15. Let λ_1 and λ_2 be defined as in (3.10). Show that (3.8) is regressive iff $\lambda_1, \lambda_2 \in \mathcal{R}$.

Theorem 3.16. *Suppose $\alpha^2 - 4\beta \neq 0$. If $\mu\beta - \alpha \in \mathcal{R}$, then a fundamental system of (3.8) is given by*

$$e_{\lambda_1}(\cdot, t_0) \quad \text{and} \quad e_{\lambda_2}(\cdot, t_0),$$

where $t_0 \in \mathbb{T}^\kappa$ and λ_1, λ_2 are given in (3.10). The solution of the initial value problem

$$(3.11) \qquad y^{\Delta\Delta} + \alpha y^\Delta + \beta y = 0, \quad y(t_0) = y_0, \quad y^\Delta(t_0) = y_0^\Delta$$

is given by

$$y_0 \frac{e_{\lambda_1}(\cdot, t_0) + e_{\lambda_2}(\cdot, t_0)}{2} + \frac{\alpha y_0 + 2y_0^\Delta}{\sqrt{\alpha^2 - 4\beta}} \frac{e_{\lambda_2}(\cdot, t_0) - e_{\lambda_1}(\cdot, t_0)}{2}.$$

Proof. Since λ_1 and λ_2 given in (3.10) are solutions of the characteristic equation (3.9), we find that (note $\lambda_1, \lambda_2 \in \mathcal{R}$ according to Exercise (3.15)) both $e_{\lambda_1}(\cdot, t_0)$ and $e_{\lambda_2}(\cdot, t_0)$ solve the dynamic equation (3.8). Moreover, the Wronskian of these two solutions is found to be

$$
\det \begin{pmatrix} e_{\lambda_1}(t, t_0) & e_{\lambda_2}(t, t_0) \\ \lambda_1 e_{\lambda_1}(t, t_0) & \lambda_2 e_{\lambda_2}(t, t_0) \end{pmatrix} = (\lambda_2 - \lambda_1) e_{\lambda_1}(t, t_0) e_{\lambda_2}(t, t_0)
$$

$$
= \sqrt{\alpha^2 - 4\beta}\, e_{\lambda_1 \oplus \lambda_2}(t, t_0),
$$

which does not vanish unless $\alpha^2 - 4\beta = 0$. Having now obtained a fundamental system $y_1 = e_{\lambda_1}(\cdot, t_0)$ and $y_2 = e_{\lambda_2}(\cdot, t_0)$ of equation (3.8), we may use Theorem

3.7 to obtain a solution of the IVP (3.11), namely $y(t) = c_1 y_1(t) + c_2 y_2(t)$, where

$$
\begin{aligned}
c_1 &= \frac{y_0 y_2^{\Delta}(t_0) - y_0^{\Delta} y_2(t_0)}{W(y_1, y_2)(t_0)} \\
&= \frac{y_0 \lambda_2 - y_0^{\Delta}}{\sqrt{\alpha^2 - 4\beta}} \\
&= \frac{y_0}{2} - \frac{\alpha y_0 + 2 y_0^{\Delta}}{2\sqrt{\alpha^2 - 4\beta}}
\end{aligned}
$$

and

$$
\begin{aligned}
c_2 &= \frac{y_1(t_0) y_0^{\Delta} - y_1^{\Delta}(t_0) y_0}{W(y_1, y_2)(t_0)} \\
&= \frac{y_0^{\Delta} - \lambda_1 y_0}{\sqrt{\alpha^2 - 4\beta}} \\
&= \frac{y_0}{2} + \frac{\alpha y_0 + 2 y_0^{\Delta}}{2\sqrt{\alpha^2 - 4\beta}}.
\end{aligned}
$$

Hence it is possible to give the solution of (3.11) in the above claimed form.   □

When $\alpha = 0$ and $\beta < 0$, we are interested in the *hyperbolic functions* defined as follows.

Definition 3.17 (Hyperbolic Functions). If $p \in C_{rd}$ and $-\mu p^2 \in \mathcal{R}$, then we define the hyperbolic functions $\cosh_p$ and $\sinh_p$ by

$$
\cosh_p = \frac{e_p + e_{-p}}{2} \quad \text{and} \quad \sinh_p = \frac{e_p - e_{-p}}{2}.
$$

Note that the regressivity condition on $-\mu p^2$ is equivalent to

$$
0 \neq 1 - \mu^2 p^2 = (1 - \mu p)(1 + \mu p)
$$

and hence is equivalent to both p and $-p$ being regressive.

In the following, if f is a function in two variables, we mean by f^{Δ} the derivative with respect to the first variable.

Lemma 3.18. *Let $p \in C_{rd}$. If $-\mu p^2 \in \mathcal{R}$, then we have*

$$
\cosh_p^{\Delta} = p \sinh_p \quad \text{and} \quad \sinh_p^{\Delta} = p \cosh_p
$$

and

$$
\cosh_p^2 - \sinh_p^2 = e_{-\mu p^2}.
$$

Proof. Using Definition 3.17, the first two formulas are easily verified, while the last formula follows from

$$
\begin{aligned}
\cosh_p^2 - \sinh_p^2 &= \left(\frac{e_p + e_{-p}}{2} \right)^2 - \left(\frac{e_p - e_{-p}}{2} \right)^2 \\
&= \frac{e_p^2 + 2 e_p e_{-p} + e_{-p}^2}{4} - \frac{e_p^2 - 2 e_p e_{-p} + e_{-p}^2}{4} \\
&= e_p e_{-p} \\
&= e_{-p \oplus p} \\
&= e_{-\mu p^2},
\end{aligned}
$$

where we have used Theorem 2.36 (vi).   □

Exercise 3.19. If $p \in C_{rd}$ and $-\mu p^2 \in \mathcal{R}$, simplify

    (i) $\cosh_p(t, s) + \sinh_p(t, s)$;

    (ii) $\cosh_p(t, s) - \sinh_p(t, s)$;

    (iii) $\dfrac{\cosh_p(s,t_0)\cosh_p(t,t_0) - \sinh_p(s,t_0)\sinh_p(t,t_0)}{\cosh_p^2(s,t_0) - \sinh_p^2(s,t_0)}$;

    (iv) $\dfrac{\cosh_p(s,t_0)\sinh_p(t,t_0) - \sinh_p(s,t_0)\cosh_p(t,t_0)}{\cosh_p^2(s,t_0) - \sinh_p^2(s,t_0)}$.

Exercise 3.20. Show that if $\gamma > 0$ with $-\gamma^2\mu \in \mathcal{R}$, then $\cosh_\gamma(\cdot, t_0)$ and $\sinh_\gamma(\cdot, t_0)$ are solutions of

$$(3.12) \qquad\qquad y^{\Delta\Delta} - \gamma^2 y = 0.$$

Theorem 3.21. *If $\gamma \in \mathbb{R} \setminus \{0\}$ with $-\gamma^2\mu \in \mathcal{R}$, then a general solution of (3.12) is given by*

$$y(t) = c_1 \cosh_\gamma(t, t_0) + c_2 \sinh_\gamma(t, t_0),$$

where c_1 and c_2 are constants.

Proof. From Exercise 3.20, $\cosh_\gamma(t, t_0)$ and $\sinh_\gamma(t, t_0)$ are solutions of (3.12). It remains to show that $\cosh_\gamma(t, t_0)$ and $\sinh_\gamma(t, t_0)$ form a fundamental set of solutions. This follows from

$$
\begin{aligned}
W\left(\cosh_\gamma(\cdot, t_0), \sinh_\gamma(\cdot, t_0)\right)(t_0) &= \det \begin{pmatrix} \cosh_\gamma(t_0, t_0) & \sinh_\gamma(t_0, t_0) \\ \gamma\sinh_\gamma(t_0, t_0) & \gamma\cosh_\gamma(t_0, t_0) \end{pmatrix} \\
&= \det \begin{pmatrix} 1 & 0 \\ 0 & \gamma \end{pmatrix} \\
&= \gamma \\
&\neq 0.
\end{aligned}
$$

$\square$

Exercise 3.22. Show that if $\gamma > 0$ and $-\gamma^2\mu \in \mathcal{R}$, then the solution of the IVP

$$y^{\Delta\Delta} - \gamma^2 y = 0, \quad y(t_0) = y_0, \quad y^\Delta(t_0) = y_0^\Delta$$

is given by

$$y(t) = y_0 \cosh_\gamma(t, t_0) + \frac{y_0^\Delta}{\gamma} \sinh_\gamma(t, t_0).$$

Exercise 3.23. Let α be constant such that $|\alpha h| \neq 1$. For $\mathbb{T} = h\mathbb{Z}$ find $\cosh_\alpha(\cdot, 0)$ and $\sinh_\alpha(\cdot, 0)$. Is it true that

$$[\cosh_\alpha(t, 0)]^2 - [\sinh_\alpha(t, 0)]^2 = 1$$

for $t \in h\mathbb{Z}$?

Theorem 3.24. *Suppose $\alpha^2 - 4\beta > 0$. Define*

$$p = -\frac{\alpha}{2} \quad \text{and} \quad q = \frac{\sqrt{\alpha^2 - 4\beta}}{2}.$$

If p and $\mu\beta - \alpha$ are regressive, then a fundamental system of (3.8) is given by

$$\cosh_{q/(1+\mu p)}(\cdot, t_0)e_p(\cdot, t_0) \quad \text{and} \quad \sinh_{q/(1+\mu p)}(\cdot, t_0)e_p(\cdot, t_0),$$

where $t_0 \in \mathbb{T}$, and the Wronskian of these two solutions is

$$qe_{\mu\beta-\alpha}(\cdot, t_0).$$

The solution of the IVP (3.11) is given by

$$\left[y_0 \cosh_{q/(1+\mu p)}(\cdot, t_0) + \frac{y_0^\Delta - py_0}{q} \sinh_{q/(1+\mu p)}(\cdot, t_0) \right] e_p(\cdot, t_0).$$

Proof. In this proof we use the convention that

$$e_p = e_p(\cdot, t_0)$$

and similarly for cosh and sinh. We apply Theorem 3.16 to find two solutions of (3.8) as

$$e_{p+q} \quad \text{and} \quad e_{p-q}.$$

By Theorem 3.1 we can construct two other solutions of (3.8) by

$$y_1 = \frac{e_{p+q} + e_{p-q}}{2} \quad \text{and} \quad y_2 = \frac{e_{p+q} - e_{p-q}}{2}.$$

We use the formulas

$$p \oplus \left(\frac{q}{1+\mu p} \right) = p + \frac{q}{1+\mu p} + \frac{\mu pq}{1+\mu p} = p + q$$

and

$$p \oplus \left(-\frac{q}{1+\mu p} \right) = p - \frac{q}{1+\mu p} - \frac{\mu pq}{1+\mu p} = p - q$$

to obtain, by using Theorem 2.36 (vi),

$$
\begin{aligned}
y_1 &= \frac{e_{p+q} + e_{p-q}}{2} \\
&= \frac{e_{p \oplus (q/(1+\mu p))} + e_{p \oplus (-q/(1+\mu p))}}{2} \\
&= \frac{e_p e_{q/(1+\mu p)} + e_p e_{-q/(1+\mu p)}}{2} \\
&= e_p \frac{e_{q/(1+\mu p)} + e_{-q/(1+\mu p)}}{2} \\
&= e_p \cosh_{q/(1+\mu p)}
\end{aligned}
$$

and similarly

$$y_2 = \frac{e_{p+q} - e_{p-q}}{2} = e_p \sinh_{q/(1+\mu p)} .$$

Next, we find by using Theorem 2.36 (ii) and Lemma 3.18 that

$$
\begin{aligned}
y_1^\Delta &= e_p^\Delta \cosh_{q/(1+\mu p)} + e_p^\sigma \cosh_{q/(1+\mu p)}^\Delta \\
&= pe_p \cosh_{q/(1+\mu p)} + e_p^\sigma \frac{q}{1+\mu p} \sinh_{q/(1+\mu p)} \\
&= pe_p \cosh_{q/(1+\mu p)} + qe_p \sinh_{q/(1+\mu p)}
\end{aligned}
$$

and similarly that

$$y_2^\Delta = pe_p \sinh_{q/(1+\mu p)} + qe_p \cosh_{q/(1+\mu p)} .$$

Use

$$y_1(t_0) = 1, \quad y_2(t_0) = 0, \quad y_1^\Delta(t_0) = p, \quad y_2^\Delta(t_0) = q$$

to verify that the function given in the statement of the theorem indeed solves
(3.11). Finally, we find the Wronskian of y_1 and y_2 as

$$
\det \begin{pmatrix} e_p \cosh_{q/(1+\mu p)} & e_p \sinh_{q/(1+\mu p)} \\ pe_p \cosh_{q/(1+\mu p)} + qe_p \sinh_{q/(1+\mu p)} & pe_p \sinh_{q/(1+\mu p)} + qe_p \cosh_{q/(1+\mu p)} \end{pmatrix}
$$

$$
\begin{aligned}
&= \det \begin{pmatrix} e_p \cosh_{q/(1+\mu p)} & e_p \sinh_{q/(1+\mu p)} \\ qe_p \sinh_{q/(1+\mu p)} & qe_p \cosh_{q/(1+\mu p)} \end{pmatrix} \\
&= qe_p^2 \left[\cosh_{q/(1+\mu p)}^2 - \sinh_{q/(1+\mu p)}^2 \right] \\
&= qe_p^2 e_{-\mu q^2/(1+\mu p)^2} \\
&= qe_{p(2+\mu p) \oplus (-\mu q^2/(1+\mu p)^2)} \\
&= qe_{2p+\mu(p^2-q^2)} \\
&= qe_{\mu\beta-\alpha},
\end{aligned}
$$

where we have used Lemma 3.18. $\square$

When $\alpha = 0$ and $\beta > 0$, we are interested in the *trigonometric functions* defined
as follows.

Definition 3.25 (Trigonometric Functions). If $p \in C_{rd}$ and $\mu p^2 \in R$, then we
define the trigonometric functions $\cos_p$ and $\sin_p$ by

$$
\cos_p = \frac{e_{ip} + e_{-ip}}{2} \quad \text{and} \quad \sin_p = \frac{e_{ip} - e_{-ip}}{2i}.
$$

Note that μp^2 is regressive iff both ip and $-ip$ are regressive, so $\cos_p$ and $\sin_p$
in Definition 3.25 are well defined.

The proofs of Lemma 3.26 and Theorem 3.32 are similar to the proofs of Lemma
3.18 and Theorem 3.24 and hence will be omitted.

Lemma 3.26. Let $p \in C_{rd}$. If $\mu p^2 \in R$, then we have

$$
\cos_p^\Delta = -p\sin_p \quad \text{and} \quad \sin_p^\Delta = p\cos_p
$$

and

$$
\cos_p^2 + \sin_p^2 = e_{\mu p^2}.
$$

We remark that for $p \in \mathbb{R}$, the regressivity condition on μp^2 is always satisfied.

Exercise 3.27. State and prove the four formulas for the trigonometric functions
analogous to the formulas in Exercise 3.19 for the hyperbolic functions. In particular, also show *Euler's formula*

(3.13) $e_{ip}(t, t_0) = \cos_p(t, t_0) + i \sin_p(t, t_0).$

Exercise 3.28. Find $\sin_\alpha(\cdot, 0)$ and $\cos_\alpha(\cdot, 0)$ for constant $\alpha \in R$ for the time scales
$\mathbb{T} = \mathbb{R}$ and $\mathbb{T} = \mathbb{Z}$, respectively.

Exercise 3.29. Find $e_1(t, 0)$, $\sin_1(t, 0)$, and $\cos_1(t, 0)$ if $\mathbb{T} = \mathbb{Z}^2 = \{k^2 : k \in \mathbb{Z}\}$.

3.3. REDUCTION OF ORDER

Exercise 3.30. Show that the identity

$$[\sin_p(t, t_0)]^2 + [\cos_p(t, t_0)]^2 = 1$$

need not hold.

Theorem 3.31. *Assume $\gamma > 0$ and $t_0 \in \mathbb{T}^\kappa$. Then*

$$y(t) = c_1 \cos_\gamma(t, t_0) + c_2 \sin_\gamma(t, t_0)$$

is a general solution of

(3.14) $$y^{\Delta\Delta} + \gamma^2 y = 0.$$

Proof. Note that the equation (3.14) is regressive because $1 + \gamma^2 \mu^2(t) \neq 0$ for $t \in \mathbb{T}^\kappa$. Using Lemma 3.26, we can easily show that $\cos_\gamma(t, t_0)$ and $\sin_\gamma(t, t_0)$ are solutions of (3.14). The reader is asked to verify

$$W\left(\cos_\gamma(\cdot, t_0), \sin_\gamma(\cdot, t_0)\right)(t_0) = \det \begin{pmatrix} 0 & 1 \\ \gamma & -\gamma \end{pmatrix} = \gamma \neq 0,$$

and then we get that $\cos_\gamma(\cdot, t_0)$ and $\sin_\gamma(\cdot, t_0)$ form a fundamental set of solutions of (3.14). It follows that

$$y(t) = c_1 \cos_\gamma(t, t_0) + c_2 \sin_\gamma(t, t_0)$$

is a general solution of (3.14). $\square$

Theorem 3.32. *Suppose $\alpha^2 - 4\beta < 0$. Define*

$$p = -\frac{\alpha}{2} \quad \text{and} \quad q = \frac{\sqrt{4\beta - \alpha^2}}{2}.$$

If p and $\mu\beta - \alpha$ are regressive, then a fundamental system of (3.8) is given by

$$\cos_{q/(1+\mu p)}(\cdot, t_0) e_p(\cdot, t_0) \quad \text{and} \quad \sin_{q/(1+\mu p)}(\cdot, t_0) e_p(\cdot, t_0),$$

where $t_0 \in \mathbb{T}$, and the Wronskian of these two solutions is

$$q e_{\mu\beta - \alpha}(\cdot, t_0).$$

The solution of the initial value problem (3.11) is given by

$$\left[y_0 \cos_{q/(1+\mu p)}(\cdot, t_0) + \frac{y_0^\Delta - p y_0}{q} \sin_{q/(1+\mu p)}(\cdot, t_0)\right] e_p(\cdot, t_0).$$

Exercise 3.33. Prove Theorem 3.32.

3.3. Reduction of Order

Now we consider the dynamic equation (3.8) in the case that $\alpha^2 - 4\beta = 0$. From (3.10) we have $\lambda_1 = \lambda_2 = p$, where

$$p = -\frac{\alpha}{2}.$$

In this case $\alpha = -2p$, $\beta = p^2$, and so the dynamic equation (3.8) is of the form

(3.15) $$y^{\Delta\Delta} - 2p y^\Delta + p^2 y = 0.$$

Hence one solution y_1 of (3.8) is given by

$$y_1(t) = e_p(t, t_0),$$

where $t_0 \in \mathbb{T}$. We will now find a second linearly independent solution of (3.15) using the so-called method of *reduction of order*. We will look for a second linearly independent solution of the form

$$(3.16) \qquad\qquad y(t) = v(t)e_p(t, t_0).$$

By the product rule and Theorem 2.36 (ii),

$$
\begin{aligned}
y^\Delta(t) &= v^\Delta(t)e_p^\sigma(t, t_0) + v(t)e_p^\Delta(t, t_0) \\
(3.17)\qquad &= v^\Delta(t)[1 + \mu(t)p]e_p(t, t_0) + pe_p(t, t_0)v(t).
\end{aligned}
$$

Now we must be careful because there are many time scales where the graininess function μ is not differentiable (see Example 1.56). We assume v is a function such that $v^\Delta(1 + \mu p)$ is differentiable. Then from (3.17) we get using the product rule that

$$
\begin{aligned}
y^{\Delta\Delta}(t) &= \left\{v^\Delta(t)[1 + \mu(t)p]\right\}^\Delta e_p^\sigma(t, t_0) + v^\Delta(t)[1 + \mu(t)p]e_p^\Delta(t, t_0) \\
&\qquad + pe_p^\sigma(t, t_0)v^\Delta(t) + pe_p^\Delta(t, t_0)v(t) \\
&= \left\{v^\Delta(t)[1 + \mu(t)p]\right\}^\Delta e_p^\sigma(t, t_0) + v^\Delta(t)[1 + \mu(t)p]pe_p(t, t_0) \\
&\qquad + p[1 + \mu(t)p]e_p(t, t_0)v^\Delta(t) + p^2 e_p(t, t_0)v(t) \\
(3.18)\qquad &= \left\{v^\Delta(t)[1 + \mu(t)p]\right\}^\Delta e_p^\sigma(t, t_0) \\
&\qquad + 2p[1 + \mu(t)p]e_p(t, t_0)v^\Delta(t) + p^2 e_p(t, t_0)v(t).
\end{aligned}
$$

Using (3.16), (3.17), and (3.18) we obtain

$$
\begin{aligned}
y^{\Delta\Delta}(t) - 2py^\Delta(t) + p^2 y(t) &= \left\{v^\Delta(t)[1 + \mu(t)p]\right\}^\Delta e_p^\sigma(t, t_0) \\
&\quad + 2p[1 + \mu(t)p]e_p(t, t_0)v^\Delta(t) + p^2 e_p(t, t_0)v(t) \\
&\quad - 2pv^\Delta(t)[1 + \mu(t)p]e_p(t, t_0) - 2p^2 e_p(t, t_0)v(t) \\
&\quad + p^2 v(t)e_p(t, t_0) \\
&= \left\{v^\Delta(t)[1 + \mu(t)p]\right\}^\Delta e_p^\sigma(t, t_0).
\end{aligned}
$$

Hence for $y(t) = v(t)e_p(t, t_0)$ to be a solution we want to choose v so that

$$\left\{v^\Delta(t)[1 + \mu(t)p]\right\}^\Delta = 0.$$

We will get that the above equation is true if we choose v so that

$$v^\Delta(t)[1 + \mu(t)p] = 1.$$

Recall in the steps above that we assumed $v^\Delta(t)[1 + \mu(t)p]$ to be differentiable, and that is true in this case. We then solve for $v^\Delta(t)$ to get

$$v^\Delta(t) = \frac{1}{1 + \mu(t)p}.$$

Hence if we take

$$v(t) = \int_{t_0}^t \frac{1}{1 + \mu(\tau)p}\Delta\tau,$$

we can check that all of the above steps are valid and so

$$(3.19) \qquad\qquad y(t) = e_p(t, t_0)\int_{t_0}^t \frac{1}{1 + \mu(\tau)p}\Delta\tau$$

is a solution of (3.8). This leads to the following theorem.

Theorem 3.34. *Suppose $\alpha^2 - 4\beta = 0$. Define*

$$p = -\frac{\alpha}{2}.$$

If $p \in \mathcal{R}$, then a fundamental system of (3.8) is given by

$$e_p(t, t_0) \quad and \quad e_p(t, t_0) \int_{t_0}^{t} \frac{1}{1 + p\mu(\tau)} \Delta\tau,$$

where $t_0 \in \mathbb{T}$, and the Wronskian of these two solutions is

$$e_{\mu\alpha^2/4}(\cdot, t_0).$$

The solution of the initial value problem (3.11) is given by

$$e_p(t, t_0) \left[y_0 + (y_0^\Delta - py_0) \int_{t_0}^{t} \frac{\Delta\tau}{1 + p\mu(\tau)} \right].$$

Exercise 3.35. Prove Theorem 3.34.

Exercise 3.36. Under the hypothesis of Theorem 3.34, show directly by substitution into the dynamic equation (3.15) that y given by (3.19) solves (3.15).

Example 3.37. Given that $\mathbb{T} = \mathbb{R}$, use (3.19) with $t_0 = 0$ to find a solution of

$$y'' - 2py' + p^2 y = 0.$$

Since $\mu(t) \equiv 0$, we get that

$$
\begin{aligned}
y(t) &= e_p(t, 0) \int_0^t \frac{1}{1 + \mu(\tau)p} \Delta\tau \\
&= e^{pt} \int_0^t 1 \Delta\tau \\
&= t e^{pt}
\end{aligned}
$$

is a solution, which of course is a well-known result.

Example 3.38. Given that $\mathbb{T} = \mathbb{Z}$, use (3.19) with $t_0 = 0$ to find a solution of

(3.20) $$\Delta^2 y(t) - 2p\Delta y(t) + p^2 y(t) = 0.$$

Since $\mu(t) \equiv 1$, we get that

$$
\begin{aligned}
y(t) &= e_p(t, 0) \int_0^t \frac{1}{1 + \mu(\tau)p} \Delta\tau \\
&= (1 + p)^t \int_0^t \frac{1}{1 + p} \Delta\tau \\
&= t(1 + p)^{t-1}
\end{aligned}
$$

solves (3.20). Since any constant times a solution is a solution, we get that

$$y(t) = t(1 + p)^t$$

is a solution.

Exercise 3.39. By directly solving (3.20) in Example 3.38, show that you get that

$$y(t) = t(1 + p)^t$$

is a solution.

In the next exercise we use a slight variation of the method of reduction of order to find a second linearly independent solution of (3.15). We also call this the method of reduction of order and we will use this method several times later in this chapter.

Exercise 3.40. In this exercise we outline another method for finding the solution (3.19) of the dynamic equation (3.15). The reader should fill in the missing steps. Suppose y is the solution of (3.15) satisfying the initial conditions

$$y(t_0) = 0 \quad \text{and} \quad y^\Delta(t_0) = 1$$

and consider

$$
\begin{aligned}
W(e_p(\cdot, t_0), y)(t) &= e_p(t, t_0)y^\Delta(t) - e_p^\Delta(t, t_0)y(t) \\
&= e_p(t, t_0)y^\Delta(t) - pe_p(t, t_0)y(t) \\
&= (y^\Delta(t) - py(t))e_p(t, t_0).
\end{aligned}
$$

In particular,

$$W(e_p(\cdot, t_0), y)(t_0) = y^\Delta(t_0) - py(t_0) = 1.$$

On the other hand, by Abel's theorem (Theorem 3.10), we have

$$W(e_p(\cdot, t_0), y)(t) = W(e_p(\cdot, t_0), y)(t_0)e_{\mu\beta-\alpha}(t, t_0) = e_{\mu\beta-\alpha}(t, t_0).$$

Hence y is a solution of the first order linear equation

$$(y^\Delta - py)e_p(t, t_0) = e_{\mu\beta-\alpha}(t, t_0)$$

or, equivalently according to Theorem 2.36 (vii),

$$y^\Delta - py = e_{(\mu\beta-\alpha)\ominus p}(t, t_0) = e_p(t, t_0).$$

Since $y(t_0) = 0$, we have by the variation of constants formula given in Theorem 2.77 that (observe Example 2.37)

$$y(t) = \int_{t_0}^t e_p(t, \sigma(\tau))e_p(\tau, t_0)\Delta\tau = e_p(t, t_0)\int_{t_0}^t \frac{1}{1 + \mu(\tau)p}\Delta\tau.$$

3.4. Method of Factoring

A second order linear equation can sometimes be factored and hence reduced to two first order equations, which we can solve by Theorem 2.77. We now show how to use the *method of factoring* to solve certain second order dynamic equations with variable coefficients. Suppose an equation is in the factored form

(3.21) $$\left(y^\Delta - py\right)^\Delta (t) - q(t)[y^\Delta(t) - p(t)y(t)] = 0,$$

where we assume that $p, q \in \mathcal{R}$. To solve this dynamic equation, let y be a solution of (3.21) and let

(3.22) $$v(t) := y^\Delta(t) - p(t)y(t).$$

With this substitution, (3.21) is equivalent to

$$v^\Delta(t) - q(t)v(t) = 0.$$

Solving this equation we get that

$$v(t) = c_2 e_q(t, t_0).$$

Substituting this into equation (3.22) we get the dynamic equation

$$y^\Delta - p(t)y = c_2 e_q(t, t_0).$$

Using the variation of constants formula from Theorem 2.77, we get that

$$(3.23) \qquad y(t) = c_1 e_p(t, t_0) + c_2 e_p(t, t_0) \int_{t_0}^t e_p(t, \sigma(\tau)) e_q(\tau, t_0) \Delta \tau$$

is a general solution of (3.21).

Example 3.41. Solve the dynamic equation

$$(3.24) \qquad y^{\Delta\Delta} - 2(t+1)y^\Delta + 4ty = 0$$

on the time scale $\mathbb{T} = \mathbb{R}$. In this case (3.24) can be written in the factored form

$$(y' - 2y)' - 2t(y' - 2y) = 0.$$

Instead of using formula (3.23) we let y be a solution of (3.24) and set

$$v(t) = y'(t) - 2y(t).$$

Solving $v' - 2tv = 0$, we get

$$v(t) = c_2 e^{t^2}.$$

This leads to the equation

$$y' - 2y = c_2 e^{t^2}.$$

Multiplying by the integrating factor e^{-2t} we obtain

$$\left(e^{-2t}y\right)' = c_2 e^{-2t+t^2}.$$

Integrating both sides of this last equation from 0 to t leads to

$$y(t) = c_1 e^{2t} + c_2 e^{2t} \int_0^t e^{-2\tau+\tau^2} d\tau$$

as a general solution of (3.24).

Example 3.42. Solve the dynamic equation

$$(3.25) \qquad y^{\Delta\Delta} - (t+3)y^\Delta + 3ty = 0$$

on the time scale $\mathbb{T} = \mathbb{N}$. In this case our dynamic equation can be written in the factored form

$$\Delta\left[\Delta y(t) - 3y(t)\right] - t[\Delta y(t) - 3y(t)] = 0.$$

Let y be a solution of (3.25) and set

$$v(t) = \Delta y(t) - 3y(t).$$

Then solving $\Delta v - tv = 0$, we get

$$v(t) = t! c_2.$$

This leads to the equation

$$\Delta y - 3y = t! c_2.$$

We rewrite this equation in the form

$$y(t+1) - 4y(t) = t! c_2.$$

Multiplying by the summation factor $\frac{1}{4^{t+1}}$ we get that

$$\Delta\left(\frac{y}{4^t}\right) = c_2 \frac{t!}{4^{t+1}}.$$

Summing both sides of this last equation and simplifying leads to

$$y(t) = c_1 4^t + c_2 4^{t-1} \sum_{\tau=0}^{t-1} \left(\frac{\tau!}{4^\tau} \right)$$

as a general solution of (3.25).

The next result is helpful when trying to write a second order linear equation in the factored form (3.21).

Theorem 3.43. *Consider the equation*

(3.26) $$y^{\Delta\Delta} - a(t)y^\Delta + b(t)y = 0.$$

If either one of the two conditions

   (i) $a = p^\sigma + q$ *and* $b = pq - p^\Delta$;
   (ii) $a = p + q$, $b = pq$, *and* p *constant*

is satisfied, then (3.26) can be written in the factored form (3.21).

Proof. Since

$$(y^\Delta - py)^\Delta - q(y^\Delta - py) = y^{\Delta\Delta} - p^\sigma y^\Delta - p^\Delta y - qy^\Delta + pqy$$
$$= y^{\Delta\Delta} - (p^\sigma + q)y^\Delta + (pq - p^\Delta)y,$$

our claim follows immediately.   □

Exercise 3.44. Factor the following dynamic equations using Theorem 3.43. Then use the method of factoring to solve the equation:

   (i) $y^{\Delta\Delta} - (5 + t)y^\Delta + 5ty = 0$;
   (ii) $y^{\Delta\Delta} - (6 + t)y^\Delta + (5 + 5t)y = 0$;
   (iii) $y^{\Delta\Delta} - (t + \sigma(t))y^\Delta + (t^2 - 1)y = 0$;
   (iv) $y^{\Delta\Delta} - 3(t + 1)y^\Delta + (9t - 3)y = 0$ on $\mathbb{T} = \mathbb{R}$;
   (v) $y^{\Delta\Delta} - (t + 6)y^\Delta + (5t - 1)y = 0$ on $\mathbb{T} = \mathbb{N}_0$.

3.5. Nonconstant Coefficients

In general, there is no method to solve second order dynamic equations with arbitrary nonconstant coefficients. However, if one solution is known, it is possible to use the reduction of order method as discussed in Section 3.3 to find a second linearly independent solution. We illustrate this procedure with the dynamic equation

(3.27) $$x^{\Delta\Delta} - q^{\circledS}(t)x^\sigma = 0,$$

where $q \in \mathbb{R}$ is constant and (compare Definition 2.18)

$$q^{\circledS}(t) = -q(\ominus q)(t) = \frac{q^2}{1 + q\mu(t)}.$$

Of course if $\mu(t)$ is independent of t (e.g, if $\mathbb{T} = h\mathbb{Z}$ for some $h > 0$ or $\mathbb{T} = \mathbb{R}$), then this is a problem with constant coefficients and can be treated using the methods from Sections 3.2 and 3.3.

We claim that $x_1 = e_q(\cdot, t_0)$ is a solution of (3.27). To see this consider

$$
\begin{aligned}
x_1^{\Delta\Delta}(t) - q^{\textcircled{2}}(t)x_1^{\sigma}(t) &= q^2 e_q(t, t_0) - \frac{q^2}{1 + q\mu(t)} e_q(\sigma(t), t_0) \\
&= q^2 e_q(t, t_0) - \frac{q^2}{1 + q\mu(t)}[1 + \mu(t)q]e_q(t, t_0) \\
&= 0,
\end{aligned}
$$

where we have applied Theorem 2.36 (ii). We now use the reduction of order technique to find the solution x_2 of the IVP

$$
x^{\Delta\Delta} - q^{\textcircled{2}}(t)x^{\sigma} = 0, \quad x_2(t_0) = 0, \quad x_2^{\Delta}(t_0) = 1.
$$

By Theorem 3.14 we know that the Wronskian of x_1 and x_2 is constant. Hence

$$
\begin{aligned}
W(x_1, x_2)(t) &\equiv W(x_1, x_2)(t_0) \\
&= \det \begin{pmatrix} x_1(t_0) & x_2(t_0) \\ x_1^{\Delta}(t_0) & x_2^{\Delta}(t_0) \end{pmatrix} \\
&= \det \begin{pmatrix} 1 & 0 \\ q & 1 \end{pmatrix} \\
&= 1
\end{aligned}
$$

for $t \in \mathbb{T}^\kappa$. This implies that

$$
W(x_1, x_2)(t) = x_1(t)x_2^{\Delta}(t) - x_1^{\Delta}(t)x_2(t) = 1
$$

for $t \in \mathbb{T}^\kappa$. It follows that

$$
e_q(t, t_0)x_2^{\Delta}(t) - qe_q(t, t_0)x_2(t) = 1.
$$

Consequently, x_2 solves the dynamic equation

(3.28) $$ x^{\Delta} = qx + e_q(t_0, t), $$

where we have used Theorem 2.36 (iv). Since $x_2(t_0) = 0$, we get by the variation of parameters formula in Theorem 2.77 applied to equation (3.28) that

$$
\begin{aligned}
x_2(t) &= \int_{t_0}^{t} e_q(t, \sigma(\tau))e_q(t_0, \tau)\Delta\tau \\
&= e_q(t, t_0) \int_{t_0}^{t} e_q(t_0, \sigma(\tau))e_q(t_0, \tau)\Delta\tau \\
&= e_q(t, t_0) \int_{t_0}^{t} \frac{e_q(t_0, \tau)}{e_q(\sigma(\tau), t_0)}\Delta\tau \\
&= e_q(t, t_0) \int_{t_0}^{t} \frac{e_q(t_0, \tau)}{[1 + q\mu(\tau)]e_q(\tau, t_0)}\Delta\tau \\
&= e_q(t, t_0) \int_{t_0}^{t} \frac{e_q^2(t_0, \tau)}{1 + q\mu(\tau)}\Delta\tau.
\end{aligned}
$$

Since x_1 and x_2 form a fundamental system, we have proved the following theorem.

Theorem 3.45. *Let $t_0 \in \mathbb{T}$. Then a general solution of the dynamic equation* (3.27) *is given by*

$$x(t) = \alpha e_q(t, t_0) + \beta e_q(t, t_0) \int_{t_0}^{t} \frac{e_q^2(t_0, \tau)}{1 + q\mu(\tau)} \Delta \tau,$$

where α and β are arbitrary constants.

Exercise 3.46. Prove that the solution of the initial value problem

$$x^{\Delta\Delta} - q^{\circledS}(t)x^{\sigma} = 0, \quad x(t_0) = x_0, \quad x^{\Delta}(t_0) = x_0^{\Delta}$$

is given by

$$x(t) = e_q(t, t_0) \left[x_0 + (x_0^{\Delta} - qx_0) \int_{t_0}^{t} \frac{e_q^2(t_0, \tau)}{1 + q\mu(\tau)} \Delta \tau \right].$$

Exercise 3.47. Verify directly by substitution into equation (3.27) that

$$x_2(t) = e_q(t, t_0) \int_{t_0}^{t} \frac{e_q^2(t_0, \tau)}{1 + q\mu(\tau)} \Delta \tau$$

is a solution of (3.27).

Exercise 3.48. Use Theorem 3.45 to find a general solution of

$$x^{\Delta\Delta} - 4^{\circledS}(t)x^{\sigma} = 0$$

and simplify your answer when $\mathbb{T} = \mathbb{R}$ and $\mathbb{T} = \mathbb{Z}$.

For the remainder of this section we now consider again a general second order dynamic equation with nonconstant coefficients of the form (3.6), where $p \in \mathcal{R}$.

Theorem 3.49. *Suppose that $z \in \mathcal{R}$ solves*

(3.29) $$z^{\Delta} + z^{\circledS} + p(t)z^{\sigma} + q(t) = 0.$$

Then $e_z(\cdot, t_0)$ solves (3.6).

Proof. Suppose z solves (3.29) and let $x = e_z = e_z(\cdot, t_0)$. Then

$$x^{\Delta} = ze_z \quad \text{and hence} \quad (x^{\Delta})^{\sigma} = z^{\sigma}e_z^{\sigma}$$

and

$$
\begin{aligned}
x^{\Delta\Delta} &= z^{\Delta}e_z^{\sigma} + ze_z^{\Delta} \\
&= z^{\Delta}e_z^{\sigma} + z^2 e_z \\
&= z^{\Delta}e_z^{\sigma} + \frac{z^2}{1 + \mu z}e_z^{\sigma} \\
&= z^{\Delta}e_z^{\sigma} + z^{\circledS}e_z^{\sigma} \\
&= \left(z^{\Delta} + z^{\circledS} \right) e_z^{\sigma}.
\end{aligned}
$$

Altogether we have

$$
\begin{aligned}
x^{\Delta\Delta} + p(x^{\Delta})^{\sigma} + qx^{\sigma} &= \left(z^{\Delta} + z^{\circledS} \right) e_z^{\sigma} + pz^{\sigma}e_z^{\sigma} + qe_z^{\sigma} \\
&= \left(z^{\Delta} + z^{\circledS} + pz^{\sigma} + q \right) e_z^{\sigma} \\
&= 0
\end{aligned}
$$

since z solves (3.29). $\qquad\square$

We now will find a second (linearly independent) solution x_2 of (3.6). Let x_2 be the solution of (3.6) satisfying

$$x_2(t_0) = 0 \quad \text{and} \quad x_2^\Delta(t_0) = 1.$$

Using Theorem 3.13 we obtain

$$
\begin{aligned}
e_z(\cdot, t_0)\left(x_2^\Delta - zx_2\right) &= W(x_1, x_2) \\
&= e_{\ominus p}(\cdot, t_0)W(x_1, x_2)(t_0) \\
&= e_{\ominus p}(\cdot, t_0).
\end{aligned}
$$

Therefore, by the quotient rule (Theorem 1.20 (v)),

$$
\begin{aligned}
\left(\frac{x_2}{e_z(\cdot, t_0)}\right)^\Delta &= \frac{x_2^\Delta e_z(\cdot, t_0) - x_2 e_z^\Delta(\cdot, t_0)}{e_z(\cdot, t_0)e_z^\sigma(\cdot, t_0)} \\
&= \frac{W(x_1, x_2)}{e_z(\cdot, t_0)e_z^\sigma(\cdot, t_0)} \\
&= \frac{e_{\ominus p}(\cdot, t_0)}{e_z(\cdot, t_0)e_z^\sigma(\cdot, t_0)}
\end{aligned}
$$

and hence

$$
\begin{aligned}
x_2(t) &= e_z(t, t_0)\int_{t_0}^t \frac{e_{\ominus p}(\tau, t_0)}{e_z(\tau, t_0)e_z^\sigma(\tau, t_0)}\Delta\tau \\
&= e_z(t, t_0)\int_{t_0}^t \frac{e_{z\oplus z\oplus p}(t_0, \tau)}{1 + \mu(\tau)z(\tau)}\Delta\tau.
\end{aligned}
$$

Hence it is easy to verify the following result.

Theorem 3.50. *If z solves equation (3.29), then*

$$
e_z(t, t_0)\left[x_0 + (x_0^\Delta - x_0 z(t_0))\int_{t_0}^t \frac{e_{z\oplus z\oplus p}(t_0, \tau)}{1 + \mu(\tau)z(\tau)}\Delta\tau\right]
$$

is the solution of the IVP

$$x^{\Delta\Delta} + p(t)x^{\Delta^\sigma} + q(t)x^\sigma = 0, \quad x(t_0) = x_0, \quad x^\Delta(t_0) = x_0^\Delta.$$

In order to apply Theorem 3.49 it is crucial to find a solution of equation (3.29). As is readily checked, this is an easy task if (3.6) is of the form (3.8). Equation (3.27) is another example in which (3.29) can be solved explicitly.

3.6. Hyperbolic and Trigonometric Functions II

In this section we consider an equation of the form

(3.30) $$x^{\Delta\Delta} = (p^\Delta - p^2 + q^2)x^\sigma.$$

Throughout we assume that p and q are functions satisfying

(3.31) $$2p + \mu(p^2 - q^2) = 0.$$

Condition (3.31) as well as many of the subsequent calculations are contained in the paper [233] by Zdeněk Pospíšil.

Example 3.51. If $\mathbb{T} = \mathbb{R}$, then condition (3.31) reads

$$p(t) = 0 \quad \text{for all} \quad t \in \mathbb{R}.$$

In this case (3.30) becomes

$$x'' = q^2(t)x$$

which has solutions

$$\exp\left(\int_0^t q(\tau)d\tau\right) \quad \text{and} \quad \exp\left(-\int_0^t q(\tau)d\tau\right)$$

and also

$$\cosh\left(\int_0^t q(\tau)d\tau\right) \quad \text{and} \quad \sinh\left(\int_0^t q(\tau)d\tau\right).$$

Next, if $\mathbb{T} = \mathbb{Z}$, condition (3.31) means

$$2p(t) + p^2(t) - q^2(t) = 0 \quad \text{for all} \quad t \in \mathbb{Z}$$

which is equivalent to

$$(1 + p(t))^2 - q^2(t) = 1 \quad \text{for all} \quad t \in \mathbb{Z}.$$

In this case equation (3.30) becomes

$$\Delta^2 x = \left[\Delta p(t) - p^2(t) + q^2(t)\right] x^\sigma$$

which is equivalent to

$$\Delta^2 x = (p(t+1) + p(t))x^\sigma.$$

If q is constant, then this equation has the two solutions

$$\cosh\left(\sum_{\tau=0}^t \log(1 + p(\tau) + q)\right) \quad \text{and} \quad \sinh\left(\sum_{\tau=0}^t \log(1 + p(\tau) + q)\right),$$

as the reader may verify directly by doing some calculations.

Lemma 3.52. *Suppose p and q satisfy (3.31). Then*

(i) $[1 + \mu(p+q)][1 + \mu(p-q)] = 1;$
(ii) $p + q$ and $p - q$ are regressive;
(iii) $(p-q)^\oslash = (p+q)^\oslash = q^2 - p^2;$
(iv) $(p-q) \oplus (p+q) = 0;$
(v) $\ominus(p-q) = p+q.$

Proof. We have

$$\begin{aligned}
[1 + \mu(p+q)][1 + \mu(p-q)] &= 1 + \mu(p+q) + \mu(p-q) + \mu^2(p^2 - q^2) \\
&= 1 + 2\mu p + \mu^2(p^2 - q^2) \\
&= 1 + \mu\left[2p + \mu(p^2 - q^2)\right] \\
&= 1,
\end{aligned}$$

where we have used (3.31) for the last equality. This calculation also shows that both $p + q$ and $p - q$ are regressive functions. Hence we can calculate

$$
\begin{aligned}
(p+q)^{\oslash} &= \frac{(p+q)^2}{1+\mu(p+q)} \\
&= (p+q)^2 \left[1+\mu(p-q)\right] \\
&= (p+q)\left[p+q+\mu(p^2-q^2)\right] \\
&= (p+q)(q-p) \\
&= q^2 - p^2,
\end{aligned}
$$

where we have used again (3.31) for the second last equality. Similarly,

$$
\begin{aligned}
(p-q)^{\oslash} &= \frac{(p-q)^2}{1+\mu(p-q)} \\
&= (p-q)^2 \left[1+\mu(p+q)\right] \\
&= (p-q)\left[p-q+\mu(p^2-q^2)\right] \\
&= (p-q)(-q-p) \\
&= q^2 - p^2.
\end{aligned}
$$

Hence $(p+q)^{\oslash}$ and $(p-q)^{\oslash}$ indeed have the same value. Finally we have

$$(p-q) \oplus (p+q) = p - q + p + q + \mu(p-q)(p+q) = 2p + \mu(p^2 - q^2) = 0$$

because of condition (3.31), and this implies $\ominus(p-q) = p+q$ by the definition of $\ominus$ as being the additive inverse operation with respect to $\oplus$. $\square$

Theorem 3.53. *Suppose p is differentiable and q is constant. If (3.31) is satisfied, then two solutions of (3.30) are given by*

$$e_{p-q}(\cdot, t_0) \quad and \quad e_{p+q}(\cdot, t_0),$$

whose Wronskian is equal to $2q$.

Proof. For this proof we omit the second argument $t_0 \in \mathbb{T}$, which is fixed throughout. Let $x_1 = e_{p-q}$ and $x_2 = e_{p+q}$. Note that our assumptions imply that both $p-q$ and $p+q$ are rd-continuous. By Lemma 3.52 (ii), $p-q$ and $p+q$ are regressive. Altogether $p-q, p+q \in \mathcal{R}$, and hence the functions x_1 and x_2 are well defined and satisfy

$$x_1^{\Delta} = (p-q)x_1 \quad and \quad x_2^{\Delta} = (p+q)x_2.$$

Moreover we have by Theorem 2.36 (ii) that

$$x_1^{\sigma} = \left[1+\mu(p-q)\right]x_1 \quad and \quad x_2^{\sigma} = \left[1+\mu(p+q)\right]x_2.$$

Thus we can calculate

$$
\begin{aligned}
x_1^{\Delta\Delta} &= (x_1^{\Delta})^{\Delta} \\
&= ((p-q)x_1)^{\Delta} \\
&= (p-q)^{\Delta}x_1^{\sigma} + (p-q)x_1^{\Delta} \\
&= p^{\Delta}x_1^{\sigma} + (p-q)^2 x_1 \\
&= p^{\Delta}x_1^{\sigma} + \frac{(p-q)^2}{1+\mu(p-q)}x_1^{\sigma} \\
&= p^{\Delta}x_1^{\sigma} + (p-q)^{\circledsymbol{2}}x_1^{\sigma} \\
&= \left(p^{\Delta} + (p-q)^{\circledsymbol{2}}\right)x_1^{\sigma} \\
&= (p^{\Delta} + q^2 - p^2)x_1^{\sigma},
\end{aligned}
$$

where we have used Lemma 3.52 (iii) in the last equation. Similarly,

$$
\begin{aligned}
x_2^{\Delta\Delta} &= (x_2^{\Delta})^{\Delta} \\
&= ((p+q)x_2)^{\Delta} \\
&= (p+q)^{\Delta}x_2^{\sigma} + (p+q)x_2^{\Delta} \\
&= p^{\Delta}x_2^{\sigma} + (p+q)^2 x_2 \\
&= p^{\Delta}x_2^{\sigma} + \frac{(p+q)^2}{1+\mu(p+q)}x_2^{\sigma} \\
&= p^{\Delta}x_2^{\sigma} + (p+q)^{\circledsymbol{2}}x_2^{\sigma} \\
&= \left(p^{\Delta} + (p+q)^{\circledsymbol{2}}\right)x_2^{\sigma} \\
&= (p^{\Delta} + q^2 - p^2)x_2^{\sigma},
\end{aligned}
$$

where the last equation is true again because of Lemma 3.52 (iii). Hence both functions x_1 and x_2 are indeed solutions of (3.30). Finally, we evaluate the Wronskian of x_1 and x_2 as

$$
\begin{aligned}
W(x_1, x_2) &= \det \begin{pmatrix} x_1 & x_2 \\ x_1^{\Delta} & x_2^{\Delta} \end{pmatrix} \\
&= \det \begin{pmatrix} x_1 & x_2 \\ (p-q)x_1 & (p+q)x_2 \end{pmatrix} \\
&= x_1(p+q)x_2 - x_2(p-q)x_1 \\
&= (p+q-p+q)x_1 x_2 \\
&= 2q x_1 x_2 \\
&= 2q,
\end{aligned}
$$

where the last equation holds because of

$$
x_1 x_2 = e_{p-q}e_{p+q} = e_{(p+q)\oplus(p-q)} = e_0 = 1
$$

by Theorem 2.36 (vi) and Lemma 3.52 (iv). $\square$

Now Theorem 3.53 gives motivation to introduce the following (alternative) *hyperbolic functions*. Note that they are in general different from the ones introduced in Definition 3.17.

Definition 3.54. If $p, q \in C_{rd}$ satisfy (3.31), then we define the hyperbolic functions ch_{pq} and sh_{pq} by

$$ch_{pq} = \frac{e_{p+q} + e_{p-q}}{2} \quad \text{and} \quad sh_{pq} = \frac{e_{p+q} - e_{p-q}}{2}.$$

Recall that the hyperbolic functions from Definition 3.17 did not satisfy the identity $\cosh^2 - \sinh^2 = 1$. However, the hyperbolic functions from Definition 3.54 do satisfy such an identity. On the other hand, while the hyperbolic functions from Definition 3.17 are derivatives of each other, this is not the case with the hyperbolic functions from Definition 3.54. As is typical for the calculus on time scales, a certain "trade-off" exists. We have the following result, where $^\Delta$ again denotes the derivative with respect to the first variable.

Lemma 3.55. *If $p, q \in C_{rd}$ satisfy (3.31), then*

$$ch_{pq}^\Delta = p\, ch_{pq} + q\, sh_{pq} \quad and \quad sh_{pq}^\Delta = q\, ch_{pq} + p\, sh_{pq}$$

and

$$ch_{pq}^2 - sh_{pq}^2 = 1.$$

Proof. We have

$$
\begin{aligned}
ch_{pq}^\Delta &= \frac{1}{2}\left[(p+q)e_{p+q} + (p-q)e_{p-q}\right] \\
&= p\frac{e_{p+q} + e_{p-q}}{2} + q\frac{e_{p+q} - e_{p-q}}{2} \\
&= p\, ch_{pq} + q\, sh_{pq}
\end{aligned}
$$

and

$$
\begin{aligned}
sh_{pq}^\Delta &= \frac{1}{2}\left[(p+q)e_{p+q} - (p-q)e_{p-q}\right] \\
&= q\frac{e_{p+q} + e_{p-q}}{2} + p\frac{e_{p+q} - e_{p-q}}{2} \\
&= q\, ch_{pq} + p\, sh_{pq}.
\end{aligned}
$$

Next,

$$
\begin{aligned}
ch_{pq}^2 - sh_{pq}^2 &= \left(\frac{e_{p+q} + e_{p-q}}{2}\right)^2 - \left(\frac{e_{p+q} - e_{p-q}}{2}\right)^2 \\
&= \frac{e_{p+q}^2 + 2e_{p+q}e_{p-q} + e_{p-q}^2 - (e_{p+q}^2 - 2e_{p+q}e_{p-q} + e_{p-q}^2)}{4} \\
&= e_{p+q}e_{p-q} \\
&= e_{(p+q)\oplus(p-q)} \\
&= e_0 \\
&= 1
\end{aligned}
$$

because of Lemma 3.52 (iv). $\square$

Theorem 3.56. *Suppose p is differentiable and $q \neq 0$ is constant. If (3.31) is satisfied, then the solution of the initial value problem*

$$x^{\Delta\Delta} = (p^{\Delta} - p^2 + q^2)x^{\sigma}, \quad x(t_0) = x_0, \quad x^{\Delta}(t_0) = x_0^{\Delta}$$

is given by

$$x(t) = x_0 \, \mathrm{ch}_{pq}(t, t_0) + \frac{x_0^{\Delta} - p x_0}{q} \, \mathrm{sh}_{pq}(t, t_0).$$

Proof. Since ch_{pq} and sh_{pq} are linear combinations of e_{p+q} and e_{p-q}, they solve (3.30) by Theorem 3.53. Hence it only remains to check the initial conditions, but since

$$\mathrm{ch}_{pq}(t_0, t_0) = 1 \quad \text{and} \quad \mathrm{sh}_{pq}(t_0, t_0) = 0,$$

we can use Lemma 3.55 to easily check that the initial conditions are satisfied. $\square$

Exercise 3.57. In this exercise we present several formulas that are satisfied by the functions ch and sh. The reader is asked to prove these results:

   (i) $e_{p+q} = \mathrm{ch}_{pq} + \mathrm{sh}_{pq}$;
   (ii) $e_{p-q} = \mathrm{ch}_{pq} - \mathrm{sh}_{pq}$;
   (iii) $|\mathrm{ch}_{pq}| \geq 1$;
   (iv) parity condition for sh: $\mathrm{sh}_{pq}(t, s) = -\mathrm{sh}_{pq}(s, t)$;
   (v) parity condition for ch: $\mathrm{ch}_{pq}(t, s) = \mathrm{ch}_{pq}(s, t)$;
   (vi) difference formula for sh:

$$\mathrm{sh}_{pq}(t, s) = \mathrm{sh}_{pq}(t, r) \, \mathrm{ch}_{pq}(r, s) - \mathrm{ch}_{pq}(t, r) \, \mathrm{sh}_{pq}(s, r);$$

   (vii) difference formula for ch:

$$\mathrm{ch}_{pq}(t, s) = \mathrm{ch}_{pq}(t, r) \, \mathrm{ch}_{pq}(s, r) - \mathrm{sh}_{pq}(t, r) \, \mathrm{sh}_{pq}(s, r);$$

   (viii) sum formula for sh:

$$\mathrm{sh}_{pq}(t, r) = \mathrm{sh}_{pq}(t, s) \, \mathrm{ch}_{pq}(s, r) + \mathrm{ch}_{pq}(t, s) \, \mathrm{sh}_{pq}(s, r);$$

   (ix) sum formula for ch:

$$\mathrm{ch}_{pq}(t, r) = \mathrm{ch}_{pq}(t, s) \, \mathrm{ch}_{pq}(s, r) + \mathrm{sh}_{pq}(t, s) \, \mathrm{sh}_{pq}(s, r).$$

Alternatively, instead of p and q, we can use a function $\alpha \in \mathcal{R}$ as follows.

Theorem 3.58. *Let $\alpha \in \mathcal{R}$. Suppose α and $\ominus\alpha$ are different and differentiable with the same derivative. Then a fundamental system of*

$$(3.32) \qquad\qquad x^{\Delta\Delta} = (\alpha^{\Delta} + \alpha^{\circledcirc})x^{\sigma}$$

is given by

$$e_{\alpha}(t, t_0) \quad \text{and} \quad e_{\ominus\alpha}(t, t_0).$$

If we introduce

$$\mathrm{ch}_{\alpha} = \frac{e_{\alpha} + e_{\ominus\alpha}}{2} \quad \text{and} \quad \mathrm{sh}_{\alpha} = \frac{e_{\alpha} - e_{\ominus\alpha}}{2},$$

then the solution of the IVP

$$x^{\Delta\Delta} = (\alpha^{\Delta} + \alpha^{\circledcirc})x^{\sigma}, \quad x(t_0) = x_0, \quad x^{\Delta}(t_0) = x_0^{\Delta}$$

is given by

$$x(t) = x_0 \, \mathrm{ch}_{\alpha}(t, t_0) + \frac{2x_0^{\Delta} - x_0[\alpha + (\ominus\alpha)]}{\alpha - (\ominus\alpha)} \, \mathrm{sh}_{\alpha}(t, t_0).$$

Proof. For $x_1 = e_\alpha$ and $x_2 = e_{\ominus\alpha}$ we have

$$x_1^\Delta = \alpha x_1 \quad \text{and} \quad x_2^\Delta = (\ominus\alpha)x_2$$

so that

$$
\begin{aligned}
x_1^{\Delta\Delta} &= \alpha^\Delta x_1^\sigma + \alpha x_1^\Delta \\
&= \alpha^\Delta x_1^\sigma + \alpha^2 x_1 \\
&= \alpha^\Delta x_1^\sigma + \frac{\alpha^2}{1+\mu\alpha} x_1^\sigma \\
&= \left(\alpha^\Delta + \frac{\alpha^2}{1+\mu\alpha}\right) x_1^\sigma \\
&= \left(\alpha^\Delta + \alpha^{\circledcirc}\right) x_1^\sigma
\end{aligned}
$$

and similarly

$$
\begin{aligned}
x_2^{\Delta\Delta} &= (\ominus\alpha)^\Delta x_2^\sigma + (\ominus\alpha)x_2^\Delta \\
&= \alpha^\Delta x_2^\sigma + (\ominus\alpha)^2 x_2 \\
&= \alpha^\Delta x_2^\sigma + \frac{(\ominus\alpha)^2}{1+\mu(\ominus\alpha)} x_2^\sigma \\
&= \left(\alpha^\Delta + \frac{(\ominus\alpha)^2}{1+\mu(\ominus\alpha)}\right) x_2^\sigma \\
&= \left(\alpha^\Delta + (\ominus\alpha)^{\circledcirc}\right) x_2^\sigma \\
&= \left(\alpha^\Delta + \alpha^{\circledcirc}\right) x_2^\sigma,
\end{aligned}
$$

where we have used (2.6) from Theorem 2.19. Hence both x_1 and x_2 are solutions of (3.32). The functions ch and sh defined as above satisfy

$$\text{ch}^2 - \text{sh}^2 = e_\alpha e_{\ominus\alpha} = e_{\alpha\oplus(\ominus\alpha)} = e_0 = 1,$$

and it is easy to see that the function given in the statement of the theorem solves the initial value problem under consideration. $\qquad\square$

Exercise 3.59. Discuss the connection between the condition on α given in Theorem 3.58 and condition (3.31). Also derive all formulas from Exercise 3.57 in terms of α.

Now we consider the equation

(3.33) $$x^{\Delta\Delta} = (p^\Delta - p^2 - q^2)x^\sigma.$$

Associated with (3.33) we employ the condition

(3.34) $$2p + \mu(p^2 + q^2) = 0.$$

For the purpose of studying (3.33) we introduce the (alternative) *trigonometric functions*. Note that they are in general different from the ones introduced in Definition 3.25.

Definition 3.60. If $p, q \in C_{rd}$ satisfy (3.34), then we define the trigonometric functions c_{pq} and s_{pq} by

$$c_{pq} = \frac{e_{p+iq} + e_{p-iq}}{2} \quad \text{and} \quad s_{pq} = \frac{e_{p+iq} - e_{p-iq}}{2i}.$$

Lemma 3.61. *If* $p, q \in C_{rd}$ *satisfy (3.34), then*

$$c_{pq}^{\Delta} = p\, c_{pq} - q\, s_{pq} \quad and \quad s_{pq}^{\Delta} = q\, c_{pq} + p\, s_{pq}$$

and

$$c_{pq}^2 + s_{pq}^2 = 1.$$

Exercise 3.62. Prove Lemma 3.61 as well as the formulas corresponding to Exercise 3.57 for the trigonometric functions defined by Definition 3.60. Use your result to find solutions of (3.33) and to solve arbitrary initial value problems for (3.33).

3.7. Euler–Cauchy Equations

In this section we consider the *Euler–Cauchy dynamic equation*

$$(3.35) \qquad t\sigma(t)y^{\Delta\Delta} + aty^{\Delta} + by = 0 \quad \text{with} \quad a, b \in \mathbb{R}$$

on a time scale $\mathbb{T}$. Throughout this section we assume that $\mathbb{T}$ is a time scale with

$$\mathbb{T} \subset (0, \infty).$$

We assume that the regressivity condition

$$(3.36) \qquad t\sigma(t) - at\mu(t) + b\mu^2(t) \neq 0 \quad \text{for all} \quad t \in \mathbb{T}^\kappa$$

is satisfied. The associated *characteristic equation* of (3.35) is defined by

$$(3.37) \qquad \lambda^2 + (a-1)\lambda + b = 0.$$

Theorem 3.63. *If the regressivity condition (3.36) is satisfied and the characteristic equation (3.37) has two distinct roots* λ_1 *and* λ_2, *then a fundamental system of (3.35) is given by*

$$(3.38) \qquad e_{\lambda_1/t}(\cdot, t_0) \quad and \quad e_{\lambda_2/t}(\cdot, t_0),$$

where $t_0 \in \mathbb{T}$.

Proof. Let $t_0 \in \mathbb{T}$ and put

$$y(t) = e_{\lambda/t}(t, t_0),$$

where we assume that $\lambda/t \in \mathcal{R}$ so that the above exponential function exists. Then we have

$$y^{\Delta}(t) = \frac{\lambda}{t}y(t) \quad \text{and hence} \quad ty^{\Delta}(t) = \lambda y(t).$$

Furthermore we find by Theorem 1.20

$$\begin{aligned}
y^{\Delta\Delta}(t) &= -\frac{\lambda}{t\sigma(t)}y(t) + \frac{\lambda}{\sigma(t)}y^{\Delta}(t) \\
&= -\frac{\lambda}{t\sigma(t)}y(t) + \frac{\lambda}{\sigma(t)}\frac{\lambda}{t}y(t) \\
&= \frac{\lambda^2 - \lambda}{t\sigma(t)}y(t)
\end{aligned}$$

and hence

$$t\sigma(t)y^{\Delta\Delta}(t) = (\lambda^2 - \lambda)y(t).$$

Therefore we have that

$$
\begin{aligned}
t\sigma(t)y^{\Delta\Delta}(t) + aty^{\Delta}(t) + by(t) &= (\lambda^2 - \lambda)y(t) + a\lambda y(t) + by(t) \\
&= (\lambda^2 - \lambda + a\lambda + b)y(t) \\
&= (\lambda^2 + (a-1)\lambda + b)\, y(t).
\end{aligned}
$$

Now assume that λ_1 and λ_2 are distinct roots of the characteristic equation (3.37). It follows that

$$
\lambda_1 + \lambda_2 = 1 - a \quad \text{and} \quad \lambda_1\lambda_2 = b.
$$

We obtain

$$
\begin{aligned}
\left(1 + \mu(t)\frac{\lambda_1}{t}\right)\left(1 + \mu(t)\frac{\lambda_2}{t}\right) &= \frac{1}{t^2}\left(t^2 + (\lambda_1 + \lambda_2)t\mu(t) + \mu^2(t)\lambda_1\lambda_2\right) \\
&= \frac{1}{t^2}\left(t^2 + (1-a)t\mu(t) + b\mu^2(t)\right) \\
&= \frac{1}{t^2}\left(t(t + \mu(t)) - at\mu(t) + b\mu^2(t)\right) \\
&= \frac{t\sigma(t) - at\mu(t) + b\mu^2(t)}{t^2} \\
&\neq 0
\end{aligned}
$$

for all $t \in \mathbb{T}^\kappa$. Hence $\lambda_1/t, \lambda_2/t \in \mathcal{R}$, and the exponential functions (3.38) are well-defined solutions of (3.35). Note that

$$
\begin{aligned}
W\left(e_{\frac{\lambda_1}{t}}(\cdot, t_0), e_{\frac{\lambda_2}{t}}(\cdot, t_0)\right)(t) &= \det\begin{pmatrix} e_{\frac{\lambda_1}{t}}(t, t_0) & e_{\frac{\lambda_2}{t}}(t, t_0) \\ \frac{\lambda_1}{t}e_{\frac{\lambda_1}{t}}(t, t_0) & \frac{\lambda_2}{t}e_{\frac{\lambda_2}{t}}(t, t_0) \end{pmatrix} \\
&= \frac{1}{t}(\lambda_2 - \lambda_1)e_{\frac{\lambda_1}{t}\oplus\frac{\lambda_2}{t}}(t, t_0) \\
&\neq 0
\end{aligned}
$$

for $t \in \mathbb{T}^\kappa$ since $\lambda_1 \neq \lambda_2$. Hence the exponential functions (3.38) form a fundamental system of solutions of (3.35). $\qquad\Box$

Remark 3.64. Under the assumptions and with the notation of Theorem 3.63, if

$$
\frac{\lambda_1}{t}, \frac{\lambda_2}{t} \in \mathcal{R}^+,
$$

then (3.38) form a fundamental set of *positive* solutions of (3.35). This follows from Theorem 2.48 (i).

Example 3.65. Solve the Euler–Cauchy dynamic equation

$$
t\sigma(t)y^{\Delta\Delta} - 4ty^{\Delta} + 6y = 0
$$

on a general time scale $\mathbb{T} \subset (0, \infty)$, and simplify the answer when $\mathbb{T} = [1, \infty)$ and when $\mathbb{T} = \mathbb{N}$, respectively. The characteristic equation is

$$
\lambda^2 - 5\lambda + 6 = 0,
$$

and so the characteristic roots are

$$
\lambda_1 = 2 \quad \text{and} \quad \lambda_2 = 3.
$$

Hence a general solution is

$$
y(t) = \alpha e_{\frac{2}{t}}(t, t_0) + \beta e_{\frac{3}{t}}(t, t_0).
$$

Note that $2/t, 3/t \in \mathcal{R}^+$. Taking $t_0 = 1$, we have by Exercise 2.60 that

$$y(t) = \alpha t^2 + \beta t^3$$

is a general solution if $\mathbb{T} = [1, \infty)$ and that

$$y(t) = \alpha(t+1)^{(2)} + \beta(t+2)^{(3)}$$

is a general solution if $\mathbb{T} = \mathbb{N}$.

Next we consider the Euler–Cauchy dynamic equation in the double root case.

Theorem 3.66. *Assume that $\alpha \in \mathbb{R}$ and $t_0 \in \mathbb{T}$. If the regressivity condition*

$$(3.39) \qquad t\sigma(t) + (2\alpha - 1)t\mu(t) + \alpha^2\mu^2(t) \neq 0 \quad \text{for all} \quad t \in \mathbb{T}^\kappa$$

holds, then a general solution of the Euler–Cauchy dynamic equation

$$(3.40) \qquad t\sigma(t)y^{\Delta\Delta} + (1 - 2\alpha)ty^{\Delta} + \alpha^2 y = 0$$

is given by

$$y(t) = c_1 e_{\frac{\alpha}{t}}(t, t_0) + c_2 e_{\frac{\alpha}{t}}(t, t_0) \int_{t_0}^{t} \frac{1}{\tau + \alpha\mu(\tau)} \Delta\tau \quad \text{for} \quad t \in \mathbb{T}.$$

Proof. Note that

$$
\begin{aligned}
\left(1 + \mu(t)\frac{\alpha}{t}\right)\left(1 + \mu(t)\frac{\alpha - 1}{\sigma(t)}\right) &= 1 + \left[\frac{\alpha}{t} + \frac{\alpha - 1}{\sigma(t)}\right]\mu(t) + \frac{\alpha(\alpha - 1)}{t\sigma(t)}\mu^2(t) \\
&= 1 + \left[\frac{\alpha\sigma(t) + (\alpha - 1)t}{t\sigma(t)}\right]\mu(t) + \frac{\alpha(\alpha - 1)}{t\sigma(t)}\mu^2(t) \\
&= 1 + \left[\frac{\alpha\mu(t) + 2\alpha t - t}{t\sigma(t)}\right]\mu(t) + \frac{\alpha(\alpha - 1)}{t\sigma(t)}\mu^2(t) \\
&= \frac{t\sigma(t) + (2\alpha - 1)t\mu(t) + \alpha^2\mu^2(t)}{t\sigma(t)} \\
&\neq 0
\end{aligned}
$$

for $t \in \mathbb{T}^\kappa$. This implies that the exponentials

$$e_{\frac{\alpha}{t}}(t, t_0) \quad \text{and} \quad e_{\frac{\alpha-1}{\sigma(t)}}(t, t_0)$$

are well defined. The characteristic equation of (3.40) is

$$\lambda^2 - 2\alpha + \alpha^2 = 0,$$

and so the characteristic roots are

$$\lambda_1 = \lambda_2 = \alpha.$$

Hence one nontrivial solution of (3.40) is

$$y(t) = e_{\frac{\alpha}{t}}(t, t_0).$$

A simple way to complete the proof of this theorem is to work Exercise 3.67 to get a second linearly independent solution of (3.40). Instead we will use the method of reduction of order to show how we got the second linearly independent solution given in Exercise 3.67. Note that (3.40) can be written in the form

$$y^{\Delta\Delta} + p(t)y^{\Delta} + q(t)y = 0,$$

where

$$p(t) = \frac{1 - 2\alpha}{\sigma(t)} \quad \text{and} \quad q(t) = \frac{\alpha^2}{t\sigma(t)}.$$

It follows that

$$
\begin{aligned}
-p(t) + \mu(t)q(t) &= \frac{2\alpha - 1}{\sigma(t)} + \mu(t)\frac{\alpha^2}{t\sigma(t)} \\
&= \frac{2\alpha - 1}{\sigma(t)} + [\sigma(t) - t]\frac{\alpha^2}{t\sigma(t)} \\
&= \frac{2\alpha - 1}{\sigma(t)} + \frac{\alpha^2}{t} - \frac{\alpha^2}{\sigma(t)} \\
&= \frac{\alpha^2}{t} - \frac{(\alpha - 1)^2}{\sigma(t)}.
\end{aligned}
$$

Now let u be a solution of (3.40) such that $u(t_0) = 0$ and

$$W\left(e_{\frac{\alpha}{t}}(\cdot, t_0), u\right)(t) = e_{-p+\mu q}(t, t_0) = e_{\frac{\alpha^2}{t} - \frac{(\alpha-1)^2}{\sigma(t)}}(t, t_0),$$

which we can do by Abel's theorem (Theorem 3.10). It follows that

$$
\begin{aligned}
\frac{W\left(e_{\frac{\alpha}{t}}(\cdot, t_0), u\right)(t)}{e_{\frac{\alpha}{t}}(t, t_0)e_{\frac{\alpha}{t}}^{\sigma}(t, t_0)} &= \frac{e_{\frac{\alpha^2}{t} - \frac{(\alpha-1)^2}{\sigma(t)}}(t, t_0)}{e_{\frac{\alpha}{t}}(t, t_0)e_{\frac{\alpha}{t}}^{\sigma}(t, t_0)} \\
&= \frac{e_{\frac{\alpha^2}{t} - \frac{(\alpha-1)^2}{\sigma(t)}}(t, t_0)}{[1 + \mu(t)\frac{\alpha}{t}]e_{\frac{\alpha}{t}}^2(t, t_0)} \\
&= \frac{t_0}{t + \alpha\mu(t)},
\end{aligned}
$$

where the last equation follows from Exercise 2.41. By the quotient rule we get that u satisfies the dynamic equation

$$\left(\frac{u}{e_{\frac{\alpha}{t}}(\cdot, t_0)}\right)^{\Delta}(t) = \frac{t_0}{t + \alpha\mu(t)}.$$

Integrating both sides from t_0 to t and using $u(t_0) = 0$ we obtain

$$\frac{u(t)}{e_{\frac{\alpha}{t}}(t, t_0)} = t_0 \int_{t_0}^{t} \frac{1}{\tau + \alpha\mu(\tau)}\Delta\tau.$$

It follows that

$$u(t) = t_0 e_{\frac{\alpha}{t}}(t, t_0) \int_{t_0}^{t} \frac{1}{\tau + \alpha\mu(\tau)}\Delta\tau$$

is a solution of (3.40), and since

$$W\left(e_{\frac{\alpha}{t}}(\cdot, t_0), u\right)(t) \neq 0,$$

we have that

$$e_{\frac{\alpha}{t}}(t, t_0) \quad \text{and} \quad u(t)$$

form a fundamental set of solutions of (3.40). But this implies that

$$e_{\frac{\alpha}{t}}(t, t_0) \quad \text{and} \quad e_{\frac{\alpha}{t}}(t, t_0) \int_{t_0}^{t} \frac{1}{\tau + \alpha\mu(\tau)}\Delta\tau$$

form a fundamental set of solutions of (3.40). Hence

$$y(t) = c_1 e_{\frac{\alpha}{t}}(t, t_0) + c_2 e_{\frac{\alpha}{t}}(t, t_0) \int_{t_0}^t \frac{1}{\tau + \alpha\mu(\tau)} \Delta\tau$$

is a general solution of (3.40).   □

Exercise 3.67. Show by direct substitution into (3.40) that

$$y(t) = e_{\frac{\alpha}{t}}(t, t_0) \int_{t_0}^t \frac{1}{\tau + \alpha\mu(\tau)} \Delta\tau$$

is a solution of (3.40) for $t \in \mathbb{T}$.

Exercise 3.68. Simplify the expression

$$e_{\frac{\alpha}{t}}(t, t_0) \int_{t_0}^t \frac{1}{\tau + \alpha\mu(\tau)} \Delta\tau$$

for $t \in \mathbb{T}$ and $t_0 = 1$ when $\mathbb{T} = [1, \infty)$ and $\mathbb{T} = \mathbb{N}$, respectively.

Exercise 3.69. Under the hypotheses of Theorem 3.40 we showed at the beginning of the proof of Theorem 3.40 that the exponential $e_{\frac{\alpha-1}{\sigma(t)}}(t, t_0)$ exists. Show by substitution that

$$te_{\frac{\alpha-1}{\sigma(t)}}(t, t_0)$$

is a solution of equation (3.40) for $t \in \mathbb{T}$. Then show that the functions

$$e_{\frac{\alpha}{t}}(t, t_0) \quad \text{and} \quad te_{\frac{\alpha-1}{\sigma(t)}}(t, t_0)$$

are linearly dependent on $\mathbb{T}$. Finally find a relationship between these two functions.

Exercise 3.70. Show that if $\alpha \in \mathbb{R}$ is constant with $\alpha/t \in \mathcal{R}^+$, then the regressivity condition (3.39) is satisfied.

Example 3.71. Solve the Euler–Cauchy dynamic equation

$$t\sigma(t)y^{\Delta\Delta} - 9ty^{\Delta} + 25y = 0$$

on a time scale $\mathbb{T} \subset (0, \infty)$. Simplify the answer in the cases when $\mathbb{T} = [1, \infty)$ and $\mathbb{T} = \mathbb{N}$. The characteristic equation is

$$\lambda^2 - 10\lambda + 25 = 0.$$

Hence

$$\lambda_1 = \lambda_2 = 5$$

are the characteristic roots. A general solution is given by

$$y(t) = c_1 e_{\frac{5}{t}}(t, t_0) + c_2 e_{\frac{5}{t}}(t, t_0) \int_{t_0}^t \frac{1}{\tau + 5\mu(\tau)} \Delta\tau$$

for $t, t_0 \in \mathbb{T}$. If $\mathbb{T} = [1, \infty)$ and $t_0 = 1$, we get that

$$y(t) = c_1 t^5 + c_2 t^5 \ln t$$

is a general solution. If $\mathbb{T} = \mathbb{N}$ and $t_0 = 1$, we get that

$$y(t) = c_1(t+4)^{(5)} + c_2(t+4)^{(5)} \sum_{\tau=1}^{t-1} \frac{1}{\tau + 5}$$

is a general solution.

Exercise 3.72. Solve the following Euler–Cauchy dynamic equations on a general time scale $\mathbb{T} \subset (0, \infty)$. Simplify your answers when $\mathbb{T} = [1, \infty)$ and when $\mathbb{T} = \mathbb{N}$.

  (i) $t\sigma(t)y^{\Delta\Delta} - 5ty^{\Delta} + 8y = 0$;
  (ii) $t\sigma(t)y^{\Delta\Delta} + ty^{\Delta} - 4y = 0$;
  (iii) $2t\sigma(t)y^{\Delta\Delta} + 7ty^{\Delta} - 5y = 0$;
  (iv) $t\sigma(t)y^{\Delta\Delta} - 3ty^{\Delta} + 4y = 0$;
  (v) $t\sigma(t)y^{\Delta\Delta} - 5ty^{\Delta} + 9y = 0$.

3.8. Variation of Parameters

In this section we consider the nonhomogeneous equation

$$(3.41) \qquad L_2 y(t) = g(t), \quad \text{where} \quad L_2 y(t) = y^{\Delta\Delta} + p(t)y^{\Delta} + q(t)y.$$

We try to find a particular solution of (3.41) of the form

$$y(t) = \alpha(t)y_1(t) + \beta(t)y_2(t),$$

where α and β are functions to be determined, and where y_1, y_2 is a fundamental system of solutions of the homogeneous equation $L_2 y = 0$. Assume that such a y is a solution of (3.41). Note that

$$
\begin{aligned}
y^{\Delta}(t) &= \alpha^{\Delta}(t)y_1^{\sigma}(t) + \alpha(t)y_1^{\Delta}(t) + \beta^{\Delta}(t)y_2^{\sigma}(t) + \beta(t)y_2^{\Delta}(t) \\
&= \alpha(t)y_1^{\Delta}(t) + \beta(t)y_2^{\Delta}(t),
\end{aligned}
$$

provided α and β satisfy

$$(3.42) \qquad \alpha^{\Delta}(t)y_1^{\sigma}(t) + \beta^{\Delta}(t)y_2^{\sigma}(t) = 0.$$

Assuming that α and β can be picked so that (3.42) is satisfied, we proceed to calculate

$$y^{\Delta\Delta}(t) = \alpha^{\Delta}(t)y_1^{\Delta^{\sigma}}(t) + \alpha(t)y_1^{\Delta\Delta}(t) + \beta^{\Delta}(t)y_2^{\Delta^{\sigma}}(t) + \beta(t)y_2^{\Delta\Delta}(t).$$

Hence

$$
\begin{aligned}
L_2 y(t) &= y^{\Delta\Delta}(t) + p(t)y^{\Delta}(t) + q(t)y(t) \\
&= \alpha(t)L_2 y_1(t) + \beta(t)L_2 y_2(t) + \alpha^{\Delta}(t)y_1^{\Delta^{\sigma}}(t) + \beta^{\Delta}(t)y_2^{\Delta^{\sigma}}(t) \\
&= \alpha^{\Delta}(t)y_1^{\Delta^{\sigma}}(t) + \beta^{\Delta}(t)y_2^{\Delta^{\sigma}}(t),
\end{aligned}
$$

since y_1 and y_2 solve (3.41). Therefore α and β need to satisfy the additional equation

$$\alpha^{\Delta}(t)y_1^{\Delta^{\sigma}}(t) + \beta^{\Delta}(t)y_2^{\Delta^{\sigma}}(t) = g(t).$$

This equation, together with equation (3.42), leads to the following system of equations for α and β:

$$
\begin{pmatrix} y_1^{\sigma}(t) & y_2^{\sigma}(t) \\ y_1^{\Delta^{\sigma}}(t) & y_2^{\Delta^{\sigma}}(t) \end{pmatrix}
\begin{pmatrix} \alpha^{\Delta}(t) \\ \beta^{\Delta}(t) \end{pmatrix}
=
\begin{pmatrix} 0 \\ g(t) \end{pmatrix}.
$$

We solve for α^Δ and β^Δ (recall that y_1 and y_2 form a fundamental system for (3.8)) to obtain

$$\begin{pmatrix} \alpha^\Delta(t) \\ \beta^\Delta(t) \end{pmatrix} = \frac{1}{W^\sigma(y_1, y_2)(t)} \begin{pmatrix} y_2^{\Delta^\sigma}(t) & -y_2^\sigma(t) \\ -y_1^{\Delta^\sigma}(t) & y_1^\sigma(t) \end{pmatrix} \begin{pmatrix} 0 \\ g(t) \end{pmatrix},$$

and therefore

$$\alpha^\Delta(t) = -\frac{y_2^\sigma(t)g(t)}{W^\sigma(y_1, y_2)(t)} \quad \text{and} \quad \beta^\Delta(t) = \frac{y_1^\sigma(t)g(t)}{W^\sigma(y_1, y_2)(t)}.$$

Integrating directly yields the following result.

Theorem 3.73 (Variation of Parameters). *Let $t_0 \in \mathbb{T}^\kappa$. Suppose that y_1 and y_2 form a fundamental set of solutions of the homogeneous equation $L_2 y = 0$. Then the solution of the initial value problem*

$$L_2 y(t) = g(t), \quad y(t_0) = y_0, \quad y^\Delta(t_0) = y_0^\Delta$$

is given by

$$y(t) = \alpha_0 y_1(t) + \beta_0 y_2(t) + \int_{t_0}^t \frac{y_1(\sigma(\tau))y_2(t) - y_2(\sigma(\tau))y_1(t)}{W(y_1, y_2)(\sigma(\tau))} g(\tau)\Delta\tau,$$

where the constants α_0 and β_0 are

$$\alpha_0 = \frac{y_2^\Delta(t_0)y_0 - y_2(t_0)y_0^\Delta}{W(y_1, y_2)(t_0)}$$

and

$$\beta_0 = \frac{y_1(t_0)y_0^\Delta - y_1^\Delta(t_0)y_0}{W(y_1, y_2)(t_0)}.$$

Example 3.74. Use the method of variation of parameters to solve

(3.43) $$y^{\Delta\Delta} - 5y^\Delta + 6y = e_4(t, t_0).$$

A general solution of the corresponding homogeneous equation is

$$u(t) = c_1 e_2(t, t_0) + c_2 e_3(t, t_0).$$

Hence there is a solution of the nonhomogeneous equation (3.43) of the form

$$y_p(t) = \alpha(t)e_2(t, t_0) + \beta(t)e_3(t, t_0),$$

where $\alpha(t)$ and $\beta(t)$ are chosen to satisfy the system of equations

(3.44) $$\alpha^\Delta(t)e_2^\sigma(t, t_0) + \beta^\Delta(t)e_3^\sigma(t, t_0) = 0$$
(3.45) $$2\alpha^\Delta(t)e_2^\sigma(t, t_0) + 3\beta^\Delta(t)e_3^\sigma(t, t_0) = e_4(t, t_0).$$

Multiplying equation (3.44) by 3 and subtracting from equation (3.45) leads to

$$\alpha^\Delta(t) = -\frac{e_4(t, t_0)}{e_2^\sigma(t, t_0)}.$$

Integrating and applying Theorem 2.38 we see that we can take

$$\alpha(t) = -\frac{1}{2}e_{\frac{2}{1+2\mu(t)}}(t, t_0).$$

Multiplying equation (3.44) by 2 and subtracting from equation (3.45) leads to

$$\beta^\Delta(t) = \frac{e_4(t, t_0)}{e_3^\sigma(t, t_0)}.$$

Integrating and applying Theorem 2.38 we see that we can take

$$\beta(t) = e_{\frac{1}{1+3\mu(t)}}(t, t_0).$$

It follows that

$$
\begin{aligned}
y_p(t) &= \alpha(t)e_2(t, t_0) + \beta(t)e_3(t, t_0) \\
&= -\frac{1}{2}e_{\frac{2}{1+2\mu(t)}}(t, t_0)e_2(t, t_0) + e_{\frac{1}{1+3\mu(t)}}(t, t_0)e_3(t, t_0) \\
&= -\frac{1}{2}e_{\frac{2}{1+2\mu(t)}\oplus 2}(t, t_0) + e_{\frac{1}{1+3\mu(t)}\oplus 3}(t, t_0) \\
&= -\frac{1}{2}e_4(t, t_0) + e_4(t, t_0) \\
&= \frac{1}{2}e_4(t, t_0),
\end{aligned}
$$

where we used Exercise 2.29. Hence a general solution of (3.43) is given by

$$y(t) = c_1 e_2(t, t_0) + c_2 e_3(t, t_0) + \frac{1}{2}e_4(t, t_0).$$

At the end of this section we will see that it is easier to solve (3.43) by the annihilator method. We now give an example where the annihilator method does not apply.

Example 3.75. Use the method of variation of parameters to solve

$$t\sigma(t)y^{\Delta\Delta} - 2ty^{\Delta} + 2y = e_{\frac{3}{t}}(t, t_0)$$

for $t \in \mathbb{T} \cap (0, \infty)$, where $t_0 \in \mathbb{T} \cap (0, \infty)$. A general solution of the corresponding homogeneous equation is

$$u(t) = c_1 e_{\frac{1}{t}}(t, t_0) + c_2 e_{\frac{2}{t}}(t, t_0).$$

Hence there is a solution of the nonhomogeneous equation of the form

$$y_p(t) = \alpha(t)e_{\frac{1}{t}}(t, t_0) + \beta(t)e_{\frac{2}{t}}(t, t_0),$$

where $\alpha(t)$ and $\beta(t)$ are chosen to satisfy the system of equations

(3.46) $$\alpha^{\Delta}(t)e_{\frac{1}{t}}^{\sigma}(t, t_0) + \beta^{\Delta}(t)e_{\frac{2}{t}}^{\sigma}(t, t_0) = 0$$

(3.47) $$\frac{1}{\sigma(t)}\alpha^{\Delta}(t)e_{\frac{1}{t}}^{\sigma}(t, t_0) + \frac{2}{\sigma(t)}\beta^{\Delta}(t)e_{\frac{2}{t}}^{\sigma}(t, t_0) = \frac{1}{t\sigma(t)}e_{\frac{3}{t}}(t, t_0).$$

Multiplying equation (3.46) by $\frac{2}{\sigma(t)}$ and subtracting from equation (3.47) leads to

$$\alpha^{\Delta}(t) = -\frac{1}{t}\frac{e_{\frac{3}{t}}(t, t_0)}{e_{\frac{1}{t}}^{\sigma}(t, t_0)}.$$

Integrating and applying Theorem 2.38 we see that we can take

$$\alpha(t) = -\frac{1}{2}e_{\frac{2}{1+\mu(t)}}(t, t_0).$$

Multiplying equation (3.46) by $\frac{1}{\sigma(t)}$ and subtracting from equation (3.47) leads to

$$\beta^{\Delta}(t) = \frac{1}{t}\frac{e_{\frac{3}{t}}(t, t_0)}{e_{\frac{2}{t}}^{\sigma}(t, t_0)}.$$

Integrating and applying Theorem 2.38 we see that we can take

$$\beta(t) = e_{\frac{1}{t+2\mu(t)}}(t, t_0).$$

It follows that

$$
\begin{aligned}
y_p(t) &= \alpha(t)e_{\frac{1}{t}}(t, t_0) + \beta(t)e_{\frac{2}{t}}(t, t_0)\\
&= -\frac{1}{2}e_{\frac{2}{t+\mu(t)}}(t, t_0)e_{\frac{1}{t}}(t, t_0) + e_{\frac{1}{t+2\mu(t)}}(t, t_0)e_{\frac{2}{t}}(t, t_0)\\
&= -\frac{1}{2}e_{\frac{1}{t}\oplus\frac{2}{t+\mu(t)}}(t, t_0) + e_{\frac{2}{t}\oplus\frac{1}{t+2\mu(t)}}(t, t_0)\\
&= -\frac{1}{2}e_{\frac{3}{t}}(t, t_0) + e_{\frac{3}{t}}(t, t_0)\\
&= \frac{1}{2}e_{\frac{3}{t}}(t, t_0).
\end{aligned}
$$

Hence a general solution is given by

$$y(t) = c_1 e_{\frac{1}{t}}(t, t_0) + c_2 e_{\frac{2}{t}}(t, t_0) + \frac{1}{2}e_{\frac{3}{t}}(t, t_0).$$

Exercise 3.76. Use the method of variation of parameters to solve each of the following dynamic equations:

(i) $y^{\Delta\Delta} - 6y^{\Delta} + 8y = e_3(t, t_0)$, where $t, t_0 \in \mathbb{T}$;
(ii) $y^{\Delta\Delta} - 3y^{\Delta} + 2y = e_2(t, t_0)$, where $t, t_0 \in \mathbb{T}$;
(iii) $t\sigma(t)y^{\Delta\Delta} - 4ty^{\Delta} + 4y = e_{\frac{6}{t}}(t, t_0)$, where $t, t_0 \in \mathbb{T} \cap (0, \infty)$;
(iv) $t\sigma(t)y^{\Delta\Delta} + ty^{\Delta} - y = e_{\frac{4}{t}}(t, t_0)$, where $t, t_0 \in \mathbb{T} \cap (0, \infty)$.

3.9. Annihilator Method

Sometimes, instead of using the variation of parameters technique, it is easier to use the so-called *annihilator method* (also called the method of *undetermined coefficients*). We now discuss this annihilator method for solving equations of the form

(3.48) $$y^{\Delta\Delta} + \alpha y^{\Delta} + \beta y = f(t),$$

where α and β are constants and f is a function that can be annihilated as defined below. First let D denote the delta derivative operator and $D^0 = I$ denote the identity operator. By $D^n f(t)$ we mean $f^{\Delta^n}(t)$.

Definition 3.77. We say $f : \mathbb{T} \to \mathbb{R}$ can be annihilated provided we can find an operator of the form $\alpha_n D^n + \alpha_{n-1}D^{n-1} + \cdots + \alpha_0 I$ such that

$$(\alpha_n D^n + \alpha_{n-1}D^{n-1} + \cdots + \alpha_0 I)f(t) = 0,$$

where α_i, $0 \le i \le n$, are constants, not all zero.

Example 3.78. Since

$$
\begin{aligned}
(D - 4I)e_4(t, t_0) &= De_4(t, t_0) - 4Ie_4(t, t_0)\\
&= 4e_4(t, t_0) - 4e_4(t, t_0)\\
&= 0,
\end{aligned}
$$

$D - 4I$ is an annihilator for $e_4(t, t_0)$.

See Table 3.1 for a list of some functions and their annihilators.

Table 3.1. Annihilators

function	annihilator
1	D
t	D^2
$e_\alpha(t, t_0)$	$D - \alpha I$
$e_\alpha(t, t_0) \int_{t_0}^t \frac{1}{1+\tau\mu(\tau)} \Delta\tau$	$(D - \alpha I)^2$
$\sin_\alpha(t, t_0),\ \alpha > 0$	$D^2 + \alpha^2 I$
$\cos_\alpha(t, t_0),\ \alpha > 0$	$D^2 + \alpha^2 I$
$\sinh_\alpha(t, t_0)$	$D^2 - \alpha^2 I$
$\cosh_\alpha(t, t_0)$	$D^2 - \alpha^2 I$

Exercise 3.79. Show that if α and β are constants, then
$$(D - \alpha I)(D - \beta I) = (D - \beta I)(D - \alpha I).$$

In the next example we show how to solve an equation of the form (3.48) by the method of annihilators.

Example 3.80. Solve the equation (3.43) by the method of annihilators. (This is the equation in Example 3.74 that we solved by the variation of parameters method.) Note that (3.43) can be written in the form
$$(D - 2I)(D - 3I)y = (D^2 - 5D + 6I)y = e_4(t, t_0).$$
Multiplying both sides by the annihilator $D - 4I$ of $e_4(t, t_0)$, we get that if y is a solution of (3.43), then y satisfies
$$(D - 4I)(D - 2I)(D - 3I)y = 0.$$
Hence
$$y(t) = \alpha e_2(t, t_0) + \beta e_3(t, t_0) + \gamma e_4(t, t_0).$$
Since $\alpha e_2(t, t_0) + \beta e_3(t, t_0)$ is a solution of the homogeneous equation corresponding to (3.43), we get that (3.43) has a solution of the form
$$y_p(t) = \gamma e_4(t, t_0).$$
To determine γ we substitute $y_p(t)$ into equation (3.43):
$$\begin{aligned} y_p^{\Delta\Delta}(t) - 5y_p^\Delta(t) + 6y_p(t) &= 16\gamma e_4(t, t_0) - 20\gamma e_4(t, t_0) + 6\gamma e_4(t, t_0) \\ &= 2\gamma e_4(t, t_0). \end{aligned}$$
Hence if we take $\gamma = \frac{1}{2}$, then we get that
$$y_p(t) = \frac{1}{2} e_4(t, t_0)$$

is a solution of (3.43). This implies that a general solution of equation (3.43) is

$$y(t) = c_1 e_2(t, t_0) + c_2 e_3(t, t_0) + \frac{1}{2} e_4(t, t_0).$$

When the method of annihilators can be used, it is usually easier to apply than the method of variation of parameters as in the case of this example.

Exercise 3.81. Use the method of annihilators to solve each of the following dynamic equations:

    (i) $y^{\Delta\Delta} - 6y^\Delta + 8y = e_3(t, t_0)$, where $t, t_0 \in \mathbf{T}$;
    (ii) $y^{\Delta\Delta} - 3y^\Delta + 2y = e_2(t, t_0)$, where $t, t_0 \in \mathbf{T}$;
    (iii) $y^{\Delta\Delta} + y^\Delta - 2y = 2 + t$, where $t, t_0 \in \mathbf{T} \cap (0, \infty)$;
    (iv) $y^{\Delta\Delta} + y = e_3(t, t_0)$, where $t, t_0 \in \mathbf{T}$.

3.10. Laplace Transform

Much of the work in this section can be found in the paper by Bohner and Peterson [**91**]. Throughout this section we assume that the time scale $\mathbf{T}_0$ is such that

$$0 \in \mathbf{T}_0 \quad \text{and} \quad \sup \mathbf{T}_0 = \infty.$$

Note that if we assume that $z \in \mathcal{R}$ is constant, then $\ominus z \in \mathcal{R}$ by Exercise 2.26 and $e_{\ominus z}(t, 0)$ is well defined on $\mathbf{T}_0$. With this in mind we make the following definition.

Definition 3.82. Assume that $x : \mathbf{T}_0 \to \mathbb{R}$ is regulated. Then the *Laplace transform* of x is defined by

$$(3.49) \qquad\qquad \mathcal{L}\{x\}(z) := \int_0^\infty x(t) e_{\ominus z}^\sigma(t, 0) \Delta t$$

for $z \in \mathcal{D}\{x\}$, where $\mathcal{D}\{x\}$ consists of all complex numbers $z \in \mathcal{R}$ for which the improper integral exists.

Hilger [**164**] defines Laplace transforms for the case where $\mathbf{T}_0$ has step-size h and derives several complicated formulas. His definition of the Laplace transform is not the same as the one given here.

Exercise 3.83. Show that if $\mathbf{T}_0$ is the real interval $[0, \infty)$, then the Laplace transform is the familiar formula for the continuous case. Show that if $\mathbf{T}_0 = \mathbb{N}_0$, then

$$(3.50) \qquad\qquad (z + 1)\mathcal{L}\{x\}(z) = \mathcal{Z}\{x\}(z + 1),$$

where $\mathcal{Z}\{x\}$ is the Z-transform of x, which is defined by

$$\mathcal{Z}\{x\}(z) = \sum_{t=0}^\infty \frac{x(t)}{z^t}$$

for those complex values of z for which this infinite sum converges.

The proof of the next result is left to the reader.

Theorem 3.84 (Linearity). *Assume f and g are regulated on $\mathbf{T}_0$ and α and β are constants. Then*

$$\mathcal{L}\{\alpha x + \beta y\}(z) = \alpha \mathcal{L}\{x\}(z) + \beta \mathcal{L}\{y\}(z)$$

for $z \in \mathcal{D}\{x\} \cap \mathcal{D}\{y\}$.

The following easy result is needed frequently throughout this section.

Lemma 3.85. *If $z \in \mathbb{C}$ is regressive, then*

$$e^{\sigma}_{\ominus z}(t,0) = \frac{e_{\ominus z}(t,0)}{1 + \mu(t)z} = -\frac{(\ominus z)(t)}{z} e_{\ominus z}(t,0).$$

Proof. By Theorem 2.36 (ii) we have

$$
\begin{aligned}
e^{\sigma}_{\ominus z}(t,0) &= (1 + \mu(t)(\ominus z)(t))e_{\ominus z}(t,0) \\
&= \left(1 - \frac{\mu(t)z}{1 + \mu(t)z}\right) e_{\ominus z}(t,0) \\
&= \frac{1}{1 + \mu(t)z} e_{\ominus z}(t,0) \\
&= -\frac{(\ominus z)(t)}{z} e_{\ominus z}(t,0).
\end{aligned}
$$

This proves our claim.   □

Example 3.86. We now shall use Lemma 3.85 to find the Laplace transform of $x(t) \equiv 1$:

$$
\begin{aligned}
\mathcal{L}\{1\}(z) &= \int_0^\infty 1 \cdot e^{\sigma}_{\ominus z}(t,0)\Delta t \\
&= -\frac{1}{z} \int_0^\infty (\ominus z)(t)e_{\ominus z}(t,0)\Delta t \\
&= -\frac{1}{z} [e_{\ominus z}(t,0)]_{t=0}^{t\to\infty} \\
&= \frac{1}{z}
\end{aligned}
$$

for all complex values of $z \in \mathcal{R}$ such that

$$\lim_{t\to\infty} e_{\ominus z}(t,0) = 0$$

holds. Therefore we get the formula

$$\mathcal{L}\{1\}(z) = \frac{1}{z}.$$

The following two results are derived using integration by parts.

Theorem 3.87. *Assume $x : \mathbb{T}_0 \to \mathbb{C}$ is such that x^{Δ} is regulated. Then*

(3.51) $$\mathcal{L}\{x^{\Delta}\}(z) = z\mathcal{L}\{x\}(z) - x(0)$$

for those regressive $z \in \mathbb{C}$ satisfying

(3.52) $$\lim_{t\to\infty} \{x(t)e_{\ominus z}(t,0)\} = 0.$$

Proof. Integration by parts (see Theorem 1.77 (v)) and Lemma 3.85 directly yield

$$
\begin{aligned}
\mathcal{L}\{x^{\Delta}\}(z) &= \int_0^{\infty} x^{\Delta}(t)e_{\ominus z}^{\sigma}(t,0)\Delta t \\
&= [x(t)e_{\ominus z}(t,0)]_{t=0}^{t\to\infty} - \int_0^{\infty} x(t)(\ominus z)(t)e_{\ominus z}(t,0)\Delta t \\
&= -x(0) + z\int_0^{\infty} x(t)e_{\ominus z}^{\sigma}(t,0)\Delta t \\
&= -x(0) + z\mathcal{L}\{x\}(z),
\end{aligned}
$$

provided (3.52) holds. □

Exercise 3.88. Show that under suitable assumptions

(3.53) $\mathcal{L}\{x^{\Delta\Delta}\}(z) = z^2\mathcal{L}\{x\}(z) - zx(0) - x^{\Delta}(0).$

Write down the formula for $\mathcal{L}\{x^{\Delta^n}\}$.

Theorem 3.89. *Assume $x : \mathbb{T}_0 \to \mathbb{C}$ is regulated. If*

$$
X(t) := \int_0^t x(\tau)\Delta\tau
$$

for $t \in \mathbb{T}_0$, then

$$
\mathcal{L}\{X\}(z) = \frac{1}{z}\mathcal{L}\{x\}(z)
$$

for those regressive $z \in \mathbb{C} \setminus \{0\}$ satisfying

(3.54) $\displaystyle\lim_{t\to\infty} \left\{ e_{\ominus z}(t,0) \int_0^t x(\tau)\Delta\tau \right\} = 0.$

Proof. We use integration by parts (see Theorem 1.77 (vi)) and Lemma 3.85 to obtain

$$
\begin{aligned}
\mathcal{L}\{X\}(z) &= \int_0^{\infty} X(t)e_{\ominus z}^{\sigma}(t,0)\Delta t \\
&= -\frac{1}{z}\int_0^{\infty} X(t)(\ominus z)(t)e_{\ominus z}(t,0)\Delta t \\
&= -\frac{1}{z}\left\{ [X(t)e_{\ominus z}(t,0)]_{t=0}^{t\to\infty} - \int_0^{\infty} X^{\Delta}(t)e_{\ominus z}^{\sigma}(t,0)\Delta t \right\} \\
&= -\frac{1}{z}\left\{ -X(0) - \int_0^{\infty} x(t)e_{\ominus z}^{\sigma}(t,0)\Delta t \right\} \\
&= \frac{1}{z}\mathcal{L}\{x\}(z),
\end{aligned}
$$

provided (3.54) holds. □

Theorem 3.90. *Assume $h_k(t,0)$, $k \in \mathbb{N}_0$, are defined as in Section 1.6. Then*

$$
\mathcal{L}\{h_k(\cdot,0)\}(z) = \frac{1}{z^{k+1}}
$$

for those regressive $z \in \mathbb{C}$ satisfying

(3.55) $\displaystyle\lim_{t\to\infty} \{h_k(t,0)e_{\ominus z}(t,0)\} = 0.$

Proof. Fix $k \in \mathbb{N}_0$. We will prove by mathematical induction that

(3.56)
$$\mathcal{L}\{h_i(\cdot,0)\}(z) = \frac{1}{z^{i+1}}$$

for $0 \le i \le k$. First note that (3.55) implies

$$\lim_{t\to\infty} \{h_i(t,0)e_{\ominus z}(t,0)\} = 0$$

for $0 \le i \le k$. Note that (3.56) holds for $i = 0$ by Example 3.86. Now assume that $1 \le i \le k$ and (3.56) holds with i replaced by $i-1$. Then by (1.9) and Theorem 3.89,

$$
\begin{aligned}
\mathcal{L}\{h_i(t,0)\}(z) &= \mathcal{L}\left\{\int_0^t h_{i-1}(\tau,0)\Delta\tau\right\}(z) \\
&= \frac{1}{z}\mathcal{L}\{h_{i-1}(t,0)\}(z) \\
&= \frac{1}{z^{i+1}}.
\end{aligned}
$$

The claim follows by the principle of mathematical induction. $\square$

Example 3.91. Now we find the Laplace transform of $x(t) = e_\alpha(t,0)$, where $\alpha \in \mathbb{C}$ is regressive. Again we use Lemma 3.85 to find

$$
\begin{aligned}
\mathcal{L}\{e_\alpha(\cdot,0)\}(z) &= \int_0^\infty e_\alpha(t,0)e_{\ominus z}^\sigma(t,0)\Delta t \\
&= \int_0^\infty \frac{1}{1+\mu(t)z}e_\alpha(t,0)e_{\ominus z}(t,0)\Delta t \\
&= \int_0^\infty \frac{1}{1+\mu(t)z}e_{\alpha\ominus z}(t,0)\Delta t \\
&= \frac{1}{\alpha-z}\int_0^\infty \frac{\alpha-z}{1+\mu(t)z}e_{\alpha\ominus z}(t,0)\Delta t \\
&= \frac{1}{\alpha-z}\int_0^\infty (\alpha\ominus z)(t)e_{\alpha\ominus z}(t,0)\Delta t \\
&= \frac{1}{\alpha-z}[e_{\alpha\ominus z}(t,0)]_{t=0}^{t\to\infty} \\
&= \frac{1}{z-\alpha},
\end{aligned}
$$

provided

$$\lim_{t\to\infty} e_{\alpha\ominus z}(t,0) = 0$$

holds.

Example 3.92. In this example we wish to remind the reader of the classical Laplace transform and the classical Z-transform as they are employed to solve higher order differential and difference equations with constant coefficients. Then we will see how these two examples can be unified using our Laplace transform as proposed in Definition 3.82. Finally, we apply our results to a certain 2-difference equation.

(i) We consider the continuous problem

(3.57) $x'' + 5x' + 6x = 0, \quad x(0) = 1, \quad x'(0) = -5.$

The "classical" Laplace transform is defined by

$$\mathcal{L}\{x\}(z) = \int_0^\infty x(t)e^{-zt}dt$$

whenever this integral converges, and by direct computation one can verify the formula

$$\mathcal{L}\{e^{\alpha t}\}(z) = \frac{1}{z - \alpha}.$$

Another formula which can be shown by integration of parts reads

$$\mathcal{L}\{x'\}(z) = z\mathcal{L}\{x\}(z) - x(0),$$

and one consequence of this formula is

$$\mathcal{L}\{x''\}(z) = z^2\mathcal{L}\{x\}(z) - zx(0) - x'(0).$$

We now take Laplace transforms on both sides of the equation in (3.57) and use the initial conditions to arrive at

$$\begin{aligned} 0 &= z^2\mathcal{L}\{x\}(z) - z + 5 + 5\left[z\mathcal{L}\{x\}(z) - 1\right] + 6\mathcal{L}\{x\}(z) \\ &= (z^2 + 5z + 6)\mathcal{L}\{x\}(z) - z \end{aligned}$$

so that

$$\mathcal{L}\{x\}(z) = \frac{z}{z^2 + 5z + 6} = \frac{3}{z + 3} - \frac{2}{z + 2}$$

and hence

$$x(t) = 3e^{-3t} - 2e^{-2t} \quad \text{for all} \quad t \in \mathbb{R}.$$

(ii) In [**125**, Example 5.7], the (discrete) problem

(3.58) $x(t + 2) + 3x(t + 1) + 2x(t) = 0, \quad x(0) = 1, \quad x(1) = -4$

is considered. The "classical" Z-transform

$$\mathcal{Z}\{x\}(z) = \sum_{t=0}^\infty \frac{x(t)}{z^t}$$

(whenever this infinite sum converges) is employed to find the solution of the initial value problem (3.58). Let us first collect some properties of $\mathcal{Z}$: We have

$$\mathcal{Z}\{\alpha^t\}(z) = \sum_{t=0}^\infty \left(\frac{\alpha}{z}\right)^k = \frac{1}{1 - \frac{\alpha}{z}} = \frac{z}{z - \alpha}$$

and

$$\begin{aligned} \mathcal{Z}\{x^\sigma\}(z) &= \sum_{t=0}^\infty \frac{x(t + 1)}{z^t} \\ &= z\sum_{t=0}^\infty \frac{x(t + 1)}{z^{t+1}} \\ &= z\sum_{t=1}^\infty \frac{x(t)}{z^t} \\ &= z\left[\mathcal{Z}\{x\}(z) - x(0)\right] \end{aligned}$$

and hence

$$\mathcal{Z}\{x^{\sigma\sigma}\}(z) = z\left[\mathcal{Z}\{x^\sigma\}(z) - x^\sigma(0)\right]$$
$$= z^2\mathcal{Z}\{x\}(z) - z^2 x(0) - zx(1).$$

Now we take Z-transforms of the equation in (3.58) to arrive at

$$0 = z^2\mathcal{Z}\{x\}(z) - z^2 + 4z + 3z\mathcal{Z}\{x\}(z) - 3z + 2\mathcal{Z}\{x\}(z)$$
$$= (z^2 + 3z + 2)\mathcal{Z}\{x\}(z) - (z^2 - z)$$

so that

$$\mathcal{Z}\{x\}(z) = z\frac{z-1}{(z+1)(z+2)} = z\left[\frac{-2}{z+1} + \frac{3}{z+2}\right] = -2\frac{z}{z+1} + 3\frac{z}{z+2}$$

and hence

$$x(t) = 3(-2)^t - 2(-1)^t \quad \text{for all} \quad t \in \mathbb{Z}.$$

(iii) Here we suggest the following alternate method for solving the problem (3.58): First we rewrite the equation in (3.58) as a *difference equation* rather than a *recurrence relation*. The IVP (3.58) is equivalent to

(3.59) $$\Delta\Delta x + 5\Delta x + 6x = 0, \quad x(0) = 1, \quad \Delta x(0) = -5.$$

Note that (3.59) now looks just like the discrete version of (3.57), whereas this fact was not apparent with equation (3.58). Now we apply the "$\tilde{Z}$-transform"; (see (3.50)) defined by

$$\tilde{\mathcal{Z}}\{x\}(z) = \frac{\mathcal{Z}\{x\}(z+1)}{z+1}.$$

We have that

$$\tilde{\mathcal{Z}}\{(1+\alpha)^t\}(z) = \frac{\mathcal{Z}\{(1+\alpha)^t\}(z+1)}{z+1}$$
$$= \frac{\frac{z+1}{z+1-(\alpha+1)}}{z+1}$$
$$= \frac{1}{z-\alpha}$$

and

$$\tilde{\mathcal{Z}}\{x^\Delta\}(z) = \frac{\mathcal{Z}\{x^\Delta\}(z+1)}{z+1}$$
$$= \frac{\mathcal{Z}\{x^\sigma - x\}(z+1)}{z+1}$$
$$= \frac{\mathcal{Z}\{x^\sigma\}(z+1) - \mathcal{Z}\{x\}(z+1)}{z+1}$$
$$= \frac{(z+1)\left[\mathcal{Z}\{x\}(z+1) - x(0)\right] - \mathcal{Z}\{x\}(z+1)}{z+1}$$
$$= \frac{z\mathcal{Z}\{x\}(z+1)}{z+1} - x(0)$$
$$= z\tilde{\mathcal{Z}}\{x\}(z) - x(0)$$

and also

$$\begin{aligned}
\tilde{Z}\{x^{\Delta\Delta}\}(z) &= z\tilde{Z}\{x^{\Delta}\}(z) - x^{\Delta}(0)\\
&= z\left[z\tilde{Z}\{x\}(z) - x(0)\right] - x^{\Delta}(0)\\
&= z^2\tilde{Z}\{x\}(z) - zx(0) - x^{\Delta}(0).
\end{aligned}$$

Now let us return to our example in (3.59) and apply $\tilde{Z}$-transforms to arrive at

$$\begin{aligned}
0 &= z^2\tilde{Z}\{x\}(z) - z + 5 + 5\left[z\tilde{Z}\{x\}(z) - 1\right] + 6\tilde{Z}\{x\}(z)\\
&= (z^2 + 5z + 6)\tilde{Z}\{x\}(z) - z.
\end{aligned}$$

Therefore

$$\tilde{Z}\{x\}(z) = \frac{z}{z^2 + 5z + 6} = \frac{z}{(z+2)(z+3)} = \frac{-2}{z+2} + \frac{3}{z+3}$$

so that

$$x(t) = -2(1-2)^t + 3(1-3)^t = 3(-2)^t - 2(-1)^t \quad \text{for all} \quad t \in \mathbb{Z}.$$

This method has the advantage that one has to "learn only one table" for transforms, as the Laplace transform and the $\tilde{Z}$-transform of corresponding functions are the same.

(iv) We can generalize all of the above examples by considering the dynamic problem

$$(3.60) \qquad x^{\Delta\Delta} + 5x^{\Delta} + 6x = 0, \quad x(0) = 1, \quad x^{\Delta}(0) = -5.$$

In (3.60), we now take our Laplace transform introduced in Definition 3.82 and use (3.51) from Theorem 3.87 and (3.53) from Exercise 3.88 to arrive at

$$\begin{aligned}
0 &= z^2\mathcal{L}\{x\}(z) - z + 5 + 5[z\mathcal{L}\{x\}(z) - 1] + 6\mathcal{L}\{x\}(z)\\
&= (z^2 + 5z + 6)\mathcal{L}\{x\}(z) - z
\end{aligned}$$

so that

$$\mathcal{L}\{x\}(z) = \frac{z}{z^2 + 5z + 6} = \frac{3}{z+3} - \frac{2}{z+2}$$

and hence (see Example 3.91)

$$x(t) = 3e_{-3}(t,0) - 2e_{-2}(t,0) \quad \text{for all} \quad t \in \mathbb{T}.$$

Now we see that all of the above examples are included in this example, as

$$e_{-a}(t,0) = \begin{cases} e^{-at} & \text{if} \quad \mathbb{T} = \mathbb{R}\\ (1-a)^t & \text{if} \quad \mathbb{T} = \mathbb{Z}. \end{cases}$$

(v) Finally, let us consider the problem

$$(3.61) \quad x(4t) + (10t - 3)x(2t) + (12t^2 - 10t + 2)x(t) = 0, \quad x(1) = 1, \quad x(2) = -4.$$

Using Example 1.41, we find that, with the time scale $\mathbb{T} = 2^{\mathbb{N}_0}$,

$$5x^{\Delta\Delta}(t) + 5x^{\Delta}(t) + 6x(t) = \frac{x(4t) - 3x(2t) + 2x(t)}{2t^2} + 5\frac{x(2t) - x(t)}{t} + 6x(t)$$

$$= \frac{x(4t) - 3x(2t) + 2x(t) + 10tx(2t) - 10tx(t) + 12t^2 x(t)}{2t^2}$$

$$= \frac{x(4t) + (10t - 3)x(2t) + (12t^2 - 10t + 2)x(t)}{2t^2}$$

$$= 0.$$

Thus (3.61) is equivalent to

$$x^{\Delta\Delta} + 5x^{\Delta} + 6x = 0, \quad x(1) = 1, \quad x(2) = -5.$$

This is of the same form as (3.60) and hence has the solution

$$x(t) = 3e_{-3}(t,1) - 2e_{-2}(t,1) \quad \text{for all} \quad t \in 2^{\mathbb{N}_0}.$$

Hence, using (2.29) from Example 2.55,

$$x(t) = 3\prod_{s \in \mathbb{T} \cap (0,t)} (1 - 3s) - 2\prod_{s \in \mathbb{T} \cap (0,t)} (1 - 2s) \quad \text{for all} \quad t \in 2^{\mathbb{N}_0}.$$

Example 3.93. Let $a \in \mathbb{T}$, $a > 0$, and define the *step function* u_a by

$$u_a(t) = \begin{cases} 0 & \text{if} \quad t \in \mathbb{T} \cap (-\infty, a) \\ 1 & \text{if} \quad t \in \mathbb{T} \cap [a, \infty). \end{cases}$$

We obtain

$$\int_0^\infty u_a(t)e_{\ominus z}(\sigma(t))\Delta t = \int_a^\infty e_{\ominus z}(\sigma(t))\Delta t = \frac{e_{\ominus z}(a)}{z}$$

provided $z \in \mathcal{R}$ satisfies

$$\lim_{t \to \infty} e_{\ominus z}(t) = 0$$

(for the last equal sign compare with the calculation in Example 3.86).

Example 3.94. Under "suitable" assumptions (see Example 3.91) we have by using the linearity theorem (Theorem 3.84)

$$\mathcal{L}\{\cosh_\alpha(\cdot, 0)\}(z) = \mathcal{L}\left\{\frac{e_\alpha(\cdot, 0) + e_{-\alpha}(\cdot, 0)}{2}\right\}(z)$$

$$= \frac{1}{2}\left(\frac{1}{z - \alpha} + \frac{1}{z + \alpha}\right)$$

$$= \frac{z}{(z - \alpha)(z + \alpha)}$$

$$= \frac{z}{z^2 - \alpha^2}.$$

Exercise 3.95. Show that

$$\mathcal{L}\{\sinh_\alpha(\cdot, 0)\}(z) = \frac{\alpha}{z^2 - \alpha^2}.$$

Example 3.96. Under "suitable" assumptions (see Example 3.91) we have

$$
\begin{aligned}
\mathcal{L}\{\cos_\alpha(\cdot,0)\}(z) &= \mathcal{L}\left\{\frac{e_{i\alpha}(\cdot,0) + e_{-i\alpha}(\cdot,0)}{2}\right\}(z) \\
&= \frac{1}{2}\left(\frac{1}{z - i\alpha} + \frac{1}{z + i\alpha}\right) \\
&= \frac{z}{(z - i\alpha)(z + i\alpha)} \\
&= \frac{z}{z^2 + \alpha^2}.
\end{aligned}
$$

Exercise 3.97. Show that

$$
\mathcal{L}\{\sin_\alpha(\cdot,0)\}(z) = \frac{\alpha}{z^2 + \alpha^2}.
$$

Exercise 3.98. Assume that $\mathbb{T}_0$ has constant graininess $\mu(t) \equiv h \geq 0$. Show that under suitable assuptions

$$
\mathcal{L}\{x^\sigma\}(z) = (1 + hz)\mathcal{L}\{x\}(z) - h(1 + hz)x(0)
$$

holds.

Theorem 3.99 (Shifting Theorem). *If $\alpha, \beta \in \mathcal{R}$ are constants, then*

(i) $\mathcal{L}\{e_\alpha(\cdot,0)\sin_{\frac{\beta}{1+\mu\alpha}}(\cdot,0)\}(z) = \frac{\beta}{(z-\alpha)^2+\beta^2}$;

(ii) $\mathcal{L}\{e_\alpha(\cdot,0)\cos_{\frac{\beta}{1+\mu\alpha}}(\cdot,0)\}(z) = \frac{z-\alpha}{(z-\alpha)^2+\beta^2}$;

(iii) $\mathcal{L}\{e_\alpha(\cdot,0)\sinh_{\frac{\beta}{1+\mu\alpha}}(\cdot,0)\}(z) = \frac{\beta}{(z-\alpha)^2-\beta^2}$;

(iv) $\mathcal{L}\{e_\alpha(\cdot,0)\cosh_{\frac{\beta}{1+\mu\alpha}}(\cdot,0)\}(z) = \frac{z-\alpha}{(z-\alpha)^2-\beta^2}$.

Proof. To prove (i) we note that $x(t) := e_\alpha(t,0)\sin_{\frac{\beta}{1+\mu\alpha}}(t,0)$ is the solution of the initial value problem

$$
x^{\Delta\Delta} - 2\alpha x^\Delta + (\alpha^2 + \beta^2)x = 0, \quad x(0) = 0, \quad x^\Delta(0) = \beta.
$$

It follows that $X(z) := \mathcal{L}\{e_\alpha(\cdot,0)\sin_{\frac{\beta}{1+\mu\alpha}}(\cdot,0)\}(z)$ satisfies

$$
z^2 X(z) - zx(0) - x^\Delta(0) - 2\alpha[zX(z) - x(0)] + (\alpha^2 + \beta^2)X(z) = 0.
$$

Using the initial conditions we get

$$
z^2 X(z) - \beta - 2\alpha z X(z) + (\alpha^2 + \beta^2)X(z) = 0.
$$

Solving for $X(z)$ we arrive at

$$
X(z) = \frac{\beta}{z^2 - 2\alpha z + (\alpha^2 + \beta^2)} = \frac{\beta}{(z-\alpha)^2 + \beta^2}.
$$

Parts (ii), (iii), and (iv) are proved similarly. □

Exercise 3.100. Prove Theorem 3.99 (ii), (iii), and (iv).

Example 3.101. Use Laplace transforms to solve the IVP

(3.62) $x^{\Delta\Delta} - 4x^\Delta + 13x = 0, \quad x(0) = 1, \quad x^\Delta(0) = 1.$

Assuming x is the solution of (3.62) we obtain after taking the Laplace transform of both sides of the dynamic equation in (3.62)

$$
z^2\mathcal{L}\{x\}(z) - zx(0) - x^\Delta(0) - 4[z\mathcal{L}\{x\}(z) - x(0)] + 13\mathcal{L}\{x\}(z) = 0.
$$

Using the initial conditions we get

$$z^2 \mathcal{L}\{x\}(z) - z - 1 - 4[z\mathcal{L}\{x\}(z) - 1] + 13\mathcal{L}\{x\}(z) = 0.$$

Solving for $\mathcal{L}\{x\}(z)$ we obtain

$$\mathcal{L}\{x\}(z) = \frac{z-3}{z^2 - 4z + 9} = \frac{z-2}{(z-2)^2 + 9} - \frac{1}{3}\frac{3}{(z-2)^2 + 9}.$$

Using Theorem 3.99 (or Table 3.2) we get that the solution of (3.62) is

$$x(t) = e_2(t,0)\cos_{\frac{3}{1+2\mu}}(t,0) - \frac{1}{3}e_{\frac{3}{1+2\mu}}(t,0)\sin_3(t,0).$$

Exercise 3.102. Use Laplace transforms to solve the following initial value problems on $\mathbb{T}_0$:

(i) $x^{\Delta\Delta} + x^{\Delta} - 2x = 0$, $x(0) = 1$, $x^{\Delta}(0) = 1$;

(ii) $x^{\Delta\Delta} - 9x = 0$, $x(0) = 0$, $x^{\Delta}(0) = 1$;

(iii) $x^{\Delta\Delta} + 2x^{\Delta} - 3x = 0$, $x(0) = 5$, $x^{\Delta}(0) = 1$;

(iv) $x^{\Delta\Delta} + 16x = 0$, $x(0) = 0$, $x^{\Delta}(0) = 3$;

(v) $x^{\Delta\Delta} - 6x^{\Delta} + 25x = 0$, $x(0) = 1$, $x^{\Delta}(0) = 2$;

(vi) $x^{\Delta\Delta} + 4x^{\Delta} + 13x = 0$, $x(0) = -1$, $x^{\Delta}(0) = 1$;

(vii) $x^{\Delta\Delta\Delta} + x^{\Delta} = e_1(t,0)$, $x(0) = x^{\Delta}(0) = x^{\Delta\Delta}(0) = 0$;

(viii) $x^{\Delta\Delta} + 6x^{\Delta} + 25x = 0$, $x(0) = 2$, $x^{\Delta}(0) = 1$.

Exercise 3.103. Assume f is a regulated function and

$$g(t) := \int_0^{\sigma(t)} \frac{1}{1 + \mu(\tau)z} \Delta\tau.$$

Show that if we can change the order of differentiation and integration for an appropriate integral, then

$$\mathcal{L}\{gf\}(z) = -\frac{d}{dz}\mathcal{L}\{f\}(z) \quad \text{for} \quad z \in \mathcal{D}(f).$$

Show that this gives a well-known formula for Laplace transforms in the case if $\mathbb{T}_0$ is the real interval $[0,\infty)$ and a well-known formula for Z-transforms by taking $\mathbb{T}_0 = \mathbb{N}_0$.

When $\mathbb{T}_0$ is the real interval $[0,\infty)$, the convolution of two functions is defined by

$$(f * g)(t) = \int_0^t f(t-s)g(s)ds \quad \text{for} \quad t \geq 0.$$

In general, if $s, t \in \mathbb{T}_0$, then it does not follow that $t - s \in \mathbb{T}_0$. So $f(t-s)$ might not be defined if f is defined only on the time scale $\mathbb{T}_0$, and so we have to give a different definition of the convolution of two functions on a general time scale $\mathbb{T}_0$.

Definition 3.104. Assume f is one of the functions e_α, $\cosh_\alpha$, $\sinh_\alpha$, $\cos_\alpha$, $\sin_\alpha$, h_k, $k \in \mathbb{N}_0$, and put $f(t) = f(t,0)$. If g is a regulated function on $\mathbb{T}_0$, then we define the *convolution* of f with g by

$$(f * g)(t) = \int_0^t f(t,\sigma(s))g(s)\Delta s \quad \text{for} \quad t \in \mathbb{T}_0.$$

Exercise 3.105. It can be shown that if $f(t) = t$ for $t \in \mathbb{T}_0$, then $1 * f = f * 1$. Prove this for the special case $\mathbb{T}_0 := \overline{q^{\mathbb{Z}}}$ by evaluating the integrals you obtain for this special case.

Theorem 3.106 (Convolution Theorem). *Assume f is one of the functions $e_\alpha(\cdot,0)$, $\cosh_\alpha(\cdot,0)$, $\sinh_\alpha(\cdot,0)$, $\cos_\alpha(\cdot,0)$, $\sin_\alpha(\cdot,0)$, $h_k(\cdot,0)$, $k \in \mathbb{N}_0$. If g is a regulated function on $\mathbb{T}_0$, then*

$$\mathcal{L}\{f * g\}(z) = \mathcal{L}\{f\}(z)\mathcal{L}\{g\}(z).$$

Proof. Let g be a regulated function. First assume $f(t,s) = e_\alpha(t,s)$ (here it is understood that we are assuming that $\alpha \in \mathbb{C}$ is regressive so that $e_\alpha(t,s)$ is well defined). In this case $f(t) = f(t,0) = e_\alpha(t,0)$. If y is the solution of the IVP

$$y^\Delta - \alpha y = g(t), \quad y(0) = 0,$$

then by the variation of constants formula in Theorem 2.77,

$$y(t) = \int_0^t e_\alpha(t,\sigma(s))g(s)\Delta s = (f * g)(t) \quad \text{for} \quad t \in \mathbb{T}_0.$$

Taking the Laplace transform of both sides of $y^\Delta - \alpha y = g(t)$ and using $y(0) = 0$, we obtain

$$z\mathcal{L}\{y\}(z) - \alpha\mathcal{L}\{y\}(z) = \mathcal{L}\{g\}(z).$$

By Example 3.91, this leads to the desired result

$$\mathcal{L}\{y\}(z) = \frac{1}{z - \alpha}\mathcal{L}\{g\}(z) = \mathcal{L}\{f\}(z)\mathcal{L}\{g\}(z).$$

Next let $f(t,s) = \cosh_\alpha(t,s)$ (here it is understood that we are assuming that $\alpha \in \mathbb{C}$ and $-\mu\alpha^2 \in \mathcal{R}$ so that $\cosh_\alpha(t,s)$ is well defined). In this case we have $f(t) = f(t,0) = \cosh_\alpha(t,0)$. Then

$$\begin{aligned}
(f * g)(t) &= \int_0^t f(t,\sigma(s))g(s)\Delta s \\
&= \int_0^t \cosh_\alpha(t,\sigma(s))g(s)\Delta s \\
&= \frac{1}{2}\int_0^t e_\alpha(t,\sigma(s))g(s)\Delta s + \frac{1}{2}\int_0^t e_{-\alpha}(t,\sigma(s))g(s)\Delta s \\
&= \frac{1}{2}\left(e_\alpha(\cdot,0) * g\right)(t) + \frac{1}{2}\left(e_{-\alpha}(\cdot,0) * g\right)(t).
\end{aligned}$$

Taking the Laplace transform of both sides we get

$$\begin{aligned}
\mathcal{L}\{f * g\}(z) &= \frac{1}{2}\mathcal{L}\{e_\alpha(\cdot,0) * g\}(z) + \frac{1}{2}\mathcal{L}\{e_{-\alpha}(\cdot,0) * g\}(z) \\
&= \frac{1}{2}\mathcal{L}\{e_\alpha(\cdot,0)\}(z)\mathcal{L}\{g\}(z) + \frac{1}{2}\mathcal{L}\{e_{-\alpha}(\cdot,0)\}(z)\mathcal{L}\{g\}(z) \\
&= \mathcal{L}\left\{\frac{e_\alpha(\cdot,0) + e_{-\alpha}(\cdot,0)}{2}\right\}(z)\mathcal{L}\{g\}(z) \\
&= \mathcal{L}\{\cosh_\alpha(\cdot,0)\}(z)\mathcal{L}\{g\}(z) \\
&= \mathcal{L}\{f\}(z)\mathcal{L}\{g\}(z).
\end{aligned}$$

Similarly, one can prove the convolution formulas when f is one of the functions $\sinh_\alpha(\cdot,0)$, $\cos_\alpha(\cdot,0)$, $\sin_\alpha(\cdot,0)$, and g a regulated function (see Exercise 3.107).

Next we prove the convolution formula when $f = h_k(\cdot,0)$, $k \in \mathbb{N}_0$ and g is a regulated function. Let y be the solution of the IVP

$$y^{\Delta^{k+1}} = g(t), \quad y^{\Delta^i}(0) = 0 \text{ for all } 0 \le i \le k.$$

Then by the variation of constants formula in Theorem 5.119,

$$y(t) = \int_0^t h_k(t, \sigma(s))g(s)\Delta s = (h_k(\cdot, 0) * g)(t).$$

Taking the Laplace transform of both sides of the dynamic equation $y^{\Delta^{k+1}} = g(t)$ and using the initial conditions, we get

$$z^{k+1}\mathcal{L}\{y\}(z) = \mathcal{L}\{g\}(z),$$

i.e.,

$$\mathcal{L}\{y\}(z) = \frac{1}{z^{k+1}}\mathcal{L}\{g\}(z) = \mathcal{L}\{h_k(\cdot, 0)\}(z)\mathcal{L}\{g\}(z),$$

which gives us the desired result.   □

Exercise 3.107. Prove the convolution theorem when f is any of the functions $\sinh_\alpha(\cdot, 0)$, $\cos_\alpha(\cdot, 0)$, $\sin_\alpha(\cdot, 0)$ and g is a regulated function. What assumptions on α are you making in each case?

Example 3.108. Solve the integral equation

$$x(t) = 3 + 4\int_0^t e_3(t, \sigma(s))x(s)\Delta s \quad \text{for} \quad t \in \mathbb{T}_0.$$

First note that we can write this equation in the form

$$x(t) = 3 + 4(e_3(\cdot, 0) * x)(t).$$

Taking the Laplace transform of both sides and using Example 3.86, Example 3.91, and the convolution theorem yields

$$\mathcal{L}\{x\}(z) = \frac{3}{z} + \frac{4}{z-3}\mathcal{L}\{x\}(z).$$

It follows that

$$\mathcal{L}\{x\}(z) = \frac{3z-9}{z(z-7)} = \frac{9}{7}\frac{1}{z} + \frac{12}{7}\frac{1}{(z-7)}.$$

Hence by Table 3.2,

$$x(t) = \frac{9}{7} + \frac{12}{7}\sin_7(t, 0).$$

Exercise 3.109. Solve each of the following integral equations:

(i) $x(t) = 2e_3(t, 0) - 5\int_0^t e_4(t, \sigma(s))x(s)\Delta s$;
(ii) $x(t) = \cos_2(t, 0) + 3\int_0^t \sin_3(t, \sigma(s))x(s)\Delta s$;
(iii) $x(t) = e_2(t, 0) + 4\int_0^t x(s)\Delta s$;
(iv) $x(t) = t + 4\int_0^t (t - \sigma(s))x(s)\Delta s$;
(v) $x(t) = h_3(t, 0) + 9\int_0^t h_2(t, \sigma(s))x(s)\Delta s$.

Corollary 3.110. *Assume f and g are each one of the functions $e_\alpha(\cdot, 0)$, $\cosh_\alpha(\cdot, 0)$, $\sinh_\alpha(\cdot, 0)$, $\cos_\alpha(\cdot, 0)$, $\sin_\alpha(\cdot, 0)$, $h_k(\cdot, 0)$, $k \in \mathbb{N}_0$. Then*

$$f * g = g * f.$$

Proof. We only prove three of the desired formulas, and the proof of the others is left to the reader. We first show that

$$e_\alpha(\cdot, 0) * e_\beta(\cdot, 0) = e_\beta(\cdot, 0) * e_\alpha(\cdot, 0).$$

To see this let

$$x := e_\alpha(\cdot, 0) * e_\beta(\cdot, 0) \quad \text{and} \quad y := e_\beta(\cdot, 0) * e_\alpha(\cdot, 0).$$

Note that $x(0) = 0$ and so, by the variation of constants formula, x is the solution of the IVP

$$x^\Delta - \alpha x = e_\beta(\cdot, 0), \quad x(0) = 0.$$

Likewise, since $y(0) = 0$, we have by the variation of constants formula that y is the solution of the IVP

$$y^\Delta - \beta y = e_\alpha(\cdot, 0), \quad y(0) = 0.$$

Note that

$$x^\Delta(0) = \alpha x(0) + e_\beta(0,0) = 1$$

and similarly $y^\Delta(0) = 1$. It then can be shown that x and y are both solutions of the same IVP

$$x^{\Delta\Delta} - 2(\alpha + \beta)x^\Delta + \alpha\beta x = 0, \quad x(0) = 0, \quad x^\Delta(0) = 1.$$

Since this IVP has a unique solution we get the desired result

$$(e_\alpha(\cdot, 0) * e_\beta(\cdot, 0))(t) = (e_\beta(\cdot, 0) * e_\alpha(\cdot, 0))(t)$$

for all $t \in \mathbb{T}_0$. Next consider

$$
\begin{aligned}
e_\alpha(\cdot, 0) * \cosh_\beta(\cdot, 0) &= e_\alpha(\cdot, 0) * \left(\frac{e_\beta(\cdot, 0) + e_{-\beta}(\cdot, 0)}{2} \right) \\
&= \frac{1}{2} \left(e_\alpha(\cdot, 0) * e_\beta(\cdot, 0) + e_\alpha(\cdot, 0) * e_{-\beta}(\cdot, 0) \right) \\
&= \frac{1}{2} \left(e_\beta(\cdot, 0) * e_\alpha(\cdot, 0) + e_{-\beta}(\cdot, 0) * e_\alpha(\cdot, 0) \right) \\
&= \left(\frac{e_\beta(\cdot, 0) + e_{-\beta}(\cdot, 0)}{2} \right) * e_\alpha(\cdot, 0) \\
&= \cosh_\beta(\cdot, 0) * e_\alpha(\cdot, 0).
\end{aligned}
$$

Finally we show that

$$e_\alpha(\cdot, 0) * h_n(\cdot, 0) = h_n(\cdot, 0) * e_\alpha(\cdot, 0).$$

Let

$$x := e_\alpha(\cdot, 0) * h_n(\cdot, 0) \quad \text{and} \quad y := h_n(\cdot, 0) * e_\alpha(\cdot, 0).$$

Since $x(0) = 0$, we have by the variation of constants formula from Theorem 2.77 that x is the solution of the IVP

$$x^\Delta - \alpha x = h_n(t, 0), \quad x(0) = 0.$$

Therefore,

$$x^{\Delta^{i+1}}(t) - \alpha x^{\Delta^i}(t) = h_{n-i}(t, 0)$$

and hence

$$x^{\Delta^i}(0) = 0, \quad 0 \le i \le n, \quad x^{\Delta^{n+1}}(0) = 1.$$

It follows that x is the solution of the IVP

$$x^{\Delta^{n+2}} - \alpha x^{\Delta^{n+1}} = 0, \quad x^{\Delta^i}(0) = 0, \quad 0 \le i \le n, \quad x^{\Delta^{n+1}}(0) = 1.$$

Using Theorem 1.117 it can be shown that

$$y^{\Delta^i}(0) = 0 \quad \text{for} \quad 0 \le i \le n.$$

Hence by the variation of constants formula (see Theorem 5.119), y is the solution of the IVP

$$y^{\Delta^{n+1}} = e_\alpha(t,0), \quad y^{\Delta^i}(0) = 0, \quad 0 \le i \le n.$$

This implies $y^{\Delta^{n+1}}(0) = 1$ and

$$y^{\Delta^{n+2}} - \alpha y^{\Delta^{n+1}} = 0.$$

Hence we have shown that y is the solution of the IVP

$$y^{\Delta^{n+2}} - \alpha y^{\Delta^{n+1}} = 0, \quad y^{\Delta^i}(0) = 0, \quad 0 \le i \le n, \quad y^{\Delta^{n+1}}(0) = 1.$$

Since x and y satisfy the same IVP, we get the desired result. $\square$

Exercise 3.111. Prove that

$$e_\alpha(\cdot,0) * \sinh_\beta(\cdot,0) = \sinh_\beta(\cdot,0) * e_\alpha(\cdot,0).$$

Theorem 3.112. *The following hold:*

(i) *If $\alpha \ne \beta$, then*

$$(e_\alpha(\cdot,0) * e_\beta(\cdot,0))(t) = \frac{1}{\beta - \alpha}[e_\beta(t,0) - e_\alpha(t,0)].$$

(ii) $(e_\alpha(\cdot,0) * e_\alpha(\cdot,0))(t) = e_\alpha(t,0) \int_0^t \frac{1}{1+\alpha\mu(s)} \Delta s.$

(iii) *If $\alpha \ne 0$, then*

$$(e_\alpha(\cdot,0) * h_k(\cdot,0))(t) = \frac{1}{\alpha^{k+1}} e_\alpha(t,0) - \sum_{j=0}^{k} \frac{1}{\alpha^{k+1-j}} h_j(t,0).$$

(iv) *If $\alpha^2 + \beta^2 > 0$, then*

$$(e_\alpha(\cdot,0) * \sin_\beta(\cdot,0))(t) = \frac{\beta e_\alpha(t,0)}{\alpha^2 + \beta^2} - \frac{\alpha \sin_\beta(t,0)}{\alpha^2 + \beta^2} - \frac{\beta \cos_\beta(t,0)}{\alpha^2 + \beta^2}.$$

(v) *If $\alpha^2 + \beta^2 > 0$, then*

$$(e_\alpha(\cdot,0) * \cos_\beta(\cdot,0))(t) = \frac{\alpha e_\alpha(t,0)}{\alpha^2 + \beta^2} + \frac{\beta \sin_\beta(t,0)}{\alpha^2 + \beta^2} - \frac{\alpha \cos_\beta(t,0)}{\alpha^2 + \beta^2}.$$

(vi) *If $\alpha \ne 0$, $\alpha \ne \beta$, then*

$$(\sin_\alpha(\cdot,0) * \sin_\beta(\cdot,0))(t) = \frac{\alpha}{(\alpha^2 - \beta^2)} \sin_\beta(t,0) - \frac{\beta}{(\alpha^2 - \beta^2)} \sin_\alpha(t,0).$$

(vii) *If $\alpha \ne 0$, $\alpha \ne \beta$, then*

$$(\cos_\alpha(\cdot,0) * \cos_\beta(\cdot,0))(t) = -\frac{\beta}{(\alpha^2 - \beta^2)} \sin_\beta(t,0) + \frac{\alpha}{(\alpha^2 - \beta^2)} \sin_\alpha(t,0).$$

(viii) *If $\alpha \ne 0$, $\alpha \ne \beta$, then*

$$(\sin_\alpha(\cdot,0) * \cos_\beta(\cdot,0))(t) = \frac{\alpha}{(\alpha^2 - \beta^2)} \cos_\beta(t,0) - \frac{\alpha}{(\alpha^2 - \beta^2)} \cos_\alpha(t,0).$$

(ix) *If $\alpha \ne 0$, then*

$$(\sin_\alpha(\cdot,0) * \sin_\alpha(\cdot,0))(t) = \frac{1}{\alpha} \sin_\alpha(t,0) - \frac{1}{2}t \cos_\alpha(t,0).$$

(x) *If $\alpha \ne 0$, then*

$$(\cos_\alpha(\cdot,0) * \cos_\alpha(\cdot,0))(t) = \frac{1}{\alpha} \sin_\alpha(t,0) + \frac{1}{2}t \cos_\alpha(t,0).$$

(xi) *If $k \geq 0$, then*

$$(\sin_\alpha(\cdot,0) * h_k(\cdot,0))(t)$$

$$= \begin{cases} (-1)^{\frac{(k+1)(k+2)}{2}} \frac{1}{\alpha^{k+1}} \cos_\alpha(t,0) + \sum_{j=0}^{\frac{k}{2}} (-1)^j \frac{h_{k-2j}(t,0)}{\alpha^{2j+1}} & \text{if } k \text{ is even} \\ (-1)^{\frac{(k+1)(k+2)}{2}} \frac{1}{\alpha^{k+1}} \sin_\alpha(t,0) + \sum_{j=0}^{\frac{k-1}{2}} (-1)^j \frac{h_{k-2j}(t,0)}{\alpha^{2j+1}} & \text{if } k \text{ is odd.} \end{cases}$$

(xii) *If $k \geq 0$, then*

$$(\cos_\alpha(\cdot,0) * h_k(\cdot,0))(t)$$

$$= \begin{cases} (-1)^{\frac{k(k+1)}{2}} \frac{1}{\alpha^{k+1}} \sin_\alpha(t,0) + \sum_{j=0}^{\frac{k-2}{2}} (-1)^j \frac{h_{k-2j-1}(t,0)}{\alpha^{2j+2}}, & \text{if } k \text{ is even} \\ (-1)^{\frac{k(k+1)}{2}} \frac{1}{\alpha^{k+1}} \cos_\alpha(t,0) + \sum_{j=0}^{\frac{k-1}{2}} (-1)^j \frac{h_{k-2j-1}(t,0)}{\alpha^{2j+2}}, & \text{if } k \text{ is odd.} \end{cases}$$

Proof. We will only prove (i) of this theorem. By the variation of constants formula the solution of the IVP

$$y^\Delta - \alpha y = e_\beta(t,0), \quad y(0) = 0$$

is given by

$$y(t) = \int_0^t e_\alpha(t, \sigma(s)) e_\beta(s,0) \Delta s = (e_\alpha(\cdot,0) * e_\beta(\cdot,0))(t).$$

Solving this IVP we get

$$(e_\alpha(\cdot,0) * e_\beta(\cdot,0))(t) = \frac{1}{\beta - \alpha}[e_\beta(t,0) - e_\alpha(t,0)],$$

which is the desired result. $\square$

Exercise 3.113. Prove Theorem 3.112 (ii) and (iv).

Example 3.114. Use Laplace transforms to solve the IVP

$$x^{\Delta\Delta} + 4x = e_3(t,0), \quad x(0) = x^\Delta(0) = 0.$$

Let x be the solution of the given IVP. Then taking the Laplace transform of both sides of the given dynamic equation gives

$$z^2 \mathcal{L}\{x\}(z) - zx(0) - x^\Delta(0) + 4\mathcal{L}\{x\}(z) = \frac{1}{z - 3}.$$

Hence

$$\mathcal{L}\{x\}(z) = \frac{1}{z - 3} \frac{1}{z^2 + 4} = \frac{1}{2} \frac{1}{z - 3} \frac{2}{z^2 + 4}.$$

From the convolution theorem (Theorem 3.106),

$$x(t) = \frac{1}{2}(e_3(\cdot,0) * \sin_2(\cdot,0))(t).$$

Using Theorem 3.112 we get

$$x(t) = \frac{1}{13} e_3(t,0) - \frac{3}{26} \sin_2(t,0) - \frac{1}{13} \cos_2(t,0).$$

Exercise 3.115. Use Laplace transforms, the convolution theorem, and Theorem 3.112 to solve the following initial value problems:

(i) $x^{\Delta\Delta} + 9x = \sin_2(t,0)$, $x(0) = x^\Delta(0) = 0$;
(ii) $x^{\Delta\Delta} + 16x = \sin_4(t,0)$, $x(0) = x^\Delta(0) = 0$;
(iii) $x^{\Delta\Delta} - 2x^\Delta + x = 9e_4(t,0)$, $x(0) = 0$, $x^\Delta(0) = 3$;
(iv) $x^{\Delta\Delta} - x^\Delta - 2x = 3e_2(t,0)$, $x(0) = 0$, $x^\Delta(0) = 1$;

Table 3.2. Laplace Transforms on $\mathbb{T}_0$

$x(t)$	$\mathcal{L}x(z)$
1	$\frac{1}{z}$
t	$\frac{1}{z^2}$
$h_k(t,0),\ k \geq 0$	$\frac{1}{z^{k+1}}$
$e_\alpha(t,0)$	$\frac{1}{z-\alpha}$
$\cosh_\alpha(t,0),\ \alpha > 0$	$\frac{z}{z^2-\alpha^2}$
$\sinh_\alpha(t,0),\ \alpha > 0$	$\frac{\alpha}{z^2-\alpha^2}$
$\cos_\alpha(t,0)$	$\frac{z}{z^2+\alpha^2}$
$\sin_\alpha(t,0)$	$\frac{\alpha}{z^2+\alpha^2}$
$u_a(t),\ a \in \mathbb{T}_0$	$\frac{1}{z}e_{\ominus z}(a,0)$
$e_\alpha(t,0)\sin_{\frac{\beta}{1+\mu\alpha}}(t,0)$	$\frac{\beta}{(z-\alpha)^2+\beta^2}$
$e_\alpha(t,0)\cos_{\frac{\beta}{1+\mu\alpha}}(t,0)$	$\frac{z-\alpha}{(z-\alpha)^2+\beta^2}$
$e_\alpha(t,0)\sinh_{\frac{\beta}{1+\mu\alpha}}(t,0)$	$\frac{\beta}{(z-\alpha)^2-\beta^2}$
$e_\alpha(t,0)\cosh_{\frac{\beta}{1+\mu\alpha}}(t,0)$	$\frac{z-\alpha}{(z-\alpha)^2-\beta^2}$

(v) $x^{\Delta\Delta} - 3x^\Delta = 3e_3(t,0)$, $x(0) = 0$, $x^\Delta(0) = 1$;

(vi) $x^{\Delta\Delta} + 25x = h_2(t,0)$, $x(0) = x^\Delta(0) = 0$;

(vii) $x^{\Delta\Delta} + x = h_3(t,0)$, $x(0) = x^\Delta(0) = 0$.

3.11. Notes and References

We discussed second order linear dynamic equations in this chapter. We introduced two different forms, (3.1) and (3.6) of second order linear dynamic equations. The first form (3.1) is analogous to the usual differential equations case and the second form (3.6) involves the shifts σ. However, in many situations, (3.6) is the "more natural" form. For example, Abel's theorem is "easier" if one considers (3.6) (compare Theorems 3.10 and 3.13). We defined trigonometric and hyperbolic functions that are of use when trying to solve (3.1), but on the other hand, in Section 3.6 we presented "alternative" trigonometric and hyperbolic functions that can be used to solve equations of the form (3.6). There is a certain trade-off: The first kind of trigonometric (hyperbolic) functions satisfy differential equations as in classical analysis, but their squares do not add up to 1 (the absolute value of the difference of the squares is not 1). The situation is precisely reversed for the second kind of trigonometric and hyperbolic functions.

Trigonometric and hyperbolic functions have been defined in Stefan Hilger [**164**] and [**90**]. Most of the results from Section 3.6 are contained in Zdeněk Pospíšil [**233**]. In either case we are able to explicitly give solutions of initial value problems. If the characteristic equation has a double root, a reduction of order technique, which is based on Abel's theorem, is used. Sometimes the method of factoring, previously unpublished but following closely the classical differential equations case, is of help. Nonhomogeneous equations are solved using variation of parameters. The method of undetermined coefficients, also called the annihilator method, is sometimes superior to the variation of parameters technique, and is also presented. While previously unpublished, this method also follows closely the classical differential equations case. In Section 3.7 we treat the case of Euler–Cauchy equations. All of those results are taken from the authors' paper [**90**].

Laplace transforms are introduced by Stefan Hilger in [**164**], but in a form that tries to combine the (continuous) Laplace transform and the discrete Z-transform. Jury [**178**] is a good source of information on the Z-transform, and for other basic results on this transform, see [**125, 191**]. The approach by Stefan Hilger manages to unify Laplace and Z-transforms, however, it gives rather complicated formulas and applies only to the two time scales $\mathbb{R}$ and $\mathbb{Z}$. Our approach in the last section is taken from our paper [**91**]. We are able to define the Laplace transform for arbitrary time scales, however, its discrete special case is not the exact Z-transform but a modification of it (for a discrete transform that is the same as our transform for $\mathbb{T} = \mathbb{Z}$ see the "Honors Thesis" by Rafe Donahue [**111**], written under the supervision of Paul Eloe). The advantage of our transform is that it applies to all time scales, and tables for all time scales are the same.

CHAPTER 4

Self-Adjoint Equations

4.1. Preliminaries and Examples

In this chapter we are concerned with the self-adjoint dynamic equation of second order

$$(4.1) \qquad Lx = 0, \quad \text{where} \quad Lx(t) = (px^\Delta)^\Delta(t) + q(t)x^\sigma(t).$$

Throughout we assume that $p, q \in C_{rd}$ and

$$p(t) \neq 0 \quad \text{for all} \quad t \in \mathbb{T}.$$

Define the set $\mathbb{D}$ to be the set of all functions $x : \mathbb{T} \to \mathbb{R}$ such that $x^\Delta : \mathbb{T}^\kappa \to \mathbb{R}$ is continuous and such that $(px^\Delta)^\Delta : \mathbb{T}^{\kappa^2} \to \mathbb{R}$ is rd-continuous. A function $x \in \mathbb{D}$ is then said to be a solution of (4.1) provided $Lx(t) = 0$ holds for all $t \in \mathbb{T}^{\kappa^2}$.

Exercise 4.1. Show that $L : \mathbb{D} \to C_{rd}$ defined by (4.1) is a linear operator, i.e.,

$$L(\alpha x_1 + \beta x_2) = \alpha L(x_1) + \beta L(x_2) \quad \text{for all} \quad x_1, x_2 \in \mathbb{D} \text{ and } \alpha, \beta \in \mathbb{R}.$$

Example 4.2. For the two cases $\mathbb{T} = \mathbb{R}$ and $\mathbb{T} = \mathbb{Z}$, we have the following:

(i) If $\mathbb{T} = \mathbb{R}$, then (4.1) is the self-adjoint differential equation

$$\frac{d}{dt}(px')(t) + q(t)x(t) = 0,$$

which has been studied extensively over the years, see, e.g., the books [**193, 238**] and the references given there.

(ii) If $\mathbb{T} = \mathbb{Z}$, then (4.1) is the self-adjoint difference equation

$$\Delta[p(t)\Delta x(t)] + q(t)x(t + 1) = 0.$$

See the books [**30, 191**] and the references there for results concerning this self-adjoint difference equation.

The next two exercises are due to Douglas Anderson.

Exercise 4.3. Find a self-adjoint scalar equation with $p(t) \equiv 1$ that has solution $x(t) = e^{at}$ for $t \in \overline{q^\mathbb{Z}}$.

Exercise 4.4. Find a linear solution to the second order self-adjoint equation

$$[(t + \alpha)x^\Delta]^\Delta - \left(\frac{1}{qt + \beta}\right)x^\sigma = 0$$

for $t \in \overline{q^\mathbb{Z}}$.

We next state the theorem concerning the existence, uniqueness, and extendability of solutions of initial value problems for $Lx = f(t)$. It is a special case of Corollary 7.12 (see also Example 7.19) which itself is proved using the global existence and uniqueness result from Theorem 8.20.

Theorem 4.5. *If $f \in C_{rd}$, $t_0 \in \mathbb{T}$, and x_0, x_0^Δ are given constants, then the initial value problem*

$$Lx = f(t), \quad x(t_0) = x_0, \quad x^\Delta(t_0) = x_0^\Delta$$

has a unique solution that exists on the whole time scale $\mathbb{T}$.

We now give three examples that appear in [**131**].

Example 4.6. Let $\mathbb{T} = [0, 1] \cup \mathbb{N}$, $p(t) \equiv 1$ and $q(t) \equiv \pi^2$ for all $t \in \mathbb{T}$. Then the solution of the initial value problem

$$Lx = 0, \quad x(0) = 0, \quad x^\Delta(0) = 1$$

is given by

$$x(t) = \begin{cases} \frac{1}{\pi}\sin(\pi t) & \text{if} \quad 0 \le t \le 1 \\ -1 & \text{if} \quad t = 2 \\ (2 - \pi^2)x(t-1) - x(t-2) & \text{if} \quad t \ge 3. \end{cases}$$

It is instructive to see how we got this solution. First note that $x^\Delta(0) = 1$ means that the right derivative, $D_r x(t)$, of the solution x at zero is one. Hence on the interval $[0, 1)$ we solve the initial value problem

$$x'' + \pi^2 x = 0, \quad x(0) = 0, \quad D_r x(0) = 1.$$

This gives us that

$$x(t) = \frac{1}{\pi}\sin(\pi t) \quad \text{for} \quad t \in [0, 1).$$

By the continuity of x we get that this last expression for x is valid on the closed interval $[0, 1]$. Next we note that since x^Δ is continuous,

$$\begin{aligned} x^\Delta(1) &= \lim_{t \to 1^-} x^\Delta(t) \\ &= \lim_{t \to 1^-} x'(t) \\ &= \lim_{t \to 1^-} \cos(\pi t) \\ &= -1. \end{aligned}$$

Then

$$-1 = x^\Delta(1) = \Delta x(1) = x(2) - x(1) = x(2).$$

For $t \ge 1$ the dynamic equation gives us

$$x^{\Delta\Delta}(t) = -\pi^2 x^\sigma(t),$$

which is the same as

$$\Delta^2 x(t) = -\pi^2 x(t+1) \quad \text{for} \quad t \ge 1.$$

Solving this last equation for $x(t+2)$ gives us that

$$x(t+2) = (2 - \pi^2)x(t+1) - x(t) \quad \text{for} \quad t \ge 1.$$

Therefore

$$x(t) = (2 - \pi^2)x(t-1) - x(t-2) \quad \text{for} \quad t \ge 3.$$

Figure 4.1. Graph of the Solution from Example 4.7

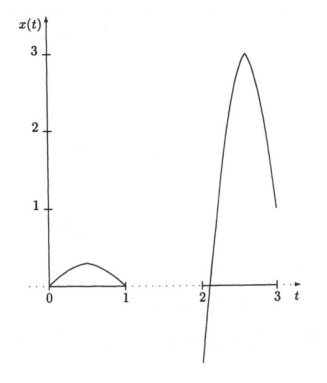

Since we know that $x(1) = 0$ and $x(2) = -1$, the above equation uniquely determines $x(t)$ for $t \geq 3$. Note that our solution "pieces" together a solution of a differential equation and a solution of a difference equation.

Example 4.7. Consider the dynamic equation

$$x^{\Delta\Delta} + \pi^2 x^\sigma = 0$$

on the time scale

$$\mathbb{T} = [0,1] \cup [2,3].$$

Let x be the solution of the above equation satisfying the initial conditions $x(0) = 0$, $x^\Delta(0) = 1$. Then

$$x(t) = \begin{cases} \frac{1}{\pi} \sin \pi t & \text{if} \quad 0 \leq t \leq 1 \\ -\cos \pi t + \frac{\pi^2 - 1}{\pi} \sin \pi t & \text{if} \quad 2 \leq t \leq 3. \end{cases}$$

To see this we get as in Example 4.6 that

$$x(t) = \frac{1}{\pi} \sin(\pi t)$$

for $0 \leq t \leq 1$ and $x(2) = -1$. Since x is a solution of the given equation, we get that

$$x^{\Delta\Delta}(1) = -\pi^2 x(2) = \pi^2.$$

This implies that

$$x^\Delta(2) - x^\Delta(1) = \pi^2.$$

Hence

$$D_r x(2) + 1 = \pi^2.$$

This gives us that $D_r x(2) = \pi^2 - 1$. Therefore on $[2,3)$ we solve the IVP

$$x'' + \pi^2 x = 0, \quad x(2) = -1, \quad x'(2) = \pi^2 - 1.$$

Solving this initial value problem we get

$$x(t) = -\cos \pi t + \frac{\pi^2 - 1}{\pi} \sin \pi t \quad \text{for} \quad t \in [2,3).$$

By continuity this last equation is also true for $t = 3$. See Figure 4.1 for a graph of the solution x.

In the next example our dynamic equation leads to a discrete problem with variable step-size.

Example 4.8. Solve the IVP

$$x^{\Delta\Delta} + x^\sigma = 0, \quad x(0) = x_0, \quad x^\Delta(0) = x_0^\Delta$$

on the time scale

$$\mathbb{T} = \bigcup_{k=0}^\infty T_k, \quad \text{where} \quad T_k = \bigcup_{n=1}^\infty \left\{ k + \frac{n-1}{n} \right\}.$$

For $t \in T_0$, this solution is determined by the IVP

$$x\left(\frac{n+1}{n+2}\right) + \frac{1 - 2n(n+1)^2}{n(n+1)^2(n+2)} x\left(\frac{n}{n+1}\right) + \frac{n}{n+2} x\left(\frac{n-1}{n}\right) = 0,$$
$$x(0) = x_0, \quad x^\Delta(0) = x_0^\Delta.$$

In Exercise 4.9 we see that the values of x on T_0 determine the values $x(1)$ and $x^\Delta(1)$. But these initial conditions at $t = 1$ and the equation

$$x\left(1 + \frac{n+1}{n+2}\right) + \frac{1 - 2n(n+1)^2}{n(n+1)^2(n+2)} x\left(1 + \frac{n}{n+1}\right) + \frac{n}{n+2} x\left(1 + \frac{n-1}{n}\right) = 0$$

determine the values of x on T_1. In particular, note that on each set T_k the solution x of the given IVP solves the same difference equation with variable step-size, where for $k \geq 1$ the initial conditions on T_k are determined by the values of the solution on T_{k-1}.

Exercise 4.9. Fill in the details for the above example. Determine the initial conditions at $t = 1$ that the solution of the IVP in Example 4.8 satisfies.

Now we consider an equation given in self-adjoint form (4.1). The p given below in Exercise 4.10 is not necessarily a differentiable function (unless the graininess of the time scale is constant). Hence (4.1) cannot be rewritten in the form (3.1) or (3.6).

Exercise 4.10. Suppose $\alpha \in \mathcal{R}$. Let $\tilde{p} \in C_{rd}$ be nonvanishing. We put

$$p = \frac{\tilde{p}}{\ominus\alpha} \quad \text{and} \quad q = \alpha\tilde{p} - \tilde{p}^\Delta.$$

Show that a fundamental system for (4.1) is given by

$$e_{\ominus\alpha}(t, t_0) \quad \text{and} \quad -p(t_0) \int_{t_0}^t \frac{\alpha(\tau)}{\tilde{p}(\tau)} e_\alpha(\tau, t) e_\alpha(\tau, t_0) \Delta\tau$$

and that the Wronskian of these two solutions is

$$p(t_0).$$

Example 4.11. Suppose $\alpha \in \mathcal{R}$. We let $\bar{p} = e_\alpha(\cdot, t_0)$ so that $q = 0$ in Exercise 4.10. Hence we consider the equation

$$(4.2) \qquad \left(\frac{e_\alpha^\sigma(\cdot, t_0)}{\alpha} x^\Delta \right)^\Delta = 0.$$

By Exercise 4.10, a fundamental system for (4.2) is given by

$$e_{\ominus\alpha}(\cdot, t_0) \quad \text{and} \quad \frac{e_{\ominus\alpha}(\cdot, t_0) - 1}{(\ominus\alpha)(t_0)}.$$

We now address the question under which circumstances an equation of the form (3.1) or (3.6) can be rewritten in self-adjoint form (4.1).

Theorem 4.12. *If $a, b \in C_{rd}$ and*

$$(4.3) \qquad 1 - a(t)\mu(t) + b(t)\mu^2(t) \neq 0 \quad \text{for all} \quad t \in \mathbb{T}^{\kappa^2},$$

then the second order dynamic equation

$$(4.4) \qquad x^{\Delta\Delta} + a(t)x^\Delta + b(t)x = 0$$

can be written in self-adjoint form (4.1), where

$$(4.5) \qquad p(t) = e_\alpha(t, t_0)$$

with $t_0 \in \mathbb{T}^\kappa$,

$$(4.6) \qquad \alpha(t) = \frac{a(t) - \mu(t)b(t)}{1 - a(t)\mu(t) + b(t)\mu^2(t)},$$

and

$$(4.7) \qquad q(t) = e_\alpha^\sigma(t, t_0)b(t) = [1 + \mu(t)\alpha(t)]p(t)b(t).$$

Proof. Assume that (4.3) holds and that α is given by (4.6). Then

$$1 + \alpha(t)\mu(t) = \frac{1}{1 - a(t)\mu(t) + b(t)\mu^2(t)} \neq 0,$$

and hence $\alpha \in \mathcal{R}$ so that p defined by (4.5) exists. Replacing $x = x^\sigma - \mu x^\Delta$ in the third term on the left side of equation (4.4), we obtain

$$x^{\Delta\Delta}(t) + [a(t) - \mu(t)b(t)] x^\Delta(t) + b(t)x^\sigma(t) = 0.$$

Multiplying both sides by $e_\alpha^\sigma(t, t_0)$, we get the equation

$$e_\alpha^\sigma(t, t_0)x^{\Delta\Delta}(t) + e_\alpha^\sigma(t, t_0) [a(t) - \mu(t)b(t)] x^\Delta(t) + e_\alpha^\sigma(t, t_0)b(t)x^\sigma(t) = 0.$$

The coefficient of $x^\Delta(t)$ is

$$\begin{aligned}
e_\alpha^\sigma(t, t_0) [a(t) - \mu(t)b(t)] &= [1 + \mu(t)\alpha(t)] e_\alpha(t, t_0) [a(t) - \mu(t)b(t)] \\
&= \frac{a(t) - \mu(t)b(t)}{1 - a(t)\mu(t) + b(t)\mu^2(t)} e_\alpha(t, t_0) \\
&= \alpha(t)e_\alpha(t, t_0) \\
&= e_\alpha^\Delta(t, t_0).
\end{aligned}$$

Hence equation (4.4) is equivalent to the equation

$$
\begin{aligned}
0 &= e_\alpha^\sigma(t, t_0) x^{\Delta\Delta}(t) + e_\alpha^\Delta(t, t_0) x^\Delta(t) + e_\alpha^\sigma(t, t_0) b(t) x^\sigma(t) \\
&= \left[e_\alpha(\cdot, t_0) x^\Delta \right]^\Delta (t) + e_\alpha^\sigma(t, t_0) b(t) x(\sigma(t)).
\end{aligned}
$$

This equation is in self-adjoint form with p given by (4.5) and q given by (4.7). $\square$

Example 4.13. Here we use Theorem 4.12 to write the dynamic equation

$$
x^{\Delta\Delta} - 5x^\Delta + 6x = 0 \quad \text{on} \quad \mathbb{T} = \mathbb{R}
$$

in self-adjoint form. Here $a(t) \equiv -5$, $b(t) \equiv 6$, and $1 - a(t)\mu(t) + b(t)\mu^2(t) = 1 \neq 0$ so Theorem 4.12 can be applied. By (4.6), $\alpha(t) = -5$. Therefore, taking $t_0 = 0$, we get from (4.5) that

$$
p(t) = e_\alpha(t, 0) = e^{-5t}
$$

and from (4.7)

$$
q(t) = [1 + \mu(t)\alpha(t)] p(t) b(t) = 6e^{-5t}.
$$

Hence the self-adjoint form of the above differential equation is

$$
\left(e^{-5t} x' \right)' + 6e^{-5t} x = 0.
$$

Example 4.14. Here we use Theorem 4.12 to write the difference equation

$$
x(t + 2) - 5x(t + 1) + 6x(t) = 0 \quad \text{on} \quad \mathbb{T} = \mathbb{Z}
$$

in self-adjoint form. We can rewrite this equation in the form

$$
\Delta^2 x(t) - 3\Delta x(t) + 2x(t) = 0.
$$

Here $a(t) \equiv -3$, $b(t) \equiv 2$, and $1 - a(t)\mu(t) + b(t)\mu^2(t) = 6 \neq 0$ so Theorem 4.12 can be applied. By (4.6), $\alpha(t) = -\frac{5}{6}$. Therefore, taking $t_0 = 0$, we get from (4.5) that

$$
p(t) = e_\alpha(t, 0) = (1 + \alpha)^t = \left(\frac{1}{6} \right)^t
$$

and from (4.7)

$$
q(t) = e_\alpha^\sigma(t) b(t) = 2 \left(\frac{1}{6} \right)^{t+1} = \frac{1}{3} \left(\frac{1}{6} \right)^t.
$$

Hence the self-adjoint form of the above difference equation is

$$
\Delta \left[\left(\frac{1}{6} \right)^t \Delta x(t) \right] + \frac{1}{3} \left(\frac{1}{6} \right)^t x(t + 1) = 0.
$$

Example 4.15. Here we use Theorem 4.12 to write the dynamic equation

$$
x^{\Delta\Delta} + 4x = 0 \quad \text{on an arbitrary time scale} \quad \mathbb{T}
$$

in self-adjoint form. In this case we have $a(t) \equiv 0$ and $b(t) \equiv 4$. This implies $1 - a(t)\mu(t) + b(t)\mu^2(t) = 1 + 4\mu^2(t) > 0$ so Theorem 4.12 can be applied. By (4.6), $\alpha(t) = \frac{-4\mu(t)}{1+4\mu^2(t)}$. Therefore, with arbitrary $t_0 \in \mathbb{T}$, we get from (4.5) that

$$
p(t) = e_\alpha(t, t_0) = e_{-4\mu/(1+4\mu^2)}(t, t_0)
$$

and from (4.7)

$$
q(t) = [1 + \mu(t)\alpha(t)] p(t) b(t) = \frac{4}{1 + 4\mu^2(t)} e_{-4\mu/(1+4\mu^2)}(t, t_0).
$$

Hence the self-adjoint form of the above dynamic equation is

$$\left(e_{-4\mu/(1+4\mu^2)}(t,t_0)x^{\Delta}\right)^{\Delta} + \frac{4}{1+4\mu^2(t)}e_{-4\mu/(1+4\mu^2)}(t,t_0)x^{\sigma} = 0.$$

Exercise 4.16. Use Theorem 4.12 to write each of the following equations in self-adjoint form:

(i) $x^{\Delta\Delta} + \frac{2}{\mu(t)}x^{\Delta} + \frac{3}{\mu^2(t)}x = 0$, where $\mu(t) \neq 0$ for all $t \in \mathbb{T}$;

(ii) $x^{\Delta\Delta} - 3x^{\Delta} + 5x = 0$, where $\mathbb{T} = \mathbb{R}$;

(iii) $x(t+2) - x(t+1) + 2x(t) = 0$, where $\mathbb{T} = \mathbb{Z}$;

(iv) $x^{\Delta\Delta} + 9x = 0$, where $\mathbb{T}$ is arbitrary.

Theorem 4.17. *If $a \in \mathcal{R}$, then the second order dynamic equation*

$$x^{\Delta\Delta} + a(t)x^{\Delta^{\sigma}} + b(t)x^{\sigma} = 0$$

can be written in self-adjoint form (4.1), where

(4.8) $$p = e_a(\cdot, t_0) \quad and \quad q = bp.$$

Proof. We rewrite

$$x^{\Delta\Delta} + ax^{\Delta^{\sigma}} + bx^{\sigma} = 0$$

equivalently as

$$\begin{aligned}
0 &= e_a(\cdot,t_0)\left\{x^{\Delta\Delta} + ax^{\Delta^{\sigma}} + bx^{\sigma}\right\} \\
&= e_a(\cdot,t_0)x^{\Delta\Delta} + e_a^{\Delta}(\cdot,t_0)x^{\Delta^{\sigma}} + be_a(\cdot,t_0)x^{\sigma} \\
&= (e_a(\cdot,t_0)x^{\Delta})^{\Delta} + be_a(\cdot,t_0)x^{\sigma} \\
&= (px^{\Delta})^{\Delta} + qx^{\sigma},
\end{aligned}$$

where p and q are defined as in (4.8). $\square$

Exercise 4.18. Use Theorem 4.17 to write each of the following equations in self-adjoint form:

(i) $x^{\Delta\Delta} + 2x^{\Delta^{\sigma}} + 3x^{\sigma} = 0$;

(ii) $x^{\Delta\Delta} + 2x^{\Delta^{\sigma}} + e_2(t_0,t)x^{\sigma} = 0$;

(iii) $x^{\Delta\Delta} + (5 + 6\mu(t))x^{\Delta^{\sigma}} + 3e_2(t_0,t)x^{\sigma} = 0$;

(iv) $x^{\Delta\Delta} + \frac{1}{1+2\mu(t)}x^{\Delta^{\sigma}} + e_2(t,t_0)x^{\sigma} = 0$.

The Prüfer transformation (established by H. Prüfer in [235]) has proved to be a useful tool in the qualitative theory of second order Sturm–Liouville differential equations (4.1) (with $\mathbb{T} = \mathbb{R}$). Now we present an extension of this transformation to the time scales case. Let x be a nontrivial solution of (4.1). Then

$$x^2(t) + (p(t)x^{\Delta}(t))^2 \neq 0 \quad \text{for all} \quad t \in \mathbb{T}$$

(note that otherwise x would be identically zero) and we can find, for each t, real numbers $\varrho(t) > 0$ and $\varphi(t)$ with $0 \leq \varphi(t) < 2\pi$ such that the equations

(4.9) $$x(t) = \varrho(t)\sin\varphi(t),$$

(4.10) $$p(t)x^{\Delta}(t) = \varrho(t)\cos\varphi(t)$$

are satisfied for all $t \in \mathbb{T}$. We call (4.9) and (4.10) the *Prüfer transformation*.

Theorem 4.19. *If x is a nontrivial solution of (4.1) and if ϱ and φ are defined by (4.9) and (4.10), then the equations*

$$(4.11) \qquad \varrho^\Delta = \varrho \left\{ \frac{1}{p} \cos\varphi(\sin\varphi)^\sigma - q\sin\varphi(\cos\varphi)^\sigma - \frac{\mu q}{p}\cos\varphi(\cos\varphi)^\sigma \right.$$
$$\left. -(\sin\varphi)^\Delta(\sin\varphi)^\sigma - (\cos\varphi)^\Delta(\cos\varphi)^\sigma \right\},$$

$$(4.12) \qquad (\sin\varphi)^\Delta(\cos\varphi)^\sigma - (\cos\varphi)^\Delta(\sin\varphi)^\sigma = \frac{1}{p}\cos\varphi(\cos\varphi)^\sigma + q\sin\varphi(\sin\varphi)^\sigma$$
$$+\frac{\mu q}{p}\cos\varphi(\sin\varphi)^\sigma$$

hold.

Proof. Use of the product rule for (4.9) yields

$$\begin{aligned}
\varrho^\Delta(\sin\varphi)^\sigma + \varrho(\sin\varphi)^\Delta &= (\varrho\sin\varphi)^\Delta \\
&= x^\Delta \\
&= \frac{1}{p}(px^\Delta) \\
&= \frac{1}{p}\varrho\cos\varphi,
\end{aligned}$$

while doing the same for (4.10) implies

$$\begin{aligned}
\varrho^\Delta(\cos\varphi)^\sigma + \varrho(\cos\varphi)^\Delta &= (\varrho\cos\varphi)^\Delta \\
&= (px^\Delta)^\Delta \\
&= -qx^\sigma \\
&= -q(x + \mu x^\Delta) \\
&= -qx - \frac{\mu q}{p}(px^\Delta) \\
&= -q\varrho\sin\varphi - \frac{\mu q}{p}\varrho\cos\varphi,
\end{aligned}$$

where we have also used that x is a solution of (4.1). Hence we obtain the two equations

$$(4.13) \qquad \varrho^\Delta(\sin\varphi)^\sigma + \varrho(\sin\varphi)^\Delta = \frac{1}{p}\varrho\cos\varphi,$$

$$(4.14) \qquad \varrho^\Delta(\cos\varphi)^\sigma + \varrho(\cos\varphi)^\Delta = -q\varrho\sin\varphi - \frac{\mu q}{p}\varrho\cos\varphi.$$

We now multiply (4.13) by $(\sin\varphi)^\sigma$ and (4.14) by $(\cos\varphi)^\sigma$ and add the resulting equations to obtain (4.11). To verify (4.12), we multiply (4.13) by $(\cos\varphi)^\sigma$ and (4.14) by $-(\sin\varphi)^\sigma$ and add the resulting equations. Dividing the obtained equation by $\varrho > 0$ directly yields (4.12). □

Remark 4.20. Theorem 4.19 suggests a method to construct solutions of the self-adjoint dynamic equation (4.1): Observe that the dynamic equation (4.12) for φ is independent of ϱ. Of course it might be difficult to solve this equation, but once a solution of (4.12) is obtained, the linear dynamic equation (4.11) for ϱ is readily solved.

The following concept of the Cauchy function will help for the study of self-adjoint equations.

Definition 4.21. We say that $x : \mathbb{T} \times \mathbb{T}^{\kappa^2} \to \mathbb{R}$ is the *Cauchy function* for (4.1) provided for each fixed $s \in \mathbb{T}^{\kappa^2}$, $x(\cdot, s)$ is the solution of the initial value problem

$$Lx(\cdot, s) = 0, \quad x(\sigma(s), s) = 0, \quad x^\Delta(\sigma(s), s) = \frac{1}{p(\sigma(s))}.$$

It is easy to verify the following example.

Example 4.22. If $q = 0$, then the Cauchy function for $(px^\Delta)^\Delta = 0$ is given by

$$x(t, s) = \int_{\sigma(s)}^t \frac{1}{p(\tau)} \Delta\tau \quad \text{for all} \quad t \in \mathbb{T}, \ s \in \mathbb{T}^{\kappa^2}.$$

In particular

$$x(t, s) = t - \sigma(s)$$

is the Cauchy function for $x^{\Delta\Delta} = 0$.

Exercise 4.23. Verify the results in Example 4.22.

We will use Theorem 1.117 (i) to prove the following variation of constants formula.

Theorem 4.24 (Variation of Constants Formula). *Assume $f \in C_{rd}$ and let $x(t, s)$ be the Cauchy function for (4.1). Then*

$$x(t) = \int_a^t x(t, s) f(s) \Delta s$$

is the solution of the initial value problem

$$Lx = f(t), \quad x(a) = 0, \quad x^\Delta(a) = 0.$$

Proof. Let $x(t, s)$ be the Cauchy function for (4.1) and set

$$x(t) = \int_a^t x(t, s) f(s) \Delta s.$$

Note that $x(a) = 0$. Using Theorem 1.117 (i), we get that

$$\begin{aligned}
x^\Delta(t) &= \int_a^t x^\Delta(t, s) f(s) \Delta s + x(\sigma(t), t) f(t) \\
&= \int_a^t x^\Delta(t, s) f(s) \Delta s
\end{aligned}$$

(where $^\Delta$ denotes the derivative with respect to the first variable) so that $x^\Delta(a) = 0$. From

$$p(t) x^\Delta(t) = \int_a^t p(t) x^\Delta(t, s) f(s) \Delta s$$

we conclude by Theorem 1.117 (i) that

$$\begin{aligned}
(px^\Delta)^\Delta(t) &= \int_a^t [px^\Delta(\cdot, s)]^\Delta(t) f(s) \Delta s + p(\sigma(t)) x^\Delta(\sigma(t), t) f(t) \\
&= \int_a^t [px^\Delta(\cdot, s)]^\Delta(t) f(s) \Delta s + f(t).
\end{aligned}$$

It follows that

$$
\begin{aligned}
Lx(t) &= (px^\Delta)^\Delta(t) + q(t)x(\sigma(t)) \\
&= \int_a^t (px^\Delta(\cdot,s))^\Delta(t)f(s)\Delta s + f(t) + q(t)\int_a^{\sigma(t)} x(\sigma(t),s)f(s)\Delta s \\
&= \int_a^t \left\{ (px^\Delta(\cdot,s))^\Delta(t) + q(t)x(\sigma(t),s) \right\} f(s)\Delta s \\
&\quad + q(t)\mu(t)x(\sigma(t),t)f(t) + f(t) \\
&= \int_a^t Lx(t,s)f(s)\Delta s + f(t) \\
&= f(t),
\end{aligned}
$$

where we have used Theorem 1.75.   □

Exercise 4.25. Use the variation of constants formula in Theorem 4.24 to solve the following initial value problems:

(i) $x^{\Delta\Delta} = 1$, $x(0) = x^\Delta(0) = 0$ for $t \in \mathbb{R}$, for $t \in \mathbb{Z}$, and for $t \in h\mathbb{Z}$;
(ii) $x^{\Delta\Delta} + x^\sigma = t$, $x(0) = x^\Delta(0) = 0$ for $t \in \mathbb{R}$.

Theorem 4.26 (Comparison Theorem for IVPs). *Assume the Cauchy function x for (4.1) satisfies $x(t,s) \geq 0$ for $t \geq \sigma(s)$. If $u, v \in \mathbb{D}$ are functions satisfying*

$$Lu(t) \geq Lv(t) \quad \text{for all } t \in [a,b], \quad u(a) = v(a), \quad u^\Delta(a) = v^\Delta(a),$$

then

$$u(t) \geq v(t) \quad \text{for all} \quad t \in [a, \sigma^2(b)].$$

Proof. Let u and v be as in the statement of this theorem and set

$$w(t) := u(t) - v(t) \quad \text{for all} \quad t \in [a, \sigma^2(b)].$$

Then

$$h(t) := Lw(t) = Lu(t) - Lv(t) \geq 0 \quad \text{for all} \quad t \in [a,b].$$

It follows that w solves the initial value problem

$$Lw(t) = h(t), \quad w(a) = w^\Delta(a) = 0.$$

Hence by the variation of constants formula (Theorem 4.24),

$$
\begin{aligned}
w(t) &= \int_a^t x(t,s)h(s)\Delta s \\
&= \int_a^{\rho(t)} x(t,s)h(s)\Delta s + \int_{\rho(t)}^t x(t,s)h(s)\Delta s \\
&= \int_a^{\rho(t)} x(t,s)h(s)\Delta s + \mu(\rho(t))x(t,\rho(t))h(\rho(t)) \\
&= \int_a^{\rho(t)} x(t,s)h(s)\Delta s \\
&\geq 0,
\end{aligned}
$$

where we also have applied Theorem 1.75.   □

Definition 4.27. If $x, y : \mathbb{T} \to \mathbb{R}$ are differentiable on $\mathbb{T}^\kappa$, then we define the *Wronskian* of x and y by

$$W(x,y)(t) = \det \begin{pmatrix} x(t) & y(t) \\ x^\Delta(t) & y^\Delta(t) \end{pmatrix} \quad \text{for} \quad t \in \mathbb{T}^\kappa.$$

Lemma 4.28. *If $x, y : \mathbb{T} \to \mathbb{R}$ are differentiable on $\mathbb{T}^\kappa$, then*

$$W(x,y)(t) = \det \begin{pmatrix} x^\sigma(t) & y^\sigma(t) \\ x^\Delta(t) & y^\Delta(t) \end{pmatrix} \quad \text{for} \quad t \in \mathbb{T}^\kappa.$$

Proof. For $t \in \mathbb{T}^\kappa$, we have by Theorem 1.16 (iv)

$$\det \begin{pmatrix} x^\sigma(t) & y^\sigma(t) \\ x^\Delta(t) & y^\Delta(t) \end{pmatrix} = \det \begin{pmatrix} x(t) + \mu(t)x^\Delta(t) & y(t) + \mu(t)y^\Delta(t) \\ x^\Delta(t) & y^\Delta(t) \end{pmatrix}$$

$$= \det \begin{pmatrix} x(t) & y(t) \\ x^\Delta(t) & y^\Delta(t) \end{pmatrix}$$

$$= W(x,y)(t),$$

which gives us the desired result. □

Definition 4.29. If $x, y : \mathbb{T} \to \mathbb{R}$ are differentiable on $\mathbb{T}^\kappa$, then the *Lagrange bracket* of x and y is defined by

$$\{x; y\}(t) = p(t)W(x,y)(t) \quad \text{for} \quad t \in \mathbb{T}^\kappa.$$

Theorem 4.30 (Lagrange Identity). *If $x, y \in \mathbb{D}$, then*

$$x^\sigma(t)Ly(t) - y^\sigma(t)Lx(t) = \{x; y\}^\Delta(t) \quad \text{for} \quad t \in \mathbb{T}^{\kappa^2}.$$

Proof. By the product rule (Theorem 1.20 (iii)), we have

$$\{x; y\}^\Delta = (xpy^\Delta - px^\Delta y)^\Delta$$
$$= x^\sigma(py^\Delta)^\Delta + x^\Delta py^\Delta - y^\sigma(px^\Delta)^\Delta - y^\Delta px^\Delta$$
$$= x^\sigma(py^\Delta)^\Delta - y^\sigma(px^\Delta)^\Delta$$
$$= x^\sigma\{(py^\Delta)^\Delta + qy^\sigma\} - y^\sigma\{(px^\Delta)^\Delta + qx^\sigma\}$$
$$= x^\sigma Ly - y^\sigma Lx$$

on $\mathbb{T}^{\kappa^2}$. □

The following two corollaries are easy consequences of Theorem 4.30.

Corollary 4.31 (Abel's Formula). *If x and y both solve (4.1), then*

$$W(x,y)(t) = \frac{C}{p(t)} \quad \text{for all} \quad t \in \mathbb{T}^\kappa,$$

where C is a constant.

Corollary 4.32. *If x and y both solve (4.1), then either*

$$W(x,y)(t) \equiv 0 \quad \text{for all} \quad t \in \mathbb{T}^\kappa$$

or

$$W(x,y)(t) \neq 0 \quad \text{for all} \quad t \in \mathbb{T}^\kappa.$$

We now state and prove a theorem that gives a formula for the Cauchy function for (4.1).

Theorem 4.33. *If u and v are linearly independent solutions of (4.1), then the Cauchy function $x(t,s)$ for (4.1) is given by*

$$(4.15) \qquad x(t,s) = \frac{u(\sigma(s))v(t) - v(\sigma(s))u(t)}{p(s)[u(s)v^\Delta(s) - u^\Delta(s)v(s)]} \quad \text{for} \quad t \in \mathbb{T},\ s \in \mathbb{T}^\kappa.$$

Proof. Let $y(t,s)$ be defined by the right-hand side of equation (4.15). Then note that for each fixed s, $y(\cdot, s)$ is a linear combination of the solutions u and v and as such is a solution of (4.1). Clearly $y(\sigma(s),s) = 0$. Also note that

$$y^\Delta(t,s) = \frac{u(\sigma(s))v^\Delta(t) - v(\sigma(s))u^\Delta(t)}{p(s)[u(s)v^\Delta(s) - u^\Delta(s)v(s)]}.$$

Using Abel's formula (Corollary 4.31) we get that

$$y^\Delta(\sigma(s),s) = \frac{W(u,v)(\sigma(s))}{p(s)W(u,v)(s)} = \frac{1}{p(\sigma(s))}.$$

From the uniqueness of solutions of initial value problems (Theorem 4.5) we have that for each fixed s,

$$x(t,s) = y(t,s),$$

which gives us the desired result. $\square$

Exercise 4.34. Let $a \in \mathbb{T}$. Show that $u(t) \equiv 1$ and $v(t) = \int_a^t \frac{1}{p(\tau)}\Delta\tau$ are linearly independent solutions of $(px^\Delta)^\Delta = 0$. Then use Theorem 4.33 to show that the Cauchy function for $(px^\Delta)^\Delta = 0$ for any time scale $\mathbb{T}$ is given by

$$x(t,s) = \int_{\sigma(s)}^t \frac{1}{p(\tau)}\Delta\tau.$$

4.2. The Riccati Equation

Definition 4.35. We say that a solution x of (4.1) has a *generalized zero* at t if

$$x(t) = 0$$

or (and in this case we also sometimes say that the generalized zero is contained in the real interval $(\rho(t),t)$) if t is left-scattered and

$$p(\rho(t))x(\rho(t))x(t) < 0.$$

We say that (4.1) is *disconjugate* on an interval $[a,b]$, if there is no nontrivial solution of (4.1) with two (or more) generalized zeros in $[a,b]$.

Note that disconjugacy of (4.1) on $[a, b]$ implies that the solution $\tilde{x}$ of the IVP

$$Lx = 0, \quad x(a) = 0, \quad x^\Delta(a) = \frac{1}{p(a)}$$

(the so-called *principal solution* of (4.1) at a) satisfies

$$p(t)\tilde{x}(t)\tilde{x}(\sigma(t)) > 0 \quad \text{for all} \quad t \in (a, b]^\kappa.$$

Assume x is a solution of (4.1) with no generalized zeros. If we make the *Riccati substitution*

(4.16)
$$z = \frac{px^\Delta}{x}$$

on $\mathbb{T}^\kappa$, then

(4.17)
$$p + \mu z = p + \mu \frac{px^\Delta}{x} = \frac{p(x + \mu x^\Delta)}{x} = \frac{px^\sigma}{x} > 0$$

on $\mathbb{T}^\kappa$. We have by Theorem 1.20 (v)

$$
\begin{aligned}
z^\Delta &= \frac{x\left(px^\Delta\right)^\Delta - p\left(x^\Delta\right)^2}{xx^\sigma} \\
&= \frac{-qxx^\sigma - p\left(x^\Delta\right)^2}{xx^\sigma} \\
&= -q - \frac{x}{px^\sigma}\left(\frac{px^\Delta}{x}\right)^2 \\
&= -q - \frac{x}{px^\sigma}z^2 \\
&= -q - \frac{z^2}{p + \mu z}
\end{aligned}
$$

on $\mathbb{T}^\kappa$, where we used (4.17) for the last equation. Hence z is a solution of the *Riccati equation*

(4.18) $Rz = 0,$ where $Rz(t) := z^\Delta(t) + q(t) + \dfrac{z^2(t)}{p(t) + \mu(t)z(t)}$

on $\mathbb{T}^\kappa$ satisfying (see (4.17))

(4.19)
$$p(t) + \mu(t)z(t) > 0$$

for all $t \in \mathbb{T}^\kappa$. Hence we have proved the following result.

Theorem 4.36. *Let x be a solution of (4.1). If x has no generalized zeros on $\mathbb{T}$, then z defined as in (4.16) is a solution of the Riccati equation (4.18) on $\mathbb{T}^\kappa$, and (4.19) holds for all $t \in \mathbb{T}^\kappa$.*

Note that if $\mathbb{T} = \mathbb{R}$, then $\mu(t) \equiv 0$ and the Riccati operator defined in (4.18) is the well-studied Riccati operator defined by

$$Rz(t) = z'(t) + q(t) + \frac{1}{p(t)}z^2(t).$$

On the other hand if $\mathbb{T} = \mathbb{Z}$, then $\mu(t) \equiv 1$ and the Riccati operator in (4.18) is the well-known operator (see, for example, [**191**, Chapter 6]) defined by

$$Rz(t) = \Delta z(t) + q(t) + \frac{z^2(t)}{p(t) + z(t)}.$$

Example 4.37. Here we will solve the Riccati equation

$$(4.20) \qquad z^\Delta + 3 \left(\frac{1}{8}\right)^{t+1} + \frac{z^2}{\left(\frac{1}{8}\right)^t + \mu(t)z} = 0$$

for $t \in \mathbb{T} = \mathbb{Z}$. Since $\mathbb{T} = \mathbb{Z}$, the self-adjoint equation corresponding to (4.20) is the difference equation

$$\Delta \left[\left(\frac{1}{8}\right)^t \Delta x(t) \right] + 3 \left(\frac{1}{8}\right)^{t+1} x(t+1) = 0.$$

Expanding this equation and simplifying we get the difference equation

$$x(t+2) - 6x(t+1) + 8x(t) = 0.$$

A general solution of this equation is

$$x(t) = c_1 2^t + c_2 4^t.$$

Hence

$$\begin{aligned} z(t) &= \frac{p(t)x^\Delta(t)}{x(t)} \\ &= \frac{p(t)\Delta x(t)}{x(t)} \\ &= \frac{\left(\frac{1}{8}\right)^t (c_1 2^t + 3c_2 4^t)}{c_1 2^t + c_2 4^t} \\ &= \left(\frac{1}{8}\right)^t \frac{c_1 + 3c_2 2^t}{c_1 + c_2 2^t} \end{aligned}$$

is a solution of (4.20). Letting $c_1 = 0$ we obtain that

$$z(t) = 3 \left(\frac{1}{8}\right)^t$$

solves (4.20). For $c_1 \neq 0$ we get that

$$z(t) = \left(\frac{1}{8}\right)^t \frac{1 + 3c2^t}{1 + c2^t}$$

is a solution of (4.20).

Example 4.38. Here we will solve the Riccati equation

$$(4.21) \qquad z^\Delta + 9e^{-6t} + \frac{z^2}{e^{-6t} + \mu(t)z} = 0$$

for $t \in \mathbb{T} = \mathbb{R}$. Since $\mathbb{T} = \mathbb{R}$, the self-adjoint equation corresponding to (4.21) is the differential equation

$$\left(e^{-6t} x'\right)' + 9e^{-6t} x = 0.$$

Expanding this and simplifying we get the differential equation

$$x'' - 6x' + 9x = 0.$$

A general solution of this equation is

$$x(t) = c_1 e^{3t} + c_2 t e^{3t}.$$

Hence

$$z(t) = \frac{p(t)x^\Delta(t)}{x(t)}$$

$$= \frac{p(t)x'(t)}{x(t)}$$

$$= \frac{e^{-6t}\left(3c_1e^{3t} + c_2e^{3t} + 3c_2te^{3t}\right)}{c_1e^{3t} + c_2te^{3t}}$$

$$= e^{-6t}\frac{3c_1 + (1+3t)c_2}{c_1 + c_2t}$$

solves (4.21). Letting $c_1 = 0$ we get that

$$z(t) = \frac{1+3t}{t}e^{-6t}$$

is a solution of (4.21). For $c_1 \neq 0$ we obtain that

$$z(t) = \frac{e^{-6t}\left[3 + (1+3t)c\right]}{1 + ct}$$

is a solution of (4.21).

Exercise 4.39. Solve each of the following Riccati equations with $\mathbb{T} = \mathbb{Z}$ (write your answer in terms of one arbitrary constant):

(i) $z^\Delta + 4\left(\frac{1}{9}\right)^{t+1} + \frac{z^2}{\left(\frac{1}{9}\right)^t + \mu(t)z} = 0$;

(ii) $z^\Delta + \frac{z^2}{1+\mu(t)z} = 0$.

Exercise 4.40. Solve each of the following Riccati equations with $\mathbb{T} = \mathbb{R} \cap [2, \infty)$ (write your answer in terms of one arbitrary constant):

(i) $z^\Delta + 1 + \frac{z^2}{1+\mu(t)z} = 0$;

(ii) $z^\Delta + \frac{4}{t^5} + \frac{z^2}{\left(\frac{1}{t^3}\right) + \mu(t)z} = 0$;

(iii) $z^\Delta - 12e^{4t} + \frac{z^2}{e^{4t} + \mu(t)z} = 0$;

(iv) $z^\Delta + 6e^{-5t} + \frac{z^2}{e^{-5t} + \mu(t)z} = 0$;

(v) $z^\Delta + 4e^{-4t} + \frac{z^2}{e^{-4t} + \mu(t)z} = 0$.

Exercise 4.41. Solve each of the following Riccati equations on an arbitrary time scale (write your answer in terms of one arbitrary constant):

(i) $z^\Delta - \frac{16}{1+4\mu(t)} + \frac{z^2}{1+z\mu(t)} = 0$;

(ii) $z^\Delta + \frac{4}{1+4\mu^2(t)}e_{-4\mu/(1+4\mu^2)}(t, t_0) + \frac{z^2}{e_{-4\mu/(1+4\mu^2)}(t, t_0) + z\mu(t)} = 0$.

Theorem 4.42. *Assume $p > 0$. Then (4.1) has a positive solution on $\mathbb{T}$ iff the Riccati equation (4.18) has a solution z on $\mathbb{T}^\kappa$ satisfying (4.19) on $\mathbb{T}^\kappa$.*

Proof. First assume that (4.1) has a positive solution on $\mathbb{T}$. Then by Theorem 4.36, the Riccati equation (4.18) has a solution z on $\mathbb{T}^\kappa$ satisfying (4.19) on $\mathbb{T}^\kappa$.

Conversely assume that z is a solution of (4.18) on $\mathbb{T}^\kappa$ such that (4.19) holds on $\mathbb{T}^\kappa$. Since by (4.19)

$$1 + \mu(t)\frac{z(t)}{p(t)} = \frac{p(t) + \mu(t)z(t)}{p(t)} > 0 \quad \text{for all} \quad t \in \mathbb{T}^\kappa,$$

we have $z/p \in \mathcal{R}^+$, and hence

$$x := e_{\frac{z}{p}}(\cdot, t_0) > 0$$

by Theorem 2.35 and Theorem 2.48 (i), where $t_0 \in \mathbb{T}$. Now $x^\Delta = (z/p)x$ implies that

$$px^\Delta = zx \quad \text{on} \quad \mathbb{T}^\kappa.$$

Taking the derivative of both sides we get that

$$\begin{aligned}
(px^\Delta)^\Delta &= (xz)^\Delta \\
&= x^\sigma z^\Delta + zx^\Delta \\
&= x^\sigma \left(-q - \frac{z^2}{p + \mu z} \right) + \frac{z^2 x}{p} \\
&= -qx^\sigma - (x + \mu x^\Delta) \frac{z^2}{p + \mu z} + \frac{z^2 x}{p} \\
&= -qx^\sigma - (x + \mu x \frac{z}{p}) \frac{z^2}{p + \mu z} + \frac{z^2 x}{p} \\
&= -qx^\sigma.
\end{aligned}$$

Hence $x = e_{\frac{z}{p}}(\cdot, t_0)$ is a positive solution of (4.1). $\qquad\square$

Theorem 4.43. *Assume $x \in \mathbb{D}$ has no generalized zeros in $\mathbb{T}$ and z is defined by the Riccati substitution (4.16) for $t \in \mathbb{T}^\kappa$. Then (4.19) holds for $t \in \mathbb{T}^\kappa$ and*

$$Lx(t) = x^\sigma(t)Rz(t) \quad \text{for} \quad t \in \mathbb{T}^{\kappa^2}.$$

Proof. We have on $\mathbb{T}^\kappa$

$$\begin{aligned}
x^\sigma z^\Delta &= x^\sigma \left(\frac{px^\Delta}{x} \right)^\Delta \\
&= \frac{x (px^\Delta)^\Delta - p (x^\Delta)^2}{x} \\
&= (px^\Delta)^\Delta - \frac{p}{x} (x^\Delta)^2 \\
&= Lx - qx^\sigma - x^\sigma \frac{x}{px^\sigma} z^2 \\
&= Lx - qx^\sigma - x^\sigma \frac{z^2}{p + \mu z}
\end{aligned}$$

by (4.17). Also (4.19) follows. Solving for Lx we get the desired result. $\qquad\square$

Definition 4.44. Let $\omega = \sup \mathbb{T}$ and if $\omega < \infty$ assume $\rho(\omega) = \omega$. Further assume $a \in \mathbb{T}$ and $q \in C_{rd}$. We say that (4.1) is oscillatory on $[a, \omega)$ provided every nontrivial real-valued solution has infinitely many generalized zeros in $[a, \omega)$.

Theorem 4.45 (Wintner's Theorem). *Assume $\sup \mathbb{T} = \infty$, $a \in \mathbb{T}$, $\mu(t) \geq K > 0$ and $0 < p(t) \leq M$ for all $t \in [a, \infty)$, and*

$$\int_a^\infty q(t)\Delta t = \infty.$$

Then (4.1) is oscillatory on $[a, \infty)$.

Proof. Assume (4.1) is nonoscillatory on $[a, \infty)$. Then there exists $t_0 \geq a$ such that (4.1) has a positive solution x on $[t_0, \infty)$. We then define z by the Riccati substitution (4.16) on $[t_0, \infty)$. By Theorem 4.36, z is a solution of the Riccati equation (4.18) on $[t_0, \infty)$ and (4.19) is satisfied on $[t_0, \infty)$. Integrating both sides of the Riccati equation from t_0 to t we get that

$$
\begin{aligned}
z(t) &= z(t_0) - \int_{t_0}^{t} q(\tau)\Delta\tau - \int_{t_0}^{t} \frac{z^2(\tau)}{p(\tau) + \mu(\tau)z(\tau)}\Delta\tau \\
&\leq z(t_0) - \int_{t_0}^{t} q(\tau)\Delta\tau.
\end{aligned}
$$

Letting $t \to \infty$ we get that

(4.22)
$$
\lim_{t\to\infty} z(t) = -\infty.
$$

But from (4.19) we have that

$$
z(t) \geq -\frac{p(t)}{\mu(t)} \geq -\frac{M}{K} \quad \text{for} \quad t \geq t_0,
$$

and this contradicts (4.22) so the proof is complete.   □

The remaining results in this section are special cases of results due to Erbe and Peterson [135].

Theorem 4.46. *Assume* $\sup \mathbb{T} = \infty$, $p > 0$, *and let* $a \in \mathbb{T}$. *Suppose that for each* $t_0 \in [a, \infty)$ *there exist* $a_0 \in [t_0, \infty)$ *and* $b_0 \in (a_0, \infty)$ *such that* $\mu(a_0) > 0$, $\mu(b_0) > 0$, *and*

(4.23)
$$
\int_{a_0}^{b_0} q(s)\Delta s \geq \frac{p(a_0)}{\mu(a_0)} + \frac{p(b_0)}{\mu(b_0)}.
$$

Then (4.1) *is oscillatory on* $[a, \infty)$.

Proof. Assume (4.1) is nonoscillatory on $[a, \infty)$. Then there is $t_0 \in [a, \infty)$ such that (4.1) has a positive solution x on $[t_0, \infty)$. We define z by the Riccati substitution (4.16) on $[t_0, \infty)$. Then by Theorem 4.42 we get that (4.19) holds on $[t_0, \infty)$ and that z is a solution of the Riccati equation (4.18) on $[t_0, \infty)$. By hypothesis there exist $a_0 \in [t_0, \infty)$ and $b_0 \in (a_0, \infty)$ such that $\mu(a_0) > 0$, $\mu(b_0) > 0$, and (4.23) holds. Since z is a solution of the Riccati equation (4.18) on $[t_0, \infty)$, we have that

$$
z^\Delta(t) = -q(t) - \frac{z^2(t)}{p(t) + \mu(t)z(t)}
$$

for $t \in [t_0, \infty)$. Integrating both sides of this last equation from a_0 to b_0 we obtain

$$
\begin{aligned}
z(b_0) &= z(a_0) - \int_{a_0}^{b_0} q(t)\Delta t - \int_{a_0}^{b_0} \frac{z^2(t)}{p(t) + \mu(t)z(t)} \Delta t \\
&\leq z(a_0) - \int_{a_0}^{b_0} q(t)\Delta t - \int_{a_0}^{\sigma(a_0)} \frac{z^2(t)}{p(t) + \mu(t)z(t)} \Delta t \\
&= z(a_0) - \frac{\mu(a_0)z^2(a_0)}{p(a_0) + \mu(a_0)z(a_0)} - \int_{a_0}^{b_0} q(t)\Delta t \\
&= \frac{p(a_0)z(a_0)}{p(a_0) + \mu(a_0)z(a_0)} - \int_{a_0}^{b_0} q(t)\Delta t \\
&< \frac{p(a_0)}{\mu(a_0)} - \int_{a_0}^{b_0} q(t)\Delta t,
\end{aligned}
$$

where in the last step we used $p(a_0) + \mu(a_0)z(a_0) > 0$. Using $p(b_0) + \mu(b_0)z(b_0) > 0$, we get that

$$
\int_{a_0}^{b_0} q(t)\Delta t < \frac{p(a_0)}{\mu(a_0)} - z(b_0) < \frac{p(a_0)}{\mu(a_0)} + \frac{p(b_0)}{\mu(b_0)}
$$

which contradicts (4.23).   □

Corollary 4.47. *Assume $a \in \mathbb{T}$, $p > 0$, and $\sup \mathbb{T} = \infty$. Assume there are right-scattered points in $\mathbb{T}$ in every neighborhood of ∞. A necessary condition for (4.1) to be nonoscillatory on $[a, \infty)$ is that there exists $t_0 \geq a$ such that*

$$
\int_{a_0}^{b_0} q(s)\Delta s < \frac{p(a_0)}{\mu(a_0)} + \frac{p(b_0)}{\mu(b_0)}
$$

holds for all $t_0 \leq a_0 < b_0$ with $\mu(a_0) > 0$ and $\mu(b_0) > 0$.

Proof. This result follows immediately from Theorem 4.46.   □

Example 4.48. In this example we show that the q-difference equation

$$
(4.24) \qquad\qquad x^{\Delta\Delta} + \frac{c}{q(q-1)t^2}x^{\sigma} = 0,
$$

where $c > 1$ is a constant, is oscillatory on $\mathbb{T} = q^{\mathbb{N}_0}$, where $q > 1$ is a constant. Let $t_0 \in [1, \infty)$ be given and pick $k_0 \in \mathbb{N}$ so that $a_0 := q^{k_0} > t_0$. Let $\tilde{q}(t) := \frac{c}{q(q-1)t^2}$

for $t \in \mathbb{T}$ and consider

$$
\begin{aligned}
\int_{a_0}^{\infty} \tilde{q}(t)\Delta t &= \int_{q^{k_0}}^{\infty} \tilde{q}(t)\Delta t \\
&= \sum_{j=k_0}^{\infty} \tilde{q}(q^j)\mu(q^j) \\
&= \sum_{j=k_0}^{\infty} \frac{c}{q(q-1)q^{2j}}(q-1)q^j \\
&= \frac{c}{q^{k_0+1}} \sum_{j=0}^{\infty} \left(\frac{1}{q}\right)^j \\
&= \frac{c}{(q-1)a_0} \\
&> \frac{1}{\mu(a_0)}.
\end{aligned}
$$

Since $\lim_{k\to\infty} \mu(t_k) = \infty$, we can now pick $b_0 \in (a_0, \infty)$ sufficiently large so that

$$
\int_{a_0}^{b_0} \tilde{q}(t)\Delta t > \frac{1}{\mu(a_0)} + \frac{1}{\mu(b_0)}.
$$

Hence from Theorem 4.46 we get that the q-difference equation (4.24) is oscillatory on $\mathbb{T}$.

Example 4.49. Let $\mathbb{T} = \mathbb{P}_{\frac{1}{2},\frac{1}{2}}$ and $q \in C_{\mathrm{rd}}$. Assume that for each $t_0 \geq 0$ there exist $k_0 \in \mathbb{N}$ and $l_0 \in \mathbb{N}$ such that $k_0 \geq t_0$ and

$$
(4.25) \qquad \sum_{j=1}^{l_0} \int_{k_0+j}^{k_0+j+\frac{1}{2}} q(t)dt + \frac{1}{2}\sum_{j=0}^{l_0-1} q\left(k_0 + j + \frac{1}{2}\right) \geq 4.
$$

Then

$$
(4.26) \qquad x^{\Delta\Delta} + q(t)x^{\sigma} = 0
$$

is oscillatory on $\mathbb{T}$. To see that this follows from Theorem 4.46, let

$$
a_0 = k_0 + \frac{1}{2} \quad \text{and} \quad b_0 = k_0 + l_0 + \frac{1}{2}.
$$

Note that a_0 and b_0 are right-scattered and

$$
\begin{aligned}
\int_{a_0}^{b_0} q(t)\Delta t &= \sum_{j=1}^{l_0} \int_{k_0+j}^{k_0+j+\frac{1}{2}} q(t)\Delta t + \sum_{j=0}^{l_0-1} \int_{k_0+j+\frac{1}{2}}^{\sigma(k_0+j+\frac{1}{2})} q(t)\Delta t \\
&= \sum_{j=1}^{l_0} \int_{k_0+j}^{k_0+j+\frac{1}{2}} q(t)dt + \frac{1}{2}\sum_{j=0}^{l_0-1} q\left(k_0 + j + \frac{1}{2}\right) \\
&\geq 4 = \frac{p(a_0)}{\mu(a_0)} + \frac{p(b_0)}{\mu(b_0)},
\end{aligned}
$$

where we used (4.25), and hence from Theorem 4.46 we get that (4.26) is oscillatory on $\mathbb{T}$. Note that there is no requirement on

$$
\liminf_{t\to\infty} \int_{t_0}^{t} q(\tau)\Delta\tau.
$$

Corollary 4.50. *Let $p > 0$. Assume there exists a strictly increasing sequence $\{t_k\}_{k \in \mathbb{N}} \subset \mathbb{T}$ with $\lim_{k \to \infty} t_k = \infty$. Suppose there are positive constants K and M such that $\mu(t_k) \geq K$ and $0 < p(t_k) \leq M$ for all $k \in \mathbb{N}$. If*

$$\text{(4.27)} \qquad \lim_{k \to \infty} \int_{t_1}^{t_k} q(t)\Delta t = \infty,$$

then (4.1) is oscillatory on $[t_1, \infty)$.

Proof. Let $t_0 \in [a, \infty)$. Then choose k_0 sufficiently large so that $a_0 := t_{k_0} \geq t_0$. Since (4.27) holds we can choose k_1 sufficiently large so that $b_0 := t_{k_1} > a_0$ and

$$\int_{a_0}^{b_0} q(t)\Delta t \geq \frac{2M}{K} \geq \frac{p(a_0)}{\mu(a_0)} + \frac{p(b_0)}{\mu(b_0)}.$$

It follows from Theorem 4.46 that (4.1) is oscillatory on $[a, \infty)$. $\qquad \square$

Theorem 4.51. *Assume $a \in \mathbb{T}$ and $\sup \mathbb{T} = \infty$. Assume for any $t_0 \in [a, \infty)$ that there exists a strictly increasing sequence $\{t_k\}_{k=1}^{\infty} \subset [t_0, \infty)$ with $\lim_{k \to \infty} t_k = \infty$, and that there are constants K_1 and K_2 such that $0 < K_1 \leq \mu(t_k) \leq K_2$ for $k \in \mathbb{N}$ and*

$$\text{(4.28)} \qquad \lim_{k \to \infty} \int_{t_1}^{t_k} q(t)\Delta t \geq \frac{1}{\mu(t_1)}.$$

Then (4.26) is oscillatory on $[a, \infty)$.

Proof. Assume that (4.26) is nonoscillatory on $[a, \infty)$. This implies that there is $t_0 \in [a, \infty)$ such that $x^{\Delta\Delta} + q(t)x^{\sigma} = 0$ has a positive solution x on $[t_0, \infty)$. We define z by the Riccati substitution (4.16) (with $p = 1$) and get by Theorem 4.42 that $1 + \mu(t)z(t) > 0$ for $t \geq t_0$ and that z is a solution of the Riccati equation

$$z^{\Delta} + q(t) + \frac{z^2}{1 + \mu(t)z} = 0 \quad \text{for} \quad t \geq t_0.$$

Corresponding to t_0, let $\{t_k\}_{k=1}^{\infty} \subset [t_0, \infty)$ be the sequence given in the statement of this theorem. Integrating both sides of the Riccati equation from t_1 to t_k, $k > 1$, and proceeding as in the proof of Theorem 4.46, we obtain

$$
\begin{aligned}
z(t_k) &= z(t_1) - \int_{t_1}^{t_k} q(t)\Delta t - \int_{t_1}^{t_k} \frac{z^2(t)}{1 + \mu(t)z(t)}\Delta t \\
&\leq z(t_1) - \int_{t_1}^{t_k} q(t)\Delta t - \int_{t_1}^{\sigma(t_1)} \frac{z^2(t)}{1 + \mu(t)z(t)}\Delta t \\
&= z(t_1) - \frac{z^2(t_1)}{1 + \mu(t_1)z(t_1)}\mu(t_1) - \int_{t_1}^{t_k} q(t)\Delta t \\
&= \frac{z(t_1)}{1 + \mu(t_1)z(t_1)} - \int_{t_1}^{t_k} q(t)\Delta t.
\end{aligned}
$$

Taking the limit of both sides as $k \to \infty$ and using $\lim_{k \to \infty} z(t_k) = 0$ (which we prove below), we get

$$\frac{1}{\mu(t_1)} \leq \frac{z(t_1)}{1 + \mu(t_1)z(t_1)}.$$

Since multiplying this inequality by $(1 + \mu(t_1)z(t_1))\mu(t_1) > 0$ implies

$$1 + \mu(t_1)z(t_1) \leq \mu(t_1)z(t_1),$$

we get a contradiction. Hence it remains to prove that

$$\lim_{k \to \infty} z(t_k) = 0.$$

To do this we first show that

$$\lim_{k \to \infty} F(t_k) = 0, \quad \text{where} \quad F = \frac{z^2}{1 + \mu z}.$$

Before we prove this, note that (4.28) implies that there is a constant M such that

$$\int_{t_1}^{t_k} q(t)\Delta t > M \quad \text{for all} \quad k \in \mathbb{N}.$$

Now consider

$$
\begin{aligned}
\sum_{j=2}^{k-1} F(t_j)\mu(t_j) &= \sum_{j=2}^{k-1} \int_{t_j}^{\sigma(t_j)} F(t)\Delta t \\
&\le \int_{\sigma(t_1)}^{t_k} F(t)\Delta t \\
&= \int_{t_1}^{t_k} F(t)\Delta t - \mu(t_1)F(t_1) \\
&= z(t_1) - \int_{t_1}^{t_k} q(t)\Delta t - z(t_k) - \mu(t_1)F(t_1) \\
&\le -z(t_k) - \int_{t_1}^{t_k} q(t)\Delta t + \frac{1}{\mu(t_1)} \\
&\le \frac{1}{\mu(t_k)} + \frac{1}{\mu(t_1)} - M \\
&\le \frac{2}{K_1} - M
\end{aligned}
$$

for $k \ge 3$. It follows that

$$\sum_{j=2}^{\infty} F(t_j)\mu(t_j) \quad \text{converges.}$$

This implies that

$$\lim_{k \to \infty} F(t_k)\mu(t_k) = 0.$$

Since $0 < K_1 \le \mu(t_k) \le K_2$ for $k \in \mathbb{N}$, it follows that

$$\lim_{k \to \infty} F(t_k) = \lim_{k \to \infty} \frac{z^2(t_k)}{1 + \mu(t_k)z(t_k)} = 0.$$

We only consider the case where all $z(t_k) \ge 0$. Then

$$0 \le \frac{z^2(t_k)}{1 + K_2 z(t_k)} \le \frac{z^2(t_k)}{1 + \mu(t_k)z(t_k)}$$

implies

$$\lim_{k \to \infty} \frac{z^2(t_k)}{1 + K_2 z(t_k)} = 0.$$

This in turn implies that $\lim_{k \to \infty} z(t_k) = 0$, which completes the proof. $\qquad \square$

Applying Theorem 4.51 one gets the following example.

Example 4.52. If $\mathbb{T} = h\mathbb{N}_0$, where $h > 0$, $c \geq \frac{1}{h^2}$, and the function q has the values

$$c, \; -\frac{c}{2}, \; -\frac{c}{2}, \; \frac{c}{3}, \; \frac{c}{3}, \; \frac{c}{3}, \; -\frac{c}{4}, \; -\frac{c}{4}, \; -\frac{c}{4}, \; -\frac{c}{4}, \; \frac{c}{5}, \ldots$$

at $0, h, 2h, 3h, \ldots$, respectively, then (4.26) is oscillatory on $\mathbb{T}$.

4.3. Disconjugacy

Theorem 4.53. *Assume $p > 0$. If (4.1) is disconjugate on $[a, \sigma^2(b)]$, then (4.1) has a positive solution on $[a, \sigma^2(b)]$. Conversely if (4.1) has a positive solution on a subinterval $\mathcal{I} \subset \mathbb{T}$, then (4.1) is disconjugate on $\mathcal{I}$.*

Proof. Assume (4.1) is disconjugate on $[a, \sigma^2(b)]$. Let u and v be solutions of (4.1) satisfying the initial conditions

$$u(a) = 0, \quad u^\Delta(a) = 1 \quad \text{and} \quad v(\sigma^2(b)) = 0, \quad v^\Delta(\sigma(b)) = -1.$$

Since (4.1) is disconjugate on $[a, \sigma^2(b)]$, $u(t) > 0$ for all $t \in (a, \sigma^2(b)]$ and $v(t) > 0$ for all $t \in [a, \sigma^2(b))$. It follows that

$$x(t) = u(t) + v(t)$$

is the desired positive solution.

Conversely assume (4.1) has a positive solution u on a subinterval $\mathcal{I} \subset \mathbb{T}$. If we assume (4.1) is not disconjugate on $\mathcal{I}$, then (4.1) has a nontrivial solution v with at least two generalized zeros in $\mathcal{I}$. Without loss of generality there are points $t_1 < t_2$ in $\mathcal{I}$ such that

$$v(t_1) \leq 0, \quad v(t_2) \leq 0$$

and $v(t) > 0$ on (t_1, t_2), where $(t_1, t_2) \neq \emptyset$. Note that

$$\left(\frac{v}{u}\right)^\Delta (t) = \frac{W(u, v)(t)}{u(t)u^\sigma(t)} = \frac{C}{p(t)u(t)u^\sigma(t)}$$

is of one sign on $\mathcal{I}^\kappa$. Hence $\frac{v}{u}$ is strictly monotone on $\mathcal{I}$. But

$$\frac{v}{u}(t_1) \leq 0, \quad \frac{v}{u}(t_0) > 0, \quad \frac{v}{u}(t_2) \leq 0,$$

where $t_0 \in (t_1, t_2)$, which contradicts the fact that $\frac{v}{u}$ is monotone. Hence the proof is complete. $\qquad \square$

We define $\mathbb{A}$ to be the set of functions

$$\mathbb{A} = \{u \in C^1_{\mathrm{prd}}([a, \sigma^2(b)], \mathbb{R}) : u(a) = u(\sigma^2(b)) = 0\}.$$

Here, C^1_{prd} denotes the set of all continuous functions whose derivatives are *piecewise* rd-continuous. Then we define the quadratic functional $\mathcal{F}$ on $\mathbb{A}$ by

$$\mathcal{F}(u) = \int_a^{\sigma^2(b)} \left\{ p(t) \left[u^\Delta(t)\right]^2 - q(t)[u(\sigma(t))]^2 \right\} \Delta t.$$

Definition 4.54. We say that $\mathcal{F}$ is positive definite on $\mathbb{A}$ provided $\mathcal{F}(u) \geq 0$ for all $u \in \mathbb{A}$ and $\mathcal{F}(u) = 0$ iff $u = 0$.

We next state and prove the "completing the square" formula.

Theorem 4.55 (Picone's Identity). *Assume z is a solution of the Riccati equation (4.18) on $[a, \sigma^2(b)]$ with $p(t) + \mu(t)z(t) > 0$ on $[a, \sigma(b)]$. Let $u \in \mathbb{A}$. Then we have for all $t \in [a, \sigma(b)]$,*

$$(zu^2)^\Delta(t) = p(t)\left[u^\Delta(t)\right]^2 - q(t)u^2(\sigma(t))$$
$$- \left\{\frac{z(t)u(\sigma(t))}{\sqrt{p(t) + \mu(t)z(t)}} - \sqrt{p(t) + \mu(t)z(t)}u^\Delta(t)\right\}^2 .$$

Proof. We have

$$\begin{aligned}
\left(zu^2\right)^\Delta(t) &= u^2(\sigma(t))z^\Delta(t) + u(\sigma(t))u^\Delta(t)z(t) + u^\Delta(t)u(t)z(t) \\
&= u^2(\sigma(t))\left(-q(t) - \frac{z^2(t)}{p(t) + \mu(t)z(t)}\right) \\
&\quad + u(\sigma(t))u^\Delta(t)z(t) + u^\Delta(t)\left[u(\sigma(t)) - \mu(t)u^\Delta(t)\right]z(t) \\
&= p(t)\left[u^\Delta(t)\right]^2 - q(t)u^2(\sigma(t)) - \frac{z^2(t)u^2(\sigma(t))}{p(t) + \mu(t)z(t)} \\
&\quad + 2u(\sigma(t))u^\Delta(t)z(t) - \left[p(t) + \mu(t)z(t)\right]\left[u^\Delta(t)\right]^2 \\
&= p(t)\left[u^\Delta(t)\right]^2 - q(t)u^2(\sigma(t)) \\
&\quad - \left\{\frac{z(t)u(\sigma(t))}{\sqrt{p(t) + \mu(t)z(t)}} - \sqrt{p(t) + \mu(t)z(t)}u^\Delta(t)\right\}^2
\end{aligned}$$

for all $t \in [a, \sigma(b)]$.   □

Theorem 4.56. *Assume x solves (4.1) on $[a, \sigma^2(b)]$ and let*

$$u(t) = \begin{cases} 0 & \text{if} \quad a \le t \le c \\ x(t) & \text{if} \quad c < t \le d \\ 0 & \text{if} \quad d < t \le \sigma^2(b), \end{cases}$$

where $a \le c \le \sigma(c) < d \le \sigma(b)$. Then

$$\mathcal{F}(u) = C + D,$$

where

$$C := \begin{cases} -p(c)x(c)x^\Delta(c) & \text{if} \quad \mu(c) = 0 \\ \frac{p(c)x(c)x^\sigma(c)}{\mu(c)} & \text{if} \quad \mu(c) > 0 \end{cases}$$

and

$$D := \begin{cases} p(d)x(d)x^\Delta(d) & \text{if} \quad \mu(d) = 0 \\ \frac{p(d)x(d)x^\sigma(d)}{\mu(d)} & \text{if} \quad \mu(d) > 0. \end{cases}$$

Proof. Let u be as in the statement of this theorem and consider

$$
\begin{aligned}
\mathcal{F}(u) &= \int_a^{\sigma^2(b)} \left\{ p(t) \left[u^\Delta(t) \right]^2 - q(t) u^2(\sigma(t)) \right\} \Delta t \\
&= \int_c^{\sigma(d)} p(t) \left[u^\Delta(t) \right]^2 \Delta t - \int_c^d q(t) u^2(\sigma(t)) \Delta t \\
&= \int_c^{\sigma(c)} p(t) \left[u^\Delta(t) \right]^2 \Delta t + \int_d^{\sigma(d)} p(t) \left[u^\Delta(t) \right]^2 \Delta t - \int_c^{\sigma(c)} q(t) u^2(\sigma(t)) \Delta t \\
&\quad + \int_{\sigma(c)}^d \left\{ p(t) \left[x^\Delta(t) \right]^2 - q(t) x^2(\sigma(t)) \right\} \Delta t \\
&= \mu(c) p(c) (u^\Delta(c))^2 + \mu(d) p(d) (u^\Delta(d))^2 - \mu(c) q(c) u^2(\sigma(c)) \\
&\quad + \left[p(t) x^\Delta(t) x(t) \right]_{t=\sigma(c)}^{t=d} - \int_{\sigma(c)}^d Lx(t) x^\sigma(t) \Delta t \\
&= C + D
\end{aligned}
$$

(we used Theorem 1.75 and Theorem 1.77 (vi)), where

$$
C = \mu(c) p(c) (u^\Delta(c))^2 - \mu(c) q(c) u^2(\sigma(c)) - p(\sigma(c)) x^\Delta(\sigma(c)) x(\sigma(c))
$$

and

$$
D = \mu(d) p(d) (u^\Delta(d))^2 + p(d) x^\Delta(d) x(d).
$$

If $\mu(c) = 0$, then $C = -p(c) x(c) x^\Delta(c)$ as desired. If $\mu(c) > 0$, then

$$
\begin{aligned}
C &= p(c) \frac{x^2(\sigma(c))}{\mu^2(c)} \mu(c) - q(c) x^2(\sigma(c)) \mu(c) - p(\sigma(c)) x^\Delta(\sigma(c)) x(\sigma^2(c)) \\
&= \frac{p(c) x^2(\sigma(c))}{\mu(c)} - \left\{ \frac{p(\sigma(c)) x^\Delta(\sigma(c)) - p(c) x^\Delta(c)}{\mu(c)} + q(c) x^\sigma(c) \right\} x^\sigma(c) \mu(c) \\
&\quad - p(c) x^\Delta(c) x^\sigma(c) \\
&= \frac{p(c) x^2(\sigma(c))}{\mu(c)} - Lx(c) x^\sigma(c) \mu(c) - p(c) \frac{x^\sigma(c) - x(c)}{\mu(c)} x^\sigma(c) \\
&= \frac{p(c) x(c) x^\sigma(c)}{\mu(c)}.
\end{aligned}
$$

If $\mu(d) = 0$, then $D = p(d) x(d) x^\Delta(d)$. Finally, if $\mu(d) > 0$, then

$$
\begin{aligned}
D &= p(d) \left[u^\Delta(d) \right]^2 \mu(d) + p(d) x^\Delta(d) x(d) \\
&= p(d) \frac{x^2(d)}{\mu^2(d)} \mu(d) + p(d) \frac{x^\sigma(d) - x(d)}{\mu(d)} x(d) \\
&= \frac{p(d) x(d) x^\sigma(d)}{\mu(d)}.
\end{aligned}
$$

This proves our claim.   □

Theorem 4.57 (Jacobi's Condition). *Assume $p > 0$. The self-adjoint equation* (4.1) *is disconjugate on $[a, \sigma^2(b)]$ iff $\mathcal{F}$ is positive definite on $\mathbb{A}$.*

Proof. Assume that (4.1) is disconjugate on $[a, \sigma^2(b)]$. Then, by Theorem 4.53, there is a solution x without generalized zeros in $[a, \sigma^2(b)]$. If we define z by the Riccati substitution (4.16), then by Theorem 4.36, z is a solution of the Riccati equation (4.18) on $[a, \sigma(b)]$ satisfying (4.19). By making appropriate definitions we can assume z is a solution of the Riccati equation (4.18) on $[a, \sigma^2(b)]$. Hence if $u \in \mathbf{A}$, then we have by Picone's identity (Theorem 4.55) that

$$\left(zu^2\right)^\Delta (t) = p(t)\left[u^\Delta(t)\right]^2 - q(t)u^2(\sigma(t))$$
$$- \left\{\frac{z(t)u^\sigma(t)}{\sqrt{p(t)+\mu(t)z(t)}} - \sqrt{p(t)+\mu(t)z(t)}u^\Delta(t)\right\}^2.$$

Integrating from a to $\sigma^2(b)$ and using $u(a) = u(\sigma^2(b)) = 0$ we get that

$$\mathcal{F}(u) = \int_a^{\sigma^2(b)} \left\{\frac{z(t)u^\sigma(t)}{\sqrt{p(t)+\mu(t)z(t)}} - \sqrt{p(t)+\mu(t)z(t)}u^\Delta(t)\right\}^2 \Delta t.$$

It follows that $\mathcal{F}(u) \geq 0$ for all $u \in \mathbf{A}$. Now assume that $\mathcal{F}(u) = 0$. Then

$$\frac{z(t)u^\sigma(t)}{\sqrt{p(t)+\mu(t)z(t)}} = \sqrt{p(t)+\mu(t)z(t)}u^\Delta(t) \quad \text{for all} \quad t \in [a, \sigma(b)].$$

It follows that u solves the initial value problem

$$u^\Delta(t) = \frac{z(t)}{p(t)+\mu(t)z(t)}u^\sigma(t), \quad u(a) = 0.$$

By Theorem 2.71 (note that $-z/(p + \mu z) \in \mathcal{R}$), this implies that $u(t) = 0$ for all $t \in [a, \sigma^2(b)]$, and hence $\mathcal{F}$ is positive definite on $\mathbf{A}$.

Conversely assume $\mathcal{F}$ is positive definite on $\mathbf{A}$. Assume (4.1) is not disconjugate on $[a, \sigma^2(b)]$. Then there are points $a \leq c \leq \sigma(c) < d \leq \sigma(b)$ and a solution x of (4.1) such that $x(c) = 0$ if $\mu(c) = 0$ and $p(c)x(c)x^\sigma(c) \leq 0$ if $\mu(c) > 0$, $x(d) = 0$ if $\mu(d) = 0$ and $p(d)x(d)x^\sigma(d) \leq 0$ if $\mu(d) > 0$, and $x(t) \neq 0$ in (c, d). Let u be defined as in Theorem 4.56. Then since $\mathcal{F}$ is positive definite on $\mathbf{A}$, $\mathcal{F}(u) > 0$. But by Theorem 4.56,

$$\mathcal{F}(u) = C + D \leq 0,$$

which is a contradiction. Hence (4.1) is in fact disconjugate on $[a, \sigma^2(b)]$. $\square$

Exercise 4.58. Show that if p is positive, then

$$\left(p(t)x^\Delta\right)^\Delta = 0$$

is disconjugate on any time scale $\mathbb{T}$. Is this also true for arbitrary $p \neq 0$?

Theorem 4.59 (Polya Factorization). *If (4.1) has a solution u with no generalized zeros in $\mathbb{T}$, then for any $x \in \mathbb{D}$ we get the* Polya factorization

$$Lx(t) = \rho_1^\sigma(t)\left\{\rho_2(\rho_1 x)^\Delta\right\}^\Delta(t) \quad \text{for all} \quad t \in \mathbb{T}^{\kappa^2},$$

where

$$\rho_1 := \frac{1}{u} \quad \text{and} \quad \rho_2 := puu^\sigma > 0.$$

Proof. Assume that u is a solution of (4.1) with no generalized zeros in $\mathbb{T}$. Then

$$\rho_2(t) = p(t)u(t)u^\sigma(t) > 0 \quad \text{for all} \quad t \in \mathbb{T}^\kappa.$$

Let $x \in \mathbb{D}$. Then by the Lagrange identity (Theorem 4.30),

$$
\begin{aligned}
Lx(t) &= \frac{1}{u^{\sigma}(t)} \{u; x\}^{\Delta}(t) \\
&= \frac{1}{u^{\sigma}(t)} \{pW(u, x)\}^{\Delta}(t) \\
&= \frac{1}{u^{\sigma}(t)} \left\{ puu^{\sigma} \left(\frac{x}{u}\right)^{\Delta} \right\}^{\Delta}(t) \\
&= \rho_1^{\sigma}(t) \{\rho_2(\rho_1 x)^{\Delta}\}^{\Delta}(t)
\end{aligned}
$$

for $t \in \mathbb{T}^{\kappa^2}$, where ρ_1 is as in the statement of the theorem. $\qquad\square$

Theorem 4.60 (Trench Factorization). *Assume $a \in \mathbb{T}$, $p > 0$, and let $\omega := \sup \mathbb{T}$. If $\omega < \infty$, then we assume $\rho(\omega) = \omega$. If (4.1) has a positive solution on $[a, \omega)$, then for any $x \in \mathbb{D}$ we get the* Trench factorization

$$
Lx(t) = \gamma_1^{\sigma}(t) \{\gamma_2(\gamma_1 x)^{\Delta}\}^{\Delta}(t) \quad \text{for all} \quad t \in [a, \omega),
$$

where γ_1 and γ_2 are positive functions on $[a, \omega)$ and

(4.29)
$$
\int_a^{\omega} \frac{1}{\gamma_2(t)} \Delta t = \infty.
$$

Proof. Since (4.1) has a positive solution on $[a, \omega)$, Lx has a Polya factorization on $[a, \omega)$ by Theorem 4.59. Hence there are positive functions ρ_1 and ρ_2 on $[a, \omega)$ such that

$$
Lx(t) = \rho_1^{\sigma}(t) \{\rho_2(\rho_1 x)^{\Delta}\}^{\Delta}(t) = \frac{1}{\alpha_1^{\sigma}(t)} \left\{ \frac{1}{\alpha_2} \left(\frac{x}{\alpha_1}\right)^{\Delta} \right\}^{\Delta}(t)
$$

for $t \in [a, \omega)$, where

$$
\alpha_i(t) := \frac{1}{\rho_i(t)} \quad \text{for } i = 1, 2, \quad t \in [a, \omega).
$$

If

$$
\int_a^{\omega} \alpha_2(t) \Delta t = \infty,
$$

then we have what we want. Assume

(4.30)
$$
\int_a^{\omega} \alpha_2(t) \Delta t < \infty.
$$

In this case we set

$$
\beta_1(t) = \alpha_1(t) \int_t^{\omega} \alpha_2(s) \Delta s \quad \text{and} \quad \beta_2(t) = \frac{\alpha_2(t)}{\int_t^{\omega} \alpha_2(s) \Delta s \int_{\sigma(t)}^{\omega} \alpha_2(s) \Delta s}
$$

for $t \in [a, \omega)$. Then by (4.30)

$$
\begin{aligned}
\int_a^{\omega} \beta_2(t) \Delta t &= \lim_{b \to \omega, b \in \mathbb{T}} \int_a^b \frac{\alpha_2(t)}{\int_t^{\omega} \alpha_2(s) \Delta s \int_{\sigma(t)}^{\omega} \alpha_2(s) \Delta s} \Delta t \\
&= \lim_{b \to \omega, b \in \mathbb{T}} \int_a^b \left\{ \frac{1}{\int_t^{\omega} \alpha_2(s) \Delta s} \right\}^{\Delta} \Delta t \\
&= \infty.
\end{aligned}
$$

For $x \in \mathbb{D}$, note that

$$\left(\frac{x}{\beta_1}\right)^\Delta (t) = \left\{\frac{\frac{x(t)}{\alpha_1(t)}}{\int_t^\omega \alpha_2(s)\Delta s}\right\}^\Delta = \frac{\int_t^\omega \alpha_2(s)\Delta s \left(\frac{x}{\alpha_1}\right)^\Delta (t) - \frac{x(t)}{\alpha_1(t)}(-\alpha_2(t))}{\int_t^\omega \alpha_2(s)\Delta s \int_{\sigma(t)}^\omega \alpha_2(s)\Delta s}$$

for $t \in [a, \omega)$. Hence

$$\frac{1}{\beta_2(t)} \left(\frac{x}{\beta_1}\right)^\Delta (t) = \left[\frac{1}{\alpha_2(t)} \left(\frac{x}{\alpha_1}\right)^\Delta (t)\right] \int_t^\omega \alpha_2(s)\Delta s + \frac{x(t)}{\alpha_1(t)}$$

for $t \in [a, \omega)$. Taking the derivative of both sides we get

$$\left\{\frac{1}{\beta_2} \left(\frac{x}{\beta_1}\right)^\Delta\right\}^\Delta (t) = \left\{\frac{1}{\alpha_2} \left(\frac{x}{\alpha_1}\right)^\Delta\right\}^\Delta (t) \int_{\sigma(t)}^\omega \alpha_2(s)\Delta s$$

for $t \in [a, \omega)$. It follows that

$$\frac{1}{\beta_1^\sigma(t)} \left\{\frac{1}{\beta_2} \left(\frac{x}{\beta_1}\right)^\Delta\right\}^\Delta (t) = \frac{1}{\alpha_1^\sigma(t)} \left\{\frac{1}{\alpha_2} \left(\frac{x}{\alpha_1}\right)^\Delta\right\}^\Delta (t) = Lx(t)$$

for $t \in [a, \omega)$. If

$$\gamma_i(t) := \frac{1}{\beta_i(t)} \quad \text{for } i = 1, 2, \quad t \in [a, \omega),$$

then

$$Lx(t) = \gamma_1^\sigma(t)\{\gamma_2(\gamma_1 x)^\Delta\}^\Delta(t)$$

for $t \in [a, \omega)$, where (4.29) is satisfied. $\square$

Theorem 4.61. *Assume $a \in \mathbb{T}$, $p > 0$, and let $\omega := \sup \mathbb{T}$. If $\omega < \infty$, then we assume $\rho(\omega) = \omega$. If (4.1) has a positive solution on $[a, \omega)$, then there is a positive solution u, called a recessive solution at ω, such that for any second linearly independent solution v, called a dominant solution at ω,*

$$\lim_{t \to \omega^-} \frac{u(t)}{v(t)} = 0, \quad \int_a^\omega \frac{1}{p(t)u(t)u^\sigma(t)} \Delta t = \infty, \quad \text{and} \quad \int_b^\omega \frac{1}{p(t)v(t)v^\sigma(t)} \Delta t < \infty,$$

where $b < \omega$ is sufficiently close. Furthermore

(4.31)
$$\frac{p(t)v^\Delta(t)}{v(t)} > \frac{p(t)u^\Delta(t)}{u(t)}$$

for $t < \omega$ sufficiently close.

Proof. Since we are assuming that (4.1) has a positive solution on $[a, \omega)$, Lx has a Trench factorization on $[a, \omega)$ by Theorem 4.60. Hence

$$Lx(t) = \gamma_1^\sigma(t) \left\{\gamma_2 (\gamma_1 x)^\Delta\right\}^\Delta (t) \quad \text{for all} \quad t \in [a, \omega),$$

where γ_1 and γ_2 are positive functions on $[a, \omega)$, and (4.29) holds. If $u(t) = \frac{1}{\gamma_1(t)}$, then we get from the Trench factorization that u is a positive solution of (4.1) on $[a, \omega)$. Let

$$v_0(t) = \frac{1}{\gamma_1(t)} \int_a^t \frac{1}{\gamma_2(s)} \Delta s.$$

Since $\gamma_2(t)(\gamma_1 v)^\Delta(t) = 1$ and $v(a) = 0$, we get by the Trench factorization that v_0 is a solution of (4.1) on $[a, \omega)$. Note that

$$\lim_{t \to \omega^-} \frac{u(t)}{v_0(t)} = \lim_{t \to \omega^-} \frac{1}{\int_a^t \frac{1}{\gamma_2(s)} \Delta s} = 0$$

by (4.29). Consider

$$\left(\frac{v_0}{u}\right)^\Delta(t) = \frac{W(u, v_0)(t)}{u(t)u^\sigma(t)} = \frac{C}{p(t)u(t)u^\sigma(t)},$$

where C is constant by Abel's formula (Corollary 4.31). Note that $C \neq 0$ since u and v_0 are linearly independent. Integrating both sides of this last equality from a to t we obtain

$$\frac{v_0(t)}{u(t)} = \int_a^t \frac{C}{p(s)u(s)u^\sigma(s)} \Delta s.$$

Letting $t \to \omega^-$ we get one of the desired results

$$\int_a^\omega \frac{1}{p(t)u(t)u^\sigma(t)} \Delta t = \infty.$$

Now let v be any solution of (4.1) such that u and v are linearly independent. Then

$$v(t) = c_1 u(t) + c_2 v_0(t),$$

where $c_2 \neq 0$. It follows that

$$\lim_{t \to \omega^-} \frac{u(t)}{v(t)} = \lim_{t \to \omega^-} \frac{u(t)}{c_1 u(t) + c_2 v_0(t)} = \lim_{t \to \omega^-} \frac{\frac{u(t)}{v_0(t)}}{c_1 \frac{u(t)}{v_0(t)} + c_2} = 0.$$

Let v be a fixed solution of (4.1) such that u and v are linearly independent. Pick $t_0 \in [a, \omega)$ so that $v(t)v^\sigma(t) > 0$ on $[t_0, \omega)$. Then consider for $t \in [t_0, \omega)$

$$\left(\frac{u}{v}\right)^\Delta(t) = \frac{v(t)u^\Delta(t) - u(t)v^\Delta(t)}{v(t)v^\sigma(t)} = \frac{W(v, u)(t)}{v(t)v^\sigma(t)} = \frac{C_1}{p(t)v(t)v^\sigma(t)},$$

where $C_1 \neq 0$ is constant by Abel's formula. Integrating both sides of this last equality from t_0 to t we obtain

$$\frac{u(t)}{v(t)} - \frac{u(t_0)}{v(t_0)} = \int_{t_0}^t \frac{C_1}{p(s)v(s)v^\sigma(s)} \Delta s.$$

Letting $t \to \omega^-$ we get another one of the desired results

$$\int_{t_0}^\omega \frac{1}{p(t)v(t)v^\sigma(t)} \Delta t < \infty.$$

We now show that (4.31) holds for $t < \omega$ sufficiently close. Above we saw that $v(t)v^\sigma(t) > 0$ on $[t_0, \omega)$. Note that the expression

$$\frac{p(t)v^\Delta(t)}{v(t)}$$

is the same if $v(t)$ is replaced by $-v(t)$. Hence without loss of generality we can assume $v > 0$ on $[t_0, \omega)$. For $t \in [t_0, \omega)$, consider

$$\frac{p(t)v^\Delta(t)}{v(t)} - \frac{p(t)u^\Delta(t)}{u(t)} = \frac{p(t)W(u, v)(t)}{v(t)u(t)} = \frac{C_2}{v(t)u(t)},$$

where $C_2 = p(t)W(u,v)(t)$ is a (nonzero) constant by Abel's formula. It remains to show that $C_2 > 0$. Since

$$\lim_{t\to\omega^-} \frac{v(t)}{u(t)} = \infty \quad \text{and} \quad \left(\frac{v}{u}\right)^\Delta (t) = \frac{W(u,v)(t)}{u(t)u^\sigma(t)} = \frac{C_2}{p(t)u(t)u^\sigma(t)},$$

we get the desired result that $C_2 > 0$, and this completes the proof. $\square$

Example 4.62. We will find solutions u and v of

$$\left[\left(\frac{1}{6}\right)^t x^\Delta\right]^\Delta + 2\left(\frac{1}{6}\right)^{t+1} x^\sigma = 0 \quad \text{for} \quad t \in \mathbb{T} = \mathbb{N}$$

and show directly that the conclusions of Theorem 4.61 are satisfied for these two solutions. Since $\mathbb{T} = \mathbb{N}$, our dynamic equation on this time scale becomes the difference equation

$$\Delta\left[\left(\frac{1}{6}\right)^t \Delta x(t)\right] + 2\left(\frac{1}{6}\right)^{t+1} x(t+1) = 0.$$

Expanding this equation and simplifying we get the recurrence relation

$$x(t+2) - 5x(t+1) + 6x(t) = 0.$$

We take $u(t) = 2^t$ and $v(t) = 3^t$ as our two solutions. Since u is positive, we know that the conclusions of Theorem 4.61 are true. To verify this directly, note that

$$\lim_{t\to\infty} \frac{u(t)}{v(t)} = \lim_{t\to\infty} \frac{2^t}{3^t} = \lim_{t\to\infty} \left(\frac{2}{3}\right)^t = 0.$$

Next consider

$$\int_0^\infty \frac{1}{p(t)u(t)u^\sigma(t)} \Delta t = \sum_{t=0}^\infty \frac{1}{p(t)u(t)u^\sigma(t)} = \frac{1}{2}\sum_{t=0}^\infty \left(\frac{3}{2}\right)^t = \infty.$$

Also

$$\int_0^\infty \frac{1}{p(t)v(t)v^\sigma(t)} \Delta t = \sum_{t=0}^\infty \frac{1}{p(t)v(t)v^\sigma(t)} = \frac{1}{3}\sum_{t=0}^\infty \left(\frac{2}{3}\right)^t < \infty.$$

Finally note that

$$\frac{p(t)v^\Delta(t)}{v(t)} = 2\left(\frac{1}{6}\right)^t > \left(\frac{1}{6}\right)^t = \frac{p(t)u^\Delta(t)}{u(t)} \quad \text{for} \quad t \in \mathbb{N}.$$

Exercise 4.63. Find solutions u and v of the following dynamic equations on the given time scales and show directly that the conclusions of Theorem 4.61 are satisfied for these two solutions:

(i) $\left(e^{-4t}x^\Delta\right)^\Delta + 4e^{-4t}x^\sigma = 0$, where $\mathbb{T} = [0,\infty)$;
(ii) $x^{\Delta\Delta} + x^\sigma = 0$, where $\mathbb{T} = [\varepsilon,\pi]$ with $0 < \varepsilon < \pi$;
(iii) $\left[\left(\frac{1}{8}\right)^t x^\Delta\right]^\Delta + 3\left(\frac{1}{8}\right)^{t+1} x^\sigma = 0$, where $\mathbb{T} = \mathbb{N}$;
(iv) $\left(\frac{1}{t^4}x^\Delta\right)^\Delta + \frac{6}{t^6}x^\sigma = 0$, where $\mathbb{T} = [1,\infty)$.

Theorem 4.64 (Leighton–Wintner Theorem). *Assume $a \in \mathbb{T}$, $p > 0$, $\sup \mathbb{T} = \infty$, and*

(4.32)
$$\int_a^\infty \frac{1}{p(t)} \Delta t = \int_a^\infty q(t)\Delta t = \infty.$$

Then (4.1) is oscillatory on $[a,\infty)$.

Proof. Assume (4.1) is nonoscillatory on $[a, \infty)$. Then (4.1) is disconjugate in a neighborhood of ∞. By Theorem 4.61, there is a dominant solution x at ∞ satisfying

$$(4.33) \qquad \int_T^\infty \frac{1}{p(t)x(t)x^\sigma(t)} \Delta t < \infty$$

for $T \in \mathbb{T}$ sufficiently large. Define z by the Riccati substitution (4.16) on $[T, \infty)$. Then, by Theorem 4.36, z satisfies (4.19) for $t \in [T, \infty)$ and z is a solution of the Riccati equation (4.18) on $[T, \infty)$. It follows that

$$z^\Delta(t) \le -q(t) \quad \text{for all} \quad t \ge T.$$

Integrating both sides of this last inequality from T to t we get that

$$z(t) - z(T) \le - \int_T^t q(\tau) \Delta \tau.$$

It follows that $\lim_{t\to\infty} z(t) = -\infty$. But then there exists $T_1 \in \mathbb{T}$ sufficiently large so that $T_1 \ge T$ and

$$z(t) = \frac{p(t)x^\Delta(t)}{x(t)} < 0 \quad \text{for all} \quad t \ge T_1.$$

Without loss of generality we can assume that $x(t) > 0$ on $[T_1, \infty)$, and then it follows that x is a positive decreasing function on $[T_1, \infty)$. But then we get that

$$\int_{T_1}^\infty \frac{1}{p(t)x(t)x^\sigma(t)} \Delta t \ge \frac{1}{x^2(T_1)} \int_{T_1}^\infty \frac{1}{p(t)} \Delta t = \infty,$$

which contradicts (4.33). $\qquad\qquad\qquad\qquad\qquad\qquad\qquad\qquad\qquad\qquad\square$

Exercise 4.65. Show that if $\mathbb{T} = q^{\mathbb{N}_0}$, where $q > 1$, then the q-difference equation

$$x^{\Delta\Delta} + \frac{1}{t \ln(2t)} x^\sigma = 0$$

is oscillatory on $\mathbb{T}$.

Exercise 4.66. Show that the Leighton–Wintner theorem (Theorem 4.64) cannot be applied to the q-difference equation (4.24) even though this equation is oscillatory on $\mathbb{T} = q^{\mathbb{N}_0}$.

4.4. Boundary Value Problems and Green's Function

In this section we are concerned with Green's functions for a general two-point boundary value problem (abbreviated by BVP) for (4.1). Many of the results given in this section can be found in Erbe and Peterson [**131, 133**].

Theorem 4.67 (Existence and Uniqueness of Solutions for General Two Point BVPs). *Assume that the homogeneous boundary value problem*

$$(4.34) \qquad Lx = 0, \quad \alpha x(a) - \beta x^\Delta(a) = \gamma x(\sigma^2(b)) + \delta x^\Delta(\sigma(b)) = 0$$

has only the trivial solution. Then the nonhomogeneous BVP

$$(4.35) \qquad Lx = h(t), \quad \alpha x(a) - \beta x^\Delta(a) = A, \quad \gamma x(\sigma^2(b)) + \delta x^\Delta(\sigma(b)) = B,$$

where A and B are given constants and $h \in C_{\mathrm{rd}}$, has a unique solution.

Proof. Let u and v be linearly independent solutions of (4.1). Then a general solution of (4.1) is given by

$$x(t) = c_1 u(t) + c_2 v(t).$$

Then x satisfies the boundary conditions in (4.34) iff

$$c_1[\alpha u(a) - \beta u^\Delta(a)] + c_2[\alpha v(a) - \beta v^\Delta(a)] = 0$$

and

$$c_1[\gamma u(\sigma^2(b)) + \delta u^\Delta(\sigma(b))] + c_2[\gamma v(\sigma^2(b)) + \delta v^\Delta(\sigma(b))] = 0,$$

i.e., iff

$$M \begin{pmatrix} c_1 \\ c_2 \end{pmatrix} = 0 \quad \text{with} \quad M = \begin{pmatrix} \alpha u(a) - \beta u^\Delta(a) & \alpha v(a) - \beta v^\Delta(a) \\ \gamma u(\sigma^2(b)) + \delta u^\Delta(\sigma(b)) & \gamma v(\sigma^2(b)) + \delta v^\Delta(\sigma(b)) \end{pmatrix}.$$

Since we are assuming that the homogeneous BVP (4.34) has only the trivial solution, we get that

$$c_1 = c_2 = 0$$

is the unique solution of the above linear system. Hence

(4.36) $\det M \neq 0.$

Now let w be a solution of the nonhomogeneous equation $Lx = h(t)$. Then a general solution of $Lx = h(t)$ is

$$x(t) = a_1 u(t) + a_2 v(t) + w(t).$$

Then x satisfies the boundary conditions in (4.35) iff

$$M \begin{pmatrix} a_1 \\ a_2 \end{pmatrix} = \begin{pmatrix} A - [\alpha w(a) - \beta w^\Delta(a)] \\ B - [\gamma w(\sigma^2(b)) + \delta w^\Delta(\sigma(b))] \end{pmatrix}.$$

Using (4.36), we get that this system has a unique solution a_1, a_2. But this implies that the BVP (4.35) has a unique solution.   □

In the next example we give a BVP of the type (4.34) which does not have just the trivial solution. In Exercise 4.69 we give a necessary and sufficient condition for some boundary value problems of the form (4.34) to have only the trivial solution.

Example 4.68. Find all solutions of the BVP

(4.37) $(p(t)x^\Delta)^\Delta = 0, \quad x^\Delta(a) = x^\Delta(\sigma(b)) = 0,$

where $a < b$. The BVP (4.37) is equivalent to a BVP of the form (4.34), where $q(t) \equiv 0$, $\alpha = \gamma = 0$, $\beta = \delta = 1$. A general solution of the equation in (4.37) is

$$x(t) = c_1 + c_2 \int_a^t \frac{1}{p(s)} \Delta s.$$

The boundary conditions lead to the equations

$$x^\Delta(a) = \frac{c_2}{p(a)} = 0 \quad \text{and} \quad x^\Delta(\sigma(b)) = \frac{c_2}{p(\sigma(b))} = 0.$$

Thus we can take $c_2 = 0$ and there is no restriction on c_1. Hence $x(t) = c_1$ solves (4.37) for any constant c_1. In particular, (4.37) has nontrivial solutions.

Exercise 4.69. Let

$$D = \alpha\gamma \int_a^{\sigma^2(b)} \frac{1}{p(s)} \Delta s + \frac{\beta\gamma}{p(a)} + \frac{\alpha\delta}{p(\sigma(b))}.$$

Show that the BVP (4.34) with $q = 0$ has only the trivial solution iff $D \neq 0$. Show that $D = 0$ for the BVP in Example 4.68.

Theorem 4.70 (Green's Function for General Two Point BVPs). *Assume that the BVP (4.34) has only the trivial solution. For each fixed $s \in [a, b]$, let $u(\cdot, s)$ be the unique solution of the BVP*

$$Lu(\cdot, s) = 0, \quad \alpha u(a, s) - \beta u^\Delta(a, s) = 0,$$

(4.38) $\gamma u(\sigma^2(b), s) + \delta u^\Delta(\sigma(b), s) = -\gamma x(\sigma^2(b), s) - \delta x^\Delta(\sigma(b), s),$

where $x(t, s)$ is the Cauchy function for (4.1). Then we define Green's function $G : [a, \sigma^2(b)] \times [a, b] \to \mathbb{R}$ for the BVP (4.34) by

$$G(t, s) = \begin{cases} u(t, s) & \text{if} \quad t \leq s \\ v(t, s) & \text{if} \quad t \geq \sigma(s), \end{cases}$$

where $v(t, s) := u(t, s) + x(t, s)$ for $t \in [a, \sigma^2(b)]$, $s \in [a, b]$. Then for each fixed $s \in [a, b]$, $v(\cdot, s)$ is a solution of (4.1) and satisfies the second boundary condition in (4.34). If $h \in C_{rd}$, then

$$x(t) := \int_a^{\sigma(b)} G(t, s) h(s) \Delta s$$

is the solution of the nonhomogeneous BVP (4.35) with $A = B = 0$.

Proof. The existence and uniqueness of $u(t, s)$ is guaranteed by Theorem 4.67. Since for each fixed $s \in [a, b]$, $u(\cdot, s)$ and $x(\cdot, s)$ are solutions of (4.1), we have for each fixed $s \in [a, b]$ that $v(\cdot, s) = u(\cdot, s) + x(\cdot, s)$ is also a solution of (4.1). It follows from (4.38) that $v(\cdot, s)$ satisfies the second boundary condition in (4.34) for each fixed $s \in [a, b]$. First note that

$$\begin{aligned}
x(t) &= \int_a^{\sigma(b)} G(t, s) h(s) \Delta s \\
&= \int_a^t G(t, s) h(s) \Delta s + \int_t^{\sigma(b)} G(t, s) h(s) \Delta s \\
&= \int_a^t v(t, s) h(s) \Delta s + \int_t^{\sigma(b)} u(t, s) h(s) \Delta s \\
&= \int_a^t [u(t, s) + x(t, s)] h(s) \Delta s + \int_t^{\sigma(b)} u(t, s) h(s) \Delta s \\
&= \int_a^{\sigma(b)} u(t, s) h(s) \Delta s + \int_a^t x(t, s) h(s) \Delta s \\
&= \int_a^{\sigma(b)} u(t, s) h(s) \Delta s + z(t),
\end{aligned}$$

where, by the variation of constants formula in Theorem 4.24, z is the solution of the IVP

$$Lz = h(t), \quad z(a) = z^\Delta(a) = 0.$$

It follows that

$$Lx(t) = \int_a^{\sigma(b)} Lu(t, s)h(s)\Delta s + Lz(t) = Lz(t) = h(t).$$

Hence x is a solution of the nonhomogeneous equation $Lx = h(t)$. It remains to show that x satisfies the two boundary conditions in (4.34). Now

$$
\begin{aligned}
\alpha x(a) - \beta x^\Delta(a) &= \int_a^{\sigma(b)} [\alpha u(a, s) - \beta u^\Delta(a, s)]h(s)\Delta s + \alpha z(a) - \beta z^\Delta(a) \\
&= \int_a^{\sigma(b)} [\alpha u(a, s) - \beta u^\Delta(a, s)]h(s)\Delta s \\
&= 0,
\end{aligned}
$$

since for each fixed s, $u(\cdot, s)$ satisfies the first boundary condition in (4.34). Hence x satisfies the first boundary condition in (4.34). From earlier in this proof,

$$
\begin{aligned}
x(t) &= \int_a^{\sigma(b)} u(t, s)h(s)\Delta s + \int_a^t x(t, s)h(s)\Delta s \\
&= \int_a^{\sigma(b)} v(t, s)h(s)\Delta s - \int_t^{\sigma(b)} x(t, s)h(s)\Delta s \\
&= \int_a^{\sigma(b)} v(t, s)h(s)\Delta s + \int_{\sigma(b)}^t x(t, s)h(s)\Delta s \\
&= \int_a^{\sigma(b)} v(t, s)h(s)\Delta s + w(t),
\end{aligned}
$$

where, by the variation of constants formula in Theorem 4.24, w solves

$$Lw = h(t), \quad w(\sigma(b)) = w^\Delta(\sigma(b)) = 0.$$

Note that

$$w(\sigma^2(b)) = w(\sigma(b)) + \mu(\sigma(b))w^\Delta(\sigma(b)) = 0.$$

Then

$$
\begin{aligned}
\gamma x(\sigma^2(b)) + \delta x^\Delta(\sigma(b)) &= \int_a^{\sigma(b)} \left[\gamma v(\sigma^2(b), s) + \delta v^\Delta(\sigma(b), s)\right] h(s)\Delta s \\
&\quad + \gamma w(\sigma^2(b)) + \delta w^\Delta(\sigma(b)) \\
&= \int_a^{\sigma(b)} \left[\gamma v(\sigma^2(b), s) + \delta v^\Delta(\sigma(b), s)\right] h(s)\Delta s \\
&= 0,
\end{aligned}
$$

since for each fixed s, $v(\cdot, s)$ satisfies the second boundary condition in (4.34). Hence x satisfies the second boundary condition in (4.34). The uniqueness of solutions of the nonhomogeneous BVP (4.35) (with $A = B = 0$) follows from Theorem 4.67. □

Exercise 4.71. Show that if the BVP (4.34) has only the trivial solution, then Green's function for the BVP (4.34) satisfies the *jump condition*

$$G^\Delta(s^+, s) - G^\Delta(s^-, s) = \frac{1}{p(s)}$$

for $s \in (a, \sigma^2(b))$ such that $\sigma(s) = s$.

Exercise 4.72. Show that if $\mathbb{T}$ is a time scale consisting of only isolated points and if the BVP (4.34) has only the trivial solution, then Green's function for (4.34) satisfies for each fixed $s \in [a, b]$,

$$LG(t, s) = \frac{\delta_{ts}}{\mu(t)}$$

for $t \in [a, b]$, where δ_{ts} is the Kronecker delta function.

In the next theorem we give another form of Green's function for the BVP (4.34) and note that Green's function satisfies a symmetry condition on the square $[a, b] \times [a, b]$.

Theorem 4.73. *Assume that the BVP* (4.34) *has only the trivial solution. Let ϕ be the solution of the IVP*

$$L\phi = 0, \quad \phi(a) = \beta, \quad \phi^\Delta(a) = \alpha,$$

and let ψ be the solution of

$$L\psi = 0, \quad \psi(\sigma^2(b)) = \delta, \quad \psi^\Delta(\sigma(b)) = -\gamma.$$

Then Green's function for the BVP (4.34) *is given by*

$$(4.39) \qquad G(t, s) = \begin{cases} \frac{1}{c}\phi(t)\psi(\sigma(s)) & \text{if} \quad t \leq s \\ \frac{1}{c}\psi(t)\phi(\sigma(s)) & \text{if} \quad t \geq \sigma(s), \end{cases}$$

where $c := p(t)W(\phi, \psi)(t)$ is a constant. Furthermore, Green's function satisfies the symmetry condition

$$G(t, \sigma(s)) = G(\sigma(s), t) \quad \text{for} \quad s, t \in [a, b].$$

Proof. Let ϕ, ψ, and c be as in the statement of this theorem. By Abel's theorem (Corollary 4.31), $c = p(t)W(\phi, \psi)(t)$ is constant. We use Theorem 4.70 to prove that G defined by (4.39) is Green's function for the BVP (4.34). Note that

$$\alpha\phi(a) - \beta\phi^\Delta(a) = \alpha\beta - \beta\alpha = 0$$

and

$$\gamma\psi(\sigma^2(b)) + \delta\psi^\Delta(\sigma(b)) = \gamma\delta - \delta\gamma = 0.$$

Hence ϕ and ψ satisfy the first and second boundary condition in (4.34), respectively. Let

$$u(t, s) := \frac{1}{c}\phi(t)\psi(\sigma(s)) \quad \text{and} \quad v(t, s) := \frac{1}{c}\psi(t)\phi(\sigma(s))$$

for $t \in [a, \sigma^2(b)]$, $s \in [a, b]$. Note that for each fixed $s \in [a, b]$, $u(\cdot, s)$ and $v(\cdot, s)$ are solutions of (4.1) on $[a, \sigma^2(b)]$. Also for each fixed $s \in [a, b]$,

$$\alpha u(a, s) - \beta u^\Delta(a, s) = \frac{1}{c}\psi(\sigma(s))[\alpha\phi(a) - \beta\phi^\Delta(a)] = 0$$

and

$$\gamma v(\sigma^2(b), s) + \delta v^\Delta(\sigma(b), s) = \frac{1}{c}\phi(\sigma(s))[\gamma\psi(\sigma^2(b)) + \delta\psi^\Delta(\sigma(b))] = 0.$$

Hence for each fixed $s \in [a, b]$, $u(\cdot, s)$ and $v(\cdot, s)$ satisfy the first and second boundary condition in (4.34), respectively. Let

$$k(t, s) := v(t, s) - u(t, s) = \frac{1}{c}[\phi(\sigma(s))\psi(t) - \phi(t)\psi(\sigma(s))].$$

It follows that for each fixed s, $k(\cdot, s)$ is a solution of (4.1). Also $k(\sigma(s), s) = 0$ and

$$
\begin{aligned}
k^\Delta(\sigma(s), s) &= \frac{1}{c}[\phi(\sigma(s))\psi^\Delta(\sigma(s)) - \phi^\Delta(\sigma(s))\psi(\sigma(s))] \\
&= \frac{\phi(\sigma(s))\psi^\Delta(\sigma(s)) - \phi^\Delta(\sigma(s))\psi(\sigma(s))}{p(\sigma(s))W(\phi, \psi)(\sigma(s))} \\
&= \frac{1}{p(\sigma(s))}.
\end{aligned}
$$

Therefore $k(t, s) = x(t, s)$ is the Cauchy function for (4.1) and we have

$$
v(t, s) = u(t, s) + x(t, s).
$$

It remains to prove that for each fixed s, $u(\cdot, s)$ satisfies (4.38). To see this consider

$$
\begin{aligned}
\gamma u(\sigma^2(b), s) + \delta u^\Delta(\sigma(b), s) &= \gamma v(\sigma^2(b), s) + \delta v^\Delta(\sigma(b), s) \\
&\quad - [\gamma x(\sigma^2(b), s) + \delta x^\Delta(\sigma(b), s)] \\
&= -[\gamma x(\sigma^2(b), s) + \delta x^\Delta(\sigma(b), s)].
\end{aligned}
$$

Hence by Theorem 4.70, $G(t, s)$ defined by (4.39) is Green's function for (4.34). It follows from (4.39) that G satisfies the condition $G(t, \sigma(s)) = G(\sigma(s), t)$ for $s, t \in [a, b]$. $\square$

In the following corollary we essentially give the special case of Theorem 4.70, where the boundary conditions are *conjugate (Dirichlet) boundary conditions*.

Corollary 4.74 (Green's Function for the Conjugate Problem). *Assume the BVP*

$$(4.40) \qquad\qquad Lx = 0, \quad x(a) = x(\sigma^2(b)) = 0.$$

has only the trivial solution. Let $x(t, s)$ be the Cauchy function for (4.1). For each fixed $s \in \mathbb{T}$, let $u(\cdot, s)$ be the unique solution of the BVP

$$
Lu(\cdot, s) = 0, \quad u(a, s) = 0, \quad u(\sigma^2(b), s) = -x(\sigma^2(b), s).
$$

Then

$$
G(t, s) = \begin{cases} u(t, s) & \text{if} \quad t \le s \\ v(t, s) & \text{if} \quad t \ge \sigma(s), \end{cases}
$$

where $v(t, s) = u(t, s) + x(t, s)$, is Green's function for the BVP (4.40). Moreover, for each fixed $s \in [a, b]$, $v(\cdot, s)$ is a solution of (4.1) and $v(\sigma^2(b), s) = 0$.

Proof. This corollary follows from Theorem 4.70 with $\alpha = \gamma = 1$ and $\beta = \delta = 0$. $\square$

Corollary 4.75. *Green's function for the BVP (4.40) with $q = 0$ is given by*

$$
G(t, s) = \begin{cases} -\dfrac{\int_a^t \frac{1}{p(\tau)}\Delta\tau \int_{\sigma(s)}^{\sigma^2(b)} \frac{1}{p(\tau)}\Delta\tau}{\int_a^{\sigma^2(b)} \frac{1}{p(\tau)}\Delta\tau} & \text{if} \quad t \le s \\[6mm] -\dfrac{\int_a^{\sigma(s)} \frac{1}{p(\tau)}\Delta\tau \int_t^{\sigma^2(b)} \frac{1}{p(\tau)}\Delta\tau}{\int_a^{\sigma^2(b)} \frac{1}{p(\tau)}\Delta\tau} & \text{if} \quad t \ge \sigma(s). \end{cases}
$$

Proof. It is easy to check that the BVP

$$
(p(t)x^\Delta)^\Delta = 0, \quad x(a) = 0, \quad x(\sigma^2(b)) = 0
$$

has only the trivial solution. By Example 4.22, the Cauchy function for $(px^\Delta)^\Delta = 0$ is given by

$$x(t,s) = \int_{\sigma(s)}^{t} \frac{1}{p(\tau)} \Delta\tau.$$

By Corollary 4.74, $u(\cdot, s)$ from the statement of Corollary 4.74 solves $(px^\Delta)^\Delta = 0$ for each fixed $s \in [a, b]$ and satisfies

$$(4.41) \qquad u(a,s) = 0 \quad \text{and} \quad u(\sigma^2(b), s) = -x(\sigma^2(b), s) = -\int_{\sigma(s)}^{\sigma^2(b)} \frac{1}{p(\tau)} \Delta\tau.$$

Since

$$x_1(t) = 1 \quad \text{and} \quad x_2(t) = \int_{a}^{t} \frac{1}{p(\tau)} \Delta\tau$$

are solutions of $(px^\Delta)^\Delta = 0$,

$$u(t,s) = \alpha(s) \cdot 1 + \beta(s) \int_{a}^{t} \frac{1}{p(\tau)} \Delta\tau.$$

Using the boundary conditions (4.41), it can be shown that

$$u(t,s) = -\frac{\int_{a}^{t} \frac{1}{p(\tau)} \Delta\tau \int_{\sigma(s)}^{\sigma^2(b)} \frac{1}{p(\tau)} \Delta\tau}{\int_{a}^{\sigma^2(b)} \frac{1}{p(\tau)} \Delta\tau}.$$

Hence $G(t, s)$ has the desired form for $t \le s$. By Corollary 4.74 for $t \ge \sigma(s)$,

$$G(t,s) = x(t,s) + u(t,s).$$

Therefore

$$
\begin{aligned}
G(t,s) &= \int_{\sigma(s)}^{t} \frac{1}{p(\tau)} \Delta\tau - \frac{\int_{a}^{t} \frac{1}{p(\tau)} \Delta\tau \int_{\sigma(s)}^{\sigma^2(b)} \frac{1}{p(\tau)} \Delta\tau}{\int_{a}^{\sigma^2(b)} \frac{1}{p(\tau)} \Delta\tau} \\[2ex]
&= \frac{\int_{\sigma(s)}^{t} \frac{1}{p(\tau)} \Delta\tau \int_{a}^{\sigma^2(b)} \frac{1}{p(\tau)} \Delta\tau - \int_{a}^{t} \frac{1}{p(\tau)} \Delta\tau \int_{\sigma(s)}^{\sigma^2(b)} \frac{1}{p(\tau)} \Delta\tau}{\int_{a}^{\sigma^2(b)} \frac{1}{p(\tau)} \Delta\tau} \\[2ex]
&= \frac{\int_{\sigma(s)}^{t} \frac{1}{p(\tau)} \Delta\tau \left[\int_{a}^{t} \frac{1}{p(\tau)} \Delta\tau + \int_{t}^{\sigma^2(b)} \frac{1}{p(\tau)} \Delta\tau\right] - \int_{a}^{t} \frac{1}{p(\tau)} \Delta\tau \int_{\sigma(s)}^{\sigma^2(b)} \frac{1}{p(\tau)} \Delta\tau}{\int_{a}^{\sigma^2(b)} \frac{1}{p(\tau)} \Delta\tau} \\[2ex]
&= \frac{\int_{a}^{t} \frac{1}{p(\tau)} \Delta\tau \left[-\int_{t}^{\sigma^2(b)} \frac{1}{p(\tau)} \Delta\tau\right] + \int_{\sigma(s)}^{t} \frac{1}{p(\tau)} \Delta\tau \int_{t}^{\sigma^2(b)} \frac{1}{p(\tau)} \Delta\tau}{\int_{a}^{\sigma^2(b)} \frac{1}{p(\tau)} \Delta\tau} \\[2ex]
&= -\frac{\int_{t}^{\sigma^2(b)} \frac{1}{p(\tau)} \Delta\tau \left[\int_{a}^{t} \frac{1}{p(\tau)} \Delta\tau - \int_{\sigma(s)}^{t} \frac{1}{p(\tau)} \Delta\tau\right]}{\int_{a}^{\sigma^2(b)} \frac{1}{p(\tau)} \Delta\tau} \\[2ex]
&= -\frac{\int_{a}^{\sigma(s)} \frac{1}{p(\tau)} \Delta\tau \int_{t}^{\sigma^2(b)} \frac{1}{p(\tau)} \Delta\tau}{\int_{a}^{\sigma^2(b)} \frac{1}{p(\tau)} \Delta\tau},
\end{aligned}
$$

which is the desired result. $\square$

For the special case of Corollary 4.75 where $\mathbb{T} = \mathbb{Z}$, Green's function is given in [**30**, Example 1.26]. See [**30**, Section 7.5] for the vector case. The following corollary follows immediately from Corollary 4.75 by taking $p(t) \equiv 1$.

Corollary 4.76. *The Green function for the BVP*

$$x^{\Delta\Delta} = 0, \quad x(a) = x(\sigma^2(b)) = 0$$

is given by

$$G(t,s) = \begin{cases} -\dfrac{(t-a)\left(\sigma^2(b)-\sigma(s)\right)}{\sigma^2(b)-a} & \text{if} \quad t \leq s \\ -\dfrac{(\sigma(s)-a)\left(\sigma^2(b)-t\right)}{\sigma^2(b)-a} & \text{if} \quad t \geq \sigma(s). \end{cases}$$

Example 4.77. In this example we use an appropriate Green function to solve the BVP

$$(4.42) \qquad x^{\Delta\Delta} = 1, \quad x(0) = x(50) = 0, \quad \text{where} \quad \mathbb{T} = \mathbb{Z}.$$

From Corollary 4.76 we get that Green's function for (4.42) is given by

$$G(t,s) = \begin{cases} \dfrac{t(s-49)}{50} & \text{if} \quad t \leq s \\ \dfrac{(s+1)(t-50)}{50} & \text{if} \quad t \geq \sigma(s). \end{cases}$$

Hence the solution of (4.42) is given by

$$\begin{aligned} x(t) &= \int_0^{50} G(t,s) \cdot 1 \Delta s \\ &= \int_0^t \frac{(s+1)(t-50)}{50} \Delta s + \int_t^{50} \frac{t(s-49)}{50} \Delta s \\ &= \frac{t-50}{50} \sum_{s=0}^{t-1}(s+1) + \frac{t}{50} \sum_{s=t}^{49}(s-49) \\ &= \frac{t(t+1)(t-50)}{100} - \frac{t(t-49)(t-50)}{100} \\ &= \frac{t(t-50)}{2}. \end{aligned}$$

Exercise 4.78. Use an appropriate Green function to solve each of the following boundary value problems on the indicated time scales:

   (i) $x^{\Delta\Delta} = 1$, $x(0) = x(50) = 0$, where $\mathbb{T} = \mathbb{R}$;
   (ii) $\left(3^t x^{\Delta}\right)^{\Delta} = 2^t$, $x(0) = x(10) = 0$, where $\mathbb{T} = \mathbb{Z}$;
   (iii) $x^{\Delta\Delta} = 1$, $x(0) = x(1) = 0$, where $\mathbb{T} = \overline{2^{\mathbb{Z}}}$;
   (iv) $\left(e^t x^{\Delta}\right)^{\Delta} = e^{2t}$, $x(0) = x(1) = 0$, where $\mathbb{T} = \mathbb{R}$.

Exercise 4.79. Let $G(t,s)$ be Green's function for the BVP

$$x^{\Delta\Delta} = 0, \quad x(a) = x(\sigma^2(b)) = 0.$$

Prove that if $\mathbb{T} = \mathbb{R}$ or if $\mathbb{T} = \mathbb{Z}$, then

$$\int_a^{\sigma(b)} |G(t,s)| \Delta s \leq \frac{(\sigma^2(b) - a)^2}{8}$$

for $t \in [a, \sigma^2(b)]$.

Theorem 4.80. *If $p > 0$ and (4.1) is disconjugate on $[a, \sigma^2(b)]$, then Green's function for the conjugate BVP (4.40) exists and satisfies*

$$G(t, s) < 0 \quad \text{for} \quad t \in (a, \sigma^2(b)), \ s \in (a, b).$$

Proof. Since (4.1) is disconjugate on $[a, \sigma^2(b)]$, Green's function

$$G(t, s) := \begin{cases} u(t, s) & \text{if} \quad t \leq s \\ v(t, s) & \text{if} \quad t \geq \sigma(s) \end{cases}$$

for (4.40) exists by Corollary 4.74. For each fixed $s \in [a, b]$, $u(\cdot, s)$ is the solution of (4.1) satisfying

$$u(a, s) = 0, \quad u(\sigma^2(b), s) = -x(\sigma^2(b), s) < 0,$$

where the last inequality is true since $x(\sigma(s), s) = 0$ and (4.1) is disconjugate on $[a, \sigma^2(b)]$. But then it follows that $u(t, s) < 0$ for $t \in (a, \sigma^2(b))$. Also for each fixed $s \in [a, b]$, $v(\cdot, s)$ is the solution of (4.1) satisfying

$$v(\sigma^2(b), s) = 0, \quad v(\sigma(s), s) = u(\sigma(s), s) + x(\sigma(s), s) = u(\sigma(s), s) < 0.$$

Now $v(t, s) < 0$ for $t \in [a, \sigma^2(b))$, and therefore $G(t, s) < 0$ for $t \in (a, \sigma^2(b))$ and $s \in (a, b)$. $\qquad\square$

Theorem 4.81 (Comparison Theorem for Conjugate BVPs). *Let $p > 0$ and assume that (4.1) is disconjugate on $[a, \sigma^2(b)]$. If $u, v \in \mathbb{D}$ satisfy*

$$Lu(t) \leq Lv(t) \quad \text{for all} \ t \in [a, b], \quad u(a) \geq v(a), \quad u(\sigma^2(b)) \geq v(\sigma^2(b)),$$

then

$$u(t) \geq v(t) \quad \text{for all} \quad t \in [a, \sigma^2(b)].$$

Proof. Let u and v be as in the statement of the theorem and set

$$w(t) = u(t) - v(t) \quad \text{for} \quad t \in [a, \sigma^2(b)].$$

If

$$h(t) := Lw(t) \quad \text{for} \quad t \in [a, b],$$

then

$$h(t) = Lu(t) - Lv(t) \leq 0 \quad \text{for} \quad t \in [a, b].$$

It follows that w solves the BVP

$$Lw = h(t) \text{ for } t \in [a, b], \quad w(a) = A, \quad w(\sigma^2(b)) = B,$$

where

$$A := u(a) - v(a) \geq 0 \quad \text{and} \quad B := u(\sigma^2(b)) - v(\sigma^2(b)) \geq 0.$$

Hence

$$w(t) = \phi(t) + \int_a^{\sigma(b)} G(t, s) h(s) \Delta s,$$

where ϕ solves the BVP

$$L\phi = 0 \text{ on } [a, b], \quad \phi(a) = A, \quad \phi(\sigma^2(b)) = B.$$

From disconjugacy,

$$\phi(t) \geq 0 \quad \text{for all} \quad t \in [a, \sigma^2(b)]$$

and

$$G(t, s) \leq 0 \quad \text{for all} \quad (t, s) \in [a, \sigma^2(b)] \times [a, b].$$

Hence $w(t) \geq 0$, which gives the desired result. $\qquad\square$

Next we consider the *focal* BVP

(4.43) $Lx = 0, \quad x(a) = x^\Delta(\sigma(b)) = 0.$

Corollary 4.82 (Green's Function for Focal BVPs). *Assume that the BVP (4.43) has only the trivial solution. For each fixed $s \in [a, b]$, let $u(\cdot, s)$ be the solution of the BVP*

$$Lu(\cdot, s) = 0, \quad u(a, s) = 0, \quad u^\Delta(\sigma(b), s) = -x^\Delta(\sigma(b), s),$$

where $x(t, s)$ is the Cauchy function of (4.1). Then

$$G(t, s) = \begin{cases} u(t, s) & \text{if} \quad t \leq s \\ u(t, s) + x(t, s) & \text{if} \quad t \geq \sigma(s) \end{cases}$$

is Green's function for the focal BVP (4.43).

Proof. This follows from Theorem 4.70 with $\alpha = \delta = 1$ and $\beta = \gamma = 0$.   □

Corollary 4.83. *The Green function for the focal BVP (4.43) with $q = 0$ is given by*

$$G(t, s) = \begin{cases} -\int_a^t \frac{1}{p(\tau)} \Delta\tau & \text{if} \quad t \leq s \\ -\int_a^{\sigma(s)} \frac{1}{p(\tau)} \Delta\tau & \text{if} \quad t \geq \sigma(s). \end{cases}$$

Proof. It is easy to see that (4.43) with $q = 0$ has only the trivial solution. Hence we can apply Corollary 4.82 to find the focal Green's function $G(t, s)$. For $t \leq s$, $G(t, s) = u(t, s)$, where for each fixed s, $u(\cdot, s)$ solves the BVP

$$Lu(\cdot, s) = 0, \quad u(a, s) = 0, \quad u^\Delta(\sigma(b), \sigma(s)) = -x^\Delta(\sigma(b), s),$$

and where $x(t, s)$ is the Cauchy function for $(px^\Delta)^\Delta = 0$. Solving this BVP, we get that

$$u(t, s) = -\int_a^t \frac{1}{p(\tau)} \Delta\tau,$$

which is the desired expression for $G(t, s)$ if $t \leq s$. If $t \geq \sigma(s)$, then

$$\begin{aligned} G(t, s) &= u(t, s) + x(t, s) \\ &= -\int_a^t \frac{1}{p(\tau)} \Delta\tau + \int_{\sigma(s)}^t \frac{1}{p(\tau)} \Delta\tau \\ &= -\int_a^{\sigma(s)} \frac{1}{p(\tau)} \Delta\tau, \end{aligned}$$

where we used Example 4.22 for the second equal sign.   □

Corollary 4.84. *The Green function for the focal BVP (4.43) with $p = 1$ and $q = 0$ is given by*

$$G(t, s) = \begin{cases} a - t & \text{if} \quad t \leq s \\ a - \sigma(s) & \text{if} \quad t \geq \sigma(s). \end{cases}$$

Proof. Put $p(t) \equiv 1$ in Corollary 4.83.   □

Definition 4.85. We say that (4.1) is *disfocal* on $[a, \sigma^2(b)]$ provided there is no nontrivial solution x with a generalized zero in $[a, \sigma^2(b)]$ followed by a generalized zero of x^Δ in $[a, \sigma(b)]$.

The proof of the following result is similar to the proof of some earlier results in this section and hence will be omitted.

Corollary 4.86 (Comparison Theorem for Focal BVPs). *Let $p > 0$. Assume that* (4.1) *is disfocal on* $[a, \sigma^2(b)]$. *If* $u, v \in \mathbb{D}$ *satisfy*

$$Lu(t) \leq Lv(t) \quad \text{for all } t \in [a, b], \quad u(a) \geq v(a), \quad u^\Delta(\sigma(b)) \geq v^\Delta(\sigma(b)),$$

then

$$u(t) \geq v(t) \quad \text{for all} \quad t \in [a, \sigma^2(b)].$$

Finally we consider the periodic BVP

$$(4.44) \qquad Lx = 0, \quad x(a) = x(a + \sigma^2(T)), \quad x^\Delta(a) = x^\Delta(a + \sigma(T)).$$

Definition 4.87. Let $T \in \mathbb{T} \cap [0, \infty)$. We say that a time scale $\mathbb{T}$ is *periodic* with period $\sigma^2(T)$ provided $t \in \mathbb{T}$ implies $t + \sigma^2(T) \in \mathbb{T}$.

We assume throughout the rest of this section that $\mathbb{T}$ is periodic with period $\sigma^2(T)$.

Theorem 4.88. *Assume that the homogeneous periodic BVP* (4.44) *has only the trivial solution. Then the nonhomogeneous BVP*

$$(4.45) \quad Lx = h(t), \quad x(a) - x(a + \sigma^2(T)) = A, \quad x^\Delta(a) - x^\Delta(a + \sigma(T)) = B,$$

where A and B are given constants and $h \in C_{rd}$, has a unique solution.

Proof. Let x_1 and x_2 be linearly independent solutions of (4.1). Then

$$x(t) = c_1 x_1(t) + c_2 x_2(t)$$

is a general solution of (4.1). Note that x satisfies the boundary conditions in (4.44) iff c_1 and c_2 are constants satisfying

$$M \begin{pmatrix} c_1 \\ c_2 \end{pmatrix} = 0 \quad \text{with} \quad M = \begin{pmatrix} x_1(a) - x_1(a + \sigma^2(T)) & x_2(a) - x_2(a + \sigma^2(T)) \\ x_1^\Delta(a) - x_1^\Delta(a + \sigma(T)) & x_2^\Delta(a) - x_2^\Delta(a + \sigma(T)) \end{pmatrix}.$$

Since we are assuming that (4.44) has only the trivial solution, it follows that

$$c_1 = c_2 = 0$$

is the unique solution of the above linear system. Hence

$$(4.46) \qquad\qquad\qquad \det M \neq 0.$$

Now we show that (4.45) has a unique solution. Let u_0 be a fixed solution of $Lu = h(t)$. Then a general solution of $Lu = h(t)$ is given by

$$u(t) = a_1 x_1(t) + a_2 x_2(t) + u_0(t).$$

It follows that u satisfies the boundary conditions in (4.45) iff a_1 and a_2 are constants satisfying the system of equations

$$M \begin{pmatrix} a_1 \\ a_2 \end{pmatrix} = \begin{pmatrix} A - u_0(a) + u_0(a + \sigma^2(T)) \\ B - u_0^\Delta(a) + u_0^\Delta(a + \sigma(T)) \end{pmatrix}.$$

This system has a unique solution because of (4.46), and hence (4.45) has a unique solution. $\square$

Theorem 4.89 (Green's Function for Periodic BVPs). *Assume that the homogeneous BVP* (4.44) *has only the trivial solution. For each fixed* $s \in \mathbb{T}$, *let* $u(\cdot, s)$ *be the solution of the BVP*

(4.47)
$$
\begin{cases}
Lu(\cdot, s) = 0 \\[2mm]
u(a, s) - u(a + \sigma^2(T), s) = x(a + \sigma^2(T), s) \\[2mm]
u^{\Delta}(a, s) - u^{\Delta}(a + \sigma(T), s) = x^{\Delta}(a + \sigma(T), s),
\end{cases}
$$

where $x(t, s)$ *is the Cauchy function for* (4.1). *Define*

(4.48)
$$
G(t, s) := \begin{cases}
u(t, s) & \text{if} \quad t \leq s \\
v(t, s) & \text{if} \quad t \geq \sigma(s),
\end{cases}
$$

where $v(t, s) := u(t, s) + x(t, s)$. *If* $h \in C_{\mathrm{rd}}$, *then*

$$
x(t) := \int_a^{\sigma(T)} G(t, s) h(s) \Delta s
$$

is the unique solution of the nonhomogeneous periodic BVP (4.45) *with* $A = B = 0$.
Furthermore for each fixed $s \in [a, b]$, $v(\cdot, s)$ *is a solution of* (4.1) *and*

$$
u(a, s) = v(a + \sigma^2(T), s), \quad u^{\Delta}(a, s) = v^{\Delta}(a + \sigma(T), s).
$$

Proof. The existence and uniqueness of $u(t, s)$ is guaranteed by Theorem 4.88. Since $v(t, s) = u(t, s) + x(t, s)$, we have for each fixed s that $v(\cdot, s)$ is a solution of (4.1). Using the boundary conditions in (4.47), it is easy to see that for each fixed s, $u(a, s) = v(a + \sigma^2(T), s)$, $u^{\Delta}(a, s) = v^{\Delta}(a + \sigma(T), s)$. Let G be as in (4.48) and consider

$$
\begin{aligned}
x(t) &= \int_a^{\sigma(T)} G(t, s) h(s) \Delta s \\
&= \int_a^t G(t, s) h(s) \Delta s + \int_t^{\sigma(T)} G(t, s) h(s) \Delta s \\
&= \int_a^t v(t, s) h(s) \Delta s + \int_t^{\sigma(T)} u(t, s) h(s) \Delta s \\
&= \int_a^t [u(t, s) + x(t, s)] h(s) \Delta s + \int_t^{\sigma(T)} u(t, s) h(s) \Delta s \\
&= \int_a^{\sigma(T)} u(t, s) h(s) \Delta s + \int_a^t x(t, s) h(s) \Delta s \\
&= \int_a^{\sigma(T)} u(t, s) h(s) \Delta s + z(t),
\end{aligned}
$$

where, by the variation of constants formula in Theorem 4.24, z solves

$$
Lz = h(t), \quad z(a) = z^{\Delta}(a) = 0.
$$

Hence

$$
Lx(t) = \int_a^{\sigma(T)} Lu(t, s) h(s) \Delta s + Lz(t) = Lz(t) = h(t).
$$

Thus x is a solution of $Lx = h(t)$. Note that

$$
\begin{aligned}
x(a) &= \int_a^{\sigma(T)} u(a,s)h(s)\Delta s + z(a) \\
&= \int_a^{\sigma(T)} v(a+\sigma^2(T),s)h(s)\Delta s \\
&= \int_a^{\sigma(T)} G(a+\sigma^2(T),s)h(s)\Delta s \\
&= x(a+\sigma^2(T))
\end{aligned}
$$

and

$$
\begin{aligned}
x^\Delta(a) &= \int_a^{\sigma(T)} u^\Delta(a,s)h(s)\Delta s + z^\Delta(a) \\
&= \int_a^{\sigma(T)} v^\Delta(a+\sigma(T),s)h(s)\Delta s \\
&= \int_a^{\sigma(T)} G^\Delta(a+\sigma(T),s)h(s)\Delta s \\
&= x^\Delta(a+\sigma(T)).
\end{aligned}
$$

Hence x satisfies the periodic boundary conditions in (4.44). □

Example 4.90. Using Theorem 4.89 we will solve the periodic BVP

$$(4.49) \qquad x^{\Delta\Delta} + x = \sin_2(t,0), \quad x(0) = x(\sigma^2(\pi)), \quad x^\Delta(0) = x^\Delta(\sigma(\pi))$$

when $\mathbb{T} = \mathbb{R}$. It is easy to show that the homogeneous BVP

$$x'' + x = 0, \quad x(0) = x(\pi), \quad x'(0) = x'(\pi)$$

has only the trivial solution, and hence we can use Theorem 4.89 to solve (4.49). By Theorem 4.89, Green's function G is given by (4.48), where for each fixed $s \in [a,b]$, $u(\cdot,s)$ is the solution of

$$(4.50) \quad u'' + u = 0, \quad u(0,s) - u(\pi,s) = x(\pi,s), \quad u'(0,s) - u'(\pi,s) = x'(\pi,s),$$

where $x(t,s)$ is the Cauchy function for $x'' + x = 0$ and $v(t,s) := u(t,s) + x(t,s)$. The Cauchy function for $x'' + x = 0$ is given by

$$x(t,s) = \sin(t-s).$$

Since for each fixed s, $u(\cdot,s)$ is a solution of $u'' + u = 0$,

$$u(t,s) = A(s)\cos t + B(s)\sin t.$$

From the boundary conditions in (4.50) we get

$$2A(s) = \sin(\pi - s) = \sin s \quad \text{and} \quad 2B(s) = \cos(\pi - s) = -\cos s.$$

It follows that

$$u(t,s) = \frac{1}{2}\sin s \cos t - \frac{1}{2}\cos s \sin t = -\frac{1}{2}\sin(t-s).$$

Therefore

$$v(t,s) = u(t,s) + x(t,s) = \frac{1}{2}\sin(t-s).$$

Hence by Theorem 4.89 the solution of the BVP (4.49) is given by

$$
\begin{aligned}
x(t) &= \int_0^\pi G(t,s)\sin(2s)ds \\
&= \int_0^t v(t,s)\sin(2s)ds + \int_t^\pi u(t,s)\sin(2s)ds \\
&= \frac{1}{2}\int_0^t \sin(t-s)\sin(2s)ds - \frac{1}{2}\int_t^\pi \sin(t-s)\sin(2s)ds \\
&= \frac{1}{2}\sin t\int_0^t \cos s\sin(2s)ds - \frac{1}{2}\cos t\int_0^t \sin(s)\sin(2s)ds \\
&\quad -\frac{1}{2}\sin t\int_t^\pi \cos s\sin(2s)ds + \frac{1}{2}\cos t\int_t^\pi \sin(s)\sin(2s)ds \\
&= \frac{1}{2}\sin t\left\{-\frac{1}{3}\sin(2t)\sin t - \frac{2}{3}\cos(2t)\cos t + \frac{2}{3}\right\} \\
&\quad -\frac{1}{2}\cos t\left\{\frac{1}{3}\cos t\sin(2t) - \frac{2}{3}\sin t\cos(2t)\right\} \\
&\quad -\frac{1}{2}\sin t\left\{\frac{1}{3}\sin(2t)\sin t + \frac{2}{3}\cos(2t)\cos t + \frac{2}{3}\right\} \\
&\quad +\frac{1}{2}\cos t\left\{-\frac{1}{3}\cos t\sin(2t) + \frac{2}{3}\sin t\cos(2t)\right\} \\
&= -\frac{1}{3}\sin^2 t\sin(2t) - \frac{2}{3}\sin t\cos t\cos(2t) + \frac{1}{3}\sin t \\
&\quad -\frac{1}{3}\cos^2 t\sin(2t) + \frac{2}{3}\sin t\cos t\cos(2t) - \frac{1}{3}\sin t \\
&= -\frac{1}{3}\sin(2t).
\end{aligned}
$$

Exercise 4.91. Use an appropriate Green function to solve each of the following periodic boundary value problems on the indicated time scales:

(i) $x^{\Delta\Delta} + x = \cos_2(t,0)$, $x(0) = x(\sigma^2(\pi))$, $x^\Delta(0) = x^\Delta(\sigma(\pi))$, where $\mathbb{T} = \mathbb{R}$;

(ii) $x^{\Delta\Delta} + x = 4$, $x(0) = x(\sigma^2(\pi/2))$, $x^\Delta(0) = x^\Delta(\sigma(\pi/2))$, where $\mathbb{T} = \mathbb{R}$.

4.5. Eigenvalue Problems

In this section we follow [14] and consider an eigenvalue problem of the form

(4.51) $$L(x) + \lambda x^\sigma = 0, \quad R_a(x) = R_b(x) = 0,$$

where

$$L(x) = x^{\Delta\Delta} + qx^\sigma$$

such that $q : \mathbb{T} \to \mathbb{R}$ is rd-continuous, and

$$R_a(x) = \alpha x(\rho(a)) + \beta x^\Delta(\rho(a)), \quad R_b(x) = \gamma x(\rho(b)) + \delta x^\Delta(\rho(b))$$

such that $\alpha, \beta, \gamma, \delta \in \mathbb{R}$ with $(\alpha^2 + \beta^2)(\gamma^2 + \delta^2) \neq 0$.

Definition 4.92. A number $\lambda \in \mathbb{R}$ is called an *eigenvalue* of (4.51) provided there exists a nontrivial solution x of the boundary value problem (4.51). Such an x is then called an *eigenfunction* corresponding to the eigenvalue λ.

Results in this section concern the *existence* of eigenvalues, *oscillation proper-ties* of eigenfunctions (i.e., we are interested in the number of generalized zeros a certain eigenfunction possesses), *comparison results* between eigenvalues of (4.51) and another eigenvalue problem

$$(4.52) \qquad L(x) + \lambda x^\sigma = 0, \quad R_a(x) = x(b) = 0,$$

and results on the so-called *Rayleigh quotient*

$$R(x) = -\frac{\langle L(x), x^\sigma \rangle}{\|x^\sigma\|^2}, \quad \text{where} \quad x \neq 0$$

such as *Rayleigh's principle*. All of these results will be derived using the so-called *extended Picone's identity* (see Theorem 4.102).

Definition 4.93. We define the inner product of x and y on $[\rho(a), b]$ by

$$\langle x, y \rangle = \int_{\rho(a)}^{b} x(t) y(t) \Delta t,$$

and we say that x and y are orthogonal (we write $x \perp y$) on $[\rho(a), b]$ provided $\langle x, y \rangle = 0$. The norm of x is defined by

$$\|x\| = \sqrt{\langle x, x \rangle}.$$

Theorem 4.94 (Green's Theorem). *If $x, y \in \mathbb{D}$, then*

$$\langle x^\sigma, Ly \rangle - \langle y^\sigma, Lx \rangle = \{x; y\}(b) - \{x; y\}(a).$$

Proof. This result follows from integrating both sides of the Lagrange identity in Theorem 4.30 from a to b and using the definition of the inner product. $\qquad \square$

It is easy to see that $W(x, y)(\rho(a)) = 0$ if $R_a(x) = R_a(y) = 0$ and $W(x, y)(b) = 0$ if $R_b(x) = R_b(y) = 0$. Hence the problem (4.51) is self-adjoint and all eigenvalues are real. In what follows we shall always assume without loss of generality that

$$(4.53) \qquad \begin{cases} \beta > \alpha\mu(\rho(a)) \text{ if } \beta \neq \alpha\mu(\rho(a)); \quad \alpha = -1 \text{ if } \beta = \alpha\mu(\rho(a)), \\ \delta > 0 \text{ if } \delta \neq 0; \quad \gamma = 1 \text{ if } \delta = 0. \end{cases}$$

We will also assume throughout that

$$(4.54) \qquad \begin{cases} \text{if } a \text{ is right-scattered, then it is also left-scattered,} \quad \text{and} \\ \text{if } b \text{ is left-scattered, then it is also right-scattered.} \end{cases}$$

Now we shall provide a characterization for the eigenvalues of (4.51). For this, we denote the unique solutions (see Theorem 4.5) of the initial value problems

$$L(x) + \lambda x^\sigma = 0, \quad x(\rho(a)) = \beta, \quad x^\Delta(\rho(a)) = -\alpha$$

by $x(\cdot, \lambda)$, where $\lambda \in \mathbb{R}$, and we put $\Lambda(\lambda) = R_b(x(\cdot, \lambda))$. With this notation, we have the following.

Theorem 4.95. *λ is an eigenvalue of (4.51) if and only if $\Lambda(\lambda) = 0$.*

Proof. If $R_b(x(\cdot, \lambda)) = 0$, then $x = x(\cdot, \lambda)$ satisfies

$$L(x) + \lambda x^\sigma = 0, \quad R_a(x) = R_b(x) = 0,$$

i.e., λ is an eigenvalue of (4.51). Conversely, let $\lambda \in \mathbb{R}$ be an eigenvalue of (4.51) with corresponding eigenfunction x. Then because of the unique solvability of initial value problems (observe $R_a(x) = 0$),

$$x = cx(\cdot, \lambda) \quad \text{with} \quad c = \frac{\beta x(\rho(a)) - \alpha x^\Delta(\rho(a))}{\alpha^2 + \beta^2}.$$

Hence $R_b(x(\cdot, \lambda)) = 0$. □

The proof of Theorem 4.95 also shows that all eigenvalues of (4.51) are *simple*. Furthermore, in view of Theorem 4.94, eigenfunctions x and y corresponding to different eigenvalues are always orthogonal in the sense

$$x^\sigma \perp y^\sigma, \quad \text{i.e.,} \quad \langle x^\sigma, y^\sigma \rangle = 0.$$

For the remainder of this section we assume that

(4.55)
$$\lim_{\nu \to \lambda} \langle x^\sigma(\cdot, \nu), x^\sigma(\cdot, \lambda) \rangle = \|x^\sigma(\cdot, \lambda)\|^2.$$

Theorem 4.96. *We have for all $t \in [\rho(a), b]$ and $\lambda \in \mathbb{R}$*

$$x(t, \lambda) \frac{\partial}{\partial \lambda} x^\Delta(t, \lambda) - x^\Delta(t, \lambda) \frac{\partial}{\partial \lambda} x(t, \lambda) = -\int_{\rho(a)}^t \{x^\sigma(\tau, \lambda)\}^2 \, \Delta\tau.$$

Proof. Let $\lambda, \nu \in \mathbb{R}$ with $\nu \neq \lambda$. Then

$$\begin{aligned}
&\left[x(t, \nu) \left\{ x^\Delta(t, \lambda) - x^\Delta(t, \nu) \right\} - x^\Delta(t, \nu) \left\{ x(t, \lambda) - x(t, \nu) \right\} \right]^\Delta \\
&= \left[x(t, \nu) x^\Delta(t, \lambda) - x^\Delta(t, \nu) x(t, \lambda) \right]^\Delta \\
&= x^\sigma(t, \nu) x^{\Delta\Delta}(t, \lambda) - x^{\Delta\Delta}(t, \nu) x^\sigma(t, \lambda) \\
&= (\nu - \lambda) x^\sigma(t, \nu) x^\sigma(t, \lambda).
\end{aligned}$$

We divide both sides by $\lambda - \nu$, integrate from $\rho(a)$ to t (note $x(\rho(a), \nu) = x(\rho(a), \lambda)$ and $x^\Delta(\rho(a), \nu) = x^\Delta(\rho(a), \lambda)$), and let $\nu \to \lambda$ (observe (4.55)) to obtain our desired equation. □

From Theorem 4.96 it follows that $\frac{x^\Delta(t,\lambda)}{x(t,\lambda)}$ is for each $t \in (a, b]$ strictly decreasing in λ whenever $x(t, \lambda) \neq 0$. Note that Theorem 4.96 implies $\frac{\partial}{\partial \lambda} x(b, \lambda) \neq 0$ in case of $x(b, \lambda) = 0$, and it also implies $\frac{\partial}{\partial \lambda} \left\{ \gamma + \delta \frac{x^\Delta(b,\lambda)}{x(b,\lambda)} \right\} \neq 0$. This together with Theorem 4.95 (observe $\Lambda(\lambda) = x(b, \lambda) \left\{ \gamma + \delta \frac{x^\Delta(b,\lambda)}{x(b,\lambda)} \right\}$) shows that all eigenvalues of (4.51) are isolated.

One concern in this section is to give the so-called *oscillation number* for any eigenfunction of (4.51). We define $n(\lambda)$ to be the number of generalized zeros of $x(\cdot, \lambda)$ in the open interval (a, b). The following result is needed later, and since its proof is analogous to the proof for the continuous case (in view of Theorem 1.16 (i), $x(t, \lambda)$ is continuous in t, whereas its continuity with respect to λ follows from [185, Section 2.6]), it is omitted here.

Lemma 4.97. $\limsup_{\nu \to \lambda} n(\nu) \leq n(\lambda) + 1$ *for all $\lambda \in \mathbb{R}$; and if equality holds, then $x(b, \lambda) = 0$.*

Exercise 4.98. Prove Lemma 4.97.

Lemma 4.99. *Put* $\lambda^* = -\left\{1 + \sup_{t \in [\rho(a),b]} q(t)\right\}$. *Let* $\lambda \leq \lambda^*$ *and* x *be a nontrivial solution of* $L(x) + \lambda x^\sigma = 0$. *Then we have the following:*

(i) *If* x *has for some* $t_1 \in [\rho(a), b]$ *a generalized zero in* $\mathcal{I} = \{t_1\} \cup (t_1, \sigma(t_1))$, *then* x^Δ *has no generalized zero in* $\mathcal{I}$;

(ii) *If* x *has a generalized zero in* $\mathcal{I}$, *then* xx^Δ *has a generalized zero in* $\mathcal{I}$ *also;*

(iii) *The function* xx^Δ *is strictly increasing on* $[\rho(a), b]$;

(iv) *If* $(xx^\Delta)(t_2) \geq 0$ *for some* $t_2 \in [\rho(a), b]$, *then* x *has no generalized zeros in* $(t_2, b]$;

(v) $[\rho(a), b]$ *contains at most one generalized zero of* x;

(vi) *If* $n(\lambda) = 0$, *then* $\lim_{\nu \to -\infty} n(\nu) = 0$.

Proof. First note that $\lambda^* > -\infty$ because of Theorem 1.65 and that $\lambda + q(t) \leq -1$ for all $\lambda \leq \lambda^*$. Assertion (i) follows from

$$
\begin{aligned}
x^\Delta(t_1)x^\Delta(\sigma(t_1)) &= x^\Delta(t_1)\left\{x^\Delta(t_1) + \mu(t_1)x^{\Delta\Delta}(t_1)\right\} \\
&= \left(x^\Delta(t_1)\right)^2 - (\lambda + q(t_1))\,\mu(t_1)x(\sigma(t_1))x^\Delta(t_1) \\
&= \left(x^\Delta(t_1)\right)^2 - (\lambda + q(t_1))\,x(\sigma(t_1))\left\{x(\sigma(t_1)) - x(t_1)\right\} \\
&= \left(x^\Delta(t_1)\right)^2 - (\lambda + q(t_1))\,(x(\sigma(t_1)))^2 + (\lambda + q(t_1))\,x(t_1)x(\sigma(t_1)) \\
&\geq \left(x^\Delta(t_1)\right)^2 + (x(\sigma(t_1)))^2 - x(t_1)x(\sigma(t_1)) > 0.
\end{aligned}
$$

For (ii), $(xx^\sigma)(t_1) \leq 0$ implies $(x^\Delta x^{\Delta^\sigma})(t_1) > 0$ by (i), and these two inequalities of course show $\left(xx^\Delta(xx^\Delta)^\sigma\right)(t_1) \leq 0$. Next, Theorem 1.76 and

$$
\begin{aligned}
(xx^\Delta)^\Delta(t) &= \left(x^\Delta(t)\right)^2 + x^{\Delta\Delta}(t)x^\sigma(t) \\
&= \left(x^\Delta(t)\right)^2 - (\lambda + q(t))\,(x^\sigma(t))^2 \geq \left(x^\Delta(t)\right)^2 + (x^\sigma(t))^2 > 0
\end{aligned}
$$

proves (iii). The identity

$$
x(t)x^\sigma(t) = x(t)\left\{x(t) + \mu(t)x^\Delta(t)\right\} = x^2(t) + \mu(t)x(t)x^\Delta(t)
$$

takes care of (iv), while (v) follows from (ii), (iii) and (iv). Finally assume $n(\lambda) = 0$ and put $\kappa = \min_{t \in [\rho(a), \sigma(b)]} x^2(t, \lambda)$. Then $\kappa > 0$ and

$$
\begin{aligned}
x(t, \nu)x^\Delta(t, \nu) &= -\alpha\beta + \int_{\rho(a)}^t \left\{\left(x^\Delta(\tau, \nu)\right)^2 - (\nu + q(\tau))\,(x^\sigma(\tau, \nu))^2\right\} \Delta\tau \\
&\geq -\alpha\beta + (\lambda^* - \nu)\kappa(t - \rho(a)) > 0
\end{aligned}
$$

for all sufficiently small ν so that (vi) follows from (iv). $\qquad\square$

Theorem 4.100. *We have* $\lim_{\lambda \to -\infty} n(\lambda) = 0$ *and* $\lim_{\lambda \to -\infty} \Lambda(\lambda) = \infty$.

Proof. Let λ^* be as in Lemma 4.99 and $\lambda \leq \lambda^*$. Then $n(\lambda^*) \in \{0, 1\}$ according to Lemma 4.99 (v). If $n(\lambda^*) = 0$, then $\lim_{\nu \to -\infty} n(\nu) = 0$ because of Lemma 4.99 (vi). Hence we assume $n(\lambda^*) = 1$. Thus $\alpha\beta > 0$ (observe $x(\rho(a), \lambda^*)x^\Delta(\rho(a), \lambda^*) = -\alpha\beta$ and use Lemma 4.99 (iv)), and $[\rho(a), a]$ does not contain a generalized zero of $x(\cdot, \lambda^*)$ (observe Lemma 4.99 (v)). Therefore there exists some $\underset{\sim}{t} \in (\rho(a), b]$ such that $[\rho(a), \underset{\sim}{t}]$ contains no generalized zero of $x(\cdot, \lambda^*)$. Put $\mathcal{I} = (\rho(a), \underset{\sim}{t}]$. Hence $\mathcal{I} \neq \emptyset$. Therefore

$$
\frac{x^\Delta(t, \lambda^*)}{x(t, \lambda^*)} < 0 \quad \text{for all} \quad t \in \mathcal{I}
$$

because of Lemma 4.99 (iv).

We now assume that

$$\frac{x^\Delta(t,\lambda)}{x(t,\lambda)} \in \left(\frac{x^\Delta(t,\lambda^*)}{x(t,\lambda^*)}, 0\right) \quad \text{for all } t \in \mathcal{I} \text{ and all } \lambda < \lambda^*.$$

This implies for all $t \in \mathcal{I}$ and all $\lambda < \lambda^*$

$$1 + \mu(t)\frac{x^\Delta(t,\lambda)}{x(t,\lambda)} \geq 1 + \mu(t)\frac{x^\Delta(t,\lambda^*)}{x(t,\lambda^*)} = \frac{x^\sigma(t,\lambda^*)}{x(t,\lambda^*)} > 0$$

and hence for all $\lambda < \lambda^*$

$$\frac{x^\Delta(\underset{\sim}{t},\lambda)}{x(\underset{\sim}{t},\lambda)} = -\frac{\beta}{\alpha} + \int_{\rho(a)}^{\underset{\sim}{t}} \left\{-(\lambda+q(t)) - \left(\frac{x^\Delta(t,\lambda)}{x(t,\lambda)}\right)^2 \frac{1}{1 + \mu(t)\frac{x^\Delta(t,\lambda)}{x(t,\lambda)}}\right\} \Delta t$$

$$\geq -\frac{\beta}{\alpha} + \int_{\rho(a)}^{\underset{\sim}{t}} \left\{(\lambda^* - \lambda) - (\lambda^* + q(t))\right.$$

$$\left. - \left(\frac{x^\Delta(t,\lambda^*)}{x(t,\lambda^*)}\right)^2 \frac{1}{1 + \mu(t)\frac{x^\Delta(t,\lambda^*)}{x(t,\lambda^*)}}\right\} \Delta t$$

$$= \frac{x^\Delta(\underset{\sim}{t},\lambda^*)}{x(\underset{\sim}{t},\lambda^*)} + (\lambda^* - \lambda)\left(\underset{\sim}{t} - \rho(a)\right)$$

which tends to infinity as $\lambda \to -\infty$, a contradiction.

Hence and because $\left(\frac{x^\Delta}{x}\right)(t,\cdot)$ is strictly decreasing according to Theorem 4.96, there must exist an $s \in \mathcal{I}$ and $\underset{\sim}{\lambda} < \lambda^*$ such that $x^\Delta(s,\underset{\sim}{\lambda}) = 0$. Then $x(s,\underset{\sim}{\lambda}) \neq 0$ and $x(\cdot,\underset{\sim}{\lambda})$ has no generalized zero in (s,b) according to Lemma 4.99 (iv). If $x(\cdot,\underset{\sim}{\lambda})$ had a generalized zero in $[\rho(a),s)$, then $(xx^\Delta)(t_1,\underset{\sim}{\lambda}) \geq 0$ for some $t_1 \in [\rho(a),s)$ because of Lemma 4.99 (ii), and this contradicts Lemma 4.99 (iii). Hence $n(\underset{\sim}{\lambda}) = 0$ and $\lim_{\nu\to-\infty} n(\nu) = 0$ due to Lemma 4.99 (vi).

Finally, using the above (note also $-x^\Delta(b,\lambda) = \alpha + \int_{\rho(a)}^b (\lambda+q(t))x^\sigma(t)\Delta t$) and from the proof of Lemma 4.99 (vi) it follows that

$$\lim_{\lambda\to-\infty} x(b,\lambda) = \lim_{\lambda\to-\infty} x^\Delta(b,\lambda) = \lim_{\lambda\to-\infty} \frac{x^\Delta(b,\lambda)}{x(b,\lambda)} = \infty.$$

Hence

$$\lim_{\lambda\to-\infty} \Lambda(\lambda) = \lim_{\lambda\to-\infty} \left\{x(b,\lambda)\left(\gamma + \delta\frac{x^\Delta(b,\lambda)}{x(b,\lambda)}\right)\right\} = \infty$$

according to our assumption (4.53).   □

From the results obtained so far, we know that the eigenvalues of (4.51) may be arranged as

$$-\infty < \lambda_1 < \lambda_2 < \cdots.$$

We shall write $\lambda_p = \infty$ for all $p > k$ if only k eigenvalues exist.

Besides the elementary results presented above, the proof of our main results requires appropriate applications of the extended Picone identity on time scales, which we shall derive now. Below we put for $\alpha \in \mathbb{R}$

$$
\alpha^\dagger = \begin{cases} \frac{1}{\alpha} & \text{if} \quad \alpha \neq 0 \\ 0 & \text{if} \quad \alpha = 0. \end{cases}
$$

Theorem 4.101 (Picone's Identity). *Let* $t \in [\rho(a), \rho(b)]$. *Suppose*

$$
\operatorname{Ker} x(t) \subset \operatorname{Ker} y(t) \quad \text{and} \quad \operatorname{Ker} x^\sigma(t) \subset \operatorname{Ker} y^\sigma(t),
$$

where $y \in C_{rd}^2$ *and* x *solves* $L(x) + \lambda x^\sigma = 0$. *Then at* t,

$$
-\left\{ L(y) + \lambda y^\sigma \right\} y^\sigma = \left[y x^\dagger W(y, x) \right]^\Delta + (x x^\sigma)^\dagger W^2(y, x).
$$

Proof. First suppose that $(x x^\sigma)(t) \neq 0$. Then by Theorem 4.30 we get

$$
\begin{aligned}
\left[y x^\dagger W(y, x) \right]^\Delta &= \frac{y^\sigma}{x^\sigma} W^\Delta(y, x) + \left(\frac{y}{x} \right)^\Delta W(y, x) \\
&= \frac{y^\sigma}{x^\sigma} \left\{ L(x) y^\sigma - L(y) x^\sigma \right\} - \frac{W^2(y, x)}{x x^\sigma} \\
&= - \left\{ L(y) + \lambda y^\sigma \right\} y^\sigma - (x x^\sigma)^\dagger W^2(y, x)
\end{aligned}
$$

at t. Next, suppose that t is right-scattered with $(x x^\sigma)(t) = 0$. Then we have $(y y^\sigma)(t) = 0$ and hence at t,

$$
W(y, x) = y x^\Delta - x y^\Delta = y \frac{x^\sigma - x}{\mu} - x \frac{y^\sigma - y}{\mu} = \frac{y x^\sigma - x y^\sigma}{\mu} = 0.
$$

If t is left-scattered, then $y x^\dagger W(y, x)$ is obviously continuous at t, and if t is left-dense, then it is also continuous at t, since according to L'Hôpital's rule (Theorem 1.119),

$$
\lim_{s \to t^-} \left[y x^\dagger W(y, x) \right](s) = \frac{y^\Delta(t)}{x^\Delta(t)} W(y, x)(t) = 0 = \left[y x^\dagger W(y, x) \right](t).
$$

Hence we may apply Theorem 4.30 to obtain at t,

$$
\begin{aligned}
\left[y x^\dagger W(y, x) \right]^\Delta &= \frac{\left[y x^\dagger W(y, x) \right]^\sigma - y x^\dagger W(y, x)}{\mu} \\
&= \frac{\left[y x^\dagger W(y, x) \right]^\sigma}{\mu} \\
&= y^\sigma (x^\sigma)^\dagger W^\Delta(y, x) \\
&= y^\sigma (x^\sigma)^\dagger \left\{ L(x) y^\sigma - L(y) x^\sigma \right\} \\
&= y^\sigma (x^\sigma)^\dagger x^\sigma \left\{ -\lambda y^\sigma - L(y) \right\} \\
&= -y^\sigma \left\{ L(y) + \lambda y^\sigma \right\} \\
&= -y^\sigma \left\{ L(y) + \lambda y^\sigma \right\} - (x x^\sigma)^\dagger W^2(y, x).
\end{aligned}
$$

Finally, if t is right-dense with $x(t) = x(\sigma(t)) = 0$, then we have $y(t) = y(\sigma(t)) = 0$, and L'Hôpital's rule yields

$$
\begin{aligned}
\left[yx^\dagger W(y,x)\right]^\Delta &= \lim_{s \to t} \frac{\left[yx^\dagger W(y,x)\right](\sigma(t)) - \left[yx^\dagger W(y,x)\right](t)}{\sigma(t) - s} \\
&= \lim_{s \to t} \frac{\left[yx^\dagger W(y,x)\right](s)}{s - t} \\
&= \lim_{s \to t} \left[yx^\dagger\right](s) \lim_{s \to t} \frac{W(y,x)(s) - W(y,x)(t)}{s - t} \\
&= \frac{y^\Delta(t)}{x^\Delta(t)} W^\Delta(y,x)(t) \\
&= 0 \\
&= -\left\{L(y) + \lambda y^\sigma\right\}(t) - (xx^\sigma)^\dagger W^2(y,x)(t).
\end{aligned}
$$

This proves our result. $\square$

With Theorem 4.101 it is now possible to prove the following key result:

Theorem 4.102 (Extended Picone's Identity). *Let $k \in \mathbb{N}$. Suppose x is a solution of $L(x) + \lambda x^\sigma = 0$. Let x_ν be solutions of $L(x_\nu) + \rho_\nu x_\nu^\sigma = 0$ with $\rho_\nu \in \mathbb{R}$, $1 \le \nu \le k$. Next, let $y \in C_{rd}^2$ satisfy $y^\sigma \perp x_\nu^\sigma$, $1 \le \nu \le k$, and suppose $x_\nu^\sigma \perp x_\mu^\sigma$, $1 \le \nu < \mu \le k$, $\|x_\nu^\sigma\| = 1$, $1 \le \nu \le k$. Put $\tilde{x}_1 = \sum_{\nu=1}^k \beta_\nu x_\nu$ with $\beta_\nu \in \mathbb{R}$, $1 \le \nu \le k$, and $\tilde{x} = y + \tilde{x}_1$. Suppose $\operatorname{Ker} x(t) \subset \operatorname{Ker} y(t)$ holds for all $t \in [\rho(a), b]$. Then*

$$
-\langle L(y) + \lambda y^\sigma, y^\sigma \rangle = \int_{\rho(a)}^b \left[(xx^\sigma)^\dagger W^2(\tilde{x},x)\right](t)\Delta t + \sum_{\nu=1}^k \beta_\nu^2 (\lambda - \rho_\nu) + T(b) - T(\rho(a)),
$$

where

$$
T = \tilde{x}y^\dagger W(\tilde{x},x) + W(\tilde{x}_1, y).
$$

Proof. We apply Theorem 4.101 and Lemma 4.94 to obtain

$$
\begin{aligned}
&\left[\tilde{x}x^\dagger W(\tilde{x},x)\right](\rho(a)) - \left[\tilde{x}x^\dagger W(\tilde{x},x)\right](b) - \int_{\rho(a)}^b \left[(xx^\sigma)^\dagger W^2(\tilde{x},x)\right](t)\Delta t \\
&= \langle L(\tilde{x}) + \lambda \tilde{x}^\sigma, \tilde{x}^\sigma \rangle \\
&= \left\langle L(y) + \sum_{\nu=1}^k \beta_\nu L(x_\nu) + \lambda y^\sigma + \lambda \tilde{x}_1^\sigma, y^\sigma + \tilde{x}_1^\sigma \right\rangle \\
&= \langle L(y) + \lambda y^\sigma, y^\sigma \rangle + \langle L(y), \tilde{x}_1^\sigma \rangle - \sum_{\nu=1}^k \beta_\nu \rho_\nu \langle x_\nu^\sigma, \tilde{x}_1^\sigma \rangle + \lambda \langle \tilde{x}_1^\sigma, \tilde{x}_1^\sigma \rangle \\
&= \langle L(y) + \lambda y^\sigma, y^\sigma \rangle + \langle L(\tilde{x}_1), y \rangle + W(\tilde{x}_1, y)(b) - W(\tilde{x}_1, y)(\rho(a)) \\
&\quad + \sum_{\nu=1}^k \beta_\nu^2 (\lambda - \rho_\nu),
\end{aligned}
$$

and hence our claim follows. $\square$

Let $\lambda_1^* < \lambda_2^* < \cdots$ denote the eigenvalues of (4.52) with corresponding normalized eigenfunctions x_ν^* (i.e., $\|x_\nu^*\| = 1$). In all the proofs of the results in the remainder of this section we use the notation from Theorem 4.102.

Theorem 4.103. *If* $\lambda \in (\lambda_k^*, \lambda_{k+1}^*]$, *then*

$$n(\lambda) = k.$$

Moreover, define a $k \times k$*-matrix* $Z^* = (z_{\mu\nu}^*)$ *by*

$$z_{\mu\nu}^* = \begin{cases} x_\nu^*(t_\mu) & \text{if } t_\mu \text{ is a zero of } x(\cdot, \lambda) \\ W(x, x_\nu^*)(t_\mu) & \text{if } t_\mu \text{ is a node of } x(\cdot, \lambda), \end{cases}$$

where $t_1 < t_2 < \cdots < t_k$ *are the generalized zeros of* $x(\cdot, \lambda)$. *Then* Z^* *is invertible.*

Proof. Let $y = 0$, $\rho_\nu = \lambda_\nu^*$, $x_\nu = x_\nu^*$, $x = x(\cdot, \lambda)$. Then $T(\rho(a)) = T(b) = 0$. Suppose that x has no generalized zeros in (a, b). Then by Theorem 4.102 (observe $\tilde{x}(b) = 0$ and $\tilde{x}(\rho(a)) = 0$ if $x(\rho(a)) = 0$),

$$0 < \sum_{\nu=1}^k (\lambda - \lambda_\nu^*) = - \int_{\rho(a)}^b \left[(xx^\sigma)^\dagger W^2(\tilde{x}, x) \right](t) \Delta t \leq 0,$$

a contradiction. Therefore x has at least one generalized zero in (a, b), say that $t_1 < t_2 < \cdots < t_p$ are all of them. Define a $p \times k$-matrix Z^* with $z_{\mu\nu}^*$ as above. Let $(\beta_1, \beta_2, \ldots, \beta_k)^T \in \operatorname{Ker} Z^*$ and $\tilde{x} = \sum_{\nu=1}^k \beta_\nu x_\nu^*$. Hence $\tilde{x}(t_\mu) = 0$ for all zeros t_μ of y and $W(\tilde{x}, x)(t_\mu) = 0$ for all nodes t_μ of x. Therefore, by Theorem 4.102,

$$0 \leq \sum_{\nu=1}^k \beta_\nu^2 (\lambda - \lambda_\nu^*) = - \int_{\rho(a)}^b \left[(xx^\sigma)^\dagger W^2(\tilde{x}, x) \right](t) \Delta t \leq 0$$

so that $\beta_1 = \beta_2 = \cdots = \beta_k = 0$ and hence $\operatorname{Ker} Z^* = \{0\}$. Thus $p \geq k$. Further, by Lemma 4.97 and Theorem 4.100 we have $p \leq k$. Therefore $p = k$ and Z^* is invertible. $\qquad\square$

According to Theorem 4.103, we have $x(t, \lambda_1^*) > 0$ for all $t \in (a, b)$. Hence $x^\Delta(b, \lambda_1^*) < 0$ (note that if b is left-scattered, then it is also right-scattered due to (4.54), and observe also Lemma 4.99 (i)) and

$$\Lambda(\lambda_1^*) = \delta x^\Delta(b, \lambda_1^*) \leq 0, \quad \text{hence} \quad \lambda_1 \leq \lambda_1^*$$

(observe (4.53) and Theorem 4.100). Because of Theorem 4.103 if $\delta \neq 0$, we also have

$$\operatorname{sgn} \Lambda(\lambda_{k+1}^*) = \operatorname{sgn} \delta x^\Delta(b, \lambda_{k+1}^*) = \operatorname{sgn} x^\Delta(b, \lambda_{k+1}^*) = (-1)^{k+1}$$

and hence $\lambda_{k+1} \leq \lambda_{k+1}^*$.

Theorem 4.104. *If* $\lambda_k^* < \infty$, *then*

$$\lambda_{k+1}^* \leq \inf_{\substack{y(\rho(a))=y(b)=0 \\ y^\sigma \perp x_\nu^{*\sigma},\, 1 \leq \nu \leq k}} R(y) \quad and \quad \lambda_{k+1}^* \leq \inf_{\substack{R_a(y)=R_b(y)=0 \\ y^\sigma \perp x_\nu^{*\sigma},\, 1 \leq \nu \leq k}} R(y).$$

Proof. Let $y \in C_{rd}^2$ with $y(\rho(a))R_a(y) = y(b)R_b(y) = 0$, $y \neq 0$, $y_\nu = x_\nu^*$, $\rho_\nu = \lambda_\nu^*$, $\lambda \in (\lambda_k^*, \lambda_{k+1}^*]$, $x = x(\cdot, \lambda)$. Then $T(\rho(a)) = T(b) = 0$. By Theorem 4.103 we may choose $\beta_1, \ldots, \beta_k \in \mathbb{R}$ such that $\tilde{x}(t_\mu) = 0$ for all zeros t_μ of x and $W(\tilde{x}, x)(t_\mu) = 0$ for all nodes t_μ of x (observe that Z^* is invertible). Then, by Theorem 4.102,

$$- \langle L(y) + \lambda y^\sigma, y^\sigma \rangle = \int_a^b \left[(xx^\sigma)^\dagger W^2(\tilde{x}, x) \right](t) \Delta t + \sum_{\nu=1}^k \beta_\nu^2 (\lambda - \lambda_\nu^*) \geq 0$$

so that $- \langle L(y), y^\sigma \rangle \geq \lambda \langle y^\sigma, y^\sigma \rangle$. $\qquad\square$

In what follows we denote by $|S|$ the number of elements of a set S. We also put for $\alpha \in \mathbb{R}$

$$\operatorname{def} \alpha = \begin{cases} 0 & \text{if} \quad \alpha \neq 0 \\ 1 & \text{if} \quad \alpha = 0. \end{cases}$$

Theorem 4.105. *The number of eigenvalues of* (4.51) *is*

$$|[a, b]| - \operatorname{def}(\beta - \alpha\mu(\rho(a))) - \operatorname{def}\delta.$$

Proof. First note that, if $|[a, b]| = \infty$ and $\lambda_k^* < \infty$, there exists $y \in \mathrm{C}_{\mathrm{rd}}^2$ with $y(\rho(a)) = y(b) = 0$, $y \neq 0$ which is orthogonal to all x_ν^*, $1 \leq \nu \leq k$. Hence Theorem 4.104 implies that $\lambda_{k+1}^* < \infty$. Therefore (4.52) has infinitely many eigenvalues.

Next, suppose $|[a, b]| = b - a + 1 < \infty$. Then all points of $[a, b]$ are right-scattered (observe that b is right-scattered since it is left-scattered due to (4.54)). Hence

$$x(a) = \beta - \alpha\mu(\rho(a)),$$

and the recursion formula

$$x(\sigma^2(t)) = \{a(t) + \lambda b(t)\}\, x(\sigma(t)) + c(t)x(t), \quad t \in [\rho(a), \rho(b)]$$

with

$$a = 1 + \frac{\mu^\sigma}{\mu} - q\mu\mu^\sigma, \quad b = -\mu\mu^\sigma, \quad \text{and} \quad c = -\frac{\mu^\sigma}{\mu}$$

can be easily verified so that our claim follows from Theorem 4.95. $\qquad\square$

Our main result in this section now reads as follows.

Theorem 4.106 (Oscillation Theorem). *The eigenvalues of* (4.51) *may be arranged as* $-\infty < \lambda_1 < \lambda_2 < \lambda_3 < \cdots$, *and an eigenfunction corresponding to* λ_{k+1} *has exactly k generalized zeros in the open interval* (a, b).

Proof. We let $y = 0$, $x_\nu = y_\nu^*$, $\rho_\nu = \lambda_\nu$, $k \in \mathbb{N}_0$, $\lambda = \lambda_{k+1}$ if $\lambda_{k+1} < \infty$, and $x = x(\cdot, \lambda)$. Then $T(\rho(a)) = T(b) = 0$, and as in the proof of Theorem 4.103 we apply Theorem 4.102 to find $n(\lambda) \leq k$. Hence our claim follows as before. Moreover, we observe that the analogously defined $k \times k$-matrix Z is invertible. $\quad\square$

Theorem 4.107 (Comparison Theorem). *If $k \in \mathbb{N}_0$ with $\lambda_k^* < \infty$, then*

$$\lambda_k \leq \lambda_k^* < \lambda_{k+1}.$$

Proof. This is clear from Theorem 4.106 and the remark after Theorem 4.103. $\quad\square$

Theorem 4.108 (Rayleigh's Principle). *We have*

$$\lambda_{k+1} = \min_{\substack{R_a(y) = R_b(y) = 0 \\ \langle x^\sigma, x_\nu^\sigma \rangle = 0,\, 1 \leq \nu \leq k}} R(y),$$

where x_ν are the normalized eigenfunctions corresponding to λ_ν, $1 \leq \nu \leq k$.

Proof. We let $y(\rho(a))R_a(y) = y(b)R_b(y) = 0$, $y \neq 0$, $\rho_\nu = \lambda_\nu$, $k \in \mathbb{N}_0$, $\lambda = \lambda_{k+1}$ if $\lambda_{k+1} < \infty$, and $x = x(\cdot, \lambda)$. Then $T(\rho(a)) = T(b) = 0$ and as in the proof of Theorem 4.104 we apply Theorem 4.102 to find

$$\lambda_{k+1}^* \leq \inf_{\substack{y(\rho(a))R_a(y) = y(b)R_b(y) = 0 \\ y^\sigma \perp x_\nu^\sigma,\, 1 \leq \nu \leq k}} R(y).$$

We note that $R(x_{k+1}) = \lambda_{k+1}$, and this implies our claim. $\qquad\square$

We now specialize our results for the continuous and the discrete case.

Example 4.109. If $\mathbb{T} = \mathbb{R}$, then our results are well known (see, e.g., [**193**, Chapter 0]), and they may be summarized as follows (we put $a = 0$ and $b = 1$): There exist infinitely many eigenvalues $\lambda_1 < \lambda_2 < \lambda_3 < \cdots$ of

$$x'' + (\lambda + q(t))x = 0, \quad \alpha x(0) + \beta x'(0) = \gamma x(1) + \delta x'(1) = 0,$$

and an eigenfunction corresponding to λ_{k+1} vanishes exactly k times in $(0,1)$. We can compute λ_{k+1} by taking the minimum of

$$-\frac{\int_0^1 (y''(t) + q(t)y(t))y(t)dt}{\int_0^1 y^2(t)dt}$$

over all C^2-functions y that satisfy the boundary conditions and $\int_0^1 y(t)x_\nu(t)dt = 0$, $1 \leq \nu \leq k$ (where x_ν are the eigenfunctions corresponding to λ_ν normalized with $\int_0^1 x_\nu^2(t)dt = 1$).

Example 4.110. If $\mathbb{T} = \mathbb{Z}$, then only some special cases of Theorem 4.106 (with $\beta = \delta = 0$) and Theorem 4.108 (with $\beta = \delta = k = 0$) are known (see, e.g., [**191**, Chapter 7]). Our results may be summarized as follows (with $a = 1$ and $b = N \in \mathbb{N}$): There exist $N - \text{def}(\beta - \alpha) - \text{def}\,\delta$ eigenvalues $\lambda_1 < \lambda_2 < \lambda_3 < \cdots$ of

$$\Delta^2 x + (\lambda + q(i))(\Delta x + x) = 0, \quad \alpha x(0) + \beta \Delta x(0) = \gamma x(N) + \delta \Delta x(N) = 0,$$

and an eigenfunction corresponding to λ_{k+1} has exactly k zeros or changes of sign in $(0, N)$ (where an interval $(i, i+1)$ for $i \in \mathbb{Z}$ is said to contain a change of sign of x whenever $x(i)x(i+1) < 0$). We can compute λ_{k+1} by taking the minimum of

$$-\frac{\sum_{i=1}^N (\Delta^2 y(i) + q(i-1)y(i))y(i)}{\sum_{i=1}^N y^2(i)}$$

over all sequences y that satisfy the boundary conditions and $\sum_{i=0}^N y(i)x_\nu(i) = 0$, $1 \leq \nu \leq k$ (where x_ν are the eigenfunctions corresponding to λ_ν normalized with $\sum_{i=0}^N x_\nu^2(i) = 1$).

4.6. Notes and References

The study of self-adjoint equations of second order, also known in the literature as Sturm–Liouville theory, has a long history, originating in Sturm's paper [**245**]. For the differential equations case we included the references [**53, 54, 63, 235**] in the bibliography. Good references concerning the difference equations case are [**94, 95, 141, 174, 175, 232, 252**]. Corresponding results for the time scales case are contained in [**119, 128, 131, 135, 145**].

The three examples that are given at the beginning of this chapter are taken from Lynn Erbe and Allan Peterson [**131**]. We also have included some results on how an arbitrary linear second order dynamic equation can be rewritten in self-adjoint form. As in Chapter 3, once again comparison of Theorems 4.12 and 4.17 shows that (3.6) might be considered as the more "natural" form of the equation than (3.1). Section 4.1 also contains some results concerning the Prüfer transformation, which follow closely the strategy for the discrete case presented in Martin Bohner and Ondřej Došlý [**79**]. For related results on the Prüfer transformation we refer to the original work by Prüfer [**235**] and to [**52, 236**].

In the section on Riccati equations we explain how a solution of a dynamic Riccati equation can be found using a solution of an associated self-adjoint second order equation. These results unify results on the well-know differential Riccati equation and on the first view rather unrelated results on difference Riccati equations, given, e.g., in Walter Kelley and Allan Peterson [**191**, Chapter 6]. For the case of $p(t)$ being positive for all $t \in \mathbb{T}$, related theory can be found in the "early" (from a time scales point of view) paper by Erbe and Hilger [**128**]. The sufficient criteria for oscillation of (4.1) given at the end of Section 4.2 are derived using a Riccati technique and are taken from Erbe and Peterson [**135**]. Their matrix analogues are given in the next chapter.

Most of the matrix analogues of Section 4.3 are contained in the following chapter as well. The first part of this section that leads to Jacobi's condition in Theorem 4.57, is a special case of Ravi Agarwal and Martin Bohner [**8**] and is given in detail for the scalar case in Erbe and Peterson [**136**]. Theorems 4.59 and 4.60 on the Polya and Trench factorization, respectively, are special cases of the results derived in Lynn Erbe, Ronald Mathsen, and Allan Peterson [**129**], and related results are contained in Martin Bohner and Paul Eloe [**85**]. For the classical results, see Polya [**231**] and Trench [**246**]. For the continuous case, the crucial Picone identity in Theorem 4.55 was originally established by Picone in [**230**].

Section 4.4 discusses boundary value problems and associated Green's functions: General two point boundary value problems from [**131**] and their special cases including conjugate and focal boundary conditions. Finally, periodic boundary conditions are discussed. Literature on boundary value problems for time scales, that discusses topics that are not presented in this book, can be found in [**10, 11, 16, 17, 40, 41, 59, 99, 154, 155, 156, 157, 158, 208**].

The material presented in the last section follows closely Ravi Agarwal, Martin Bohner, and Patricia Wong [**14**], and the general strategy is as in Werner Kratz [**193**, Chapter 0], where an extension of Picone's identity (see Theorem 4.102) is used to derive in an elementary way the main theorems concerning Sturm–Liouville eigenvalue problems, namely an oscillation theorem, a comparison theorem, and Rayleigh's principle. We discuss here only separated boundary conditions and leave the general nonseparated case as an open problem. Corresponding results on differential equations are rather classical and can be found in Kratz [**193**, Chapter 0]. The case of nonseparated boundary conditions is in Bohner [**63**] and goes back to Birkhoff [**60**]. Corresponding results on difference equations are contained, e.g., in Kelley and Peterson [**191**, Chapter 7] and originate in Atkinson's [**46**] work. However, no results on discrete Sturm–Liouville eigenvalue problems involving nonseparated boundary conditions, e.g., an oscillation number giving a formula for the number of generalized zeros of a given eigenfunction, can be found in the literature.

CHAPTER 5

Linear Systems and Higher Order Equations

5.1. Regressive Matrices

Definition 5.1. Let A be an $m \times n$-matrix-valued function on $\mathbb{T}$. We say that A is rd-continuous on $\mathbb{T}$ if each entry of A is rd-continuous on $\mathbb{T}$, and the class of all such rd-continuous $m \times n$-matrix-valued functions on $\mathbb{T}$ is denoted, similar to the scalar case (see Definition 1.58), by

$$C_{rd} = C_{rd}(\mathbb{T}) = C_{rd}(\mathbb{T}, \mathbb{R}^{m \times n}).$$

We say that A is differentiable on $\mathbb{T}$ provided each entry of A is differentiable on $\mathbb{T}$, and in this case we put

$$A^{\Delta} = (a_{ij}^{\Delta})_{1 \leq i \leq m, 1 \leq j \leq n}, \quad \text{where} \quad A = (a_{ij})_{1 \leq i \leq m, 1 \leq j \leq n}.$$

Theorem 5.2. *If A is differentiable at $t \in \mathbb{T}^\kappa$, then $A^{\sigma}(t) = A(t) + \mu(t)A^{\Delta}(t)$.*

Proof. We use (iv) of Theorem 1.16 to obtain

$$
\begin{aligned}
A^{\sigma} &= (a_{ij}^{\sigma}) \\
&= (a_{ij} + \mu a_{ij}^{\Delta}) \\
&= (a_{ij}) + \mu(a_{ij}^{\Delta}) \\
&= A + \mu A^{\Delta}
\end{aligned}
$$

at t. $\qquad\square$

Theorem 5.3. *Suppose A and B are differentiable $n \times n$-matrix-valued functions. Then*

  (i) $(A + B)^{\Delta} = A^{\Delta} + B^{\Delta}$;
  (ii) $(\alpha A)^{\Delta} = \alpha A^{\Delta}$ *if α is constant;*
  (iii) $(AB)^{\Delta} = A^{\Delta}B^{\sigma} + AB^{\Delta} = A^{\sigma}B^{\Delta} + A^{\Delta}B$;
  (iv) $(A^{-1})^{\Delta} = -(A^{\sigma})^{-1}A^{\Delta}A^{-1} = -A^{-1}A^{\Delta}(A^{\sigma})^{-1}$ *if AA^{σ} is invertible;*
  (v) $(AB^{-1})^{\Delta} = (A^{\Delta} - AB^{-1}B^{\Delta})(B^{\sigma})^{-1} = (A^{\Delta} - (AB^{-1})^{\sigma}B^{\Delta})B^{-1}$ *if BB^{σ} is invertible.*

Proof. We only show the first parts of (iii) and (iv) and leave the remainder of the proof as an exercise. Let $1 \leq i, j \leq n$. We use Theorem 1.20 to calculate the ijth entry of $(AB)^\Delta$:

$$\left(\sum_{k=1}^n a_{ik}b_{kj}\right)^\Delta = \sum_{k=1}^n (a_{ik}b_{kj})^\Delta$$

$$= \sum_{k=1}^n (a_{ik}^\Delta b_{kj}^\sigma + a_{ik}b_{kj}^\Delta)$$

$$= \sum_{k=1}^n a_{ik}^\Delta b_{kj}^\sigma + \sum_{k=1}^n a_{ik}b_{kj}^\Delta,$$

and this is the ijth entry of the matrix $A^\Delta B^\sigma + AB^\Delta$ (see Definition 5.1). Next, we use part (iii) to differentiate $I = AA^{-1}$:

$$0 = A^\Delta A^{-1} + A^\sigma (A^{-1})^\Delta,$$

and solving for $(A^{-1})^\Delta$ yields the first formula in (iv). $\qquad\square$

Exercise 5.4. Prove the remaining parts of Theorem 5.3.

We consider the linear system of dynamic equations

(5.1) $$y^\Delta = A(t)y,$$

where A is an $n \times n$-matrix-valued function on $\mathbb{T}$. A vector-valued $y : \mathbb{T} \to \mathbb{R}^n$ is said to be a solution of (5.1) on $\mathbb{T}$ provided $y^\Delta(t) = A(t)y(t)$ holds for each $t \in \mathbb{T}^\kappa$. In order to state the main theorem on solvability of initial value problems involving equation (5.1), we make the following definition.

Definition 5.5 (Regressivity). An $n \times n$-matrix-valued function A on a time scale $\mathbb{T}$ is called *regressive* (with respect to $\mathbb{T}$) provided

(5.2) $$I + \mu(t)A(t) \quad \text{is invertible for all} \quad t \in \mathbb{T}^\kappa,$$

and the class of all such regressive and rd-continuous functions is denoted, similar to the scalar case (see Definition 2.25), by

$$\mathcal{R} = \mathcal{R}(\mathbb{T}) = \mathcal{R}(\mathbb{T}, \mathbb{R}^{n \times n}).$$

We say that the system (5.1) is *regressive* provided $A \in \mathcal{R}$.

Exercise 5.6. Show that the $n \times n$-matrix-valued function A is regressive iff the eigenvalues $\lambda_i(t)$ of $A(t)$ are regressive for all $1 \le i \le n$.

Exercise 5.7. Show that a 2×2-matrix-valued function A is regressive iff the scalar-valued function $\operatorname{tr} A + \mu \det A$ is regressive (where $\operatorname{tr} A$ denotes the *trace* of the matrix A, i.e., the sum of the diagonal elements of A). Derive similar characterizations for the 3×3-case, the 4×4-case, and the general $n \times n$-case.

Now the existence and uniqueness theorem for initial value problems for equation (5.1) reads as follows.

Theorem 5.8 (Existence and Uniqueness Theorem). *Let $A \in \mathcal{R}$ be an $n \times n$-matrix-valued function on $\mathbb{T}$ and suppose that $f : \mathbb{T} \to \mathbb{R}^n$ is rd-continuous. Let $t_0 \in \mathbb{T}$ and $y_0 \in \mathbb{R}^n$. Then the initial value problem*

$$y^\Delta = A(t)y + f(t), \quad y(t_0) = y_0$$

has a unique solution $y : \mathbb{T} \to \mathbb{R}^n$.

Proof. This is a special case of the result given in Theorem 8.20 (it is easy to see that Definition 5.1 and Definition 8.4 define A^Δ equivalently).   □

It follows from Theorem 5.8 that the matrix initial value problem

(5.3) $$Y^\Delta = A(t)Y, \quad Y(t_0) = Y_0,$$

where Y_0 is a constant $n \times n$-matrix, has a unique (matrix-valued) solution Y.

Example 5.9. If $\mathbb{T} = \mathbb{R}$, then any matrix-valued function A on $\mathbb{T}$ satisfies condition (5.2) and hence is regressive. In this case, a matrix-valued function A is rd-continuous iff it is continuous. So the initial value problem (5.3) has a unique solution provided A is continuous. Now, if $\mathbb{T} = \mathbb{Z}$, then it is in a sense the other way around: Any matrix-valued function A on $\mathbb{T}$ is rd-continuous, since there are no left-dense or right-dense points. However, in order for a matrix-valued function A on $\mathbb{T}$ to be regressive, the matrix $I + A(t)$ needs to be invertible for each $t \in \mathbb{Z}$, which holds if and only if $A(t)$ has no eigenvalue equal to -1, for each $t \in \mathbb{Z}$. So the initial value problem (5.3) has a unique solution provided $A(t)$ has no eigenvalue equal to -1, for all $t \in \mathbb{Z}$ (see Exercise 5.6).

Before we give some properties of the exponential matrix we define the "circle plus" addition $\oplus$ and the "circle minus" subtraction $\ominus$.

Definition 5.10. Assume A and B are regressive $n \times n$-matrix-valued functions on $\mathbb{T}$. Then we define $A \oplus B$ by

$$(A \oplus B)(t) = A(t) + B(t) + \mu(t)A(t)B(t) \quad \text{for all} \quad t \in \mathbb{T}^\kappa,$$

and we define $\ominus A$ by

$$(\ominus A)(t) = -[I + \mu(t)A(t)]^{-1}A(t) \quad \text{for all} \quad t \in \mathbb{T}^\kappa.$$

Exercise 5.11. Show that if A is regressive on $\mathbb{T}$, then

$$(\ominus A)(t) = -A(t)[I + \mu(t)A(t)]^{-1} \quad \text{for all} \quad t \in \mathbb{T}^\kappa.$$

Lemma 5.12. $(\mathcal{R}(\mathbb{T}, \mathbb{R}^{n \times n}), \oplus)$ *is a group.*

Proof. Let $A, B \in \mathcal{R}$. Then $I + \mu(t)A(t)$ and $I + \mu(t)B(t)$ are invertible for all $t \in \mathbb{T}^\kappa$, and therefore

$$\begin{aligned}
I + \mu(t)(A \oplus B)(t) &= I + \mu(t)[A(t) + B(t) + \mu(t)A(t)B(t)] \\
&= I + \mu(t)A(t) + \mu(t)B(t) + \mu^2(t)A(t)B(t) \\
&= [I + \mu(t)A(t)][I + \mu(t)B(t)]
\end{aligned}$$

is also invertible for each $t \in \mathbb{T}^\kappa$, being the product of two invertible matrices. Hence (note that $\mu \in C_{rd}$)

$$A \oplus B \in \mathcal{R}.$$

Next, if $A \in \mathcal{R}$, then $\ominus A$ defined in Definition 5.10 satisfies $(A \oplus (\ominus A))(t) = 0$ for all $t \in \mathbb{T}^\kappa$ according to Exercise 5.15. It remains to show that

$$\ominus A \in \mathcal{R}.$$

But this follows since because of Exercise 5.11

$$
\begin{aligned}
I + \mu(t)(\ominus A)(t) &= I - \mu(t)A(t)[I + \mu(t)A(t)]^{-1} \\
&= [I + \mu(t)A(t)][I + \mu(t)A(t)]^{-1} - \mu(t)A(t)[I + \mu(t)A(t)]^{-1} \\
&= [I + \mu(t)A(t) - \mu(t)A(t)][I + \mu(t)A(t)]^{-1} \\
&= [I + \mu(t)A(t)]^{-1}
\end{aligned}
$$

is invertible for all $t \in \mathbb{T}^\kappa$. In Exercise 5.13, the reader is asked to check that $\oplus$ satisfies the associative law. $\square$

Exercise 5.13. Show that $\oplus$ satisfies the associative law, i.e.,

$$A \oplus (B \oplus C) = (A \oplus B) \oplus C \quad \text{for all} \quad A, B, C \in \mathcal{R}.$$

Definition 5.14. If the matrix-valued functions A and B are regressive on $\mathbb{T}$, then we define $A \ominus B$ by

$$(A \ominus B)(t) = (A \oplus (\ominus B))(t) \quad \text{for all} \quad t \in \mathbb{T}^\kappa.$$

Exercise 5.15. Let A and B be regressive matrix-valued functions. State and prove the analogues to the scalar formulas from Exercise 2.28.

If A is a matrix, then we let A^* denote its conjugate transpose.

Exercise 5.16. Show that $(A^*)^\Delta = (A^\Delta)^*$ holds for any differentiable matrix-valued function A.

Exercise 5.17. Suppose A and B are regressive matrix-valued functions. Show that

(i) A^* is regressive;
(ii) $\ominus A^* = (\ominus A)^*$;
(iii) $A^* \oplus B^* = (A \oplus B)^*$.

Definition 5.18 (Matrix Exponential Function). Let $t_0 \in \mathbb{T}$ and assume that $A \in \mathcal{R}$ is an $n \times n$-matrix-valued function. The unique matrix-valued solution of the IVP

$$Y^\Delta = A(t)Y, \quad Y(t_0) = I,$$

where I denotes as usual the $n \times n$-identity matrix, is called the *matrix exponential function* (at t_0), and it is denoted by $e_A(\cdot, t_0)$.

Example 5.19. Assume that A is a constant $n \times n$-matrix. If $\mathbb{T} = \mathbb{R}$, then

$$e_A(t, t_0) = e^{A(t-t_0)},$$

while if $\mathbb{T} = \mathbb{Z}$ and $I + A$ is invertible, then

$$e_A(t, t_0) = (I + A)^{(t-t_0)}.$$

Exercise 5.20. Let $\mathbb{T} = 2^{\mathbb{N}_0}$. If $A \in \mathcal{R}$ is a constant $n \times n$-matrix, find the matrix exponential $e_A(t, 1)$.

In the following theorem we give some properties of the matrix exponential function.

Theorem 5.21. *If $A, B \in \mathcal{R}$ are matrix-valued functions on $\mathbb{T}$, then*

(i) $e_0(t, s) \equiv I$ *and* $e_A(t, t) \equiv I$;
(ii) $e_A(\sigma(t), s) = (I + \mu(t)A(t))e_A(t, s)$;

(iii) $e_A^{-1}(t,s) = e_{\ominus A^*}^*(t,s);$

(iv) $e_A(t,s) = e_A^{-1}(s,t) = e_{\ominus A^*}^*(s,t);$

(v) $e_A(t,s)e_A(s,r) = e_A(t,r);$

(vi) $e_A(t,s)e_B(t,s) = e_{A\oplus B}(t,s)$ if $e_A(t,s)$ and $B(t)$ commute.

Proof. Part (i). Consider the initial value problem

$$Y^\Delta = 0, \quad Y(s) = I,$$

which has exactly one solution by Theorem 5.8. Since $Y(t) \equiv I$ is a solution of this IVP, we have $e_0(t,s) = Y(t) \equiv I$.

Part (ii). According to Theorem 5.2 we have

$$\begin{aligned}
e_A(\acute{\sigma}(t),s) &= e_A(t,s) + \mu(t)e_A^\Delta(t,s) \\
&= e_A(t,s) + \mu(t)A(t)e_A(t,s) \\
&= (I + \mu(t)A(t))e_A(t,s).
\end{aligned}$$

Part (iii). Let $Y(t) = (e_A^{-1}(t,s))^*$. Then

$$\begin{aligned}
Y^\Delta(t) &= -\left(e_A^{-1}(\sigma(t),s)e_A^\Delta(t,s)e_A^{-1}(t,s)\right)^* \\
&= -\left(e_A^{-1}(\sigma(t),s)A(t)e_A(t,s)e_A^{-1}(t,s)\right)^* \\
&= -\left(e_A^{-1}(\sigma(t),s)A(t)\right)^* \\
&= -\left(e_A^{-1}(t,s)(I + \mu(t)A(t))^{-1}A(t)\right)^* \\
&= \left(e_A^{-1}(t,s)(\ominus A)(t)\right)^* \\
&= (\ominus A^*)(t)(e_A^{-1}(t,s))^* \\
&= (\ominus A^*)(t)Y(t),
\end{aligned}$$

where we have used (ii) and Exercise 5.17 (ii). Also,

$$Y(s) = (e_A^{-1}(s,s))^* = (I^{-1})^* = I.$$

Hence Y solves the initial value problem

$$Y^\Delta = (\ominus A^*)(t)Y, \quad Y(s) = I,$$

which has exactly one solution according to Theorem 5.8, and therefore we have $e_{\ominus A^*}(t,s) = Y(t) = (e_A^{-1}(t,s))^*$ so that $e_A^{-1}(t,s) = e_{\ominus A^*}^*(t,s)$.

Part (iv). This is Exercise 5.22.

Part (v). Consider $Y(t) = e_A(t,s)e_A(s,r)$. Then

$$Y^\Delta(t) = e_A^\Delta(t,s)e_A(s,r) = A(t)e_A(t,s)e_A(s,r) = A(t)Y(t)$$

and $Y(r) = e_A(r,s)e_A(s,r) = I$ according to (iv). Hence Y solves the IVP

$$Y^\Delta = A(t)Y, \quad Y(r) = I,$$

which has exactly one solution according to Theorem 5.8, and therefore we have $e_A(t,r) = Y(t) = e_A(t,s)e_A(s,r)$.

Part (vi). Let $Y(t) = e_A(t,s)e_B(t,s)$ and suppose that $e_A(t,s)$ and $B(t)$ commute. We use Theorem 5.2 and Theorem 5.3 (iii) to calculate

$$
\begin{aligned}
Y^\Delta(t) &= e_A^\Delta(t,s)e_B^\sigma(t,s) + e_A(t,s)e_B^\Delta(t,s) \\
&= A(t)e_A(t,s)(I + \mu(t)B(t))e_B(t,s) + e_A(t,s)B(t)e_B(t,s) \\
&= A(t)(I + \mu(t)B(t))e_A(t,s)e_B(t,s) + B(t)e_A(t,s)e_B(t,s) \\
&= [A(t)(I + \mu(t)B(t)) + B(t)]\,e_A(t,s)e_B(t,s) \\
&= (A \oplus B)(t)e_A(t,s)e_B(t,s) \\
&= (A \oplus B)(t)Y(t).
\end{aligned}
$$

Also $Y(s) = e_A(s,s)e_B(s,s) = I \cdot I = I$. So Y solves the initial value problem

$$
Y^\Delta = (A \oplus B)(t)Y, \quad Y(s) = I,
$$

which has exactly one solution according to Theorem 5.8, and therefore we have $e_{A \oplus B}(t,s) = Y(t) = e_A(t,s)e_B(t,s)$.   □

Exercise 5.22. Prove part (iv) of Theorem 5.21.

The matrix analogue of Theorem 2.39 now reads as follows.

Theorem 5.23. *If $A \in \mathcal{R}$ and $a, b, c \in \mathbb{T}$, then*

$$
[e_A(c, \cdot)]^\Delta = -[e_A(c, \cdot)]^\sigma A
$$

and

$$
\int_a^b e_A(c, \sigma(t))A(t)\Delta t = e_A(c, a) - e_A(c, b).
$$

Proof. We use the properties given in Theorem 5.21 and Exercise 5.17 (ii) to find

$$
\begin{aligned}
e_A(c, \sigma(t))A(t) &= e_{\ominus A^*}^*(\sigma(t), c)A(t) \\
&= \{[I + \mu(t)(\ominus A^*)(t)]\,e_{\ominus A^*}(t, c)\}^* A(t) \\
&= e_{\ominus A^*}^*(t, c)\,[I + \mu(t)(\ominus A^*)(t)]^* A(t) \\
&= e_{\ominus A^*}^*(t, c)\,[I - \mu(t)\{A^*(t)[I + \mu(t)A^*(t)]^{-1}\}]^* A(t) \\
&= e_{\ominus A^*}^*(t, c)\,\{[I + \mu(t)A^*(t) - \mu(t)A^*(t)][I + \mu(t)A^*(t)]^{-1}\}^* A(t) \\
&= e_{\ominus A^*}^*(t, c)[I + \mu(t)A(t)]^{-1}A(t) \\
&= \{A^*(t)[I + \mu(t)A^*(t)]^{-1}e_{\ominus A^*}(t, c)\}^* \\
&= -\{(\ominus A^*)(t)e_{\ominus A^*}(t, c)\}^* \\
&= -\{e_{\ominus A^*}^\Delta(t, c)\}^* \\
&= -[e_A(c, \cdot)]^\Delta(t).
\end{aligned}
$$

This proves our result.   □

Together with the homogeneous equation (5.1) we also consider the nonhomogeneous equation

(5.4)
$$
y^\Delta = A(t)y + f(t),
$$

where $f : \mathbb{T} \to \mathbb{R}^n$ is a vector-valued function. We have the following result for initial value problems for equation (5.4).

Theorem 5.24 (Variation of Constants). *Let $A \in \mathcal{R}$ be an $n \times n$-matrix-valued function on $\mathbb{T}$ and suppose that $f : \mathbb{T} \to \mathbb{R}^n$ is rd-continuous. Let $t_0 \in \mathbb{T}$ and $y_0 \in \mathbb{R}^n$. Then the initial value problem*

$$(5.5) \qquad\qquad y^{\Delta} = A(t)y + f(t), \quad y(t_0) = y_0$$

has a unique solution $y : \mathbb{T} \to \mathbb{R}^n$. Moreover, this solution is given by

$$(5.6) \qquad\qquad y(t) = e_A(t, t_0)y_0 + \int_{t_0}^{t} e_A(t, \sigma(\tau))f(\tau)\Delta\tau.$$

Proof. First, y given by (5.6) is well defined and can be rewritten because of Theorem 5.21 (v) as

$$y(t) = e_A(t, t_0) \left\{ y_0 + \int_{t_0}^{t} e_A(t_0, \sigma(\tau))f(\tau)\Delta\tau \right\}.$$

We use the product rule to differentiate y:

$$\begin{aligned}
y^{\Delta}(t) &= A(t)e_A(t, t_0) \left\{ y_0 + \int_{t_0}^{t} e_A(t_0, \sigma(\tau))f(\tau)\Delta\tau \right\} \\
&\quad + e_A(\sigma(t), t_0)e_A(t_0, \sigma(t))f(t) \\
&= A(t)y(t) + f(t).
\end{aligned}$$

Obviously, $y(t_0) = y_0$. Therefore y is a solution of (5.5).

Now we show that y is the only solution of (5.5). Assume $\tilde{y}$ is another solution of (5.5) and put $v(t) = e_A(t_0, t)\tilde{y}(t)$. By Theorem 5.21 (v) we have $\tilde{y}(t) = e_A(t, t_0)v(t)$ and therefore

$$\begin{aligned}
A(t)e_A(t, t_0)v(t) + f(t) &= A(t)\tilde{y}(t) + f(t) \\
&= \tilde{y}^{\Delta}(t) \\
&= A(t)e_A(t, t_0)v(t) + e_A(\sigma(t), t_0)v^{\Delta}(t),
\end{aligned}$$

so $v^{\Delta}(t) = e_A(t_0, \sigma(t))f(t)$. Since $v(t_0)$ must be equal to y_0, this yields

$$v(t) = y_0 + \int_{t_0}^{t} e_A(t_0, \sigma(\tau))f(\tau)\Delta\tau$$

and therefore $\tilde{y} = y$, where y is given by (5.6). $\square$

In order to apply Theorem 5.21 (vi), it is important to know under what conditions the two matrices e_A and B commute. We investigate this problem next.

Theorem 5.25. *Suppose $A \in \mathcal{R}$ and C is differentiable. If C is a solution of the dynamic equation*

$$C^{\Delta} = A(t)C - C^{\sigma}A(t),$$

then

$$C(t)e_A(t, s) = e_A(t, s)C(s).$$

Proof. Fix $s \in \mathbb{T}$ and consider

$$F(t) = C(t)e_A(t, s) - e_A(t, s)C(s).$$

Then $F(s) = 0$ and

$$
\begin{aligned}
F^{\Delta}(t) &= C(\sigma(t))A(t)e_A(t,s) + C^{\Delta}(t)e_A(t,s) - A(t)e_A(t,s)C(s) \\
&= \left[C^{\Delta}(t) + C(\sigma(t))A(t) - A(t)C(t) \right] e_A(t,s) \\
&\quad + A(t) \left[C(t)e_A(t,s) - e_A(t,s)C(s) \right] \\
&= A(t)F(t).
\end{aligned}
$$

Hence F solves the initial value problem

$$
F^{\Delta} = A(t)F, \quad F(s) = 0.
$$

By the variation of constants theorem (Theorem 5.24) we have $F \equiv 0$, but this means that $C(t)e_A(t,s)$ and $e_A(t,s)C(s)$ commute. $\qquad \square$

Corollary 5.26. *Suppose $A \in \mathcal{R}$ and C is a constant matrix. If C commutes with A, then C commutes with e_A. In particular, if A is a constant matrix, then A commutes with e_A.*

Proof. This follows from Theorem 5.25 since $C^{\Delta} \equiv 0$ in this case. $\qquad \square$

As in the scalar case, along with (5.1), we consider its *adjoint equation*

$$
(5.7) \qquad\qquad x^{\Delta} = -A^*(t)x^{\sigma}.
$$

We have the following result on initial value problems involving the adjoint equation.

Theorem 5.27 (Variation of Constants). *Let $A \in \mathcal{R}$ be an $n \times n$-matrix-valued function on $\mathbb{T}$ and suppose that $f : \mathbb{T} \to \mathbb{R}^n$ is rd-continuous. Let $t_0 \in \mathbb{T}$ and $x_0 \in \mathbb{R}^n$. Then the initial value problem*

$$
(5.8) \qquad\qquad x^{\Delta} = -A^*(t)x^{\sigma} + f(t), \quad x(t_0) = x_0
$$

has a unique solution $x : \mathbb{T} \to \mathbb{R}^n$. Moreover, this solution is given by

$$
(5.9) \qquad\qquad x(t) = e_{\ominus A^*}(t,t_0)x_0 + \int_{t_0}^{t} e_{\ominus A^*}(t,\tau)f(\tau)\Delta\tau.
$$

Proof. We rewrite

$$
\begin{aligned}
x^{\Delta} &= -A^*(t)x^{\sigma} + f(t) \\
&= -A^*(t)\left[x + \mu(t)x^{\Delta} \right] + f(t) \\
&= -A^*(t)x - \mu(t)A^*(t)x^{\Delta} + f(t),
\end{aligned}
$$

i.e.,

$$
\left[I + \mu(t)A^*(t) \right] x^{\Delta} = -A^*(t)x + f(t).
$$

Since A is regressive, A^* is regressive as well (see Exercise 5.17 (i)), and hence the above is equivalent to

$$
\begin{aligned}
x^{\Delta} &= -\left[I + \mu(t)A^*(t) \right]^{-1} A^*(t)x + \left[I + \mu(t)A^*(t) \right]^{-1} f(t) \\
&= (\ominus A^*(t))x + \left[I + \mu(t)A^*(t) \right]^{-1} f(t).
\end{aligned}
$$

Hence we may apply Theorem 5.24 to obtain the solution of (5.8) as

$$
\begin{aligned}
x(t) &= e_{\ominus A^*}(t, t_0)x_0 + \int_{t_0}^{t} e_{\ominus A^*}(t, \sigma(\tau)) \left[I + \mu(\tau)A^*(\tau) \right]^{-1} f(\tau) \Delta\tau \\
&= e_{\ominus A^*}(t, t_0)x_0 + \int_{t_0}^{t} e_A^*(\sigma(\tau), t) \left[I + \mu(\tau)A^*(\tau) \right]^{-1} f(\tau) \Delta\tau \\
&= e_{\ominus A^*}(t, t_0)x_0 + \int_{t_0}^{t} \left\{ \left[I + \mu(\tau)A(\tau) \right]^{-1} e_A(\sigma(\tau), t) \right\}^* f(\tau) \Delta\tau \\
&= e_{\ominus A^*}(t, t_0)x_0 + \int_{t_0}^{t} \left\{ e_A(\tau, t) \right\}^* f(\tau) \Delta\tau \\
&= e_{\ominus A^*}(t, t_0)x_0 + \int_{t_0}^{t} e_{\ominus A^*}(t, \tau) f(\tau) \Delta\tau,
\end{aligned}
$$

where we have applied Theorem 5.21 (ii) and (iii).   □

We now present the dynamic equation version of Liouville's formula.

Theorem 5.28 (Liouville's Formula). *Let $A \in \mathcal{R}$ be a 2×2-matrix-valued function and assume that X is a solution of $X^\Delta = A(t)X$. Then X satisfies Liouville's formula*

(5.10) $\det X(t) = e_{\operatorname{tr} A + \mu \det A}(t, t_0) \det X(t_0) \quad for \quad t \in \mathbb{T}.$

Proof. Note that $A \in \mathcal{R}$ implies $\operatorname{tr} A + \mu \det A \in \mathcal{R}$ by Exercise 5.7. We put

$$
A = \begin{pmatrix} a_{11} & a_{12} \\ a_{21} & a_{22} \end{pmatrix} \quad \text{and} \quad X = \begin{pmatrix} x_{11} & x_{12} \\ x_{21} & x_{22} \end{pmatrix}.
$$

Then

$$
\begin{aligned}
(\det X)^\Delta &= (x_{11}x_{22} - x_{12}x_{21})^\Delta \\
&= x_{11}^\Delta x_{22} + x_{11}^\sigma x_{22}^\Delta - x_{12}^\Delta x_{21} - x_{12}^\sigma x_{21}^\Delta \\
&= \det \begin{pmatrix} x_{11}^\Delta & x_{12}^\Delta \\ x_{21} & x_{22} \end{pmatrix} + \det \begin{pmatrix} x_{11}^\sigma & x_{12}^\sigma \\ x_{21}^\Delta & x_{22}^\Delta \end{pmatrix} \\
&= \det \begin{pmatrix} a_{11}x_{11} + a_{12}x_{21} & a_{11}x_{12} + a_{12}x_{22} \\ x_{21} & x_{22} \end{pmatrix} \\
&\quad + \det \begin{pmatrix} x_{11} + \mu x_{11}^\Delta & x_{12} + \mu x_{12}^\Delta \\ x_{21}^\Delta & x_{22}^\Delta \end{pmatrix} \\
&= a_{11} \det X + \det \begin{pmatrix} x_{11} & x_{12} \\ x_{21}^\Delta & x_{22}^\Delta \end{pmatrix} + \mu \det \begin{pmatrix} x_{11}^\Delta & x_{12}^\Delta \\ x_{21}^\Delta & x_{22}^\Delta \end{pmatrix}
\end{aligned}
$$

$$
= a_{11} \det X + \det \begin{pmatrix} x_{11} & x_{12} \\ a_{21}x_{11} + a_{22}x_{21} & a_{21}x_{12} + a_{22}x_{22} \end{pmatrix}
$$

$$
+ \mu \det \begin{pmatrix} x_{11}^{\Delta} & x_{12}^{\Delta} \\ x_{21}^{\Delta} & x_{22}^{\Delta} \end{pmatrix}
$$

$$
= (a_{11} + a_{22}) \det X + \mu \det \begin{pmatrix} a_{11}x_{11} + a_{12}x_{21} & a_{11}x_{12} + a_{12}x_{22} \\ x_{21}^{\Delta} & x_{22}^{\Delta} \end{pmatrix}
$$

$$
= \operatorname{tr} A \det X + \mu \det \left\{ a_{11} \begin{pmatrix} x_{11} & x_{12} \\ x_{21}^{\Delta} & x_{22}^{\Delta} \end{pmatrix} + a_{12} \begin{pmatrix} x_{21} & x_{22} \\ x_{21}^{\Delta} & x_{22}^{\Delta} \end{pmatrix} \right\}
$$

$$
= \operatorname{tr} A \det X + \mu(a_{11}a_{22} \det X - a_{12}a_{21} \det X)
$$

$$
= (\operatorname{tr} A + \mu \det A) \det X.
$$

Solving this first order scalar dynamic equation (use Theorem 2.62) we obtain (5.10). $\qquad\square$

Exercise 5.29. Derive Liouville's formula for $n = 3$ and $n = 4$. Prove a formula for $n \in \mathbb{N}$ using induction.

5.2. Constant Coefficients

In this section we study the vector dynamic equation

$$
(5.11) \qquad\qquad x^{\Delta} = Ax,
$$

where $A \in \mathcal{R}$ is a real constant $n \times n$-matrix.

Theorem 5.30. *If λ_0, ξ is an eigenpair for the constant $n \times n$-matrix A, then $x(t) = e_{\lambda_0}(t, t_0)\xi$ is a solution of* (5.11) *on $\mathbb{T}$.*

Proof. Assume λ_0, ξ is an eigenpair for A. Since A is assumed to be regressive on $\mathbb{T}$, we have by Exercise 5.6 that $\lambda_0 \in \mathcal{R}$, and so

$$
x(t) := e_{\lambda_0}(t, t_0)\xi
$$

is well defined on $\mathbb{T}$. Then

$$
\begin{aligned}
x^{\Delta}(t) &= \lambda_0 e_{\lambda_0}(t, t_0)\xi \\
&= e_{\lambda_0}(t, t_0)\lambda_0 \xi \\
&= e_{\lambda_0}(t, t_0)A\xi \\
&= A e_{\lambda_0}(t, t_0)\xi \\
&= Ax(t)
\end{aligned}
$$

for $t \in \mathbb{T}^{\kappa}$. $\qquad\square$

Example 5.31. Solve the vector dynamic equation

$$(5.12) \qquad x^{\Delta} = \begin{pmatrix} -3 & -2 \\ 3 & 4 \end{pmatrix} x.$$

The eigenvalues of the coefficient matrix in (5.12) are $\lambda_1 = -2$ and $\lambda_2 = 3$. Hence, by Exercise 5.6, the vector equation (5.12) is regressive for any time scale such that $1 - 2\mu(t) \neq 0$ for all $t \in \mathbb{T}^{\kappa}$. Eigenvectors corresponding to λ_1 and λ_2 are

$$\xi_1 = \begin{pmatrix} 2 \\ -1 \end{pmatrix} \quad \text{and} \quad \xi_2 = \begin{pmatrix} 1 \\ -3 \end{pmatrix},$$

respectively. It follows that

$$x(t) = c_1 e_{-2}(t, t_0) \begin{pmatrix} 2 \\ -1 \end{pmatrix} + c_2 e_3(t, t_0) \begin{pmatrix} 1 \\ -3 \end{pmatrix}$$

is a general solution of (5.12), for any time scale $\mathbb{T}$ with $-2 \in \mathcal{R}$.

Theorem 5.32. *Assume $A \in \mathcal{R}$ is a real $n \times n$-matrix-valued function on $\mathbb{T}^{\kappa}$. If $x(t) = u(t) + iv(t)$ is a complex vector-valued solution of $x^{\Delta} = A(t)x$ on $\mathbb{T}$, where u and v are real vector-valued functions on $\mathbb{T}$, then u and v are real vector-valued solutions of $x^{\Delta} = A(t)x$ on $\mathbb{T}$.*

Proof. Let

$$x(t) = u(t) + iv(t)$$

be a complex vector-valued solution of $x^{\Delta} = A(t)x$ on $\mathbb{T}$, where u and v are real vector-valued functions on $\mathbb{T}$. Then

$$u^{\Delta}(t) + iv^{\Delta}(t) = x^{\Delta}(t) = A(t)x(t) = A(t)u(t) + iA(t)v(t)$$

for $t \in \mathbb{T}^{\kappa}$. Equating real and imaginary parts we have the desired result

$$u^{\Delta}(t) = A(t)u(t), \quad v^{\Delta}(t) = A(t)v(t)$$

for $t \in \mathbb{T}^{\kappa}$. $\qquad\qquad\qquad\qquad\qquad\qquad\qquad\qquad\qquad\qquad\square$

Example 5.33. Solve the vector dynamic equation

$$(5.13) \qquad x^{\Delta} = \begin{pmatrix} 3 & 1 \\ -13 & -3 \end{pmatrix} x.$$

The eigenvalues of the coefficient matrix are $\lambda_1 = 2i$ and $\lambda_2 = -2i$. Hence, by Exercise 5.6 the vector equation (5.13) is regressive for any time scale. An eigenvector corresponding to $\lambda_1 = 2i$ is

$$\xi_1 = \begin{pmatrix} 1 \\ -3 + 2i \end{pmatrix}.$$

It follows from Theorem 5.30 and Euler's formula (3.13) that

$$x(t) = e_{2i}(t, t_0) \begin{pmatrix} 1 \\ -3 + 2i \end{pmatrix}$$

$$= [\cos_2(t, t_0) + i \sin_2(t, t_0)] \begin{pmatrix} 1 \\ -3 + 2i \end{pmatrix}$$

$$= \begin{pmatrix} \cos_2(t, t_0) \\ -3 \cos_2(t, t_0) - 2 \sin_2(t, t_0) \end{pmatrix} + i \begin{pmatrix} \sin_2(t, t_0) \\ 2 \cos_2(t, t_0) - 3 \sin_2(t, t_0) \end{pmatrix}$$

is a complex vector solution of (5.13). By Theorem 5.32,

$$\begin{pmatrix} \cos_2(t, t_0) \\ -3 \cos_2(t, t_0) - 2 \sin_2(t, t_0) \end{pmatrix} \quad \text{and} \quad \begin{pmatrix} \sin_2(t, t_0) \\ 2 \cos_2(t, t_0) - 3 \sin_2(t, t_0) \end{pmatrix}$$

are solutions of (5.13). Therefore

$$x(t) = c_1 \begin{pmatrix} \cos_2(t, t_0) \\ -3 \cos_2(t, t_0) - 2 \sin_2(t, t_0) \end{pmatrix} + c_2 \begin{pmatrix} \sin_2(t, t_0) \\ 2 \cos_2(t, t_0) - 3 \sin_2(t, t_0) \end{pmatrix}$$

is a general solution of (5.13) for any time scale $\mathbb{T}$.

Exercise 5.34. Solve the following vector dynamic equations:

(i) $x^{\Delta} = \begin{pmatrix} 3 & -1 \\ 5 & -3 \end{pmatrix} x;$

(ii) $x^{\Delta} = \begin{pmatrix} 2 & 2 \\ 3 & -3 \end{pmatrix} x;$

(iii) $x^{\Delta} = \begin{pmatrix} 1 & -2 \\ 5 & -1 \end{pmatrix} x;$

(iv) $x^{\Delta} = \begin{pmatrix} 0 & 1 & 0 \\ 0 & 0 & 1 \\ 12 & 4 & -3 \end{pmatrix} x.$

Whenever a real constant $n \times n$-matrix A has n linearly independent eigenvectors (for example every symmetric matrix has n linearly independent eigenvectors), we can use Theorem 5.30 to solve (5.11). However not every constant $n \times n$-matrix A has n linearly independent eigenvectors. The Putzer algorithm for dynamic equations (Theorem 5.35) can be used to solve (5.11) for any time scale on which A is regressive. This result is mentioned in Ahlbrandt and Ridenhour [31].

Theorem 5.35 (Putzer Algorithm). *Let $A \in \mathcal{R}$ be a constant $n \times n$-matrix. Suppose $t_0 \in \mathbb{T}$. If $\lambda_1, \lambda_2, \dots, \lambda_n$ are the eigenvalues of A, then*

$$(5.14) \qquad e_A(t, t_0) = \sum_{i=0}^{n-1} r_{i+1}(t) P_i,$$

where $r(t) := (r_1(t), r_2(t), \dots, r_n(t))^T$ is the solution of the IVP

$$(5.15) \qquad r^\Delta = \begin{pmatrix} \lambda_1 & 0 & 0 & \cdots & 0 \\ 1 & \lambda_2 & 0 & \ddots & \vdots \\ 0 & 1 & \lambda_3 & \ddots & \vdots \\ \vdots & \ddots & \ddots & \ddots & 0 \\ 0 & \cdots & 0 & 1 & \lambda_n \end{pmatrix} r, \quad r(t_0) = \begin{pmatrix} 1 \\ 0 \\ 0 \\ \vdots \\ 0 \end{pmatrix},$$

and the P-matrices $P_0, P_1, \dots, P_n$ are recursively defined by $P_0 = I$ and

$$P_{k+1} = (A - \lambda_{k+1} I) P_k \quad \text{for} \quad 0 \le k \le n-1.$$

Proof. Let $A \in \mathcal{R}$. From Exercise 5.6 we get that the coefficient matrix in (5.15) is regressive. Using Theorem 5.8 we find that the IVP (5.15) has a unique solution $r(t) := (r_1(t), r_2(t), \dots, r_n(t))^T$. Let $P_0, P_1, \dots, P_n$ be the matrices defined in the statement of this theorem and let X be defined by the right-hand side of (5.14). Then

$$
\begin{aligned}
X^\Delta(t) - AX(t) &= \sum_{i=0}^{n-1} r_{i+1}^\Delta(t) P_i - A \sum_{i=0}^{n-1} r_{i+1}(t) P_i \\
&= r_1^\Delta(t) P_0 + \sum_{i=1}^{n-1} r_{i+1}^\Delta(t) P_i - \sum_{i=0}^{n-1} r_{i+1}(t) A P_i \\
&= \lambda_1 r_1(t) P_0 + \sum_{i=1}^{n-1} [r_i(t) + \lambda_{i+1} r_{i+1}(t)] P_i - \sum_{i=0}^{n-1} r_{i+1}(t) A P_i \\
&= \sum_{i=1}^{n-1} r_i(t) P_i - \sum_{i=0}^{n-1} (A - \lambda_{i+1} I) r_{i+1}(t) P_i \\
&= \sum_{i=1}^{n-1} r_i(t) P_i - \sum_{i=0}^{n-1} r_{i+1}(t) P_{i+1} \\
&= -r_n(t) P_n \\
&= 0,
\end{aligned}
$$

where we have used that

$$P_n = (A - \lambda_n I)P_{n-1} = \cdots = (A - \lambda_n I)(A - \lambda_{n-1}I)\cdots(A - \lambda_1 I) = 0$$

by the Cayley–Hamilton theorem. Since

$$X(t_0) = \sum_{i=0}^{n-1} r_{i+1}(t_0)P_i = r_1(t_0)P_0 = I$$

(observe the initial condition in (5.15)), we get that X solves the IVP

$$X^\Delta = AX, \quad X(t_0) = I.$$

By Theorem 5.8 and Definition 5.18, the matrix exponential $e_A(t, t_0)$ is the unique solution of this IVP, and hence

$$e_A(t, t_0) = X(t) = \sum_{i=0}^{n-1} r_{i+1}(t)P_i,$$

which is what we wanted to prove. □

Example 5.36. We now solve the vector equation

$$(5.16) \qquad\qquad x^\Delta = Ax, \quad \text{where} \quad A = \begin{pmatrix} 2 & 1 \\ -1 & 4 \end{pmatrix}$$

on any time scale $\mathbb{T}$. Then we use our answer to solve (5.16) for $\mathbb{T} = \mathbb{R}$ and $\mathbb{T} = \mathbb{Z}$. First we find the matrix exponential $e_A(t, t_0)$. The eigenvalues of the coefficient matrix in (5.16) are $\lambda_1 = \lambda_2 = 3$. It turns out that there is only one linearly independent eigenvector for A so we cannot use Theorem 5.30 to find a general solution of (5.16). Since $1 + \lambda_1 \mu(t) = 1 + \lambda_2 \mu(t) = 1 + 3\mu(t) \neq 0$, $A \in \mathcal{R}$ on any time scale $\mathbb{T}$. Hence in this example we can solve (5.16) on any time scale. The P-matrices are given by

$$P_0 = I = \begin{pmatrix} 1 & 0 \\ 0 & 1 \end{pmatrix} \quad \text{and} \quad P_1 = (A - \lambda_1 I)P_0 = A - 3I = \begin{pmatrix} -1 & 1 \\ -1 & 1 \end{pmatrix}.$$

We want to choose r_1 and r_2 so that

$$r_1^\Delta = 3r_1, \quad r_1(t_0) = 1 \quad \text{and} \quad r_2^\Delta = r_1 + 3r_2, \quad r_2(t_0) = 0.$$

Solving the first IVP for r_1 we get that

$$r_1(t) = e_3(t, t_0).$$

Solving the second IVP, i.e.,

$$r_2^\Delta = 3r_2 + e_3(t, t_0), \quad r_2(t_0) = 0,$$

we obtain by Exercise 2.79 (iii)

$$r_2(t) = e_3(t, t_0) \int_{t_0}^t \frac{\Delta s}{1 + 3\mu(s)}.$$

Using the Putzer algorithm (Theorem 5.35), we get that for any time scale

$$
\begin{aligned}
e_A(t, t_0) &= r_1(t) P_0 + r_2(t) P_1 \\
&= e_3(t, t_0) \begin{pmatrix} 1 & 0 \\ 0 & 1 \end{pmatrix} + e_3(t, t_0) \int_{t_0}^{t} \frac{\Delta s}{1 + 3\mu(s)} \begin{pmatrix} -1 & 1 \\ -1 & 1 \end{pmatrix}.
\end{aligned}
$$

A general solution of (5.16) is then given by

$$
x(t) = e_A(t, t_0)\xi,
$$

where $\xi = \begin{pmatrix} c_1 \\ c_2 \end{pmatrix}$ is an arbitrary constant vector. Therefore

$$
x(t) = c_1 e_3(t, t_0) \begin{pmatrix} 1 - f(t, t_0) \\ -f(t, t_0) \end{pmatrix} + c_2 e_3(t, t_0) \begin{pmatrix} f(t, t_0) \\ 1 + f(t, t_0) \end{pmatrix}
$$

with

$$
f(t, t_0) = \int_{t_0}^{t} \frac{\Delta s}{1 + 3\mu(s)}.
$$

If $\mathbb{T} = \mathbb{R}$ and $t_0 = 0$, then $f(t, t_0) = t$ and

$$
x(t) = c_1 e^{3t} \begin{pmatrix} 1 - t \\ -t \end{pmatrix} + c_2 e^{3t} \begin{pmatrix} t \\ 1 + t \end{pmatrix}.
$$

If $\mathbb{T} = \mathbb{Z}$ and $t_0 = 0$, then $f(t, t_0) = t/4$ and

$$
\begin{aligned}
x(t) &= c_1 4^t \begin{pmatrix} 1 - t/4 \\ -t/4 \end{pmatrix} + c_2 4^t \begin{pmatrix} t/4 \\ 1 + t/4 \end{pmatrix} \\
&= a_1 4^t \begin{pmatrix} 4 - t \\ -t \end{pmatrix} + a_2 4^t \begin{pmatrix} t \\ 4 + t \end{pmatrix},
\end{aligned}
$$

where $a_1 = c_1/4$ and $a_2 = c_2/4$.

Exercise 5.37. Use Putzer's algorithm (Theorem 5.35) to solve $x^\Delta = Ax$ (first for any time scale $\mathbb{T}$, and then, using your result, for the special time scales $\mathbb{T} = \mathbb{R}$ and $\mathbb{T} = \mathbb{Z}$) for A being each of the matrices

$$
\begin{pmatrix} 0 & 1 \\ -2 & 3 \end{pmatrix}, \quad \begin{pmatrix} 1 & 1 \\ -1 & 3 \end{pmatrix}, \quad \begin{pmatrix} 6 & -2 \\ -2 & 9 \end{pmatrix}.
$$

5.3. Self-Adjoint Matrix Equations

Let P and Q be Hermitian $n \times n$-matrix-valued functions on a time scale $\mathbb{T}$ such that $P(t)$ is invertible for all $t \in \mathbb{T}$. (A matrix M is called *Hermitian* if $M^* = M$.) In this section we are concerned with the *self-adjoint* second order matrix differential equation

$$(5.17) \qquad LX = 0, \quad \text{where} \quad LX(t) := \left(PX^{\Delta}\right)^{\Delta}(t) + Q(t)X^{\sigma}(t)$$

on $\mathbb{T}^{\kappa^2}$.

Definition 5.38. Let $\mathbb{D}$ denote the set of all $n \times n$-matrix-valued functions X defined on $\mathbb{T}$ such that X is differentiable on $\mathbb{T}^{\kappa}$ and $(PX^{\Delta})^{\Delta}$ is rd-continuous on $\mathbb{T}^{\kappa^2}$. We say that X is a solution of the self-adjoint matrix equation (5.17) on $\mathbb{T}$ provided $X \in \mathbb{D}$ and

$$LX(t) = 0 \quad \text{for all} \quad t \in \mathbb{T}^{\kappa^2}.$$

Next, putting

$$Z = \begin{pmatrix} X \\ PX^{\Delta} \end{pmatrix} \quad \text{on } \mathbb{T}^{\kappa} \quad \text{and} \quad S = \begin{pmatrix} 0 & P^{-1} \\ -Q & -\mu QP^{-1} \end{pmatrix} \quad \text{on } \mathbb{T},$$

it is easy to see that X solves (5.17) if and only if (note that $X^{\sigma} = X + \mu X^{\Delta}$ by Theorem 5.2) Z solves

$$(5.18) \qquad\qquad\qquad Z^{\Delta} = S(t)Z.$$

Now $S \in C_{\mathrm{rd}}$ is regressive (see Definition 5.5) on $\mathbb{T}$ because

$$(I + \mu S)^{-1} = \begin{pmatrix} I & \mu P^{-1} \\ -\mu Q & I - \mu^2 QP^{-1} \end{pmatrix}^{-1} = \begin{pmatrix} I - \mu^2 P^{-1}Q & -\mu P^{-1} \\ \mu Q & I \end{pmatrix}$$

holds on $\mathbb{T}^{\kappa}$. Hence, by Theorem 5.8, any IVP

$$Z^{\Delta} = S(t)Z, \quad Z(a) = Z_a,$$

i.e.,

$$LX = 0, \quad X(a) = X_a, \quad X^{\Delta}(a) = X_a^{\Delta}$$

with any choice of a $2n \times n$-matrix Z_a, i.e., $n \times n$-matrices X_a and X_a^{Δ}, has a unique solution.

Definition 5.39. The unique solution of the initial value problem

$$LX = 0, \quad X(a) = 0, \quad X^{\Delta}(a) = P^{-1}(a)$$

is called the *principal solution* of (5.17) (at a), while the unique solution of the initial value problem

$$LX = 0, \quad X(a) = -I, \quad X^{\Delta}(a) = 0$$

is said to be the *associated solution* of (5.17) (at a).

Definition 5.40. If $X, Y \in \mathbb{D}$, then we define the (generalized) *Wronskian matrix* of X and Y by

$$W(X,Y)(t) = X^*(t)P(t)Y^\Delta(t) - [P(t)X^\Delta(t)]^*Y(t)$$

for $t \in \mathbb{T}^\kappa$.

Next we state and prove the Lagrange identity for the matrix case.

Theorem 5.41 (Lagrange Identity). *If $X, Y \in \mathbb{D}$, then*

$$X^*(\sigma(t))LY(t) - [LX(t)]^*\,Y(\sigma(t)) = [W(X,Y)]^\Delta(t)$$

for $t \in \mathbb{T}^{\kappa^2}$.

Proof. Let $X, Y \in \mathbb{D}$ and calculate (use Exercise 5.16)

$$
\begin{aligned}
W^\Delta(X,Y) &= \left\{ X^*PY^\Delta - (PX^\Delta)^*Y \right\}^\Delta \\
&= (X^*)^\sigma(PY^\Delta)^\Delta + (X^*)^\Delta PY^\Delta - (PX^\Delta)^*Y^\Delta - \left\{(PX^\Delta)^\Delta\right\}^*Y^\sigma \\
&= (X^*)^\sigma(PY^\Delta)^\Delta - \left\{(PX^\Delta)^\Delta\right\}^*Y^\sigma \\
&= (X^*)^\sigma\left\{(PY^\Delta)^\Delta + QY^\sigma\right\} - \left\{(PX^\Delta)^\Delta + QX^\sigma\right\}^*Y^\sigma \\
&= (X^*)^\sigma LY - (LX)^*Y^\sigma
\end{aligned}
$$

on $\mathbb{T}^{\kappa^2}$.   □

An immediate corollary of the Lagrange identity is Abel's formula for the matrix case.

Corollary 5.42 (Abel's Formula). *If X and Y are solutions of (5.17) on $\mathbb{T}$, then*

$$W(X,Y)(t) \equiv C$$

for $t \in \mathbb{T}^\kappa$, where C is a constant matrix.

Proof. Assume X and Y are solutions of (5.17) on $\mathbb{T}$. By the Lagrange identity (Theorem 5.41) we get that

$$W^\Delta(X,Y)(t) = 0 \quad \text{for all} \quad t \in \mathbb{T}^{\kappa^2}.$$

Hence $W(X,Y)(t)$ must be a constant matrix for all $t \in \mathbb{T}^\kappa$.   □

From Abel's formula we get that if X is a solution of (5.17) on $\mathbb{T}$, then

$$W(X,X)(t) \equiv C \quad \text{for} \quad t \in \mathbb{T}^\kappa,$$

where C is a constant matrix. With this in mind we make the following definition.

Definition 5.43.   (i) If X is a solution of (5.17) satisfying

$$W(X,X)(t) \equiv 0 \quad \text{for} \quad t \in \mathbb{T}^\kappa,$$

then we say that X is a *prepared solution* (or *conjoined solution* or *isotropic solution*) of (5.17).

(ii) If X and Y are two conjoined solutions with

$$W(X,Y)(t) \equiv I \quad \text{for} \quad t \in \mathbb{T}^\kappa,$$

then we say that X and Y are *normalized conjoined bases* of (5.17).

Theorem 5.44. *Assume that X is a solution of (5.17) on $\mathbb{T}$. Then the following are equivalent:*

   (i) *X is a prepared solution;*
   (ii) *$X^*(t)P(t)X^\Delta(t)$ is Hermitian for all $t \in \mathbb{T}^\kappa$;*
   (iii) *$X^*(t_0)P(t_0)X^\Delta(t_0)$ is Hermitian for some $t_0 \in \mathbb{T}^\kappa$.*

Proof. Assume that X is a solution of (5.17) on $\mathbb{T}$. Since

$$W(X, X)(t) = X^*(t)P(t)X^\Delta(t) - [P(t)X^\Delta(t)]^*X(t)$$

for $t \in \mathbb{T}^\kappa$, it follows that X is a prepared solution of (5.17) iff $X^*(t)P(t)X^\Delta(t)$ is Hermitian for all $t \in \mathbb{T}^\kappa$. By Abel's formula,

$$W(X, X)(t) = 0 \quad \text{for all} \quad t \in \mathbb{T}^\kappa$$

iff

$$W(X, X)(t_0) = 0 \quad \text{for some} \quad t_0 \in \mathbb{T}^\kappa.$$

Using this fact it follows that (iii) is equivalent to (i) and to (ii).   □

Note that one can easily get prepared solutions of (5.17) by taking initial conditions at $t_0 \in \mathbb{T}$ so that $X^*(t_0)P(t_0)X^\Delta(t_0)$ is Hermitian.

Exercise 5.45. Show that if X is a prepared solution of (5.17) and if $z_0 \in \mathbb{C}$, then Y defined by $Y(t) = z_0 X(t)$ is also a prepared solution of (5.17).

Exercise 5.46. Let $t_0 \in \mathbb{T}$. Show that

$$x = \sin_1(\cdot, t_0) \quad \text{and} \quad y = \cos_1(\cdot, t_0)$$

are prepared solutions of $x^{\Delta\Delta} + x = 0$ but that $z = e_i(\cdot, t_0)$ is a solution which is not prepared.

In the Sturmian theory for (5.17) the matrix function $X^*(\sigma(t))P(t)X(t)$ is important. We note the following result.

Lemma 5.47. *Let X be a solution of (5.17). If X is prepared, then*

$$X^*(\sigma(t))P(t)X(t) \quad \text{is Hermitian for all} \quad t \in \mathbb{T}^\kappa.$$

Conversely, if there is $t_0 \in \mathbb{T}^\kappa$ such that $\mu(t_0) > 0$ and $X^*(\sigma(t_0))P(t_0)X(t_0)$ is Hermitian, then X is a prepared solution of (5.17). Also, if X is a nonsingular prepared solution, then

$$P(t)X(\sigma(t))X^{-1}(t), \; P(t)X(t)X^{-1}(\sigma(t)), \; \text{and} \; Z(t) := P(t)X^\Delta(t)X^{-1}(t)$$

are Hermitian for all $t \in \mathbb{T}^\kappa$.

Proof. Let X be a solution of (5.17). The relation (apply Theorem 5.2)

$$(X^*)^\sigma PX = (X + \mu X^\Delta)^* PX = X^* PX + \mu(X^\Delta)^* PX$$

proves the first two statements of this lemma. Now assume that X is a nonsingular prepared solution of (5.17). Then

(5.19) $$(X^*)^\sigma PX = X^* PX^\sigma \quad \text{and} \quad X^* PX^\Delta = (X^\Delta)^* PX$$

on $\mathbb{T}^\kappa$ by Theorem 5.44 and what we showed above. We multiply the first equation in (5.19) from the left with $(X^{-1})^*$ and from the right with X^{-1} to obtain that $PX^\sigma X^{-1}$ is Hermitian. To see that $PX(X^\sigma)^{-1}$ is Hermitian, we multiply the first equation in (5.19) with $((X^\sigma)^{-1})^*$ from the left and with $(X^\sigma)^{-1}$ from the right.

Multiplying the second equation in (5.19) with $(X^{-1})^*$ from the left and with X^{-1} from the right shows that Z is Hermitian.   □

Lemma 5.48. *Assume that X is a prepared solution of (5.17) on $\mathbb{T}$. Then the following are equivalent:*

    (i) $X^*(\sigma(t))P(t)X(t) > 0$ *on* $\mathbb{T}^\kappa$;
    (ii) $X(t)$ *is nonsingular and* $P(t)X(\sigma(t))X^{-1}(t) > 0$ *on* $\mathbb{T}^\kappa$;
    (iii) $X(t)$ *is nonsingular and* $P(t)X(t)X^{-1}(\sigma(t)) > 0$ *on* $\mathbb{T}^\kappa$.

Proof. First note that $X^*(\sigma(t))P(t)X(t) > 0$ for $t \in \mathbb{T}^\kappa$ implies that $X(t)$ is nonsingular for $t \in \mathbb{T}$. Since X is a prepared solution, we have by Lemma 5.47,

$$(5.20) \qquad PX^\sigma X^{-1} = (X^{-1})^*(X^\sigma)^*P \quad \text{and} \quad PX(X^\sigma)^{-1} = ((X^\sigma)^{-1})^*X^*P$$

on $\mathbb{T}^\kappa$. We multiply the right-hand side of the first equation in (5.19) from the right with XX^{-1} to obtain the equivalence of (i) and (ii). For the equivalence of (i) and (iii), multiply the right-hand side of the second equation in (5.20) from the right with $X^\sigma(X^\sigma)^{-1}$.   □

We shall now consider the *matrix Riccati dynamic equation*

$$(5.21) \qquad RZ = 0, \quad \text{where} \quad RZ := Z^\Delta + Q(t) + Z^* \left\{ P(t) + \mu(t)Z \right\}^{-1} Z.$$

Theorem 5.49 (Riccati Equation). *If the self-adjoint matrix equation (5.17) has a prepared solution X such that $X(t)$ is invertible for all $t \in \mathbb{T}$, then Z defined by the Riccati substitution*

$$(5.22) \qquad Z(t) = P(t)X^\Delta(t)X^{-1}(t),$$

$t \in \mathbb{T}^\kappa$, is a Hermitian solution of the matrix Riccati equation (5.21) on $\mathbb{T}^\kappa$. Conversely, if (5.21) has a Hermitian solution Z on $\mathbb{T}^\kappa$, then there exists a prepared solution X of (5.17) such that $X(t)$ is invertible for all $t \in \mathbb{T}$ and relation (5.22) holds.

Proof. First we assume that X solves (5.17) and is invertible on $\mathbb{T}$. We then define Z by the matrix Riccati substitution (5.22) so that on $\mathbb{T}^{\kappa^2}$

$$\begin{aligned}
Z^\Delta + Q + Z^*(P + \mu Z)^{-1}Z &= Z^\Delta + Q + Z\left\{I + (X^\sigma - X)X^{-1}\right\}^{-1}P^{-1}Z \\
&= (PX^\Delta)^\Delta(X^\sigma)^{-1} - PX^\Delta(X^\sigma)^{-1}X^\Delta X^{-1} + Q \\
&\quad + PX^\Delta X^{-1}(X^\sigma X^{-1})^{-1}P^{-1}PX^\Delta X^{-1} \\
&= -QX^\sigma(X^\sigma)^{-1} - PX^\Delta(X^\sigma)^{-1}X^\Delta X^{-1} + Q + PX^\Delta(X^\sigma)^{-1}X^\Delta X^{-1} \\
&= 0
\end{aligned}$$

holds according to Theorem 5.3, i.e., Z is a Hermitian (see Lemma 5.47) solution of (5.21) on $\mathbb{T}^\kappa$.

Conversely, let Z be a Hermitian solution of (5.21) on $\mathbb{T}^\kappa$, let $t_0 \in \mathbb{T}$, and put

$$X = e_{P^{-1}Z}(\cdot, t_0).$$

Note that, by Theorem 5.8, X is well defined because $I + \mu P^{-1} Z = P^{-1}(P + \mu Z)$ is invertible on $\mathbb{T}^\kappa$. Then X is invertible on $\mathbb{T}$ by Theorem 5.21, and we have

$$
\begin{aligned}
(PX^\Delta)^\Delta &= (ZX)^\Delta \\
&= Z^\Delta X^\sigma + Z X^\Delta \\
&= -QX^\sigma - Z(P + \mu Z)^{-1} Z X^\sigma + Z P^{-1} Z X \\
&= -QX^\sigma + Z(P + \mu Z)^{-1} \left\{ (P + \mu Z)P^{-1} Z X - Z X^\sigma \right\} \\
&= -QX^\sigma + Z(P + \mu Z)^{-1} \left\{ (P + \mu Z)P^{-1} Z X - Z(X + \mu X^\Delta) \right\} \\
&= -QX^\sigma + Z(P + \mu Z)^{-1} \left\{ ZX + \mu Z P^{-1} Z X - Z(X + \mu P^{-1} Z X) \right\} \\
&= -QX^\sigma
\end{aligned}
$$

on $\mathbb{T}^{\kappa^2}$. We extend X to a solution of (5.17) so that, since $(X^* P X^\Delta)(t_0) = Z(t_0)$ is Hermitian, X is indeed a prepared solution of (5.17). $\qquad\square$

Theorem 5.50. *The self-adjoint matrix equation (5.17) has a prepared solution X on $\mathbb{T}$ with $X^*(\sigma(t))P(t)X(t) > 0$ on $\mathbb{T}^\kappa$ iff the matrix Riccati equation (5.21) has a Hermitian solution Z on $\mathbb{T}^\kappa$ satisfying*

$$
P(t) + \mu(t)Z(t) > 0 \quad \text{for all} \quad t \in \mathbb{T}^\kappa.
$$

Proof. Assuming that (5.22) holds, note that

$$
\text{(5.23)} \qquad P + \mu Z = P + \mu P X^\Delta X^{-1} = P(I + (X^\sigma - X)X^{-1}) = P X^\sigma X^{-1}
$$

implies

$$
X^*(P + \mu Z)X = X^* P X^\sigma
$$

and apply Theorem 5.49. $\qquad\square$

Our next result (Picone's identity) is the main tool for proving Jacobi's condition for the self-adjoint vector equation (5.17), i.e., the vector analogue of Theorem 4.57.

Theorem 5.51 (Picone's Identity). *Let $\alpha \in \mathbb{R}^n$ and suppose X and Y are normalized conjoined bases of (5.17) such that X is invertible on $\mathbb{T}^\kappa$. We put*

$$
Z = P X^\Delta X^{-1} \quad \text{and} \quad D = X(X^\sigma)^{-1} P^{-1} \quad \text{on} \quad \mathbb{T}^\kappa.
$$

Let $t \in \mathbb{T}^\kappa$ and assume that $u : \mathbb{T} \to \mathbb{R}^n$ is differentiable at t. Then we have at t

$$
\begin{aligned}
(u^* Z u + 2\alpha^* X^{-1} u - \alpha^* X^{-1} Y \alpha)^\Delta = (u^\Delta)^* P u^\Delta - (u^\sigma)^* Q u^\sigma \\
- \{ P u^\Delta - Z u - (X^{-1})^* \alpha \}^* D \{ P u^\Delta - Z u - (X^{-1})^* \alpha \}.
\end{aligned}
$$

Proof. According to the differentiation rules from Theorem 5.3, we have at t

$$
\begin{aligned}
(X^{-1} Y)^\Delta &= -(X^\sigma)^{-1} X^\Delta X^{-1} Y + (X^\sigma)^{-1} Y^\Delta \\
&= -(X^\sigma)^{-1} P^{-1} Z Y + (X^\sigma)^{-1} P^{-1} P Y^\Delta \\
&= (X^\sigma)^{-1} P^{-1} (X^{-1})^* \left\{ X^* P Y^\Delta - (X^\Delta)^* P Y \right\} \\
&= (X^\sigma)^{-1} P^{-1} (X^{-1})^* W(X, Y) \\
&= (X^\sigma)^{-1} P^{-1} (X^{-1})^*,
\end{aligned}
$$

and (using Theorem 5.2 and (5.23) and Theorem 5.49),

$$(u^*Zu)^\Delta + (Pu^\Delta - Zu)^*X(X^\sigma)^{-1}P^{-1}(Pu^\Delta - Zu)$$

$$= (u^\Delta)^*Zu + (u^\sigma)^*Z^\Delta u^\sigma + (u^\sigma)^*Zu^\Delta + (u^\Delta)^*PX(X^\sigma)^{-1}u^\Delta$$
$$-u^*ZX(X^\sigma)^{-1}u^\Delta - (u^\Delta)^*PX(X^\sigma)^{-1}P^{-1}Zu + u^*ZX(X^\sigma)^{-1}P^{-1}Zu$$

$$= (u^\Delta)^*Zu - (u^\sigma)^*Qu^\sigma - (u^\sigma)^*ZX(X^\sigma)^{-1}P^{-1}Zu^\sigma + (u^\sigma)^*Zu^\Delta$$
$$+(u^\Delta)^*PX(X^\sigma)^{-1}u^\Delta - u^*ZX(X^\sigma)^{-1}u^\Delta - (u^\Delta)^*PX(X^\sigma)^{-1}P^{-1}Zu$$
$$+u^*ZX(X^\sigma)^{-1}P^{-1}Zu$$

$$= (u^\Delta)^*Zu - (u^\sigma)^*Qu^\sigma - (u^\sigma)^*ZX(X^\sigma)^{-1}P^{-1}Zu^\sigma + (u^\sigma)^*Zu^\Delta$$
$$+(u^\Delta)^*P(X^\sigma - \mu X^\Delta)(X^\sigma)^{-1}u^\Delta - u^*ZX(X^\sigma)^{-1}u^\Delta$$
$$-(u^\Delta)^*P(X^\sigma - \mu X^\Delta)(X^\sigma)^{-1}P^{-1}Zu + u^*ZX(X^\sigma)^{-1}P^{-1}Zu$$

$$= (u^\Delta)^*Pu^\Delta - (u^\sigma)^*Qu^\sigma + (u^\sigma)^*Zu^\Delta - \mu(u^\Delta)^*PX^\Delta(X^\sigma)^{-1}u^\Delta$$
$$-u^*ZX(X^\sigma)^{-1}u^\Delta + \mu(u^\Delta)^*PX^\Delta(X^\sigma)^{-1}P^{-1}Zu + u^*ZX(X^\sigma)^{-1}P^{-1}Zu$$
$$-(u^\sigma)^*ZX(X^\sigma)^{-1}P^{-1}Zu^\sigma$$

$$= (u^\Delta)^*Pu^\Delta - (u^\sigma)^*Qu^\sigma + (u^\sigma)^*Zu^\Delta - (u^\sigma)^*ZX(X^\sigma)^{-1}P^{-1}Zu^\sigma$$
$$-(\mu u^\Delta + u)^*ZX(X^\sigma)^{-1}u^\Delta + (\mu u^\Delta + u)^*ZX(X^\sigma)^{-1}P^{-1}Zu$$

$$= (u^\Delta)^*Pu^\Delta - (u^\sigma)^*Qu^\sigma + (u^\sigma)^*Zu^\Delta - (u^\sigma)^*ZX(X^\sigma)^{-1}P^{-1}Zu^\sigma$$
$$-(u^\sigma)^*ZX(X^\sigma)^{-1}u^\Delta + (u^\sigma)^*ZX(X^\sigma)^{-1}P^{-1}Zu$$

$$= (u^\Delta)^*Pu^\Delta - (u^\sigma)^*Qu^\sigma + (u^\sigma)^*Zu^\Delta$$
$$-(u^\sigma)^*Z(X^\sigma - \mu X^\Delta)(X^\sigma)^{-1}u^\Delta + (u^\sigma)^*ZX(X^\sigma)^{-1}P^{-1}Z(u - u^\sigma)$$

$$= (u^\Delta)^*Pu^\Delta - (u^\sigma)^*Qu^\sigma + \mu(u^\sigma)^*Z\left\{X^\Delta(X^\sigma)^{-1} - X(X^\sigma)^{-1}X^\Delta X^{-1}\right\}u^\Delta$$

$$= (u^\Delta)^*Pu^\Delta - (u^\sigma)^*Qu^\sigma$$

so that

$$\left(u^*Zu + 2\alpha^*X^{-1}u - \alpha^*X^{-1}Y\alpha\right)^\Delta - (u^\Delta)^*Pu^\Delta + (u^\sigma)^*Qu^\sigma$$
$$= -(Pu^\Delta - Zu)^*X(X^\sigma)^{-1}P^{-1}(Pu^\Delta - Zu) - 2\alpha^*(X^\sigma)^{-1}X^\Delta X^{-1}u$$
$$+2\alpha^*(X^\sigma)^{-1}u^\Delta - \alpha^*(X^\sigma)^{-1}P^{-1}(X^{-1})^*\alpha$$
$$= -\left\{Pu^\Delta - Zu - (X^{-1})^*\alpha\right\}^*X(X^\sigma)^{-1}P^{-1}\left\{Pu^\Delta - Zu - (X^{-1})^*\alpha\right\}$$
$$= -\left\{Pu^\Delta - Zu - (X^{-1})^*\alpha\right\}^*D\left\{Pu^\Delta - Zu - (X^{-1})^*\alpha\right\},$$

and hence our claimed identity follows. $\square$

Now we generalize the notion of positive definiteness from Definition 4.54 to the case of the vector equation (5.17).

Definition 5.52. The quadratic functional

$$\mathcal{F}(u) = \int_a^b \left\{(u^\Delta)^*Pu^\Delta - (u^\sigma)^*Qu^\sigma\right\}(t)\Delta t$$

is called *positive definite* (we write $\mathcal{F} > 0$) provided

$$\mathcal{F}(u) > 0 \quad \text{for all} \quad u \in C^1_{prd}\left([a,b],\mathbb{R}^n\right)\setminus\{0\} \quad \text{with} \quad u(a) = u(b) = 0.$$

Lemma 5.53. *If* $u \in C^1_{prd}$ *and*

$$\left\{(Pu^\Delta)^\Delta + Qu^\sigma\right\}(t) = 0 \quad \text{for all} \quad t \in [a,b]^{\kappa^2},$$

then

$$\int_a^{\rho(b)} \left\{ (u^\Delta)^* P u^\Delta - (u^\sigma)^* Q u^\sigma \right\}(t) \Delta t = \left\{ u^* P u^\Delta \right\}(\rho(b)) - \left\{ u^* P u^\Delta \right\}(a).$$

Proof. We have on $[a, b]^{\kappa^2}$

$$
\begin{aligned}
(u^* P u^\Delta)^\Delta &= (u^\sigma)^* (P u^\Delta)^\Delta + (u^\Delta)^* P u^\Delta \\
&= (u^\Delta)^* P u^\Delta - (u^\sigma)^* Q u^\sigma
\end{aligned}
$$

so that the assertion follows.   □

Definition 5.54. A conjoined solution of (5.17) is said to have *no focal points* in $(a, b]$ provided it satisfies

$$X \text{ invertible on } (a, b] \quad \text{and} \quad X(X^\sigma)^{-1} P^{-1} \geq 0 \text{ on } [a, b]^\kappa.$$

If a is right-dense and $X(a)$ is singular, then $X^{-1}(a)$ has to be replaced by the Moore–Penrose inverse of $X(a)$ (see Definition 7.36).

Our next result, Theorem 5.55, offers a sufficient condition for the positive definiteness of $\mathcal{F}$. A special case of this result yields one direction of Jacobi's condition, which we shall prove in Theorem 5.57.

Theorem 5.55 (Sufficient Condition for Positive Definiteness). *A sufficient condition for $\mathcal{F} > 0$ is that there exist normalized conjoined bases X and Y of (5.17) such that X has no focal points in $(a, b]$.*

Proof. We let X and Y be as above and put $D = X(X^\sigma)^{-1} P^{-1}$ on $[a, b]^\kappa$. Pick any $u \in C^1_{\text{prd}}$ with $u(a) = u(b) = 0$. First we shall consider the case that a is right-scattered so that we have $\sigma(a) > a$ and $\mu(a) > 0$. By Theorem 1.75 and Theorem 5.51 (with $\alpha = 0$) we have

$$
\begin{aligned}
\mathcal{F}(u) &= \int_a^{\sigma(a)} \left\{ (u^\Delta)^* P u^\Delta - (u^\sigma)^* Q u^\sigma \right\}(t) \Delta t \\
&\quad + \int_{\sigma(a)}^b \left\{ (u^\Delta)^* P u^\Delta - (u^\sigma)^* Q u^\sigma \right\}(t) \Delta t \\
&= \left\{ (u^\Delta)^* P u^\Delta - (u^\sigma)^* Q u^\sigma \right\}(a) \mu(a) \\
&\quad + \int_{\sigma(a)}^b \left\{ (u^* Z u)^\Delta + (P u^\Delta - Z u)^* D (P u^\Delta - Z u) \right\}(t) \Delta t \\
&= \left\{ (u^\sigma)^* P u^\sigma \mu^{-1} - (u^\sigma)^* Q u^\sigma \mu \right\}(a) + \left\{ u^* Z u \right\}(b) \\
&\quad - \left\{ u^* Z u \right\}(\sigma(a)) + \int_{\sigma(a)}^b \left\{ (P u^\Delta - Z u)^* D (P u^\Delta - Z u) \right\}(t) \Delta t \\
&\geq \left\{ (u^\sigma)^* P u^\sigma \mu^{-1} - (u^\sigma)^* Q u^\sigma \mu \right\}(a) + \left\{ u^* Z u \right\}(b) - \left\{ u^* Z u \right\}(\sigma(a)) \\
&= \left\{ (u^\sigma)^* P u^\sigma \mu^{-1} - (u^\sigma)^* Q u^\sigma \mu - (u^\sigma)^* (P X^\Delta X^{-1})^\sigma u^\sigma \right\}(a) \\
&= (u^\sigma)^*(a) \left\{ P \mu^{-1} - Q \mu - \left(P X^\Delta + \mu (P X^\Delta)^\Delta \right) (X^\sigma)^{-1} \right\}(a) u^\sigma(a) \\
&= (u^\sigma)^*(a) \left\{ P \mu^{-1} - Q \mu - \left(P X^\Delta - \mu Q X^\sigma \right) (X^\sigma)^{-1} \right\}(a) u^\sigma(a) \\
&= (u^\sigma)^*(a) (P X (X^\sigma)^{-1})(a) \mu^{-1}(a) u^\sigma(a) \\
&= (u^\sigma)^*(a) (P D P \mu^{-1})(a) u^\sigma(a)
\end{aligned}
$$

$$\geq \quad 0.$$

If however, $\mathcal{F}(u) = 0$, then because of Theorem 5.51

$$(Pu^{\Delta} - Zu)(t) = 0 \quad \text{for all} \quad t \in [\sigma(a), \rho(b)],$$

i.e., $u^{\Delta}(t) = (X^{\Delta}X^{-1}u)(t)$ for all $t \in [\sigma(a), \rho(b)]$. Since

$$I + \mu X^{\Delta}X^{-1} = I + (X^{\sigma} - X)X^{-1} = X^{\sigma}X^{-1},$$

the map $X^{\Delta}X^{-1}$ is regressive on $[\sigma(a), \rho(b)]$ so that by Theorem 5.8 the IVP

$$u^{\Delta} = (X^{\Delta}X^{-1})(t)u \quad \text{on} \quad [\sigma(a), \rho(b)], \quad u(b) = 0$$

admits only one, namely the trivial solution. Hence $u = 0$ on $[a, b]$ so that $\mathcal{F} > 0$ follows.

Next, let a be right-dense and pick a decreasing sequence $\{a_m\}_{m \in \mathbb{N}} \subset [a, b]$ with $\lim_{m \to \infty} a_m = a$. Let $m \in \mathbb{N}$. We put $\alpha_m = -\{(X^{\Delta})^* Pu\}(a_m)$ and apply Picone's identity (Theorem 5.51 with $\alpha = \alpha_m$) to obtain the following:

$$\int_{a_m}^{b} \{(u^{\Delta})^* Pu^{\Delta} - (u^{\sigma})^* Qu^{\sigma}\}(t)\Delta t = \int_{a_m}^{b} (u^* Zu + 2\alpha_m^* X^{-1}u - \alpha_m^* X^{-1}Y\alpha_m)^{\Delta}(t)\Delta t$$

$$+ \int_{a_m}^{b} \{Pu^{\Delta} - Zu - (X^{-1})^*\alpha_m\}^* D \{Pu^{\Delta} - Zu - (X^{-1})^*\alpha_m\}(t)\Delta t$$

$$\geq \int_{a_m}^{b} (u^* Zu + 2\alpha_m^* X^{-1}u - \alpha_m^* X^{-1}Y\alpha_m)^{\Delta}(t)\Delta t$$

$$= -(\alpha_m^* X^{-1}Y\alpha_m)(b) - (u^* Zu + 2\alpha_m^* X^{-1}u - \alpha_m^* X^{-1}Y\alpha_m)(a_m)$$

$$= -(\alpha_m^* X^{-1}Y\alpha_m)(b) + \{u^* Zu + u^* ZY(X^{\Delta})^* Pu\}(a_m)$$

$$= -(\alpha_m^* X^{-1}Y\alpha_m)(b) + \{u^* Zu + u^*(X^{-1})^*(X^* PY^{\Delta} - I)(X^{\Delta})^* Pu\}(a_m)$$

$$= -(\alpha_m^* X^{-1}Y\alpha_m)(b) + \{u^* PY^{\Delta}(X^{\Delta})^* Pu\}(a_m),$$

and hence by Theorem 5.51

$$\mathcal{F}(u) = \lim_{m \to \infty} \int_{a_m}^{b} \{(u^{\Delta})^* Pu^{\Delta} - (u^{\sigma})^* Qu^{\sigma}\}(t)\Delta t$$

$$\geq -\lim_{m \to \infty} \{\alpha_m^*(X^{-1}Y)(b)\alpha_m + (u^* PY^{\Delta})(a_m)\alpha_m\}$$

$$= 0.$$

Finally, if $\mathcal{F}(u) = 0$, we know that $\lim_{m \to \infty} \int_{a_m}^{b} \{z_m^* Dz_m\}(t)\Delta t$ exists and equals 0, where

$$z_m := Pu^{\Delta} - Zu - (X^{-1})^*\alpha_m \to Pu^{\Delta} - Zu =: z \quad \text{as} \quad m \to \infty$$

holds uniformly (observe also Theorem 1.16 (i)) on $[a_k, b]$ for each $k \in \mathbb{N}$. Now let $k \in \mathbb{N}$. We first note

$$\int_{a_m}^{b} \{z_m^* Dz_m\}(t)\Delta t \geq \int_{a_k}^{b} \{z_m^* Dz_m\}(t)\Delta t \quad \text{for all} \quad m \geq k.$$

Hence, by applying Theorem 5.51 we have

$$\lim_{m \to \infty} \int_{a_m}^{b} \{z_m^* Dz_m\}(t)\Delta t \geq \int_{a_k}^{b} \{z^* Dz\}(t)\Delta t.$$

Now we let $k \to \infty$ and apply Theorem 5.51 to obtain

$$0 = \lim_{m \to \infty} \int_{a_m}^{b} \{z_m^* D z_m\}(t) \Delta t \geq \int_{a}^{b} \{z^* D z\}(t) \Delta t \geq 0.$$

Hence, again by Theorem 5.51, we have $z = 0$ on $[a, b]^\kappa$ so that as in the first part of this proof $u = 0$ follows. Therefore, $\mathcal{F}$ is again positive definite and the proof of our result is complete. □

Definition 5.56. We say that equation (5.17) is *disconjugate* on $[a, b]$ if the principal solution $\tilde{X}$ of (5.17) satisfies

$$\tilde{X} \text{ invertible on } (a, b] \quad \text{and} \quad \tilde{X}(\tilde{X}^\sigma)^{-1} P^{-1} > 0 \text{ on } (a, b]^\kappa.$$

Using this terminology, (5.17) is disconjugate iff the principal solution of (5.17) has no focal points in $(a, b]$. Now we are ready to prove one of the main results (Jacobi's condition) of this section (compare with Theorem 4.57).

Theorem 5.57 (Jacobi's Condition). *$\mathcal{F} > 0$ iff (5.17) is disconjugate.*

Proof. First, if (5.17) is disconjugate, then we can apply Theorem 5.55 with $\tilde{X}$ and $\tilde{Y}$, where $\tilde{X}$ is the principal solution of (5.17) and $\tilde{Y}$ is the associated solution of (5.17) (see Definition 5.39), and hence disconjugacy implies $\mathcal{F} > 0$.

Conversely, suppose that (5.17) is not disconjugate. Then there exists $t_0 \in \mathbb{T}$ with exactly one of the following two properties:

I. $t_0 \in (a, b]$ such that $\tilde{X}(t)$ is invertible for all $t \in (a, t_0)$ and

$$\tilde{X}(t_0) \quad \text{is singular,}$$

II. $t_0 \in (a, b]^\kappa$ such that $\tilde{X}(t)$ is invertible for all $t \in (a, b]$ and

$$\tilde{D}(t_0) = \left\{ \tilde{X}(\tilde{X}^\sigma)^{-1} P^{-1} \right\}(t_0) \quad \text{is not positive definite.}$$

Let $d \in \mathbb{R}^n \setminus \{0\}$ with $\tilde{X}(t_0)d = 0$ in Case I and

$$d^*(\tilde{X}^* P \tilde{X}^\sigma)(t_0)d = (P\tilde{X}^\sigma d)^* \tilde{D}(P\tilde{X}^\sigma d)(t_0) \leq 0$$

in Case II. Putting

$$u(t) = \begin{cases} \tilde{X}(t)d & \text{for} \quad t \leq t_0 \\ 0 & \text{otherwise} \end{cases}$$

yields $u(a) = u(b) = 0$, $u(t) \neq 0$ for all $t \in (a, t_0)$, and, moreover, $u \in C^1_{\text{prd}}$ except for Case II with t_0 right-dense. Hence we assume for the moment this is not the case, and we will deal with this special case later. Now it is easy to see that both $u^\Delta(t)$ and $u^\sigma(t)$ are zero for all $t > t_0$ so that

$$\left\{ (u^\Delta)^* P u^\Delta - (u^\sigma)^* Q u^\sigma \right\}(t) = 0 \quad \text{for all} \quad t > t_0.$$

In the following we use Lemma 5.53 and Theorem 1.75, and for $\alpha \in \mathbb{R}$ we put (as before in Theorem 4.101) $\alpha^\dagger = 0$ if $\alpha = 0$ and $\alpha^\dagger = 1/\alpha$ if $\alpha \neq 0$. We have

$$
\begin{aligned}
\mathcal{F}(u) &= \int_a^{\sigma(t_0)} \left\{ (u^\Delta)^* P u^\Delta - (u^\sigma)^* Q u^\sigma \right\}(t) \Delta t \\
&= \int_a^{t_0} \left\{ (u^\Delta)^* P u^\Delta - (u^\sigma)^* Q u^\sigma \right\}(t) \Delta t \\
&\quad + \int_{t_0}^{\sigma(t_0)} \left\{ (u^\Delta)^* P u^\Delta - (u^\sigma)^* Q u^\sigma \right\}(t) \Delta t \\
&= \left\{ d^* \tilde{X}^* P \tilde{X}^\Delta d \right\}(t_0) + \left\{ (u^\Delta)^* P u^\Delta - (u^\sigma)^* Q u^\sigma \right\}(t_0) \mu(t_0) \\
&= d^* \left\{ \tilde{X}^* P \tilde{X}^\Delta + \tilde{X}^* P \tilde{X} \mu^\dagger \right\}(t_0) d \\
&= d^* \left\{ \mu^\dagger \tilde{X}^* P \tilde{X}^\sigma + (1 - \mu^\dagger \mu) \tilde{X}^* P \tilde{X}^\Delta \right\}(t_0) d \leq 0.
\end{aligned}
$$

Therefore $\mathcal{F} \not> 0$.

Now we discuss the remaining Case II with right-dense t_0. This situation implies $P(t_0) \not> 0$, i.e., there exists $d \in \mathbb{R}^n$ such that $d^* P(t_0) d < 0$. We have to show $\mathcal{F} \not> 0$, and for the sake of achieving a contradiction we assume $\mathcal{F} > 0$. First we suppose that t_0 is left-scattered. Then $t_0 \neq b$. Let $\{t_m\}_{m \in \mathbb{N}} \subset [t_0, b]$ be a decreasing sequence converging to t_0. For $m \in \mathbb{N}$ we put

$$
u_m(t) = \begin{cases} \frac{t_m - t}{\sqrt{t_m - t_0}} d & \text{if} \quad t \in (t_0, t_m] \\ 0 & \text{otherwise.} \end{cases}
$$

Hence $u_m(a) = u_m(b) = 0$, $u_m(t_0) = \sqrt{t_m - t_0} d \neq 0$, and $u \in C^1_{\mathrm{prd}}$. Therefore

$$
\begin{aligned}
0 &< \mathcal{F}(u_m) = \int_{t_0}^{t_m} \left\{ (u_m^\Delta)^* P u_m^\Delta - (u_m^\sigma)^* Q u_m^\sigma \right\}(t) \Delta t \\
&= \int_{t_0}^{t_m} \left\{ \frac{d^*}{\sqrt{t_m - t_0}} P(t) \frac{d}{\sqrt{t_m - t_0}} - d^* \frac{t_m - \sigma(t)}{\sqrt{t_m - t_0}} Q(t) \frac{t_m - \sigma(t)}{\sqrt{t_m - t_0}} d \right\} \Delta t \\
&= d^* \left\{ \frac{1}{t_m - t_0} \int_{t_0}^{t_m} P(t) \Delta t \right\} d - d^* \left\{ \int_{t_0}^{t_m} \frac{(t_m - \sigma(t))^2}{t_m - t_0} Q(t) \Delta t \right\} d \\
&\to d^* P(t_0) d < 0
\end{aligned}
$$

as $m \to \infty$, which yields our desired contradiction. The remaining case of left-dense t_0 can be treated the same way with

$$
y_m(t) = \begin{cases} \frac{t - t_m^*}{\sqrt{t_0 - t_m^*}} d & \text{if} \quad t \in [t_m^*, t_0) \\ 0 & \text{otherwise} \end{cases}
$$

or

$$
y_m(t) = \begin{cases} \frac{t - t_m^*}{\sqrt{t_0 - t_m^*}} d & \text{if} \quad t \in [t_m^*, t_0) \\ \frac{t_m - t}{\sqrt{t_m - t_0}} d & \text{if} \quad t \in (t_0, t_m] \\ 0 & \text{otherwise} \end{cases}
$$

if $t_0 = b$ or otherwise, respectively, using a strictly increasing sequence $\{t_m^*\}_{m \in \mathbb{N}}$ with $\lim_{m \to \infty} t_m^* = t_0$. $\qquad \square$

Definition 5.58. We call a solution X of (5.17) a *basis* whenever

$$\text{rank} \begin{pmatrix} X(a) \\ P(a)X^{\Delta}(a) \end{pmatrix} = n.$$

Concerning conjoined bases of (5.17) we have the following version of Sturm's separation result on time scales.

Theorem 5.59 (Sturm's Separation Theorem). *Suppose there exists a conjoined basis of (5.17) with no focal points in $(a, b]$. Then equation (5.17) is disconjugate on $[a, b]$.*

Proof. Let X be a conjoined basis of (5.17) with no focal points in $(a, b]$. Since X is a basis, we note that the matrix

$$K = X^*(a)X(a) + (X^{\Delta})^*(a)P^2(a)X^{\Delta}(a) \quad \text{is invertible.}$$

Let Y be the solution of (5.17) satisfying

$$Y(a) = -P(a)X^{\Delta}(a)K^{-1}, \quad Y^{\Delta}(a) = P^{-1}(a)X(a)K^{-1}.$$

Then in view of the Wronskian identity (Corollary 5.42), Y satisfies

$$\begin{aligned}
\left\{ Y^*PY^{\Delta} - (Y^{\Delta})^*PY \right\}(t) &\equiv \left\{ Y^*PY^{\Delta} - (Y^{\Delta})^*PY \right\}(a) \\
&= -(K^{-1})^*(X^{\Delta})^*(a)P(a)X(a)K^{-1} + (K^{-1})^*X^*(a)P(a)X^{\Delta}(a)K^{-1} \\
&= (K^{-1})^* \left\{ X^*PX^{\Delta} - (X^{\Delta})^*PX \right\}(a)K^{-1} \\
&= 0
\end{aligned}$$

and

$$\begin{aligned}
\left\{ X^*PY^{\Delta} - (X^{\Delta})^*PY \right\}(t) &\equiv \left\{ X^*PY^{\Delta} - (X^{\Delta})^*PY \right\}(a) \\
&= X^*(a)X(a)K^{-1} + (X^{\Delta})^*(a)P^2(a)X^{\Delta}(a)K^{-1} \\
&= KK^{-1} \\
&= I,
\end{aligned}$$

and hence X and Y are normalized conjoined bases of (5.17). An application of Theorem 5.55 now yields that $\mathcal{F} > 0$. However, now Jacobi's condition (Theorem 5.57) in turn shows that (5.17) is disconjugate, i.e., the principal solution of (5.17) has no focal points in $(a, b]$. $\square$

Now we shall also consider the equation

(5.24) $$\left[\tilde{P}(t)X^{\Delta} \right]^{\Delta} + \tilde{Q}(t)X^{\sigma} = 0,$$

where $\tilde{P}$ and $\tilde{Q}$ satisfy the same assumptions as P and Q.

Theorem 5.60 (Sturm's Comparison Theorem). *Suppose we have for all $t \in \mathbb{T}$*

$$\tilde{P}(t) \leq P(t) \quad \text{and} \quad \tilde{Q}(t) \geq Q(t).$$

If (5.24) is disconjugate, then (5.17) is also disconjugate.

Proof. Suppose (5.24) is disconjugate. Then by Jacobi's condition (Theorem 5.57)

$$\tilde{\mathcal{F}}(u) = \int_a^b \left\{ (u^\Delta)^* \tilde{P} u^\Delta - (u^\sigma)^* \tilde{Q} u^\sigma \right\} (t) \Delta t > 0$$

for all nontrivial $u \in C_{\mathrm{prd}}^1$ with $u(a) = u(b) = 0$. For such u we also have

$$
\begin{aligned}
\mathcal{F}(u) &= \int_a^b \left\{ (u^\Delta)^* P u^\Delta - (u^\sigma)^* Q u^\sigma \right\} (t) \Delta t \\
&\geq \int_a^b \left\{ (u^\Delta)^* \tilde{P} u^\Delta - (u^\sigma)^* \tilde{Q} u^\sigma \right\} (t) \Delta t \\
&= \tilde{\mathcal{F}}(u) > 0.
\end{aligned}
$$

Hence $\mathcal{F} > 0$ and thus (5.17) is disconjugate by Theorem 5.57. $\qquad\square$

We are now concerned with oscillation properties of (5.17).

Definition 5.61. Assume $\sup \mathbb{T} = \infty$ and let $a \in \mathbb{T}$. We say that (5.17) is *nonoscillatory* on $[a, \infty)$ provided there exist a prepared solution X of (5.17) and $t_0 \in [a, \infty)$ such that

$$X^*(\sigma(t)) P(t) X(t) > 0 \quad \text{for all} \quad t \geq t_0.$$

Otherwise we say that (5.17) is *oscillatory* on $[a, \infty)$.

The oscillation results given here from Theorem 5.63 through Theorem 5.71 are taken from [135] and generalize (and improve) the oscillation results of Peterson and Ridenhour [226] for a discrete self-adjoint difference equation with step-size one. These results generalize the corresponding results for the scalar case given at the end of Section 4.2. We now introduce some notation that we use in the remainder of this section. If A is a Hermitian $n \times n$-matrix, then we let $\lambda_i(A)$ denote the ith eigenvalue of A so that

$$\lambda_{\max}(A) = \lambda_1(A) \geq \cdots \geq \lambda_n(A) = \lambda_{\min}(A).$$

We also let $\operatorname{tr} A$ denote the *trace* of an $n \times n$-matrix A, i.e., the sum of all diagonal elements of A. We make frequent use of *Weyl's theorem* [176, Theorem 4.3.1] that says if A and B are Hermitian matrices, then

$$(5.25) \qquad\qquad \lambda_i(A + B) \geq \lambda_i(A) + \lambda_{\min}(B).$$

Exercise 5.62. Show that if A and B are Hermitian matrices, then

$$\lambda_i(A - B) \leq \lambda_i(A) - \lambda_{\min}(B).$$

We now state and prove an oscillation theorem for the matrix equation

$$(5.26) \qquad\qquad X^{\Delta\Delta} + Q(t) X^\sigma = 0.$$

Theorem 5.63. *Assume* $\sup \mathbb{T} = \infty$ *and let* $a \in \mathbb{T}$. *Assume that for any* $t_0 \in [a, \infty)$ *there exist* $t_0 \leq a_0 < b_0$ *such that* $\mu(a_0) > 0$, $\mu(b_0) > 0$, *and*

$$(5.27) \qquad\qquad \lambda_{\max} \left(\int_{a_0}^{b_0} Q(t) \Delta t \right) \geq \frac{1}{\mu(a_0)} + \frac{1}{\mu(b_0)}.$$

Then (5.26) *is oscillatory on* $[a, \infty)$.

Proof. Assume (5.26) is nonoscillatory on $[a, \infty)$. Then there exist $t_0 \in [a, \infty)$ and a prepared solution X of (5.26) satisfying

$$X^*(\sigma(t))X(t) > 0 \quad \text{for all} \quad t \geq t_0.$$

If we define Z by the Riccati substitution (5.22) (with $P = I$) for $t \in [t_0, \infty)$, then by Theorem 5.50 we get that

$$I + \mu(t)Z(t) > 0 \quad \text{for all} \quad t \geq t_0,$$

and Z is a Hermitian solution of the Riccati equation (5.21) on $[t_0, \infty)$. By hypothesis, there exist $t_0 \leq a_0 < b_0$ such that $\mu(a_0) > 0$, $\mu(b_0) > 0$, and inequality (5.27) holds. Integrating both sides of the Riccati equation from a_0 to $t > a_0$, we obtain

$$
\begin{aligned}
Z(t) &= Z(a_0) - \int_{a_0}^{t} Q(s)\Delta s - \int_{a_0}^{t} Z(s)[I + \mu(s)Z(s)]^{-1}Z(s)\Delta s \\
&\leq Z(a_0) - \int_{a_0}^{t} Q(s)\Delta s - \int_{a_0}^{\sigma(a_0)} Z(s)[I + \mu(s)Z(s)]^{-1}Z(s)\Delta s \\
&= Z(a_0) - \mu(a_0)Z(a_0)[I + \mu(a_0)Z(a_0)]^{-1}Z(a_0) - \int_{a_0}^{t} Q(s)\Delta s \\
&= Z(a_0)[I + \mu(a_0)Z(a_0)]^{-1}[I + \mu(a_0)Z(a_0) - \mu(a_0)Z(a_0)] - \int_{a_0}^{t} Q(s)\Delta s \\
&= Z(a_0)[I + \mu(a_0)Z(a_0)]^{-1} - \int_{a_0}^{t} Q(s)\Delta s.
\end{aligned}
$$

It follows that

$$(5.28) \qquad Z(t) + \int_{a_0}^{t} Q(s)\Delta s \leq Z(a_0)[I + \mu(a_0)Z(a_0)]^{-1}.$$

Now let U be a unitary matrix (i.e., $U^*U = I$) such that

$$Z(a_0) = U^*DU \quad \text{with} \quad D := \text{diag}(d_1, \dots, d_n),$$

where $d_i = \lambda_i(Z(a_0))$ is the ith eigenvalue of $Z(a_0)$, $i \in \mathbb{N}$. Consider

$$
\begin{aligned}
Z(a_0)[I + \mu(a_0)Z(a_0)]^{-1} &= U^*DU[I + \mu(a_0)U^*DU]^{-1} \\
&= U^*DU\{U^*[I + \mu(a_0)D]U\}^{-1} \\
&= U^*D[I + \mu(a_0)D]^{-1}U.
\end{aligned}
$$

Since $I + \mu(a_0)Z(a_0) > 0$ implies that

$$1 + \mu(a_0)d_i > 0$$

and $h(x) := \frac{x}{1+\mu(a_0)x}$ is increasing when $1 + \mu(a_0)x > 0$, it follows that

$$\lambda_i\left(Z(a_0)[I + \mu(a_0)Z(a_0)]^{-1}\right) = \lambda_i\left(D[I + \mu(a_0)D]\right) = \frac{d_i}{1 + \mu(a_0)d_i}.$$

Using $h(x) = \frac{x}{1+\mu(a_0)x} < \frac{1}{\mu(a_0)}$ when $1 + \mu(a_0)x > 0$, we get

$$(5.29) \qquad \lambda_i\left(Z(a_0)[I + \mu(a_0)Z(a_0)]^{-1}\right) = \frac{d_i}{1 + \mu(a_0)d_i} < \frac{1}{\mu(a_0)}.$$

Hence from (5.28) we obtain

$$\lambda_i\left(Z(t) + \int_{a_0}^{t} Q(s)\Delta s\right) \leq \lambda_i\left(Z(a_0)[I + \mu(a_0)Z(a_0)]^{-1}\right) < \frac{1}{\mu(a_0)}.$$

Applying Weyl's inequality (5.25), we get

$$\frac{1}{\mu(a_0)} > \lambda_{\max}\left(Z(t) + \int_{a_0}^{t} Q(s)\Delta s\right) \geq \lambda_{\max}\left(\int_{a_0}^{t} Q(s)\Delta s\right) + \lambda_{\min}\left(Z(t)\right).$$

Letting $t = b_0$ and rearranging terms in this inequality we obtain

$$\lambda_{\max}\left(\int_{a_0}^{b_0} Q(s)\Delta s\right) < \frac{1}{\mu(a_0)} - \lambda_{\min}(Z(b_0)).$$

Since $I + \mu(b_0)Z(b_0) > 0$ implies $\lambda_{\min}\left(Z(b_0)\right) > -\frac{1}{\mu(b_0)}$, we get that

$$\lambda_{\max}\left(\int_{a_0}^{b_0} Q(s)\Delta s\right) < \frac{1}{\mu(a_0)} + \frac{1}{\mu(b_0)},$$

which is a contradiction. $\qquad\square$

Example 5.64. Let the scalar functions $q_i : \mathbb{T}^\kappa \to \mathbb{R}$ be rd-continuous on $\mathbb{T} := \mathbb{P}_{\frac{1}{2},\frac{1}{2}}$ for all $1 \leq i \leq n$, and assume that for each $t_0 \in [0,\infty)$ there exist $k_0 \in \mathbb{N}$ and $l_0 \in \mathbb{N}$ such that $k_0 \geq t_0$ and

$$(5.30) \qquad \sum_{j=1}^{l_0} \int_{k_0+j}^{k_0+j+\frac{1}{2}} q_1(t)dt + \frac{1}{2}\sum_{j=0}^{l_0-1} q_1\left(k_0+j+\frac{1}{2}\right) \geq 4.$$

Then, if

$$Q(t) := \mathrm{diag}\left(q_1(t), q_2(t), \dots, q_n(t)\right),$$

it follows that the matrix dynamic equation (5.26) is oscillatory on $\mathbb{T}$. To see that this follows from Theorem 5.63, let $t_0 \in \mathbb{T}$ be given and let

$$a_0 := k_0 + \frac{1}{2} \quad \text{and} \quad b_0 := k_0 + l_0 + \frac{1}{2}.$$

Note that a_0 and b_0 are right-scattered and

$$\lambda_{\max}\left(\int_{a_0}^{b_0} Q(t)\Delta t\right) \geq \int_{a_0}^{b_0} q_1(t)\Delta t$$

$$= \sum_{j=1}^{l_0} \int_{k_0+j}^{k_0+j+\frac{1}{2}} q_1(t)\Delta t + \sum_{j=0}^{l_0-1} \int_{k_0+j+\frac{1}{2}}^{\sigma(k_0+j+\frac{1}{2})} q_1(t)\Delta t$$

$$= \sum_{j=1}^{l_0} \int_{k_0+j}^{k_0+j+\frac{1}{2}} q_1(t)dt + \frac{1}{2}\sum_{j=0}^{l_0-1} q_1\left(k_0+j+\frac{1}{2}\right).$$

Since it can be shown that

$$\frac{1}{\mu(a_0)} + \frac{1}{\mu(b_0)} = 4,$$

this result follows from Theorem 5.63. Note that there is no requirement on

$$\liminf_{t\to\infty} \int_{t_0}^{t} q_j(\tau)\Delta\tau, \quad 1 \leq j \leq n.$$

Corollary 5.65. *Assume* $\sup \mathbb{T} = \infty$ *and let* $a \in \mathbb{T}$. *If there exists a sequence* $\{t_k\}_{k=1}^{\infty} \subset [a, \infty)$ *such that* $\lim_{k \to \infty} t_k = \infty$ *with* $\mu(t_k) \geq K > 0$ *for some* $K > 0$ *and such that*

$$(5.31) \qquad \limsup_{t_k \to \infty} \lambda_{\max} \left(\int_a^{t_k} Q(s) \Delta s \right) = +\infty,$$

then (5.26) *is oscillatory.*

Proof. Let $t_0 \in [a, \infty)$. Choose k_0 sufficiently large so that $a_0 := t_{k_0} \in [t_0, \infty)$. Using (5.31), we can pick $k_1 > k_0$ sufficiently large with $b_0 := t_{k_1}$ chosen so that

$$\lambda_{\max} \left(\int_{a_0}^{b_0} Q(s) \Delta s \right) \geq \frac{2}{K} \geq \frac{1}{\mu(a_0)} + \frac{1}{\mu(b_0)}.$$

The conclusion now follows from Theorem 5.63. $\qquad \qquad \square$

It should be noted that Corollary 5.65 is an extension to the time scales situation of the so-called "largest eigenvalue theorem" which says that the matrix differential system

$$Y'' + Q(t)Y = 0$$

is oscillatory if

$$\lim_{t \to \infty} \lambda_{\max} \left(\int_{t_0}^t Q(s) ds \right) = \infty.$$

In the discrete case where $\mathbb{T} = \mathbb{Z}$, one can replace "lim" by "lim sup", as was noted in [**95, 226**].

Theorem 5.66. *Assume* $\sup \mathbb{T} = \infty$ *and let* $a \in \mathbb{T}$. *Suppose that there is a strictly increasing sequence* $\{t_k\}_{k \in \mathbb{N}} \subset [a, \infty)$ *such that* $\mu(t_k) > 0$ *for* $k \in \mathbb{N}$, *with* $\lim_{k \to \infty} t_k = \infty$. *Further assume that there is a sequence* $\{\tau_k\}_{k \in \mathbb{N}} \subset [a, \infty)$ *such that* $\sigma(\tau_k) > \tau_k \geq \sigma(t_k)$ *for* $k \in \mathbb{N}$ *with*

$$\lambda_{\min} \left(\frac{P(t_k)}{\mu(t_k)} + \frac{P(\tau_k)}{\mu(\tau_k)} - \int_{t_k}^{\tau_k} Q(s) \Delta s \right) \leq 0$$

for $k \in \mathbb{N}$. *Then* (5.17) *is oscillatory on* $[a, \infty)$.

Proof. Assume (5.17) is nonoscillatory on $[a, \infty)$. Then there exist a prepared solution X of (5.17) and $t_0 \in [a, \infty)$ such that $X^*(\sigma(t))P(t)X(t) > 0$ on $[t_0, \infty)$. We define Z by the Riccati substitution (5.22) for $t \in [t_0, \infty)$. Then by Theorem 5.50 we get that

$$P(t) + \mu(t)Z(t) > 0 \quad \text{for all} \quad t \geq t_0,$$

and Z is a Hermitian solution of the Riccati equation (5.21) on $[t_0, \infty)$. Pick a fixed $k \in \mathbb{N}$ so that $t_k \geq t_0$. Integrating both sides of the Riccati equation from t_k to τ_k, we obtain

$$
\begin{aligned}
Z(\tau_k) &= Z(t_k) - \int_{t_k}^{\tau_k} Q(t)\Delta t - \int_{t_k}^{\tau_k} Z(t)(P(t) + \mu(t)Z(t))^{-1}Z(t)\Delta t \\
&\leq Z(t_k) - \int_{t_k}^{\tau_k} Q(t)\Delta t - \int_{t_k}^{\sigma(t_k)} Z(t)(P(t) + \mu(t)Z(t))^{-1}Z(t)\Delta t \\
&= Z(t_k) - \mu(t_k)Z(t_k)[P(t_k) + \mu(t_k)Z(t_k)]^{-1}Z(t_k) - \int_{t_k}^{\tau_k} Q(t)\Delta t \\
&= Z(t_k)[P(t_k) + \mu(t_k)Z(t_k)]^{-1}[P(t_k) + \mu(t_k)Z(t_k) - \mu(t_k)Z(t_k)] \\
&\quad - \int_{t_k}^{\tau_k} Q(t)\Delta t \\
&= Z(t_k)[P(t_k) + \mu(t_k)Z(t_k)]^{-1}P(t_k) - \int_{t_k}^{\tau_k} Q(t)\Delta t.
\end{aligned}
$$

But using (5.23) we get that

$$
\begin{aligned}
Z(t_k)[P(t_k) + \mu(t_k)Z(t_k)]^{-1}P(t_k) &= Z(t_k)X(t_k)X^{-1}(\sigma(t_k))P^{-1}(t_k)P(t_k) \\
&= P(t_k)X^{\Delta}(t_k)X^{-1}(\sigma(t_k)) \\
&= \frac{P(t_k)}{\mu(t_k)}[X^{\sigma}(t_k) - X(t_k)]X^{-1}(\sigma(t_k)) \\
&= \frac{P(t_k)}{\mu(t_k)} - \frac{1}{\mu(t_k)}P(t_k)X(t_k)X^{-1}(\sigma(t_k)) \\
&< \frac{P(t_k)}{\mu(t_k)},
\end{aligned}
$$

where in the last step we used Lemma 5.48. Hence from above

$$
Z(\tau_k) < \frac{P(t_k)}{\mu(t_k)} - \int_{t_k}^{\tau_k} Q(t)\Delta t.
$$

Using $P(\tau_k) + \mu(\tau_k)Z(\tau_k) > 0$, we obtain

$$
\frac{P(\tau_k)}{\mu(\tau_k)} + \frac{P(t_k)}{\mu(t_k)} - \int_{t_k}^{\tau_k} Q(t)\Delta t > 0,
$$

which is a contradiction. □

Exercise 5.67. Assume sup $\mathbb{T} = \infty$ and let $a \in \mathbb{T}$. Show that if for any $t_0 \in [a, \infty)$ there exist $t_0 \leq a_0 < b_0$ such that $\mu(a_0) > 0$, $\mu(b_0) > 0$, and

$$
\lambda_{\min}\left(\frac{P(a_0)}{\mu(a_0)} + \frac{P(b_0)}{\mu(b_0)} - \int_{a_0}^{b_0} Q(s)\Delta s\right) \leq 0,
$$

then (5.17) is oscillatory on $[a, \infty)$.

Corollary 5.68. *Assume* sup $\mathbb{T} = \infty$ *and let* $a \in \mathbb{T}$. *A necessary condition for* (5.17) *to be nonoscillatory on* $[a, \infty)$ *is that for any strictly increasing sequence* $\{t_k\}_{k \in \mathbb{N}} \subset [a, \infty)$ *such that* $\mu(t_k) > 0$ *for* $k \in \mathbb{N}$, *with* $\lim_{k \to \infty} t_k = \infty$, *there is* $N \in \mathbb{N}$ *such that*

$$
D_k := \frac{P(t_k)}{\mu(t_k)} + \frac{P(t_{k+1})}{\mu(t_{k+1})} - \int_{t_k}^{t_{k+1}} Q(s)\Delta s > 0 \quad for \quad k \geq N.
$$

Proof. In Theorem 5.66, take $\tau_k = t_{k+1}$ for k sufficiently large. □

Corollary 5.69. *Assume* $\sup \mathbb{T} = \infty$ *and let* $a \in \mathbb{T}$. *Assume that there is a strictly increasing sequence* $\{t_k\}_{k \in \mathbb{N}} \subset [a, \infty)$ *such that* $\mu(t_k) > 0$ *for* $k \in \mathbb{N}$, *with* $\lim_{k \to \infty} t_k = \infty$. *Furthermore assume that there are sequences* $\{s_k\}_{k \in \mathbb{N}} \subset [a, \infty)$ *and* $\{\tau_k\}_{k \in \mathbb{N}} \subset [a, \infty)$ *such that* $\sigma(\tau_k) > \tau_k \geq \sigma(s_k) > s_k \geq \sigma(t_k)$, $k \in \mathbb{N}$, *with*

$$\int_{t_k}^{s_k} Q(t) \Delta t \geq \frac{P(t_k)}{\mu(t_k)}$$

and

$$\lambda_{\min} \left(\frac{P(\tau_k)}{\mu(\tau_k)} - \int_{s_k}^{\tau_k} Q(t) \Delta t \right) \leq 0$$

for $k \in \mathbb{N}$. *Then* (5.17) *is oscillatory on* $[a, \infty)$.

Proof. It follows from Weyl's inequality (Exercise 5.62) that if A and B are Hermitian matrices, then

$$\lambda_{\min} (A - B) \leq \lambda_{\min}(A) - \lambda_{\min}(B).$$

We use this fact in the following chain of inequalities. Consider

$$\lambda_{\min} \left(\frac{P(t_k)}{\mu(t_k)} + \frac{P(\tau_k)}{\mu(\tau_k)} - \int_{t_k}^{\tau_k} Q(t) \Delta t \right)$$

$$= \lambda_{\min} \left(\left[\frac{P(\tau_k)}{\mu(\tau_k)} - \int_{s_k}^{\tau_k} Q(t) \Delta t \right] - \left[\int_{t_k}^{s_k} Q(t) \Delta t - \frac{P(t_k)}{\mu(t_k)} \right] \right)$$

$$\leq \lambda_{\min} \left(\frac{P(\tau_k)}{\mu(\tau_k)} - \int_{s_k}^{\tau_k} Q(t) \Delta t \right) - \lambda_{\min} \left(\int_{t_k}^{s_k} Q(t) \Delta t - \frac{P(t_k)}{\mu(t_k)} \right)$$

$$\leq 0.$$

Hence the result follows from Theorem 5.66. $\qquad\qquad\qquad\qquad\qquad\qquad\square$

Corollary 5.70. *Assume* $\sup \mathbb{T} = \infty$ *and let* $a \in \mathbb{T}$. *Suppose also that*

$$\lim_{t \to \infty} \lambda_{\min} \left(\int_{t_0}^{t} Q(s) \Delta s \right) = \infty \tag{5.32}$$

and that for each $T \in [a, \infty)$ *there is* $t \in [T, \infty)$ *such that* $\mu(t) > 0$ *and*

$$\lambda_{\min} \left(\frac{P(t)}{\mu(t)} - \int_{T}^{t} Q(s) \Delta s \right) \leq 0. \tag{5.33}$$

Then (5.17) *is oscillatory on* $[a, \infty)$.

Proof. Let $\{t_k\}_{k \in \mathbb{N}} \subset [a, \infty)$ be a strictly increasing sequence with $\mu(t_k) > 0$ and $\lim_{k \to \infty} t_k = \infty$. Using (5.32) we obtain that for each k there exists $\tau_k > \sigma(t_k)$ so that

$$\int_{t_k}^{\tau_k} Q(t) \Delta t \geq \frac{P(t_k)}{\mu(t_k)} \quad \text{for} \quad k \in \mathbb{N}.$$

Using (5.33) we get that for each $k \in \mathbb{N}$ there exists $s_k \geq \sigma(\tau_k)$ so that

$$\lambda_{\min} \left(\frac{P(s_k)}{\mu(s_k)} - \int_{\tau_k}^{s_k} Q(t) \Delta t \right) \leq 0.$$

It follows from Corollary 5.69 that (5.17) is oscillatory on $[a, \infty)$. $\qquad\qquad\square$

Theorem 5.71. *Assume* $\sup \mathbb{T} = \infty$ *and let* $a \in \mathbb{T}$. *Suppose for each* $t_0 \geq a$ *there is a strictly increasing sequence* $\{t_k\}_{k=1}^{\infty} \subset [t_0, \infty)$ *with* $\mu(t_k) > 0$ *and* $\lim_{k \to \infty} t_k = \infty$, *and that there are constants* K_1 *and* K_2 *such that* $0 < K_1 \leq \mu(t_k) \leq K_2$ *for* $k \in \mathbb{N}$ *with*

$$\lim_{k \to \infty} \lambda_{\max} \left(\int_{t_1}^{t_k} Q(t)\Delta t \right) \geq \frac{1}{\mu(t_1)}.$$

Further assume that there is a constant M *such that*

$$\mathrm{tr} \left(\int_{t_1}^{t_k} Q(t)\Delta t \right) \geq M \quad \text{for} \quad k \in \mathbb{N}.$$

Then equation (5.26) *is oscillatory on* $[a, \infty)$.

Proof. Assume (5.26) is nonoscillatory on $[a, \infty)$. This implies that there exist a nontrivial prepared solution X of (5.26) and $t_0 \in [a, \infty)$ such that

$$X^*(\sigma(t))X(t) > 0 \quad \text{for} \quad t \geq t_0.$$

If we define Z by the Riccati substitution (5.22) (with $P = I$), then by Theorem 5.50 we get that

$$I + \mu(t)Z(t) > 0 \quad \text{for} \quad t \geq t_0,$$

and Z is Hermitian satisfying

$$Z^{\Delta}(t) + Q(t) + F(t) = 0 \quad \text{with} \quad F = Z^*(I + \mu Z)^{-1}Z.$$

Corresponding to t_0 let $\{t_k\}_{k=1}^{\infty} \subset [t_0, \infty)$ be the sequence guaranteed in the statement of this theorem. Integrating both sides of the Riccati equation from t_1 to t_k, $k > 1$, gives

$$
\begin{aligned}
Z(t_1) &= Z(t_k) + \int_{t_1}^{t_k} Q(t)\Delta t + \int_{t_1}^{t_k} F(t)\Delta t \\
&= Z(t_k) + \int_{t_1}^{t_k} Q(t)\Delta t + \int_{t_1}^{\sigma(t_1)} F(t)\Delta t + \int_{\sigma(t_1)}^{t_k} F(t)\Delta t \\
&= Z(t_k) + \int_{t_1}^{t_k} Q(t)\Delta t + F(t_1)\mu(t_1) + \int_{\sigma(t_1)}^{t_k} F(t)\Delta t.
\end{aligned}
$$

Simplifying we get

$$Z(t_k) + \int_{t_1}^{t_k} Q(t)\Delta t + \int_{\sigma(t_1)}^{t_k} F(t)\Delta t = Z(t_1)\left[I + \mu(t_1)Z(t_1)\right]^{-1}.$$

Hence

(5.34) $$\lambda_{\max}\left(Z(t_k) + \int_{t_1}^{t_k} Q(t)\Delta t + \int_{\sigma(t_1)}^{t_k} F(t)\Delta t\right)$$

$$= \lambda_{\max}\left(Z(t_1)\left(I + \mu(t_1)Z(t_1)\right)^{-1}\right).$$

By Weyl's inequality (5.25),

$$\lambda_{\max}\left(Z(t_1)\left(I + \mu(t_1)Z(t_1)\right)^{-1}\right)$$

$$\geq \lambda_{\min}\left(Z(t_k)\right) + \lambda_{\max}\left(\int_{t_1}^{t_k} Q(t)\Delta t\right) + \lambda_{\min}\left(\int_{\sigma(t_1)}^{t_k} F(t)\Delta t\right)$$

$$(5.35) \qquad \geq \lambda_{\min}\left(Z(t_k)\right) + \lambda_{\max}\left(\int_{t_1}^{t_k} Q(t)\Delta t\right).$$

Taking the limit of both sides as $k \to \infty$ and using

$$(5.36) \qquad \lim_{k \to \infty} \lambda_{\min}\left(Z(t_k)\right) = 0,$$

which we will prove later, we obtain

$$\lambda_{\max}\left(Z(t_1)\left(I + \mu(t_1)Z(t_1)\right)^{-1}\right) \geq \lim_{k \to \infty} \lambda_{\max}\left(\int_{t_1}^{t_k} Q(t)\Delta t\right) \geq \frac{1}{\mu(t_1)}.$$

But in the proof of Theorem 5.63 we showed that $I + \mu(t_1)Z(t_1) > 0$ implies that

$$(5.37) \qquad \lambda_{\max}\left(Z(t_1)\left(I + \mu(t_1)Z(t_1)\right)^{-1}\right) < \frac{1}{\mu(t_1)},$$

and this gives us a contradiction. Hence to complete the proof of this theorem it remains to prove that (5.36) holds. In fact, we shall prove that

$$\lim_{k \to \infty} \lambda_i\left(Z(t_k)\right) = 0 \quad \text{for all} \quad 1 \leq i \leq n,$$

which includes (5.36) as a special case. To do this we first show that

$$(5.38) \qquad \lim_{k \to \infty} \lambda_i\left(F(t_k)\right) = 0 \quad \text{for all} \quad 1 \leq i \leq n.$$

Since $F(t) \geq 0$ implies $\operatorname{tr}(F(t)) \geq 0$, we have

$$\sum_{j=1}^{k} \mu(t_j)\lambda_i\left(F(t_j)\right) \leq \sum_{j=1}^{k} \mu(t_j)\operatorname{tr}\left(F(t_j)\right)$$

$$= \sum_{j=1}^{k} \int_{t_j}^{\sigma(t_j)} \operatorname{tr}\left(F(t)\right)\Delta t$$

$$\leq \int_{t_1}^{\sigma(t_k)} \operatorname{tr}\left(F(t)\right)\Delta t$$

$$\leq \operatorname{tr}\left(\int_{t_1}^{\sigma(t_k)} F(t)\Delta t\right)$$

$$\leq n\lambda_{\max}\left(\int_{t_1}^{\sigma(t_k)} F(t)\Delta t\right)$$

for all $k > 1$. We now show that the sequence

$$\left\{\lambda_{\max}\left(\int_{t_1}^{t_k} F(t)\Delta t\right)\right\}_{k \in \mathbb{N}}$$

is bounded. From (5.35) and (5.37) we get

$$\frac{1}{\mu(t_1)} > \lambda_{\min}\left(Z(t_k)\right) + \lambda_{\max}\left(\int_{t_1}^{t_k} Q(t)\Delta t\right).$$

Using (5.34) and (5.37) and applying Weyl's inequality (5.25) twice yields

$$\frac{1}{\mu(t_1)} > \lambda_{\max}\left(Z(t_k) + \int_{t_1}^{t_k} Q(t)\Delta t + \int_{\sigma(t_1)}^{t_k} F(t)\Delta t\right)$$

$$\geq \lambda_{\min}\left(Z(t_k)\right) + \lambda_{\min}\left(\int_{t_1}^{t_k} Q(t)\Delta t\right) + \lambda_{\max}\left(\int_{\sigma(t_1)}^{t_k} F(t)\Delta t\right).$$

It follows that

$$(5.39) \quad \frac{1}{\mu(t_1)} + \frac{1}{\mu(t_k)} > \lambda_{\min}\left(\int_{t_1}^{t_k} Q(t)\Delta t\right) + \lambda_{\max}\left(\int_{\sigma(t_1)}^{t_k} F(t)\Delta t\right).$$

From Theorem 5.63, we can, without loss of generality, assume that

$$(5.40) \quad \lambda_{\max}\left(\int_{t_1}^{t_k} Q(t)\Delta t\right) < \frac{1}{\mu(t_1)} + \frac{1}{\mu(t_k)}$$

holds for $k > 1$. Therefore, by assumption,

$$\begin{aligned}
M &\leq \operatorname{tr}\left(\int_{t_1}^{t_k} Q(t)\Delta t\right) \\
&= \sum_{i=1}^{n} \lambda_i\left(\int_{t_1}^{t_k} Q(t)\Delta t\right) \\
&= \lambda_{\min}\left(\int_{t_1}^{t_k} Q(t)\Delta t\right) + \sum_{i=1}^{n-1} \lambda_i\left(\int_{t_1}^{t_k} Q(t)\Delta t\right) \\
&< \lambda_{\min}\left(\int_{t_1}^{t_k} Q(t)\Delta t\right) + (n-1)\left(\frac{1}{\mu(t_1)} + \frac{1}{\mu(t_k)}\right).
\end{aligned}$$

Solving for the last term on the right-hand side of the inequality (5.39) and using the above inequality we get

$$\lambda_{\max}\left(\int_{\sigma(t_1)}^{t_k} F(t)\Delta t\right) < n\left(\frac{1}{\mu(t_1)} + \frac{1}{\mu(t_k)}\right) - M.$$

Therefore

$$\sum_{j=1}^{\infty} \mu(t_j)\operatorname{tr}\left(F(t_j)\right) < \infty$$

and so since $\lambda_i\left(F(t)\right) \leq \operatorname{tr}\left(F(t)\right)$ and $0 < K_1 \leq \mu(t_k) \leq K_2$ for $k \in \mathbb{N}$, (5.38) follows, i.e.,

$$\lim_{k\to\infty} \frac{\left(\lambda_i(Z(t_k))\right)^2}{1 + \mu(t_k)\lambda_i(Z(t_k))} = 0 \quad \text{for all} \quad 1 \leq i \leq n.$$

Similar to the argument to prove (5.29) in Theorem 5.63 we can show

$$\lambda_i\left(F(t_k)\right) = \frac{d_i^2}{1 + \mu(t_k)d_i}.$$

If we only consider the case $\lambda_i(Z(t_k)) \geq 0$, then

$$0 \leq \frac{(\lambda_i(Z(t_k)))^2}{1 + K_2\lambda_i(Z(t_k))} \leq \frac{(\lambda_i(Z(t_k)))^2}{1 + \mu(t_k)\lambda_i(Z(t_k))},$$

which implies

$$\lim_{k \to \infty} \frac{(\lambda_i(Z(t_k)))^2}{1 + K_2\lambda_i(Z(t_k))} = 0 \quad \text{for all} \quad 1 \leq i \leq n,$$

which in turn implies that $\lim_{k\to\infty} \lambda_i(Z(t_k)) = 0$ for all $1 \leq i \leq n$. This completes the proof. $\square$

5.4. Asymptotic Behavior of Solutions

The purpose of this section is to provide a unified treatment leading to a framework for discussing the asymptotic behavior of solutions of certain kinds of dynamic equations. For the two major extreme cases, $\mathbb{T} = \mathbb{R}$ and $\mathbb{T} = \mathbb{Z}$, results of this type are rather well known. The differential equations case involves results on the theory of asymptotic integration, which has been well documented in the monograph of Eastham [124] (see also the papers by Harris and Lutz [149, 150, 151]). The difference equations case involves results on the asymptotic behavior of products, and we refer to the paper by Benzaid and Lutz [57] and the monographs of Agarwal [5] and Elaydi [125]. Here, we present methods from the paper by Bohner and Lutz [88] which lead to the asymptotic behavior of solutions of dynamic equations on time scales. This means finding conditions on the coefficients of a dynamic equation so that solutions may be represented asymptotically in terms of known elementary functions, functions which satisfy rather simple, scalar equations, and functions which are asymptotic to constants as $t \to \infty$ in $\mathbb{T}$. So basically the idea is to reduce the study of solutions of a system of dynamic equations to the one-dimensional (scalar) case. For differential equations such simple equations can be explicitly solved using exponentials and quadrature, while in the difference equation case solutions can be constructed as elementary products. In the general case one is led to certain scalar dynamic equations on time scales whose solutions are exponential functions on the particular time scale.

The asymptotic behavior of solutions of the dynamic equation

$$x^\Delta = A(t)x$$

means for us a representation of a (fundamental) matrix of solutions in the form

$$e_A(t) = P(t)\,[I + E(t)]\,D(t),$$

where $P(t)$ represents an explicitly known matrix of elementary functions, $D(t)$ represents a diagonal matrix of solutions of scalar time scale equations, and $E(t)$ represents a matrix of functions satisfying $E(t) \to 0$ as $t \to \infty$. To accomplish this, we think of $P(t)$ constructed from certain explicit linear transformations $x = P(t)y$ of the original system into a system whose main term is diagonal with a "small" perturbation. The factor $[I + E(t)]$ in the representation comes from two sources. One involves further, so-called preliminary transformations, whose role is to bring the system into almost diagonal form, which means that the coefficient matrix consists of an explicit diagonal matrix (possibly different from the former one) plus an absolutely summable perturbation (with respect to the time scale). The other part contributing to $[I + E(t)]$ arises from determining the asymptotic behavior of the almost diagonal system, and this follows from an analogue for dynamic equations on time scales of an important and fundamental result of N. Levinson [214] for differential equations. For Levinson's result, two kinds of assumptions are required; the first is the summability condition on the perturbation and the second is a so-called dichotomy, or separation condition on the elements of the diagonal matrix. The purpose of the dichotomy condition is to insure that solutions of the diagonal system have sufficiently different asymptotic behavior which small perturbations cannot influence. The dichotomy conditions that we find for the time scale case are natural generalizations of both Levinson's classical ones and those of Benzaid and Lutz [57] in the case of difference equations.

Throughout we assume that $\sup \mathbb{T} = \infty$. We let $n \in \mathbb{N}$, $b : \mathbb{T} \to \mathbb{R}^n$ and $f : \mathbb{T} \times \mathbb{R}^n \to \mathbb{R}^n$ be rd-continuous functions, and let A be an $n \times n$-matrix-valued function on $\mathbb{T}$ satisfying

$$(5.41) \qquad\qquad A \in \mathcal{R}.$$

We now establish a relationship between the system of linear dynamic equations

$$(5.42) \qquad\qquad y^\Delta = A(t)y + b(t)$$

and its perturbation

$$(5.43) \qquad\qquad x^\Delta = A(t)x + b(t) + f(t, x),$$

where f is thought of as "small" with respect to the linear part in a sense that will be made precise below. We fix $t_0 \in \mathbb{T}$. Subject to (5.41), according to Theorem 5.8, the matrix exponential (see Definition 5.18)

$$e_A(t) := e_A(t, t_0)$$

is the unique $n \times n$-matrix-valued solution of

$$(5.44) \qquad\qquad Y^\Delta = A(t)Y, \quad Y(t_0) = I.$$

We now assume (5.41),

$$(5.45) \qquad\qquad \int_{t_0}^\infty |f(\tau, 0)| \Delta\tau < \infty,$$

$$(5.46) \qquad \begin{cases} |f(\tau, x_1) - f(\tau, x_2)| \le \gamma(\tau)|x_1 - x_2| \text{ for all } x_1, x_2 \in \mathbb{R}^n, \tau \ge t_0 \\ \\ \text{and such that } \int_{t_0}^\infty \gamma(\tau)\Delta\tau < \infty, \end{cases}$$

and (see (5.56) below)

$$(5.47) \qquad \begin{cases} \text{there exist supplementary projections } P_1, P_2 \text{ and } K > 0 \text{ with} \\ \\ |e_A(t)P_1 e_A^{-1}(s)| \le K \quad \text{for all} \quad t \in \mathbb{T}, t > t_0, s \in [t_0, t) \\ \\ |e_A(t)P_2 e_A^{-1}(s)| \le K \quad \text{for all} \quad s \in \mathbb{T}, s > t_0, t \in [t_0, s), \end{cases}$$

and introduce an operator T defined by $(t \geq t_1)$

$$(5.48) \qquad (Tx)(t) \ = \ \int_{t_1}^{t} e_A(t) P_1 e_A^{-1}(\sigma(\tau)) f(\tau, x(\tau)) \Delta\tau$$

$$- \int_{t}^{\infty} e_A(t) P_2 e_A^{-1}(\sigma(\tau)) f(\tau, x(\tau)) \Delta\tau,$$

where x is bounded on $[t_0, \infty)$, and where $t_1 \in [t_0, \infty)$ is picked in such a manner that (observe (5.46))

$$(5.49) \qquad\qquad \theta = K \int_{t_1}^{\infty} \gamma(\tau) \Delta\tau < 1$$

holds. Note that T, given by (5.48), is well defined whenever $x : \mathbb{T} \to \mathbb{R}^n$ is bounded on $[t_0, \infty)$ (see Theorem 5.21 (iii), (5.45), (5.46), and (5.47)). The following useful result on the operator T is related to the variation of constants formula.

Lemma 5.72. *Assume (5.41) through (5.47) and define T by (5.48). Then the relationship*

$$(5.50) \qquad\qquad (Tx)^{\Delta} - A(Tx) = f(\cdot, x)$$

holds on $[t_0, \infty)$.

Proof. By the product rule (Theorem 5.3 (iii)), we find for $t \in [t_0, \infty)$

$$(Tx)^{\Delta}(t) - A(t)(Tx)(t) = e_A^{\Delta}(t) \int_{t_1}^{t} P_1 e_A^{-1}(\sigma(\tau)) f(\tau, x(\tau)) \Delta\tau$$

$$+ e_A^{\sigma}(t) P_1 e_A^{-1}(\sigma(t)) f(t, x(t)) - e_A^{\Delta}(t) \int_{t}^{\infty} P_2 e_A^{-1}(\sigma(\tau)) f(\tau, x(\tau)) \Delta\tau$$

$$+ e_A^{\sigma}(t) P_2 e_A^{-1}(\sigma(t)) f(t, x(t)) - A(t) e_A(t) \int_{t_1}^{t} P_1 e_A^{-1}(\sigma(\tau)) f(\tau, x(\tau)) \Delta\tau$$

$$+ A(t) e_A(t) \int_{t}^{\infty} P_2 e_A^{-1}(\sigma(\tau)) f(\tau, x(\tau)) \Delta\tau$$

$$= \ e_A(\sigma(t)) \left(P_1 + P_2 \right) e_A^{-1}(\sigma(t)) f(t, x(t))$$

$$= \ f(t, x(t)),$$

where we applied (5.44) and (5.56). $\square$

The first main result of this section (see also [**104, 240**] and [**57**, Theorem 1.1]) now reads as follows.

Theorem 5.73. *Assume (5.41), (5.45), (5.46), and (5.47). Then there exists a one-to-one and bicontinuous correspondence between the bounded solutions of (5.42) and (5.43). In addition, if*

$$(5.51) \qquad\qquad \lim_{t \to \infty} \{ e_A(t) P_1 \} = 0,$$

then the difference between corresponding solutions of (5.42) and (5.43) tends to zero as $t \to \infty$.

Proof. For a bounded function (more precisely, bounded on $[t_1, \infty)$) we put

$$\|x\| = \sup_{t \in [t_1, \infty)} |x(t)|.$$

If x_1 and x_2 are two bounded functions, it is easy to prove that the estimate

$$(5.52) \qquad |(Tx_1)(t) - (Tx_2)(t)| \leq \theta \|x_1 - x_2\|$$

holds for each $t \in [t_1, \infty)$, where θ is defined by (5.49). Hence, for such x_1 and x_2, we have

$$(5.53) \qquad \|Tx_1 - Tx_2\| \leq \theta \|x_1 - x_2\|.$$

Next, since

$$|(T0)(t)| \leq K \int_{t_1}^{\infty} |f(\tau, 0)| \Delta \tau$$

holds by (5.47), and since (5.53) and (5.46) imply

$$\|Tx\| \leq \|Tx - T0\| + \|T0\| \leq \theta \|x\| + \|T0\|,$$

we also have

$$(5.54) \qquad \|Tx\| \leq \theta \|x\| + K \int_{t_1}^{\infty} |f(\tau, 0)| \Delta \tau.$$

By (5.53) and (5.54), T is a contraction mapping that maps the set of bounded (on $[t_1, \infty)$) functions to itself. Hence

$$(5.55) \qquad y = x - Tx$$

establishes a one-to-one correspondence between bounded (for $t \geq t_1$) solutions of (5.42) and (5.43) (using (5.50)). Using (5.52), we have

$$(1 + \theta)^{-1} \|y_1 - y_2\| \leq \|x_1 - x_2\| \leq (1 - \theta)^{-1} \|y_1 - y_2\|$$

which shows that the correspondence is bicontinuous on $[t_1, \infty)$. Using (5.41), this correspondence can be extended to $[t_0, t_1]$. Finally, given $\varepsilon > 0$, choose $t_2(\varepsilon) > t_1$ so that

$$K \int_{t_2(\varepsilon)}^{\infty} |f(\tau, x(\tau))| \Delta \tau < \frac{\varepsilon}{2}.$$

Then

$$\begin{aligned} |x(t) - y(t)| &= |(Tx)(t)| \\ &< |e_A(t)P_1| \left| \int_{t_1}^{t_2(\varepsilon)} e_A^{-1}(\sigma(\tau)) f(\tau, x(\tau)) \Delta \tau \right| + \frac{\varepsilon}{2}, \end{aligned}$$

and letting $t \to \infty$, we see that $x(t) - y(t) = o(1)$ (i.e., $x(t) - y(t) \to 0$) as $t \to \infty$. $\square$

Remark 5.74. Condition (5.46) is a *growth condition* on the perturbation term, and (5.47) is a so-called *dichotomy condition*. We also refer to the book by S. Elaydi [125, Section 7.4] for a discussion of the dichotomy condition for difference equations. In (5.47), P_1 and P_2 are $n \times n$-matrices with (see [104, page 37])

$$(5.56) \qquad P_1^2 = P_1, \quad P_2^2 = P_2, \quad P_1 + P_2 = I, \quad P_1 P_2 = P_2 P_1 = 0.$$

Now we establish a version of Levinson's perturbation lemma (see [214] and [57, Lemma 2.1]). To do so, we apply Theorem 5.73. Let scalar functions $\lambda_k : \mathbb{T} \to \mathbb{R}$, $1 \leq k \leq n$, be given. We assume

$$(5.57) \qquad \lambda_k \in \mathcal{R} \quad \text{for all} \quad 1 \leq k \leq n.$$

Exercise 5.75. Assume (5.57). Show that the n^2 functions w_{ij} introduced by

$$(5.58) \qquad\qquad w_{ij} = e_{\lambda_j \ominus \lambda_i} \quad \text{for} \quad 1 \le i, j \le n$$

are well defined. Next, put

$$W_i = \text{diag} \{ w_{i1}, w_{i2}, \dots, w_{in} \} \quad \text{for} \quad 1 \le i \le n$$

and find a matrix Λ_i such that

$$W_i = e_{\Lambda_i}.$$

We put

$$(5.59) \qquad\qquad \Lambda(t) = \text{diag} \{ \lambda_1(t), \dots, \lambda_n(t) \}$$

and let $R \in C_{rd}$ be an $n \times n$-matrix-valued function on $\mathbb{T}$.

Theorem 5.76 (Levinson's Perturbation Lemma). *Assume* (5.57),

$$(5.60) \qquad\qquad \int_{t_0}^{\infty} \left| \frac{R(\tau)}{1 + \mu(\tau)\lambda_i(\tau)} \right| \Delta\tau < \infty, \quad 1 \le i \le n,$$

and suppose that

$$(5.61) \quad \begin{cases} \text{there exist } K, m > 0 \text{ such that each } (i,j) \text{ with } i \ne j \text{ satisfies either} \\[2mm] \text{(a)} \lim_{\tau \to \infty} w_{ij}(\tau) = 0 \quad \text{and} \quad \left| \frac{w_{ij}(s)}{w_{ij}(t)} \right| \ge m \text{ for all } t \ge t_0, s \in [t_0, t] \\[2mm] \text{or} \\[2mm] \text{(b)} \left| \frac{w_{ij}(s)}{w_{ij}(t)} \right| \le K \text{ for all } t \ge t_0, s \in [t_0, t], \end{cases}$$

where w_{ij} are given by (5.58). *Then the linear system $x^\Delta = [\Lambda(t) + R(t)]x$ has a fundamental matrix X such that*

$$(5.62) \qquad\qquad X(t) = [I + o(1)]e_\Lambda(t) \quad \text{as} \quad t \to \infty.$$

Proof. First of all we fix $i \in \{1, \dots, n\}$. We put

$$\Lambda_i(t) = \frac{\Lambda(t) - \lambda_i(t)I}{1 + \mu(t)\lambda_i(t)} \quad \text{and} \quad f_i(t, x) = \frac{R(t)x}{1 + \mu(t)\lambda_i(t)},$$

and now Theorem 5.73 will be applied to

$$(5.63) \qquad\qquad w_i^\Delta = \Lambda_i(t)w_i$$

and

$$(5.64) \qquad\qquad v_i^\Delta = \Lambda_i(t)v_i + f_i(t, v_i).$$

Condition (5.41) requires nonsingularity of

$$I + \mu(t)\Lambda_i(t) = \frac{I + \mu(t)\Lambda(t)}{1 + \mu(t)\lambda_i(t)},$$

and this is of course equivalent to (5.57), since $\Lambda(t)$ is a diagonal matrix according to (5.59). Next, f_i satisfies (5.45) trivially, and also (5.46) with $\gamma = \frac{R}{1 + \mu\lambda_i}$, and this follows from (5.60). Finally, we need to verify (5.47). For this purpose, we define

$$p_j = \begin{cases} 1 & \text{if } (i,j) \text{ satisfies (a) in (5.61)} \\ \\ 0 & \text{otherwise.} \end{cases}$$

Moreover, let $P_1 = \text{diag}\{p_1, \ldots, p_n\}$ and $P_2 = I - P_1$ so that (5.56) holds. By Exercise 5.75, (a) of (5.61) implies $\lim_{t \to 0}\{W_i(t)P_1\} = 0$ and hence (5.51) for each $i \in \{1, \ldots, n\}$. Next, if $t_0 \le s \le t$ with $s, t \in \mathbb{T}$, then $W_i(t)P_1W_i^{-1}(s)$ is a diagonal matrix with entries 0 or, if (i,j) satisfies (a) of (5.61), $\frac{w_{ij}(t)}{w_{ij}(s)}$, with $\left|\frac{w_{ij}(t)}{w_{ij}(s)}\right| \le \frac{1}{m}$. Also, if $t_0 \le t \le s$ with $s, t \in \mathbb{T}$, then $W_i(t)P_2W_i^{-1}(s)$ is a diagonal matrix with entries 0 or, if (i,j) satisfies (b) of (5.61), $\frac{w_{ij}(t)}{w_{ij}(s)}$, with $\left|\frac{w_{ij}(t)}{w_{ij}(s)}\right| \le K$. Thus (5.47) is satisfied with $e_A = W_i$ for each $i \in \{1, \ldots, n\}$.

Now, by Theorem 5.73 and since $w_i(t) = u_i$ (where $u_i \in \mathbb{R}^n$ is the ith unit vector) is trivially a bounded solution of (5.63), there exists a bounded solution v_i of (5.64) with

$$v_i = w_i + \varepsilon_i = u_i + \varepsilon_i \quad \text{and} \quad \lim_{t \to \infty} \varepsilon_i(t) = 0, \quad 1 \le i \le n.$$

We put $x_i = v_i e_{\lambda_i} = (u_i + \varepsilon_i)e_{\lambda_i}$. Observe that $e_\Lambda = \text{diag}\{e_{\lambda_1}, \ldots, e_{\lambda_n}\}$. By Theorem 5.3 (iii)

$$\begin{aligned} x_i^\Delta(t) &= (e_{\lambda_i}v_i)^\Delta(t) \\ &= e_{\lambda_i}^\Delta(t)v_i(t) + e_{\lambda_i}^\sigma(t)v_i^\Delta(t) \\ &= \lambda_i(t)e_{\lambda_i}(t)v_i(t) + \{e_{\lambda_i}(t) + \mu(t)e_{\lambda_i}^\Delta(t)\}\frac{\Lambda(t) - \lambda_i(t)I + R(t)}{1 + \mu(t)\lambda_i(t)}v_i(t) \\ &= \lambda_i(t)x_i(t) + e_{\lambda_i}(t)\{\Lambda(t) - \lambda_i(t)I + R(t)\}v_i(t) \\ &= [\Lambda(t) + R(t)]x_i(t), \end{aligned}$$

and this implies (5.62) from the statement of Theorem 5.76. □

Remark 5.77. The proof of Theorem 5.76 shows that fewer hypotheses are required provided we are interested only in the condition

$$x_i(t) = [1 + o(1)]e_{\lambda_i}(t)$$

for some fixed $i \in \{1, \ldots, n\}$. More precisely, for this we only need the "ith parts" of both conditions (5.57) and (5.60), and then each $j \in \{1, \ldots, n\}$ is required to satisfy either (a) or (b) of (5.61).

Remark 5.78. In what follows we say that an $n \times n$-matrix-valued function Λ with

$$(5.65) \qquad \Lambda(t) = \text{diag}\{\lambda_1(t), \ldots, \lambda_n(t)\} \quad \text{and} \quad \Lambda \in \mathcal{R}$$

satisfies the dichotomy condition provided (5.61) holds (with the notation (5.58)). We also say that an $n \times n$-matrix-valued function $R \in C_{rd}$ *satisfies the growth condition with respect to* Λ if (5.60) holds.

For the remainder of this section we consider a dynamic equation of the form

$$(5.66) \qquad x^\Delta = [\Lambda(t) + V(t) + R(t)]x,$$

where $\Lambda, V, R \in C_{rd}$ are $n \times n$-matrix-valued functions on $\mathbb{T}$ with (5.65) and

$$(5.67) \qquad\qquad V(t) \to 0 \quad \text{as} \quad t \to \infty.$$

Now, if either Theorem 5.73 can be directly applied to (5.66) (with $A = \Lambda + V$ and $f(t, x) = R(t)x$) or if Theorem 5.76 could be applied (with perturbation term $V + R$), then it follows that (5.66) has a solution X satisfying

$$X(t) = [I + o(1)]e_{\Lambda+V}(t) \quad \text{as} \quad t \to \infty.$$

We now consider situations where these results cannot be directly applied, and we look for certain types of preliminary linear transformations, whose role is to modify (5.66) to a form (with a possibly different Λ) so that one of these theorems is applicable to yield an asymptotic representation for solutions. For our further investigations in this direction the following auxiliary result is useful.

Lemma 5.79. *Suppose Q is a differentiable $n \times n$-matrix-valued function on $\mathbb{T}$ such that $Q(t) \to 0$ as $t \to \infty$. Then, for sufficiently large $t \in \mathbb{T}$,*

$$(5.68) \qquad P(t) = (I + Q^\sigma(t))^{-1} \left[(\Lambda(t) + V(t))(I + Q(t)) - Q^\Delta(t) \right]$$

is well defined. If there are $n \times n$-matrix-valued functions $\hat{\Lambda}, \hat{R} \in C_{rd}$ with $P = \hat{\Lambda} + \hat{R}$ such that $\hat{\Lambda}$ satisfies the dichotomy condition and such that both $\hat{R}$ and R satisfy the growth condition with respect to $\hat{\Lambda}$, then there exists a solution X of (5.66) satisfying

$$X(t) = [I + o(1)]\, e_{\hat{\Lambda}}(t) \quad \text{as} \quad t \to \infty.$$

Proof. Our assumptions make the transformation

$$w = [I + Q(t)]^{-1} x, \quad \text{i.e.,} \quad x = [I + Q(t)]\, w$$

well defined for sufficiently large $t \in \mathbb{T}$. This implies (use Theorem 5.3 (iii))

$$
\begin{aligned}
[\Lambda(t) + V(t) + R(t)]\,[I + Q(t)]\, w(t) &= [\Lambda(t) + V(t) + R(t)]\, x(t) \\
&= x^\Delta(t) \\
&= [I + Q^\sigma(t)]\, w^\Delta(t) + Q^\Delta(t) w(t)
\end{aligned}
$$

so that (use (5.68))

$$
\begin{aligned}
w^\Delta(t) &= [I + Q^\sigma(t)]^{-1} \left\{ [\Lambda(t) + V(t) + R(t)]\,[I + Q(t)] - Q^\Delta(t) \right\} w(t) \\
&= \left[P(t) + (I + Q^\sigma(t))^{-1} R(t)\,(I + Q(t)) \right] w(t).
\end{aligned}
$$

Now, Theorem 5.76 (where R there is replaced by $\hat{R} + (I + Q^\sigma)^{-1} R(I + Q)$) yields (note that both $(I + Q^\sigma(t))^{-1}$ and $(I + Q(t))$ are bounded because of $Q(t) \to 0$ as $t \to \infty$) the existence of W with

$$W^\Delta = \left[\hat{\Lambda}(t) + \hat{R}(t) + (I + Q^\sigma(t))^{-1} R(t)\,(I + Q(t)) \right] W$$

such that $W(t) = [I + o(1)]\, e_{\hat{\Lambda}}(t)$ as $t \to \infty$, and hence there exists a solution X of (5.66) with

$$
\begin{aligned}
X(t) &= [I + Q(t)]\, W(t) \\
&= [I + Q(t)]\,[I + o(1)]\, e_{\hat{\Lambda}}(t) \\
&= [I + o(1)]\, e_{\hat{\Lambda}}(t)
\end{aligned}
$$

as $t \to \infty$, i.e., (5.62) holds (with $\Lambda = \hat{\Lambda}$). $\qquad\square$

In order to apply Lemma 5.79 we now consider various options for constructing $Q(t)$ depending upon the properties of Λ and V. These are motivated by analogy with the continuous case [151] and the discrete case [57].

Theorem 5.80. *Let* $\Lambda, V, R \in C_{rd}$ *be* $n \times n$-matrix-valued functions on $\mathbb{T}$. *Suppose that*

$$Q(t) := -\int_t^\infty V(\tau)\Delta\tau \text{ is well defined for each } t \in [t_0, \infty).$$

If Λ *satisfies the dichotomy condition and if both* $\Lambda Q - Q^\sigma \Lambda + VQ$ *and* R *satisfy the growth condition with respect to* Λ, *then there exists a solution* X *of* (5.66) *satisfying*

$$X(t) = [I + o(1)]\, e_\Lambda(t) \quad as \quad t \to \infty.$$

Proof. First, since $\int_{t_0}^\infty V(\tau)\Delta\tau$ converges, we have $Q(t) \to 0$ as $t \to \infty$. For sufficiently large $t \in \mathbb{T}$ we hence may define

$$\hat{R}(t) = (I + Q^\sigma(t))^{-1}\, [\Lambda(t)Q(t) - Q^\sigma(t)\Lambda(t) + V(t)Q(t)]\,.$$

Then we have (observe $Q^\Delta(t) = V(t)$ for $t \in \mathbb{T}$ with $t \geq t_0$) for P from (5.68) on $[t_1, \infty)$ for sufficiently large $t_1 \in \mathbb{T}$

$$
\begin{aligned}
P &= (I + Q^\sigma)^{-1}\left[(\Lambda + V)(I + Q) - Q^\Delta\right] \\
&= (I + Q^\sigma)^{-1}[\Lambda + V + \Lambda Q + VQ - V - (I + Q^\sigma)\Lambda] + \Lambda \\
&= (I + Q^\sigma)^{-1}(\Lambda Q - Q^\sigma\Lambda + VQ) + \Lambda \\
&= \Lambda + \hat{R}
\end{aligned}
$$

so that our result follows from Lemma 5.79. $\qquad\qquad\square$

Theorem 5.81. *Let* $\Lambda, V, R \in C_{rd}$ *be* $n \times n$-matrix-valued functions with (5.65) and (5.67) such that $\Lambda(t)$ tends to an $n \times n$-matrix with distinct eigenvalues so that $\Lambda(t) + V(t)$ has distinct eigenvalues $\hat{\lambda}_1(t), \ldots, \hat{\lambda}_n(t)$ for sufficiently large $t \in \mathbb{T}$. Put $\hat{\Lambda}(t) = \mathrm{diag}\left\{\hat{\lambda}_1(t), \ldots, \hat{\lambda}_n(t)\right\}$. If $\hat{\Lambda}$ satisfies the dichotomy condition and if V is differentiable and both $V^\Delta(I + \mu\hat{\Lambda})$ and R satisfy the growth condition with respect to $\hat{\Lambda}$, then there exists a solution X of (5.66) satisfying*

$$X(t) = [I + o(1)]\, e_{\hat{\Lambda}}(t) \quad as \quad t \to \infty.$$

Proof. Since for sufficiently large $t \in \mathbb{T}$ the eigenvalues of $\Lambda(t) + V(t)$ are all distinct, there exist matrices $Q(t)$ such that

(5.69) $$\hat{\Lambda}(t) = (I + Q(t))^{-1}\, [\Lambda(t) + V(t)]\, (I + Q(t))\,,$$

and the explicit construction of these matrices $Q(t)$ works as in [149, Remark 2.1] to show that $Q(t) \to 0$ and $Q^\Delta(t) = O(|V^\Delta(t)|)$ as $t \to \infty$. Next, we define (for sufficiently large $t \in \mathbb{T}$)

$$\hat{R}(t) = -(I + Q^\sigma(t))^{-1}\, Q^\Delta(t)\left(I + \mu(t)\hat{\Lambda}(t)\right).$$

With this notation we have on $[t_1, \infty)$ for sufficiently large $t_1 \in \mathbb{T}$ (see (5.68))

$$
\begin{aligned}
P &= (I + Q^\sigma)^{-1} \left[(\Lambda + V)(I + Q) - Q^\Delta \right] \\
&= (I + Q^\sigma)^{-1} \left\{ \left[(I + Q) - (I + Q^\sigma) \right] (I + Q)^{-1} (\Lambda + V)(I + Q) - Q^\Delta \right\} \\
&\quad + (I + Q)^{-1} (\Lambda + V)(I + Q) \\
&= (I + Q^\sigma)^{-1} \left\{ -\mu Q^\Delta (I + Q)^{-1} (\Lambda + V)(I + Q) - Q^\Delta \right\} + \hat{\Lambda} \\
&= -(I + Q^\sigma)^{-1} Q^\Delta I + \mu \hat{\Lambda} + \hat{\Lambda} \\
&= \hat{\Lambda} + \hat{R},
\end{aligned}
$$

where we used Theorem 5.2. Now our conclusion follows from Lemma 5.79. $\qquad \square$

Remark 5.82. The assumption that $\lim_{t \to \infty} \Lambda(t)$ exists and has distinct eigenvalues can be weakened since the proof only requires that (5.69) holds with $\lim_{t \to \infty} Q(t) = 0$. If there exists a nonzero, scalar-valued function g so that $\lim_{t \to \infty} \{ g(t) \left[\Lambda(t) + V(t) \right] \}$ has distinct eigenvalues, then clearly (5.69) holds with $\hat{\Lambda}$ replaced by $g\hat{\Lambda}$. This remark is useful for cases in which $\lim_{t \to \infty} \mu(t) = \infty$ (see, e.g., Example 5.87 involving a q-difference equation).

Theorem 5.83. Let $\Lambda, R, V \in C_{rd}$ be $n \times n$-matrix-valued functions on $\mathbb{T}$. Suppose that (5.65) holds so that the w_{ij} from (5.58) are never zero, and assume that for some $p > 1$

$$
(5.70) \qquad \int_{t_0}^{\infty} \left| \frac{v_{ij}(\tau)}{1 + \mu(\tau) \lambda_j(\tau)} \right|^p \Delta \tau < \infty \qquad \text{for all} \quad 1 \le i \ne j \le n,
$$

where v_{ij} are the ijth entries of V. Let $q = \frac{p}{p-1}$ and suppose that each pair (i, j) with $i \ne j$ satisfies either

$$
(5.71) \qquad \lim_{t \to \infty} \int_{t_0}^{t} \left| \frac{w_{ji}(t)}{w_{ji}(\sigma(\tau))} \right|^q \Delta \tau = 0
$$

or

$$
(5.72) \qquad \int_{t}^{\infty} \left| \frac{w_{ji}(t)}{w_{ji}(\sigma(\tau))} \right|^q \Delta \tau < \infty \qquad \text{for all} \quad t \in [t_0, \infty).
$$

If (5.71) holds, then we put

$$
q_{ij}(t) = \int_{t_0}^{t} \frac{w_{ji}(t)}{w_{ji}(\sigma(\tau))} \frac{v_{ij}(\tau)}{1 + \mu(\tau) \lambda_j(\tau)} \Delta \tau,
$$

and if (5.72) holds, then we put

$$
q_{ij}(t) = -\int_{t}^{\infty} \frac{w_{ji}(t)}{w_{ji}(\sigma(\tau))} \frac{v_{ij}(\tau)}{1 + \mu(\tau) \lambda_j(\tau)} \Delta \tau.
$$

Let $Q(t)$ be the matrix that has $q_{ij}(t)$ as its ijth entries, $i \ne j$ and $q_{ii}(t) \equiv 0$. Finally, let $\operatorname{diag} V(t)$ be the diagonal $n \times n$-matrix that has the same diagonal entries as $V(t)$ and suppose that $\hat{\Lambda} := \Lambda + \operatorname{diag} V$ satisfies the dichotomy condition and that both $VQ - Q^\sigma \operatorname{diag} V$ and R satisfy the growth condition with respect to $\hat{\Lambda}$. Then there exists a solution X of (5.66) satisfying

$$
X(t) = [I + o(1)] \, e_{\hat{\Lambda}}(t) \qquad \text{as} \quad t \to \infty.
$$

Proof. First we show that (5.70), (5.71), and (5.72) imply $\lim_{t\to\infty} Q(t) = 0$. Our conclusion then follows from Lemma 5.79, since $P = \hat{\Lambda} + \hat{R}$ (see (5.68)) with

(5.73)
$$\hat{R}(t) = (I + Q^\sigma(t))^{-1} [V(t)Q(t) - Q^\sigma(t) \operatorname{diag} V(t)]$$

(for sufficiently large $t \in \mathbb{T}$), which will be shown below. Now, suppose the pair (i,j) satisfies (5.71). Then for $t \in [t_0, \infty)$, by Hölder's inequality (see Theorem 6.13),

$$
\begin{aligned}
|q_{ij}(t)| &\leq \int_{t_0}^t \left| \frac{w_{ji}(t)}{w_{ji}(\sigma(\tau))} \frac{v_{ij}(\tau)}{1 + \mu(\tau)\lambda_j(\tau)} \right| \Delta\tau \\
&\leq \left\{ \int_{t_0}^t \left| \frac{w_{ji}(t)}{w_{ji}(\sigma(\tau))} \right|^{\frac{p}{p-1}} \Delta\tau \right\}^{\frac{p-1}{p}} \left\{ \int_{t_0}^t \left| \frac{v_{ij}(\tau)}{1 + \mu(\tau)\lambda_j(\tau)} \right|^p \Delta\tau \right\}^{\frac{1}{p}} \\
&\leq \left\{ \int_{t_0}^t \left| \frac{w_{ji}(t)}{w_{ji}(\sigma(\tau))} \right|^{\frac{p}{p-1}} \Delta\tau \right\}^{\frac{p-1}{p}} \left\{ \int_{t_0}^\infty \left| \frac{v_{ij}(\tau)}{1 + \mu(\tau)\lambda_j(\tau)} \right|^p \Delta\tau \right\}^{\frac{1}{p}} \\
&\to 0
\end{aligned}
$$

as $t \to \infty$, because of (5.70). Next, if the pair (i,j) satisfies (5.72), then again by Hölder's inequality

$$
\begin{aligned}
|q_{ij}(t)| &\leq \int_t^\infty \left| \frac{w_{ji}(t)}{w_{ji}(\sigma(\tau))} \frac{v_{ij}(\tau)}{1 + \mu(\tau)\lambda_j(\tau)} \right| \Delta\tau \\
&\leq \left\{ \int_t^\infty \left| \frac{w_{ji}(t)}{w_{ji}(\sigma(\tau))} \right|^{\frac{p}{p-1}} \Delta\tau \right\}^{\frac{p-1}{p}} \left\{ \int_t^\infty \left| \frac{v_{ij}(\tau)}{1 + \mu(\tau)\lambda_j(\tau)} \right|^p \Delta\tau \right\}^{\frac{1}{p}} \\
&\to 0
\end{aligned}
$$

as $t \to \infty$, because of (5.70).

Now we want to show that Q is a solution of

(5.74)
$$Q^\Delta = V(t) - \operatorname{diag} V(t) + \Lambda(t)Q - Q^\sigma \Lambda(t).$$

First, if the pair (i,j) satisfies (5.71), then q_{ij} is differentiable, and it follows from the product rule (Theorem 5.3 (iii)) that

$$
\begin{aligned}
q_{ij}^\Delta(t) &= w_{ji}^\Delta(t) \int_{t_0}^t \frac{v_{ij}(\tau)}{(1 + \mu(\tau)\lambda_j(\tau)) \, w_{ji}(\sigma(\tau))} \Delta\tau + \frac{w_{ji}(\sigma(t))v_{ij}(t)}{(1 + \mu(t)\lambda_j(t)) \, w_{ji}(\sigma(t))} \\
&= \frac{\lambda_i(t) - \lambda_j(t)}{1 + \mu(t)\lambda_j(t)} w_{ji}(t) \int_{t_0}^t \frac{v_{ij}(\tau)}{(1 + \mu(\tau)\lambda_j(\tau)) \, w_{ji}(\sigma(\tau))} \Delta\tau + \frac{v_{ij}(t)}{1 + \mu(t)\lambda_j(t)} \\
&= \frac{\lambda_i(t) - \lambda_j(t)}{1 + \mu(t)\lambda_j(t)} q_{ij}(t) + \frac{v_{ij}(t)}{1 + \mu(t)\lambda_j(t)}
\end{aligned}
$$

holds. Hence q_{ij} is the unique solution of the IVP

$$q_{ij}^\Delta = \frac{\lambda_i(t) - \lambda_j(t)}{1 + \mu(t)\lambda_j(t)} q_{ij} + (1 - \delta_{ij}) \frac{v_{ij}(t)}{1 + \mu(t)\lambda_j(t)}, \quad q_{ij}(t_0) = 0,$$

where δ_{ij} stands for the usual Kronecker symbol. This initial value problem is equivalent to (multiply the equation by $1 + \mu(t)\lambda_j(t)$)

$$q_{ij}^\Delta = (1 - \delta_{ij}) \, v_{ij}(t) + \lambda_i(t)q_{ij} - \lambda_j(t)q_{ij}^\sigma, \quad q_{ij}(t_0) = 0.$$

Similarly, q_{ij} satisfies this same dynamic equation (with, however, a different initial condition) provided the pair (i,j) satisfies (5.72). Thus we arrive at a proof of (5.74).

Now, for sufficiently large $t \in \mathbb{T}$, we define $\hat{R}$ by (5.73). If $t_1 \in \mathbb{T}$ is sufficiently large, we have on $[t_1, \infty)$ (see (5.68))

$$
\begin{aligned}
P &= (I + Q^\sigma)^{-1} \left[(\Lambda + V)(I + Q) - Q^\Delta \right] \\
&= (I + Q^\sigma)^{-1} \left(\Lambda + V + \Lambda Q + VQ - Q^\Delta \right) \\
&= (I + Q^\sigma)^{-1} \left(\Lambda + V + \Lambda Q + VQ - V + \operatorname{diag} V - \Lambda Q + Q^\sigma \Lambda \right) \\
&= (I + Q^\sigma)^{-1} \left(\Lambda + VQ + \operatorname{diag} V + Q^\sigma \Lambda \right) \\
&= (I + Q^\sigma)^{-1} \left[(I + Q^\sigma)\Lambda + (I + Q^\sigma) \operatorname{diag} V + VQ - Q^\sigma \operatorname{diag} V \right] \\
&= \Lambda + \operatorname{diag} V + (I + Q^\sigma)^{-1} \left(VQ - Q^\sigma \operatorname{diag} V \right) \\
&= \hat{\Lambda} + \hat{R},
\end{aligned}
$$

where we used (5.74). Again, an application of Lemma 5.79 finishes the proof. $\square$

Remark 5.84. In the continuous case the result of Theorem 5.83 with $1 < p \leq 2$ is due to P. Hartman and A. Wintner and was generalized to the case $p > 2$ by Harris and Lutz [151]. There, the conditions (5.71) and (5.72) follow from a rather strong separation condition on $\operatorname{Re}[\lambda_i(t) - \lambda_j(t)]$ (see Example 5.85), and it also follows that Q is in the same L^p class as V. In the case of an arbitrary time scale, however, it appears difficult to find general conditions on Λ which imply (5.71) or (5.72).

We conclude this section with some examples.

Example 5.85. Let us briefly discuss the two (well-known) cases $\mathbb{T} = \mathbb{R}$ and $\mathbb{T} = \mathbb{Z}$. First, if $\mathbb{T} = \mathbb{R}$, then the functions w_{ij} from (5.58) are

$$
w_{ij}(t) = \exp \left\{ \int_{t_0}^{t} [\lambda_j(\tau) - \lambda_i(\tau)] \, d\tau \right\}
$$

provided (5.57) holds. Thus

$$
\left| \frac{w_{ij}(s)}{w_{ij}(t)} \right| = \left| \exp \left\{ \int_{t}^{s} [\lambda_j(\tau) - \lambda_i(\tau)] \, d\tau \right\} \right| = \exp \left\{ \int_{s}^{t} \operatorname{Re} [\lambda_i(\tau) - \lambda_j(\tau)] \, d\tau \right\},
$$

where Re denotes the real part of a complex number. Now, if there exist $m > 0$ and $t^* \in \mathbb{R}$ such that for each pair (i,j) with $i \neq j$

$$
|\operatorname{Re} [\lambda_i(t) - \lambda_j(t)]| \geq m \quad \text{for all} \quad t \in [t^*, \infty),
$$

then it follows that

(5.75)
$$
\begin{cases}
\text{for each pair } (i,j) \text{ with } i \neq j \text{ either} \\[2mm]
\text{(a) } \operatorname{Re} [f_i(t) - f_j(t)] \geq m \quad \text{for all} \quad t \in [t^*, \infty) \\[2mm]
\text{or} \\[2mm]
\text{(b) } \operatorname{Re} [f_i(t) - f_j(t)] \leq -m \quad \text{for all} \quad t \in [t^*, \infty),
\end{cases}
$$

where $f_k(t) = \lambda_k(t)$, provided (5.57) holds so that $\lambda_i - \lambda_j$ are continuous functions. Thus, (5.75) (a) implies (5.61) (a) and (5.75) (b) implies (5.61) (b), and hence (5.75) is a sufficient condition for the dichotomy condition (5.61).

Next, if $\mathbb{T} = \mathbb{Z}$, then the functions w_{ij} from (5.58) are (provided (5.57) holds)

$$w_{ij}(t) = \prod_{\tau=t_0}^{t-1} \frac{1+\lambda_j(\tau)}{1+\lambda_i(\tau)}.$$

Now

$$\begin{aligned}
\left|\frac{w_{ij}(s)}{w_{ij}(t)}\right| &= \prod_{\tau=s}^{t-1} \left|\frac{1+\lambda_i(\tau)}{1+\lambda_j(\tau)}\right| = \exp\left\{\sum_{\tau=s}^{t-1} [\log|1+\lambda_i(\tau)| - \log|1+\lambda_j(\tau)|]\right\} \\
&= \exp\left\{\int_s^t [\log|1+\lambda_i(\tau)| - \log|1+\lambda_j(\tau)|]\,\Delta\tau\right\} \\
&= \exp\left\{\int_s^t \mathrm{Re}\,[f_i(\tau) - f_j(\tau)]\,\Delta\tau\right\},
\end{aligned}$$

this time with $f_k(t) = \log(1+\lambda_k(t))$, so that again (5.75) is sufficient for the dichotomy condition (5.61) to hold. Alternatively we can give in this case the following sufficient condition for the dichotomy condition (5.61):

$$(5.76) \qquad \lim_{t\to\infty} \alpha_k(t) = \alpha_k,\ 1 \le k \le n \quad \text{with} \quad |\alpha_i| \ne |\alpha_j| \text{ for all } 1 \le i < j \le n,$$

where $\alpha_k(t) = 1 + \lambda_k(t)$. To see that (5.76) is really sufficient for (5.61) we note that for a certain index pair (i, j) with $i \ne j$ (5.76) implies

$$(5.77) \qquad \text{(a) } |\alpha_i| > |\alpha_j| \quad \text{or} \quad \text{(b) } |\alpha_i| < |\alpha_j|.$$

It is easy to see that (5.77) (a) implies (note $\alpha_i \ne 0$) (5.61) (a) and that (5.77) (b) is sufficient (note $\alpha_j \ne 0$) for (5.61) (b).

Example 5.86. Let us assume that $\mathbb{T}$ is a time scale that has only isolated points. Then, assuming (5.57), the functions w_{ij} from (5.58) may easily be checked to have both of the forms ($t \in \mathbb{T}, t \ge t_0$)

$$(5.78) \qquad w_{ij}(t) = \prod_{\tau\in[t_0,t)} \frac{\alpha_j(\tau)}{\alpha_i(\tau)} \quad \text{with} \quad \alpha_k(\tau) = 1 + \mu(\tau)\lambda_k(\tau)$$

and
$$(5.79)$$
$$w_{ij}(t) = \exp\left\{\int_{t_0}^t [f_j(\tau) - f_i(\tau)]\Delta\tau\right\} \quad \text{with} \quad f_k(\tau) = \frac{\log[1+\mu(\tau)\alpha_k(\tau)]}{\mu(\tau)}.$$

Hence for $s, t \in \mathbb{T}$ with $t_0 \le s \le t$, (5.78) and (5.79) imply

$$\left|\frac{w_{ij}(s)}{w_{ij}(t)}\right| = \prod_{\tau\in[s,t)} \left|\frac{\alpha_i(\tau)}{\alpha_j(\tau)}\right| = \exp\left\{\int_s^t \mathrm{Re}[f_i(\tau) - f_j(\tau)]\Delta\tau\right\}.$$

From this we see that both (5.75) and (5.76), using the above f_k and α_k from (5.79) and (5.78), are sufficient conditions for the dichotomy condition (5.61).

Example 5.87 (*q-difference equations*). Let $q > 1$ and consider $\mathbb{T} = q^{\mathbb{N}_0}$. As a special example, we consider the so-called *q-difference equation of Jacobi*

$$(5.80) \qquad\qquad ty(qt) - y(t) = 1.$$

It is well known that (5.80) has the Theta-series

$$\Theta(t) = -\sum_{n=0}^{\infty} q^{\frac{n(n-1)}{2}} t^n$$

as a formal solution, which can be shown to diverge everywhere. Recently, C. Zhang [253] has studied the summability of this series and its asymptotic properties, especially for complex t. Since we are concerned with the behavior of solutions near ∞, the formal series

$$\hat{\Theta}(t) = \sum_{n=1}^{\infty} q^{\frac{n(n+1)}{2}} t^{-n}$$

is more relevant for us. It also can be seen to diverge for all t, but one would expect that it, too, would be an asymptotic expansion for an actual solution of (5.80) as $t \to \infty$. It is possible to apply Theorem 5.73 to this equation and a perturbation $tx(qt) - x(t) = 1 - \frac{q}{t}$, which has solution $x(t) = \frac{q}{t}$. They can be expressed in the form of (5.42), resp. (5.43), with $A(t) = \frac{1-t}{t^2(q-1)}$, $b(t) = \frac{1}{t^2(q-1)}$, and $f(t, x) = -\frac{q}{t^3(q-1)}$. Since f is easily seen to satisfy (5.45) and (5.46), it follows that (5.80) has a solution y satisfying $y(t) = x(t) + o(1)$ as $t \to \infty$. This is a rather weak asymptotic statement which could be improved by modifying Theorem 5.73, but we can also achieve a better asymptotic result in the following way: Replacing t by qt to obtain $qty(q^2t) - y(qt) = 1$ and subtracting (5.80), we obtain

$$qty(q^2t) - (t+1)y(qt) + y(t) = 0,$$

a homogeneous *second order* q-difference equation having not only the Theta-series $\hat{\Theta}(t)$ as a formal solution, but other solutions having quite different asymptotic behavior. We put $x(t) = \begin{pmatrix} y(t) \\ y(qt) \end{pmatrix}$ and rewrite the above equation as

$$(5.81) \quad x^\sigma = \begin{pmatrix} y(tq) \\ \frac{1}{qt}\left\{(t+1)y(qt) - y(t)\right\} \end{pmatrix} = \tilde{A}(t)x \quad \text{with} \quad \tilde{A}(t) = \begin{pmatrix} 0 & 1 \\ -\frac{1}{qt} & \frac{t+1}{qt} \end{pmatrix}.$$

Let us introduce the transformation

$$x(t) = Pz(t) \quad \text{with} \quad P = \begin{pmatrix} 1 & q \\ 0 & 1 \end{pmatrix}$$

so that (5.81) is transformed into

$$z^\sigma = A^*(t)z \quad \text{with} \quad A^*(t) = P^{-1}\tilde{A}(t)P = \begin{pmatrix} \frac{1}{t} & \frac{q-1}{t} \\ -\frac{1}{qt} & \frac{1}{q} + \frac{1-q}{qt} \end{pmatrix}$$

or, equivalently,

$$z^\Delta = A(t)z \quad \text{with} \quad A(t) = \frac{1}{(q-1)t}\left\{A^*(t) - I\right\}.$$

Now we apply Theorem 5.81 (see in particular also Remark 5.82) since $A = \Lambda + V$ with

$$\Lambda(t) = \frac{1}{(q-1)t} \begin{pmatrix} \frac{1}{t} - 1 & 0 \\ 0 & \frac{1}{q} + \frac{1-q}{qt} - 1 \end{pmatrix} \quad \text{and} \quad V(t) = \begin{pmatrix} 0 & \frac{1}{t^2} \\ -\frac{1}{q(q-1)t^2} & 0 \end{pmatrix}.$$

Note that for $h(t) = \frac{1}{t^2}$ we have

$$h^\Delta(t) = \frac{h(qt) - h(t)}{(q-1)t} = \frac{1-q^2}{q^2(q-1)t^3} = -\frac{1+q}{q^2t^3}$$

so that

$$V^\Delta(t) = \begin{pmatrix} 0 & -\frac{1+q}{q^2t^3} \\ \frac{1+q}{q^3(q-1)t^3} & 0 \end{pmatrix}.$$

Observe that the dichotomy conditions are satisfied for $\Lambda(t)$ (use (5.76) and see the remarks in Example 5.86) and

$$V^\Delta(t)\left(I + \mu(t)\hat{\Lambda}(t)\right) = \begin{pmatrix} 0 & -\frac{1+q}{q^2t^3}\left(\frac{1}{q} + \frac{1-q}{qt}\right) \\ \frac{1+q}{q^3(q-1)t^4} & 0 \end{pmatrix} + O(t^{-5})$$

satisfies the growth condition with respect to Λ. Hence Theorem 5.81 implies that $z^\Delta = [\Lambda(t) + V(t)]z$ has a matrix solution Z satisfying

$$Z(t) = [I + o(1)]\,W(t) \quad \text{as} \quad t \to \infty,$$

where W is a fundamental solution of the diagonal system

$$w^\Delta = \Lambda(t)w, \quad w = \begin{pmatrix} w_1 \\ w_2 \end{pmatrix}.$$

The first component w_1 is a solution of the scalar equation

$$w(qt) = \frac{1}{t}w(t),$$

which has the solution

$$w_1(t) = \sqrt{t}\exp\left(-\frac{(\ln t)^2}{2\ln q}\right).$$

The second component w_2 is a solution of

$$(5.82) \qquad w(qt) = \left(\frac{1}{q} + \frac{1-q}{qt}\right)w(t),$$

and we show that there exists a solution with the asymptotic representation

$$w_2(t) = \frac{1}{t}(1 + o(1)) \quad \text{as} \quad t \to \infty.$$

To see this, observe that $\tilde{w}(t) = \frac{1}{t} + \frac{q}{t^2}$ is a solution of

$$(5.83) \qquad w(qt) = \left(\frac{1}{q} + \frac{1-q}{qt} + r(t)\right)w(t) \quad \text{with} \quad r(t) = \frac{q-1}{t^2}\left(1 + \frac{q}{t}\right)^{-1}.$$

Applying Theorem 5.76 to the equations (5.82) and (5.83), we see that (5.82) has a solution of the form

$$w_2(t) = \left(\frac{1}{t} + \frac{q}{t^2}\right)(1 + o(1)) = \frac{1}{t}(1 + o(1)) \quad \text{as} \quad t \to \infty.$$

Finally, we see that the system (5.81) has a vector solution of the form

$$\begin{pmatrix} 1 & q \\ 0 & 1 \end{pmatrix} \begin{pmatrix} 1 + o(1) & o(1) \\ o(1) & 1 + o(1) \end{pmatrix} \begin{pmatrix} 0 \\ \frac{1}{t}(1 + o(1)) \end{pmatrix} = \begin{pmatrix} \frac{q}{t}(1 + o(1)) \\ \frac{1}{t}(1 + o(1)) \end{pmatrix}$$

as $t \to \infty$. Hence there exists a solution of the scalar, second order q-difference equation of the form

$$(5.84) \qquad\qquad y(t) = \frac{q}{t}(1 + o(1)) \quad \text{as} \quad t \to \infty.$$

The general solution also involves terms coming from w_1, whose behavior near ∞ is of somewhat smaller order than the dominant terms. Note that w_1 is also a solution of the homogeneous q-difference equation corresponding to (5.80), and it plays an important role as well in explaining why the formal Theta-series solutions diverge and how to re-sum them (see [**253**]). Of course, the asymptotic result (5.84) is a much weaker statement than asserting that $\hat{\Theta}$ is an *asymptotic expansion* for an actual solution, either in the sense of Poincaré or Gevrey. However, our results could be used also in the case of nonanalytic perturbations, provided they satisfy our required conditions. Based on the asymptotic expansion results for the differential and difference equations cases and also some results such as in [**253**] in the q-difference case, it is likely that better asymptotic results than (5.84) can be achieved in the case of general time scales, but we choose not to do this here.

5.5. Higher Order Linear Dynamic Equations

Many results from this section are contained in the publications by Bohner and Eloe [**85**] and Erbe, Mathsen, and Peterson [**130**].

For $n \in \mathbb{N}_0$ and rd-continuous functions $p_i : \mathbb{T} \to \mathbb{R}$, $1 \leq i \leq n$, we consider the nth order linear dynamic equation

$$(5.85) \qquad\qquad Ly = 0, \quad \text{where} \quad Ly = y^{\Delta^n} + \sum_{i=1}^{n} p_i y^{\Delta^{n-i}}.$$

A function $y : \mathbb{T} \to \mathbb{R}$ is said to be a solution of the equation (5.85) on $\mathbb{T}$ provided y is n times differentiable (see Definition 1.27) on $\mathbb{T}^{\kappa^n}$ and satisfies $Ly(t) = 0$ for all $t \in \mathbb{T}^{\kappa^n}$. It follows with Theorem 1.16 (i) that y^{Δ^n} is an rd-continuous function on $\mathbb{T}^{\kappa^n}$.

In order to give criteria under which initial value problems involving the linear dynamic equation of nth order (5.85) have unique solutions, we start by rewriting (5.85) as a system of dynamic equations.

Lemma 5.88. *If* $z = \begin{pmatrix} z_1 \\ \vdots \\ z_n \end{pmatrix} : \mathbb{T} \to \mathbb{R}^n$ *satisfies for all* $t \in \mathbb{T}^\kappa$

$$(5.86) \qquad z^\Delta(t) = A(t)z(t), \quad where \quad A = \begin{pmatrix} 0 & 1 & 0 & \cdots & 0 \\ \vdots & 0 & 1 & \ddots & \vdots \\ \vdots & & \ddots & \ddots & 0 \\ 0 & \cdots & \cdots & 0 & 1 \\ -p_n & \cdots & \cdots & -p_2 & -p_1 \end{pmatrix},$$

then $y = z_1$ *is a solution of equation* (5.85). *Conversely, if* y *solves* (5.85) *on* $\mathbb{T}$,

then $z = \begin{pmatrix} y \\ y^\Delta \\ \vdots \\ y^{\Delta^{n-1}} \end{pmatrix} : \mathbb{T} \to \mathbb{R}^n$ *satisfies* (5.86) *for all* $t \in \mathbb{T}^{\kappa^n}$.

Proof. This is easy to see because $z^\Delta = A(t)z$ can be equivalently rewritten as $z_i^\Delta = z_{i+1}$, $1 \leq i \leq n-1$, $z_n^\Delta = -\sum_{i=1}^n p_i(t)z_{n-i+1}$. □

Definition 5.89. We say that the equation (5.85) is *regressive* provided system (5.86) is regressive (see Definition 5.5).

An immediate consequence of Lemma 5.88, Definition 5.89, and Theorem 5.8 is the following.

Corollary 5.90. *Suppose* (5.85) *is regressive. Then any initial value problem*

$$Ly = 0, \quad y^{\Delta^i}(t_0) = \alpha_i \quad for \ all \quad 0 \leq i \leq n-1,$$

with $t_0 \in \mathbb{T}$ *and* $\alpha_i \in \mathbb{R}$, $0 \leq i \leq n-1$, *has a unique solution.*

Theorem 5.91. *The equation* (5.85) *is regressive if* $p_i \in C_{rd}$ *for all* $1 \leq i \leq n$ *and*

$$(5.87) \qquad 1 + \sum_{i=1}^n (-\mu(t))^i p_i(t) \neq 0 \quad for \ all \quad t \in \mathbb{T}^\kappa.$$

Proof. It is enough to calculate $\det(I + \mu A)$, where A is given in (5.86). We multiply the $(n-i)$th column of $I + \mu A$ by $(-\mu)^i$ and add it to the nth column, for all

$1 \leq i \leq n - 1$, to obtain

$$
\det(I + \mu A) \;=\; \det
\begin{pmatrix}
1 & \mu & 0 & \cdots & & 0 \\
0 & 1 & \mu & \ddots & & \vdots \\
\vdots & \ddots & \ddots & \ddots & & 0 \\
0 & \cdots & 0 & 1 & & \mu \\
-\mu p_n & \cdots & \cdots & & -\mu p_2 & 1 - \mu p_1
\end{pmatrix}
$$

$$
=\; \det
\begin{pmatrix}
1 & \mu & 0 & \cdots & \cdots & & 0 \\
0 & 1 & \mu & \ddots & & & \vdots \\
\vdots & \ddots & \ddots & \ddots & \ddots & & \vdots \\
\vdots & & \ddots & 1 & \mu & & 0 \\
0 & \cdots & \cdots & 0 & 1 & & 0 \\
-\mu p_n & \cdots & \cdots & \cdots & -\mu p_2 & & 1 + \sum_{i=1}^{n}(-\mu)^i p_i
\end{pmatrix}
$$

$$
=\; 1 + \sum_{i=1}^{n}(-\mu)^i p_i,
$$

which never vanishes according to our assumption (5.87). □

Example 5.92. If $\mathbb{T} = \mathbb{R}$, then $\mu \equiv 0$, and (5.87) is always satisfied, while if $\mathbb{T} = \mathbb{Z}$, then $\mu \equiv 1$ and condition (5.87) becomes $1 + \sum_{i=1}^{n}(-1)^i p_i \neq 0$.

Exercise 5.93. Determine whether the following equations are regressive on the indicated time scale $\mathbb{T}$:

  (i) $y^{\Delta^2} = y$, where $\mathbb{T} = \mathbb{Z}$;
  (ii) $y^{\Delta^2} = y^\sigma$, where $\mathbb{T}$ is any time scale;
  (iii) $\sum_{\nu=0}^{n} \frac{1}{t^\nu} y^{\Delta^{n-\nu}} = 0$, where $\mathbb{T} = 2^{\mathbb{N}_0}$ and $n \in \mathbb{N}$;
  (iv) $y^{\Delta^3} - \frac{7}{2} y^{\Delta^2} + \frac{7}{2} y^\Delta - y = 0$, where $\mathbb{T}$ is any time scale.

Definition 5.94. An equation

$$(5.88) \qquad \sum_{i=0}^{n} p_i(t) y^{\Delta^{n-i}} = 0 \quad \text{with} \quad p_0(t) \neq 0 \text{ for all } t \in \mathbb{T}^{\kappa^n}$$

is said to be regressive if $y^{\Delta^n} + \sum_{i=1}^{n} \frac{p_i(t)}{p_0(t)} y^{\Delta^{n-i}} = 0$ is regressive.

Theorem 5.95. *Assume (5.88) is regressive. If $p_i \in C_{\mathrm{rd}}^1$ for all $0 \leq i \leq n$ and $y^{\Delta^n} \in C_{\mathrm{rd}}^1$, then $(Ly)^\Delta = 0$ is regressive as well.*

Proof. Let us calculate $(Ly)^\Delta$:

$$
\begin{aligned}
(Ly)^\Delta &= y^{\Delta^{n+1}} + \sum_{i=1}^{n} \left(p_i^\Delta y^{\Delta^{n-i}} + p_i^\sigma y^{\Delta^{n-i+1}} \right) \\
&= y^{\Delta^{n+1}} + \sum_{i=2}^{n+1} p_{i-1}^\Delta y^{\Delta^{n-i+1}} + \sum_{i=1}^{n} p_i^\sigma y^{\Delta^{n-i+1}} \\
&= y^{\Delta^{n+1}} + \sum_{i=1}^{n+1} \hat{p}_i y^{\Delta^{n-i+1}}
\end{aligned}
$$

with $\hat{p}_1 = p_1^\sigma$, $\hat{p}_{n+1} = p_n^\Delta$, and $\hat{p}_i = p_{i-1}^\Delta + p_i^\sigma$, $2 \le i \le n$. Now we have

$$
\begin{aligned}
1 + \sum_{i=1}^{n+1} (-\mu)^i \hat{p}_i &= 1 - \mu p_1^\sigma + (-\mu)^{n+1} p_n^\Delta + \sum_{i=2}^{n} (-\mu)^i (p_{i-1}^\Delta + p_i^\sigma) \\
&= 1 - \mu p_1^\sigma + (-\mu)^{n+1} p_n^\Delta + \sum_{i=1}^{n-1} (-\mu)^{i+1} p_i^\Delta + \sum_{i=2}^{n} (-\mu)^i p_i^\sigma \\
&= 1 - \mu p_1^\sigma + (-\mu)^{n+1} p_n^\Delta + \mu^2 p_1^\Delta + (-\mu)^n p_n^\sigma + \sum_{i=2}^{n-1} (-\mu)^i (p_i^\sigma - \mu p_i^\Delta) \\
&= 1 - \mu p_1 + (-\mu)^n p_n + \sum_{i=2}^{n-1} (-\mu)^i p_i \\
&= 1 + \sum_{i=1}^{n} (-\mu)^i p_i.
\end{aligned}
$$

Also, if p_i^Δ exists and is rd-continuous, then p_i is continuous and hence $p_i^\sigma \in C_{\mathrm{rd}}$, hence $\hat{p}_i \in C_{\mathrm{rd}}$, $1 \le i \le n$. If $Ly = 0$ is regressive, then so is $(Ly)^\Delta = 0$. □

Applying Theorem 5.95 successively, we obtain the following as a consequence.

Corollary 5.96. *Let $u_i : \mathbb{T} \to \mathbb{R} \setminus \{0\}$ for all $1 \le i \le n$ be given functions, and define operators L_i recursively by $L_1 y = u_1 y$ and $L_{i+1} y = u_{i+1} (L_i y)^\Delta$, $1 \le i \le n$. If $L_i y$ has C_{rd}^1-coefficients for all $1 \le i \le n-1$, then the equation $L_{n+1} y = 0$, i.e.,*

$$
u_{n+1} (u_n (\cdots (u_2 (u_1 y)^\Delta)^\Delta \cdots)^\Delta)^\Delta = 0,
$$

is regressive.

Let us now consider an nth order linear dynamic equation of the form

$$
(5.89) \qquad y^{\Delta^n} + \sum_{k=1}^{n} q_k(t) \left(y^{\Delta^{n-k}} \right)^\sigma = 0.
$$

Of course, if $\mathbb{T} = \mathbb{R}$, the p_k and the q_k in (5.85) and (5.89) are the same. But for an arbitrary time scale it is much easier to check if (5.89) is regressive than if (5.85) is regressive. This statement is illustrated by our next result. Compare also with Exercise 5.93 (ii).

Theorem 5.97. *If $q_k \in C_{\mathrm{rd}}$ for all $1 \le k \le n$ and*

$$
(5.90) \qquad\qquad q_1 \in \mathcal{R},
$$

then equation (5.89) is regressive.

Proof. We start by transforming equation (5.89) into an equivalent dynamical equation of the form (5.85), and then proceed to verify that condition (5.90) is equivalent to (5.87). We have

$$
y^{\Delta^n} + \sum_{k=1}^{n} q_k \left(y^{\Delta^{n-k}} \right)^\sigma = y^{\Delta^n} + \sum_{k=1}^{n} q_k \left(y^{\Delta^{n-k}} + \mu y^{\Delta^{n-k+1}} \right)
$$

$$
= y^{\Delta^n} + \sum_{k=1}^{n} q_k y^{\Delta^{n-k}} + \sum_{k=0}^{n-1} \mu q_{k+1} y^{\Delta^{n-k}}
$$

$$
= (1 + \mu q_1) y^{\Delta^n} + \sum_{k=1}^{n-1} (q_k + \mu q_{k+1}) y^{\Delta^{n-k}} + q_n y
$$

so that we may divide (5.89) by $1 + \mu q_1$ (see (5.90)) to obtain (5.85) with

$$
p_k = \frac{q_k + \mu q_{k+1}}{1 + \mu q_1} \text{ for all } 1 \le k \le n-1, \quad p_n = \frac{q_n}{1 + \mu q_1}.
$$

Let us check to see that (5.87) holds:

$$
1 + \sum_{i=1}^{n} (-\mu)^i p_i = \frac{1 + \mu q_1 + \sum_{i=1}^{n-1} (-\mu)^i (q_i + \mu q_{i+1}) + (-\mu)^n q_n}{1 + \mu q_1}
$$

$$
= \frac{1 + \mu q_1 + \sum_{i=1}^{n-1} (-\mu)^i q_i - \sum_{i=1}^{n-1} (-\mu)^{i+1} q_{i+1} + (-\mu)^n q_n}{1 + \mu q_1}
$$

$$
= \frac{1 + \mu q_1 + \sum_{i=1}^{n-1} (-\mu)^i q_i - \sum_{i=2}^{n} (-\mu)^i q_i + (-\mu)^n q_n}{1 + \mu q_1}
$$

$$
= \frac{1}{1 + \mu q_1} \ne 0.
$$

Of course $p_k \in C_{rd}$ for all $1 \le k \le n$, and so our assertion follows from Theorem 5.91. $\qquad\square$

Example 5.98. Again, for $\mathbb{T} = \mathbb{R}$, (5.90) is always satisfied, while for $\mathbb{T} = \mathbb{Z}$, (5.90) is equivalent to $q_1(t)$ being different from -1 for all $t \in \mathbb{Z}$.

The converse of Theorem 5.97 is also true: Every regressive equation (5.85) can be written in the form (5.89) such that the assumptions of Theorem 5.97 hold. This is the contents of our next result.

Theorem 5.99. *If an equation* (5.85) *is regressive, then there exist* $q_k \in C_{rd}$, $1 \le k \le n$, *with* q_1 *satisfying* (5.90) *such that* y *solves* (5.85) *if and only if it solves* (5.89). *More precisely, we have*

$$
q_k = \frac{\sum_{i=k}^{n} (-\mu)^{i-k} p_i}{1 + \sum_{i=1}^{n} (-\mu)^i p_i} \quad \textit{for all} \quad 1 \le k \le n.
$$

Proof. First of all we have for all $1 \le i \le n$

$$\sum_{k=1}^{i}(-\mu)^{i-k}\left(y^{\Delta^{n-k}}\right)^{\sigma} = \sum_{k=1}^{i}(-\mu)^{i-k}\left(y^{\Delta^{n-k}} + \mu y^{\Delta^{n-k+1}}\right)$$

$$= \sum_{k=1}^{i}(-\mu)^{i-k}y^{\Delta^{n-k}} - \sum_{k=1}^{i}(-\mu)^{i-k+1}y^{\Delta^{n-k+1}}$$

$$= \sum_{k=1}^{i}(-\mu)^{i-k}y^{\Delta^{n-k}} - \sum_{k=0}^{i-1}(-\mu)^{i-k}y^{\Delta^{n-k}}$$

$$= y^{\Delta^{n-i}} - (-\mu)^{i}y^{\Delta^{n}},$$

and therefore

$$\left\{1+\sum_{i=1}^{n}(-\mu)^{i}p_i\right\}y^{\Delta^{n}} + \sum_{k=1}^{n}\left\{\sum_{i=k}^{n}(-\mu)^{i-k}p_i\right\}\left(y^{\Delta^{n-k}}\right)^{\sigma}$$

$$= \left\{1+\sum_{i=1}^{n}(-\mu)^{i}p_i\right\}y^{\Delta^{n}} + \sum_{i=1}^{n}\left\{\sum_{k=1}^{i}(-\mu)^{i-k}\left(y^{\Delta^{n-k}}\right)^{\sigma}\right\}p_i$$

$$= \left\{1+\sum_{i=1}^{n}(-\mu)^{i}p_i\right\}y^{\Delta^{n}} + \sum_{i=1}^{n}\left\{y^{\Delta^{n-i}} - (-\mu)^{i}y^{\Delta^{n}}\right\}p_i$$

$$= y^{\Delta^{n}} + \sum_{i=1}^{n}p_iy^{\Delta^{n-i}}.$$

Since $1+\sum_{i=1}^{n}(-\mu)^{i}p_i \ne 0$ because of (5.87), our claim follows. Note also that $1+\mu q_1 = \frac{1}{1+\sum_{i=1}^{n}(-\mu)^{i}p_i} \ne 0$. $\square$

Now we want to study properties of solutions.

Definition 5.100. Let $y_k : \mathbb{T} \to \mathbb{R}$ be $(m-1)$ times differentiable functions for all $1 \le k \le m$. We then define the *Wronski determinant* $W = W(y_1,\dots,y_m)$ of the set $\{y_1,\dots,y_m\}$ by $W(y_1,\dots,y_m) = \det V(y_1,\dots,y_m)$, where

$$V(y_1,\dots,y_m) = \begin{pmatrix} y_1 & y_2 & \cdots & y_m \\ y_1^{\Delta} & y_2^{\Delta} & \cdots & y_m^{\Delta} \\ \vdots & \vdots & & \vdots \\ y_1^{\Delta^{m-1}} & y_2^{\Delta^{m-1}} & \cdots & y_m^{\Delta^{m-1}} \end{pmatrix},$$

i.e., $(V(y_1,\dots,y_m))_{ij} = y_j^{\Delta^{i-1}}$ for all $1 \le i,j \le m$.

The Wronski determinant in case of differential equations is often called the *Wronskian*, while the word *Casoratian* (see, e.g., [125, Definition 2.11]) is sometimes used in connection with difference equations.

We next give a sometimes useful alternative representation of the Wronskian, which mainly is a consequence of Theorem 1.16 (iv).

Theorem 5.101. *Let* $y_k : \mathbb{T} \to \mathbb{R}$ *be* $(m-1)$ *times differentiable for all* $1 \le k \le m$. *Then*

$$W(y_1, \dots, y_m) = \det \begin{pmatrix} y_1^\sigma & y_2^\sigma & \cdots & y_m^\sigma \\ (y_1^\Delta)^\sigma & (y_2^\Delta)^\sigma & \cdots & (y_m^\Delta)^\sigma \\ \vdots & \vdots & & \vdots \\ \left(y_1^{\Delta^{m-2}}\right)^\sigma & \left(y_2^{\Delta^{m-2}}\right)^\sigma & \cdots & \left(y_m^{\Delta^{m-2}}\right)^\sigma \\ y_1^{\Delta^{m-1}} & y_2^{\Delta^{m-1}} & \cdots & y_m^{\Delta^{m-1}} \end{pmatrix}.$$

Proof. By Theorem 1.16 (iv) we have

$$y_i^{\Delta^j} = \left(y_i^{\Delta^j}\right)^\sigma - \mu y_i^{\Delta^{j+1}} \quad \text{for all} \quad 1 \le i \le m, \, 0 \le j \le m-1.$$

First, replace y_i in the first row of $V(y_1, \dots, y_m)$ by $y_i^\sigma - \mu y_i^\Delta$ and then add μ times the second row. Continue to do the same process with the second row all the way to the $(m-1)$st row to arrive at the claimed formula for $W(y_1, \dots, y_m)$. □

Exercise 5.102. Find the following Wronski determinants:

  (i) $W\left(e_{\lambda_1}(\cdot, t_0), e_{\lambda_2}(\cdot, t_0), e_{\lambda_3}(\cdot, t_0)\right)$, where $\lambda_1, \lambda_2, \lambda_3 \in \mathcal{R}$ are constants;
 (ii) $W\left(h_0(\cdot, t_0), h_1(\cdot, t_0), \dots, h_{n-1}(\cdot, t_0)\right)$.

The next result illustrates the importance of Wronskians.

Theorem 5.103. *If there exist solutions* y_k, $1 \le k \le n$, *of the regressive equation* (5.85) *(or* (5.89)*) such that* $W(y_1, \dots, y_n)(t_0) \ne 0$ *for some* $t_0 \in \mathbb{T}^{\kappa^{n-1}}$, *then each solution* y *of* (5.85) *(or of* (5.89)*) can be written as* $y = \sum_{k=1}^n \alpha_k y_k$, *where the* $\alpha_k \in \mathbb{R}$, $1 \le k \le n$, *are uniquely determined.*

Proof. Let y be a solution of (5.85). Since $W(y_1, \dots, y_n)(t_0) \ne 0$, the matrix $V(y_1, \dots, y_n)(t_0)$ is invertible, and the vector

$$\begin{pmatrix} \alpha_1 \\ \vdots \\ \alpha_n \end{pmatrix} = \left(V(y_1, \dots, y_n)(t_0)\right)^{-1} \begin{pmatrix} y(t_0) \\ y^\Delta(t_0) \\ \vdots \\ y^{\Delta^{n-1}}(t_0) \end{pmatrix}$$

is well defined. We put $\tilde{y} = \sum_{k=1}^{n} \alpha_k y_k$. Then $\tilde{y}$ obviously also solves (5.85) (or (5.89)), and it satisfies

$$
\begin{pmatrix} \tilde{y}(t_0) \\ \tilde{y}^{\Delta}(t_0) \\ \vdots \\ \tilde{y}^{\Delta^{n-1}}(t_0) \end{pmatrix} = \begin{pmatrix} y_1(t_0) & y_2(t_0) & \cdots & y_n(t_0) \\ y_1^{\Delta}(t_0) & y_2^{\Delta}(t_0) & \cdots & y_n^{\Delta}(t_0) \\ \vdots & \vdots & & \vdots \\ y_1^{\Delta^{n-1}}(t_0) & y_2^{\Delta^{n-1}}(t_0) & \cdots & y_n^{\Delta^{n-1}}(t_0) \end{pmatrix} \begin{pmatrix} \alpha_1 \\ \alpha_2 \\ \vdots \\ \alpha_n \end{pmatrix}
$$

$$
= V(y_1, \ldots, y_n)(t_0) \begin{pmatrix} \alpha_1 \\ \vdots \\ \alpha_n \end{pmatrix}
$$

$$
= V(y_1, \ldots, y_n)(t_0)(V(y_1, \ldots, y_n)(t_0))^{-1} \begin{pmatrix} y(t_0) \\ y^{\Delta}(t_0) \\ \vdots \\ y^{\Delta^{n-1}}(t_0) \end{pmatrix}
$$

$$
= \begin{pmatrix} y(t_0) \\ y^{\Delta}(t_0) \\ \vdots \\ y^{\Delta^{n-1}}(t_0) \end{pmatrix}
$$

so that $y^{\Delta^i}(t_0) = \tilde{y}^{\Delta^i}(t_0)$ for all $0 \le i \le n-1$. Hence by Corollary 5.90, $y = \tilde{y}$. Therefore $y = \sum_{k=1}^{n} \alpha_k y_k$. If $y = \sum_{k=1}^{n} \beta_k y_k$ for some $\beta_k \in \mathbb{R}$, $1 \le k \le n$, then

$$
\begin{pmatrix} y(t_0) \\ y^{\Delta}(t_0) \\ \vdots \\ y^{\Delta^{n-1}}(t_0) \end{pmatrix} = V(y_1, \ldots, y_n)(t_0) \begin{pmatrix} \beta_1 \\ \beta_2 \\ \vdots \\ \beta_n \end{pmatrix}
$$

so that

$$
\begin{pmatrix} \beta_1 \\ \beta_2 \\ \vdots \\ \beta_n \end{pmatrix} = (V(y_1, \dots, y_n)(t_0))^{-1} \begin{pmatrix} y(t_0) \\ y^{\Delta}(t_0) \\ \vdots \\ y^{\Delta^{n-1}}(t_0) \end{pmatrix} = \begin{pmatrix} \alpha_1 \\ \alpha_2 \\ \vdots \\ \alpha_n \end{pmatrix},
$$

and hence this representation is unique. $\square$

Definition 5.104. A set of solutions $\{y_1, \dots, y_n\}$ of the regressive equation (5.85) (or (5.89)) is called a *fundamental system* for (5.85) (or (5.89)) if there is $t_0 \in \mathbb{T}^{\kappa^{n-1}}$ such that $W(y_1, \dots, y_n)(t_0) \neq 0$.

As in the theory for ordinary differential equations we will see that a fundamental system $\{y_1, \dots, y_n\}$ satisfies $W(y_1, \dots, y_n)(t) \neq 0$ for all $t \in \mathbb{T}^{\kappa^{n-1}}$. In order to show this we first derive a linear dynamic equation of first order which is satisfied by $W(y_1, \dots, y_n)$. To do so, however, we first need to be able to differentiate determinants. This is done next.

Theorem 5.105 (Derivative of Determinants). *Let $m \in \mathbb{N}$ and an $m \times m$-matrix-valued function $A = (a_{ij})_{1 \leq i, j \leq m}$ with differentiable entries $a_{ij} : \mathbb{T} \to \mathbb{R}$ be given. For $1 \leq k \leq m$ we define matrix-valued functions $B^{(k)} = (b_{ij}^{(k)})_{1 \leq i, j \leq m}$ by*

$$
b_{ij}^{(k)} = \begin{cases} a_{ij}^{\sigma} & \text{if } i < k \\ a_{ij}^{\Delta} & \text{if } i = k \\ a_{ij} & \text{if } i > k. \end{cases}
$$

Then $\det A : \mathbb{T} \to \mathbb{R}$ is differentiable, and the formula

$$
(\det A)^{\Delta} = \sum_{k=1}^{m} \det B^{(k)}
$$

holds.

Proof. We prove the statement by induction. First, for $m = 1$ the theorem holds since $(\det A)^{\Delta} = a_{11}^{\Delta} = b_{11}^{(1)} = B^{(1)}$. Now we assume that the statement holds for some $m \in \mathbb{N}$. We put

$$
V_k = (\text{adj}(a_{ij})_{1 \leq i, j \leq m+1})_{k1}
$$

and calculate

$$
\begin{aligned}
[\det(a_{ij})_{1\le i,j\le m+1}]^\Delta
&= \left[\sum_{k=1}^{m+1} a_{1k}V_k\right]^\Delta \\
&= \sum_{k=1}^{m+1} (a_{1k}V_k)^\Delta \\
&= \sum_{k=1}^{m+1} a_{1k}^\Delta V_k + \sum_{k=1}^{m+1} a_{1k}^\sigma V_k^\Delta \\
&= \det(b_{ij}^{(1)})_{1\le i,j\le m+1} + \sum_{k=1}^{m+1} a_{1k}^\sigma V_k^\Delta \\
&= \det(b_{ij}^{(1)})_{1\le i,j\le m+1} + \sum_{k=2}^{m+1} \det(b_{ij}^{(k)})_{1\le i,j\le m+1} \\
&= \sum_{k=1}^{m+1} \det(b_{ij}^{(k)})_{1\le i,j\le m+1},
\end{aligned}
$$

where we have used the induction hypothesis to find V_k^Δ (note that V_k is itself a determinant of an $m \times m$-matrix). $\qquad\square$

Exercise 5.106. Use Theorem 5.105 (and also direct computation) to differentiate the determinants of the matrices

$$
\begin{pmatrix} y_1 & y_2 \\ y_1^{\Delta^2} & y_2^{\Delta^2} \end{pmatrix}, \quad V(y_1,y_2), \quad
\begin{pmatrix} 1 & t & t^2 \\ t & 1 & t \\ t^2 & t & 1 \end{pmatrix}, \quad V(y_1,y_2,y_3).
$$

As an application of Theorem 5.105, it is now easy to differentiate Wronski determinants.

Corollary 5.107. *If* $y_i : \mathbb{T} \to \mathbb{R}$, $1 \le i \le m$, *are* m *times differentiable, then*

$$
(5.91) \quad (W(y_1,\ldots,y_m))^\Delta = \det
\begin{pmatrix}
y_1^\sigma & y_2^\sigma & \cdots & y_m^\sigma \\
(y_1^\Delta)^\sigma & (y_2^\Delta)^\sigma & \cdots & (y_m^\Delta)^\sigma \\
\vdots & \vdots & & \vdots \\
(y_1^{\Delta^{m-2}})^\sigma & (y_2^{\Delta^{m-2}})^\sigma & \cdots & (y_m^{\Delta^{m-2}})^\sigma \\
y_1^{\Delta^m} & y_2^{\Delta^m} & \cdots & y_m^{\Delta^m}
\end{pmatrix},
$$

i.e., $((W(y_1,\ldots,y_m))^\Delta)_{ij} = \begin{cases} (y_j^{\Delta^{i-1}})^\sigma & \text{if } 1 \le i \le m-1 \\ y_j^{\Delta^m} & \text{if } i = m. \end{cases}$

Proof. By Theorem 5.105, $(W(y_1,\dots,y_m))^\Delta = \sum_{k=1}^m \det B^{(k)}$, where the $B^{(k)}$ are constructed as in Theorem 5.105. But each $B^{(k)}$ for all $1 \le k \le m-1$ has two identical rows (namely the kth and the $(k+1)$st) so that $\det B^{(k)} = 0$ for all $1 \le k \le m-1$. Only $\det B^{(m)}$ remains, and $B^{(m)}$ is equal to the matrix on the right-hand side of equation (5.91). $\qquad\square$

We now can use Corollary 5.107 to prove the following result.

Theorem 5.108. *If* $y_1,\dots,y_n$ *are functions which are* n *times differentiable on* $\mathbb{T}^{\kappa^n}$ *and if* $W = W(y_1,\dots,y_n)$, *then*

$$
W^\Delta = \det
\begin{pmatrix}
y_1^\sigma & y_2^\sigma & \cdots & y_n^\sigma \\
(y_1^\Delta)^\sigma & (y_2^\Delta)^\sigma & \cdots & (y_n^\Delta)^\sigma \\
\vdots & \vdots & & \vdots \\
(y_1^{\Delta^{n-2}})^\sigma & (y_2^{\Delta^{n-2}})^\sigma & \cdots & (y_n^{\Delta^{n-2}})^\sigma \\
Ly_1 & Ly_2 & \cdots & Ly_n
\end{pmatrix}
- q_1 W^\sigma,
$$

where Ly *is given by the left-hand side of equation* (5.89).

Proof. We apply Corollary 5.107. We have $y_k^{\Delta^n} = Ly_k - \sum_{i=1}^n q_i(y_k^{\Delta^{n-i}})^\sigma$ for all $1 \le k \le n$. Replacing the $y_k^{\Delta^n}$ in the last row of the matrix on the right-hand side of (5.91), and adding q_i times the ith row to the last row, for all $1 \le i \le n-1$, we arrive at

$$
W^\Delta = \det
\begin{pmatrix}
y_1^\sigma & y_2^\sigma & \cdots & y_n^\sigma \\
(y_1^\Delta)^\sigma & (y_2^\Delta)^\sigma & \cdots & (y_n^\Delta)^\sigma \\
\vdots & \vdots & & \vdots \\
(y_1^{\Delta^{n-2}})^\sigma & (y_2^{\Delta^{n-2}})^\sigma & \cdots & (y_n^{\Delta^{n-2}})^\sigma \\
Ly_1 - q_1(y_1^{\Delta^{n-1}})^\sigma & Ly_2 - q_1(y_2^{\Delta^{n-1}})^\sigma & \cdots & Ly_n - q_1(y_n^{\Delta^{n-1}})^\sigma
\end{pmatrix}
$$

$$
= \det
\begin{pmatrix}
y_1^\sigma & y_2^\sigma & \cdots & y_n^\sigma \\
(y_1^\Delta)^\sigma & (y_2^\Delta)^\sigma & \cdots & (y_n^\Delta)^\sigma \\
\vdots & \vdots & & \vdots \\
(y_1^{\Delta^{n-2}})^\sigma & (y_2^{\Delta^{n-2}})^\sigma & \cdots & (y_n^{\Delta^{n-2}})^\sigma \\
Ly_1 & Ly_2 & \cdots & Ly_n
\end{pmatrix}
$$

$$-q_1 \det \begin{pmatrix} y_1^\sigma & y_2^\sigma & \cdots & y_n^\sigma \\ (y_1^\Delta)^\sigma & (y_2^\Delta)^\sigma & \cdots & (y_n^\Delta)^\sigma \\ \vdots & \vdots & & \vdots \\ (y_1^{\Delta^{n-2}})^\sigma & (y_2^{\Delta^{n-2}})^\sigma & \cdots & (y_n^{\Delta^{n-2}})^\sigma \\ (y_1^{\Delta^{n-1}})^\sigma & (y_2^{\Delta^{n-1}})^\sigma & \cdots & (y_n^{\Delta^{n-1}})^\sigma \end{pmatrix}$$

$$= \det \begin{pmatrix} y_1^\sigma & y_2^\sigma & \cdots & y_n^\sigma \\ (y_1^\Delta)^\sigma & (y_2^\Delta)^\sigma & \cdots & (y_n^\Delta)^\sigma \\ \vdots & \vdots & & \vdots \\ (y_1^{\Delta^{n-2}})^\sigma & (y_2^{\Delta^{n-2}})^\sigma & \cdots & (y_n^{\Delta^{n-2}})^\sigma \\ Ly_1 & Ly_2 & \cdots & Ly_n \end{pmatrix} - q_1 W^\sigma$$

so that our claimed formula follows. $\square$

Now we show that the Wronski determinant of n solutions of (5.89) satisfies a first order linear dynamic equation.

Theorem 5.109 (Abel's Formula for (5.89)). *Suppose that* (5.89) *is regressive, i.e.,* $q_k \in C_{rd}$ *for all* $1 \le k \le n$ *and* $q_1 \in \mathcal{R}$. *If* $y_1, \dots, y_n$ *are* n *solutions of* (5.89) *and if* $W = W(y_1, \dots, y_n)$, *then*

$$W^\Delta = -q_1 W^\sigma = (\ominus q_1)W.$$

In particular

$$W(t) = W(t_0)e_{\ominus q_1}(t, t_0) \quad \text{for all} \quad t \in \mathbb{T}^{\kappa^{n-1}}.$$

Proof. This follows from Theorem 5.108 since $Ly_k = 0$ for all $1 \le k \le n$ and hence $W^\Delta = -q_1 W^\sigma$. Also, since (5.89) is regressive,

$$\begin{aligned} (1 + \mu q_1)W^\Delta &= -q_1 W^\sigma + \mu q_1 W^\Delta \\ &= q_1(\mu W^\Delta - W^\sigma) \\ &= -q_1 W \end{aligned}$$

can be solved for W^Δ (because of (5.90)). Solving for W we get by Theorem 2.71 that $W(t) = W(t_0)e_{\ominus q_1}(t, t_0)$ for $t \in \mathbb{T}^{\kappa^{n-1}}$. $\square$

Corollary 5.110 (Abel's Formula for (5.85)). *Suppose* (5.85) *is regressive, i.e.,* $p_k \in C_{rd}$ *for all* $1 \le k \le n$ *and* $1 + \sum_{i=1}^n (-\mu)^i p_i$ *never vanishes. If* $y_1, \dots, y_n$ *are* n *solutions of* (5.85), *and if* $W = W(y_1, \dots, y_n)$, *then*

$$W^\Delta = -\left\{ \sum_{i=1}^n (-\mu)^{i-1} p_i \right\} W.$$

In particular

$$W(t) = W(t_0)e_{-\sum_{i=1}^n(-\mu)^{i-1}p_i}(t,t_0) \quad \text{for all} \quad t \in \mathbb{T}^{\kappa^{n-1}}.$$

Proof. We apply Theorem 5.109 and Theorem 5.99. First, by Theorem 5.99, the y_k are also solutions of (5.89) for all $1 \le k \le n$, where the q_k, $1 \le k \le n$, are given in Theorem 5.99. In particular,

$$q_1 = \frac{\sum_{i=1}^n(-\mu)^{i-1}p_i}{1+\sum_{i=1}^n(-\mu)^i p_i}.$$

By Theorem 5.109, $W^\Delta = -\frac{q_1}{1+\mu q_1}W$. Hence $1+\mu q_1 = \frac{1}{1+\sum_{i=1}^n(-\mu)^i p_i}$ and therefore $\frac{q_1}{1+\mu q_1} = \sum_{i=1}^n(-\mu)^{i-1}p_i$. Solving this equation for W we get by Theorem 2.62 that $W(t) = W(t_0)e_{-\sum_{i=1}^n(-\mu)^{i-1}p_i}(t,t_0)$ for $t \in \mathbb{T}^{\kappa^{n-1}}$. $\square$

The following result follows from Abel's formula and the fact that an exponential function is never zero.

Corollary 5.111. *A set $\{y_1,\dots,y_n\}$ is a fundamental system of (5.85) (or of (5.89)) if and only if $y_1,\dots,y_n$ are solutions of (5.85) (or of (5.89)) satisfying $W(y_1,\dots,y_n)(t) \ne 0$ for all $t \in \mathbb{T}^{\kappa^{n-1}}$.*

Exercise 5.112. Rewrite Theorem 5.109 and Corollary 5.110 for the cases $\mathbb{T} = \mathbb{R}$ and $\mathbb{T} = \mathbb{Z}$ and solve the corresponding differential and difference equations to find W.

Exercise 5.113. Consider

$$\frac{h/2}{\cosh h - 1}y^{\Delta^2} = y^\Delta + \frac{1}{h}y, \quad \mathbb{T} = h\mathbb{Z}, \quad h > 0.$$

Show that $\{\cosh t, \sinh t\}$ is a fundamental system.

Now let $f : \mathbb{T} \to \mathbb{R}$ be rd-continuous and consider the *nonhomogeneous* equation

$$(5.92) \qquad y^{\Delta^n} + \sum_{i=1}^n p_i(t)y^{\Delta^{n-i}} = f(t),$$

which is equivalent to a linear system $z^\Delta = A(t)z + b(t)$, where A is given in (5.86) and $b = (0,\dots,0,f)^T$.

Definition 5.114. We define the Cauchy function $y : \mathbb{T} \times \mathbb{T}^{\kappa^n} \to \mathbb{R}$ for the linear dynamic equation (5.85) to be for each fixed $s \in \mathbb{T}^{\kappa^n}$ the solution of the IVP

$$Ly = 0, \quad y^{\Delta^i}(\sigma(s),s) = 0, \quad 0 \le i \le n-2, \quad y^{\Delta^{n-1}}(\sigma(s),s) = 1.$$

Example 5.115. Note that

$$y(t,s) := h_{n-1}(t,\sigma(s))$$

is the Cauchy function for $y^{\Delta^n} = 0$.

Theorem 5.116. *If $\{y_1,\dots,y_n\}$ is a fundamental system of the regressive equation (5.85), then the Cauchy funtion is given by*

$$y(t,s) = \frac{W(\sigma(s),t)}{W(\sigma(s))},$$

where $W(s) := W(y_1, \ldots, y_n)(s)$ *and*

$$W(s,t) = \det \begin{pmatrix} y_1(s) & \cdots & y_n(s) \\ y_1^{\Delta}(s) & \cdots & y_n^{\Delta}(s) \\ \vdots & & \vdots \\ y_1^{\Delta^{n-2}}(s) & \cdots & y_n^{\Delta^{n-2}}(s) \\ y_1(t) & \cdots & y_n(t) \end{pmatrix},$$

i.e., the last row of $V(y_1, \ldots, y_n)(s)$ *is replaced by* $(y_1(t), \ldots, y_n(t))$.

Exercise 5.117. Prove Theorem 5.116.

Exercise 5.118. Use Theorem 5.116 to find the Cauchy functions for the following dynamic equations:

(i) $y^{\Delta\Delta\Delta} = 0$;

(ii) $y^{\Delta\Delta\Delta} - y^{\Delta} = 0$.

Theorem 5.119 (Variation of Constants). *Suppose* $\{y_1, \ldots, y_n\}$ *is a fundamental system of the regressive equation* (5.85). *Let* $f \in C_{rd}$. *Then the solution of the IVP*

$$Ly = f(t), \quad y^{\Delta^i}(t_0) = 0, \quad 0 \le i \le n-1$$

is given by

$$y(t) = \int_{t_0}^t y(t,s) f(s) \Delta s,$$

where $y(t,s)$ *is the Cauchy function for* (5.85).

Proof. With y defined as above we have

$$y^{\Delta^i}(t) = \int_{t_0}^t y^{\Delta^i}(t,s) f(s) \Delta s + y^{\Delta^i}(\sigma(t),t) f(t) = \int_{t_0}^t y^{\Delta^i}(t,s) f(s) \Delta s$$

for $0 \le i \le n-1$ and

$$y^{\Delta^n}(t) = \int_{t_0}^t y^{\Delta^n}(t,s) f(s) \Delta s + y^{\Delta^{n-1}}(\sigma(t),t) f(t) = \int_{t_0}^t y^{\Delta^n}(t,s) f(s) \Delta s + f(t).$$

It follows from these equations that

$$y^{\Delta^i}(t_0) = 0, \quad 0 \le i \le n-1$$

and

$$Ly(t) = \int_{t_0}^t Ly(t,s) f(s) \Delta s + f(t) = f(t),$$

and the proof is complete. $\square$

Exercise 5.120. Use the variation of constants formula in 5.119 to solve the following initial value problems:

(i) $y^{\Delta\Delta\Delta} = 1$, $y(t_0) = y^{\Delta}(t_0) = y^{\Delta\Delta}(t_0) = 0$;

(ii) $y^{\Delta\Delta\Delta} - y^{\Delta} = 2$, $y(t_0) = y^{\Delta}(t_0) = y^{\Delta\Delta}(t_0) = 0$.

Definition 5.121. Let $y : \mathbb{T} \to \mathbb{R}$ be $(k-1)$ times differentiable, $k \in \mathbb{N}$. We say that y has a *generalized zero* (GZ) of order greater than or equal to k at $t \in \mathbb{T}^{\kappa^{k-1}}$ provided

$$(5.93) \qquad y^{\Delta^i}(t) = 0 \quad \text{for all} \quad 0 \le i \le k-1$$

or

$$(5.94) \qquad y^{\Delta^i}(t) = 0 \text{ for all } 0 \le i \le k-2 \quad \text{and} \quad y^{\Delta^{k-1}}(\rho(t))y^{\Delta^{k-1}}(t) < 0$$

holds.

Note that in the case of (5.94), t must necessarily be left-scattered, because otherwise $\rho(t) = t$ and $0 > y^{\Delta^{k-1}}(\rho(t))y^{\Delta^{k-1}}(t) = (y^{\Delta^{k-1}}(t))^2 \ge 0$, a contradiction. Also, if t is left-scattered, we have $\sigma(\rho(t)) = t$ (see Exercise 1.4).

Lemma 5.122. *Condition (5.94) holds if and only if*

$$(5.95) \qquad y^{\Delta^i}(t) = 0 \text{ for all } 0 \le i \le k-2 \quad \text{and} \quad (-1)^{k-1}y(\rho(t))y^{\Delta^{k-1}}(t) < 0.$$

Proof. Suppose (5.94) holds. Then $\sigma(\rho(t)) = t$, $\mu(\rho(t)) > 0$, and

$$y^{\Delta^{m-1}}(\rho(t)) = \frac{y^{\Delta^{m-2}}(t) - y^{\Delta^{m-2}}(\rho(t))}{\mu(\rho(t))} = -\frac{y^{\Delta^{m-2}}(\rho(t))}{\mu(\rho(t))} \quad \text{for all} \quad 2 \le m \le k$$

so that

$$y^{\Delta^{k-1}}(\rho(t)) = \frac{(-1)^{k-1}y(\rho(t))}{(\mu(\rho(t)))^{k-1}}.$$

Therefore (5.95) holds. The proof of the converse is left to the reader in Exercise 5.123. □

Exercise 5.123. Prove that (5.95) implies (5.94).

We state the following result without its technical but easy proof.

Lemma 5.124. *Let $j \in \mathbb{N}_0$ and $t \in \mathbb{T}^{\kappa^j}$. Then*

$$y^{\Delta^i}(t) = 0, \quad 0 \le i \le j,$$

if and only if

$$y^{\Delta^i}(\sigma^l(t)) = 0, \quad 0 \le i \le j-l, \ 0 \le l \le j.$$

In this case

$$y^{\Delta^{j+1-l}}(\sigma^l(t)) = \prod_{s=0}^{l-1} \mu(\sigma^s(t))y^{\Delta^{j+1}}(t).$$

Exercise 5.125. Prove Lemma 5.124.

Remark 5.126. If y has a GZ of order greater than or equal to k at t we shall say that y has at least k GZs, counting multiplicities. Note that if y has a GZ of order greater than or equal to k at t, then, as a corollary to Lemma 5.124, y has a GZ of order greater than or equal to $k-1$ at $\sigma(t)$. In order to avoid redundancies as we count GZs, if y has a GZ of order greater than or equal to k_1 at t_1 and y has a GZ of order greater than or equal to k_2 at t_2 and $\sigma^{k_1-1}(t_1) < t_2$, we shall say that y has at least $k_1 + k_2$ GZs, counting multiplicities.

The following is a dynamic version of Rolle's theorem.

Theorem 5.127 (Rolle's Theorem). *If y has at least $k \in \mathbb{N}$ GZs on $\mathbb{T} = [a,b]$, counting multiplicities, then y^Δ has at least $k-1$ GZs on $\mathbb{T}$, counting multiplicities.*

Proof. We shall show the following two statements:

(i) If y has a GZ of order greater than or equal to $m \in \mathbb{N}$ at t, then y^Δ has a GZ of order greater than or equal to $m-1$ at t.

(ii) If y has a GZ of order greater than or equal to $m \in \mathbb{N}$ at t and if y has a GZ of order greater than or equal to 1 at s with $\sigma^{m-1}(t) < s$, then y^Δ has at least m GZs in $[t, s)$, counting multiplicities.

Once statements (i) and (ii) are established, the claim of the theorem follows by partitioning $\mathbb{T} = [a, b]$ appropriately.

Statement (i) clearly holds and hence it remains to prove (ii). In fact, taking into account (i), to prove (ii) it suffices to show the following two statements:

(iii) If $y(r) = 0$ and y^Δ has no GZ in $[r, s)$, where $r < \rho(s)$, then y has no GZ at s.

(iv) If $y(\rho(r))y(r) < 0$ and y^Δ has no GZ in $[r, s)$, where $r < \rho(s)$, then y has no GZ at s.

First, if the assumptions of (iii) hold, then $y^\Delta(\tau) > 0$ for all $\tau \in [r, s)$ or $y^\Delta(\tau) < 0$ for all $\tau \in [r, s)$ so that

$$y(\rho(s))y(s) = \left\{ \int_r^{\rho(s)} y^\Delta(\tau)\Delta\tau \right\} \left\{ \int_r^s y^\Delta(\tau)\Delta\tau \right\} > 0.$$

Second, if the assumptions of (iv) hold, then $\rho(r) < r$ and

$$y(\rho(r))y^\Delta(\rho(r)) = y(\rho(r))(y(r) - y(\rho(r)))/\mu(\rho(r)) < 0$$

and hence, since $y^\Delta(\tau)$ is of constant sign on $[\rho(r), s)$, we have

$$y(\rho(r))y^\Delta(\tau) < 0 \quad \text{for all} \quad \tau \in [\rho(r), s).$$

Then

$$y(\rho(r))y(t) = y(\rho(r)) \left\{ y(r) + \int_r^t y^\Delta(\tau)\Delta\tau \right\} < 0 \quad \text{for} \quad t \in \{\rho(s), s\}$$

so that $y^2(\rho(r))y(\rho(s))y(s) > 0$ and hence $y(\rho(s))y(s) > 0$. $\square$

5.6. Notes and References

Linear dynamic systems are discussed in this chapter. We prove the existence and uniqueness of solutions of initial value problems involving a first order linear dynamic system via the general existence and uniqueness theorem given in Chapter 8. The matrix exponential is introduced as the unique solution of a certain initial value problem, and many properties analogous to the scalar case given in Chapter 2 are also given for the matrix case. Some of the results given in the first section can also be found in Hilger [**159, 160**] and in Keller [**190**]. The second section treats the case of a constant coefficient matrix, and after some general theory involving such problems we present a Putzer algorithm, which is essentially from Calvin Ahlbrandt and Jerry Ridenhour [**31**]. Other studies on linear dynamic systems, which are not discussed in this book, include Bernd Aulbach and Christian Pötzsche [**51**] and

Stefan Siegmund [241] (see [239] for the corresponding continuous "Sacker–Sell theory").

The section on self-adjoint second order linear dynamic matrix equations is rather large. Sources for the corresponding differential equations theory are [54, 114, 115, 116, 192, 193, 238], while corresponding difference equations theory can be found in [1, 22, 26, 27, 28, 30, 37, 66, 72, 222, 225, 226, 227]. Related time scales results are contained in [8, 134, 135]. Most of the results presented in this section are matrix analogues of the corresponding results from Chapter 4.

Section 5.4 follows closely the paper by Martin Bohner and Donald Lutz [88]. We included time scales versions of the classical Levinson perturbation lemma and Hartman–Wintner's theorem, as well as an application to q-difference equations in Example 5.87. For the original differential equations theory, see Levinson [214], Wasow [250], and Harris and Lutz [149, 150, 151]. Eastham [124] features a documentation of the theory on asymptotic integration. Corresponding results for difference equations are due to Benzaid and Lutz [57] (see also [240]). For a summary of the discrete results, see Saber Elaydi [125].

Linear dynamic equations of higher order are discussed in the last section, and of course, as is known in both the differential and difference equations case, a higher order equation can always be rewritten as a system using Lemma 5.88. We study essentially two different forms of nth order linear dynamic equations, namely (5.85), which is familiar from differential equations and which might be thought of at the first view as the "right" equation, and (5.89), which has the σ operators in each term but the leading one, and which turns out once again as the more "natural" form of a dynamic equation (compare the two Abel theorems in Theorem 5.110 and Theorem 5.109). Generalized zeros of solutions of higher order dynamic equations on time scales are first defined in Martin Bohner and Paul Eloe [85] (following essentially Hartman's [153] original definition for the discrete case; see also Coppel [105] for the continuous case), where also interpolating families of functions (for the discrete case see Paul Eloe and Johnny Henderson [127]) are considered. The paper Lynn Erbe, Ronald Mathsen, and Allan Peterson [130] contains related results and higher order versions of the Polya and Trench factorizations (see also [148] for the discrete case).

CHAPTER 6

Dynamic Inequalities

6.1. Gronwall's Inequality

We start this section with a comparison theorem. Throughout we let $t_0 \in \mathbb{T}$.

Theorem 6.1. *Let* $y, f \in C_{rd}$ *and* $p \in \mathcal{R}^+$. *Then*

$$y^\Delta(t) \leq p(t)y(t) + f(t) \quad \text{for all} \quad t \in \mathbb{T}$$

implies

$$y(t) \leq y(t_0)e_p(t, t_0) + \int_{t_0}^t e_p(t, \sigma(\tau))f(\tau)\Delta\tau \quad \text{for all} \quad t \in \mathbb{T}.$$

Proof. We use the product rule and Theorem 2.36 (ii) to calculate

$$
\begin{aligned}
{[ye_{\ominus p}(\cdot, t_0)]}^\Delta(t) &= y^\Delta(t)e_{\ominus p}(\sigma(t), t_0) + y(t)(\ominus p)(t)e_{\ominus p}(t, t_0) \\
&= y^\Delta(t)e_{\ominus p}(\sigma(t), t_0) + y(t)\frac{(\ominus p)(t)}{1 + \mu(t)(\ominus p)(t)}e_{\ominus p}(\sigma(t), t_0) \\
&= \left[y^\Delta(t) - (\ominus(\ominus p))(t)y(t)\right]e_{\ominus p}(\sigma(t), t_0) \\
&= [y^\Delta(t) - p(t)y(t)]e_{\ominus p}(\sigma(t), t_0).
\end{aligned}
$$

Since $p \in \mathcal{R}^+$, we have $\ominus p \in \mathcal{R}^+$ by Lemma 2.47. This implies $e_{\ominus p} > 0$ by Theorem 2.48 (i). Now

$$
\begin{aligned}
y(t)e_{\ominus p}(t, t_0) - y(t_0) &= \int_{t_0}^t [y^\Delta(\tau) - p(\tau)y(\tau)]e_{\ominus p}(\sigma(\tau), t_0)\Delta\tau \\
&\leq \int_{t_0}^t f(\tau)e_{\ominus p}(\sigma(\tau), t_0)\Delta\tau \\
&= \int_{t_0}^t e_p(t_0, \sigma(\tau))f(\tau)\Delta\tau,
\end{aligned}
$$

and hence the assertion follows by applying Theorem 2.36. $\qquad\square$

Theorem 6.1 has two consequences, which are given next.

Theorem 6.2 (Bernoulli's Inequality). *Let* $\alpha \in \mathbb{R}$ *with* $\alpha \in \mathcal{R}^+$. *Then*

$$e_\alpha(t, s) \geq 1 + \alpha(t - s) \quad \text{for all} \quad t \geq s.$$

Proof. Since $\alpha \in \mathcal{R}^+$, we have $e_\alpha(t, s) > 0$ for all $t, s \in \mathbb{T}$. Suppose $t, s \in \mathbb{T}$ with $t \geq s$. Let $y(t) = \alpha(t - s)$. Then

$$\alpha y(t) + \alpha = \alpha^2(t - s) + \alpha \geq \alpha = y^\Delta(t).$$

Since $y(s) = 0$, we have by Theorem 6.1 (with $p(t) = f(t) \equiv \alpha$)

$$y(t) \leq \int_s^t e_\alpha(t, \sigma(\tau))\alpha\Delta\tau = e_\alpha(t, s) - 1,$$

where we used Theorem 2.39. Hence $e_\alpha(t, s) \geq 1 + y(t) = 1 + \alpha(t - s)$ follows. □

Example 6.3. If $\mathbb{T} = [0, \infty)$, then Theorem 6.2 says that for any $\alpha \in \mathbb{R}$

$$e^{\alpha t} \geq 1 + \alpha t \quad \text{for all} \quad t \geq 0$$

holds, and for $\mathbb{T} = \mathbb{N}$ we have for any $\alpha > -1$

$$(1 + \alpha)^n \geq 1 + \alpha n \quad \text{for all} \quad n \in \mathbb{N}.$$

This last inequality is the well-known Bernoulli inequality from calculus.

Theorem 6.4 (Gronwall's Inequality). *Let* $y, f \in C_{\mathrm{rd}}$ *and* $p \in \mathcal{R}^+$, $p \geq 0$. *Then*

$$y(t) \leq f(t) + \int_{t_0}^t y(\tau)p(\tau)\Delta\tau \quad \text{for all} \quad t \in \mathbb{T}$$

implies

$$y(t) \leq f(t) + \int_{t_0}^t e_p(t, \sigma(\tau))f(\tau)p(\tau)\Delta\tau \quad \text{for all} \quad t \in \mathbb{T}.$$

Proof. Define

$$z(t) = \int_{t_0}^t y(\tau)p(\tau)\Delta\tau.$$

Then $z(t_0) = 0$ and

$$z^\Delta(t) = y(t)p(t) \leq [f(t) + z(t)]p(t) = p(t)z(t) + p(t)f(t).$$

By Theorem 6.1,

$$z(t) \leq \int_{t_0}^t e_p(t, \sigma(\tau))f(\tau)p(\tau)\Delta\tau,$$

and hence the claim follows because of $y(t) \leq f(t) + z(t)$. □

If we take $\mathbb{T} = h\mathbb{N}_0$ and $t_0 = 0$ in Gronwall's inequality, we get the following example.

Example 6.5. Let $\mathbb{T} = h\mathbb{Z} \cap [0, \infty)$. If y and f are functions defined on $\mathbb{T}$ and $\gamma > 0$ is a constant such that

$$y(t) \leq f(t) + \gamma \sum_{\tau=0}^{\frac{t}{h}-1} y(\tau h) \quad \text{for all} \quad t \in \mathbb{T},$$

then

$$y(t) \leq f(t) + \gamma \sum_{\tau=0}^{\frac{t}{h}-1} f(\tau h)(1 + \gamma h)^{\frac{t-h(\tau+1)}{h}} \quad \text{for all} \quad t \in \mathbb{T}.$$

Corollary 6.6. *Let* $y \in C_{\mathrm{rd}}$ *and* $p \in \mathcal{R}^+$ *with* $p \geq 0$. *Then*

$$y(t) \leq \int_{t_0}^t y(\tau)p(\tau)\Delta\tau \quad \text{for all} \quad t \in \mathbb{T}$$

implies

$$y(t) \leq 0 \quad \text{for all} \quad t \in \mathbb{T}.$$

Proof. This is Theorem 6.4 with $f(t) \equiv 0$.   □

Corollary 6.7. *Let* $y \in C_{rd}$, $p \in \mathcal{R}^+$, $p \geq 0$, *and* $\alpha \in \mathbb{R}$. *Then*

$$y(t) \leq \alpha + \int_{t_0}^{t} y(\tau)p(\tau)\Delta\tau \quad \textit{for all} \quad t \in \mathbb{T}$$

implies

$$y(t) \leq \alpha e_p(t, t_0) \quad \textit{for all} \quad t \in \mathbb{T}.$$

Proof. In Theorem 6.4, let $f(t) \equiv \alpha$. Then by Theorem 6.4,

$$\begin{aligned}
y(t) &\leq \alpha + \int_{t_0}^{t} e_p(t, \sigma(\tau))\alpha p(\tau)\Delta\tau \\
&= \alpha \left[1 + \int_{t_0}^{t} p(\tau)e_p(t, \sigma(\tau))\Delta\tau \right] \\
&= \alpha \left[1 + e_p(t, t_0) - e_p(t, t) \right] \\
&= \alpha e_p(t, t_0),
\end{aligned}$$

where we used Theorem 2.39.   □

Corollary 6.8. *Let* $y \in C_{rd}$ *and* $\alpha, \beta, \gamma \in \mathbb{R}$ *with* $\gamma > 0$. *Then*

$$y(t) \leq \alpha + \beta(t - t_0) + \gamma \int_{t_0}^{t} y(\tau)\Delta\tau \quad \textit{for all} \quad t \in \mathbb{T}$$

implies

$$y(t) \leq \left(\alpha + \frac{\beta}{\gamma} \right) e_\gamma(t, t_0) - \frac{\beta}{\gamma} \quad \textit{for all} \quad t \in \mathbb{T}.$$

Proof. In Theorem 6.4, let $f(t) = \alpha + \beta(t - t_0)$ and $p(t) \equiv \gamma$. Note that for $w(\tau) = e_\gamma(t, \tau)$ we have $w^\Delta(\tau) = -\gamma e_\gamma(t, \sigma(\tau))$ by Theorem 2.39. By Theorem 6.4,

$$\begin{aligned}
y(t) &\leq f(t) + \int_{t_0}^{t} e_\gamma(t, \sigma(\tau))\gamma f(\tau)\Delta\tau \\
&= f(t)w(t) - \int_{t_0}^{t} w^\Delta(\tau)f(\tau)\Delta\tau \\
&= f(t_0)w(t_0) + \int_{t_0}^{t} w(\sigma(\tau))f^\Delta(\tau)\Delta\tau \\
&= \alpha e_\gamma(t, t_0) + \int_{t_0}^{t} e_\gamma(t, \sigma(\tau))\beta\Delta\tau \\
&= \alpha e_\gamma(t, t_0) + \frac{\beta}{\gamma} \int_{t_0}^{t} \gamma e_\gamma(t, \sigma(\tau))\Delta\tau \\
&= \alpha e_\gamma(t, t_0) + \frac{\beta}{\gamma} \left(e_\gamma(t, t_0) - 1 \right),
\end{aligned}$$

where we used integration by parts (Theorem 1.77 (vi)) and Theorem 2.39.   □

To conclude this section we now present nonlinear versions of the comparison Theorem 6.1 and of Gronwall's inequality (Theorem 6.4). A nonlinear version of Gronwall's inequality sometimes is called a *Bihari* type inequality. Further results on Bihari inequalities are contained in [187].

Theorem 6.9. *Let* $g : \mathbb{T} \times \mathbb{R} \to \mathbb{R}$ *be a function with*
$$g(t, x_1) \leq g(t, x_2) \text{ for all } t \in \mathbb{T} \quad \text{whenever} \quad x_1 \leq x_2.$$
Let $v, w : \mathbb{T} \to \mathbb{R}$ *be differentiable with*
$$v^\Delta(t) \leq g(t, v(t)), \ w^\Delta(t) \geq g(t, w(t)) \quad \text{for all} \quad t \in \mathbb{T}^\kappa \setminus \{t_0\},$$
where $t_0 \in \mathbb{T}$. *Then*
$$v(t_0) < w(t_0)$$
implies
$$v(t) < w(t) \quad \text{for all} \quad t \geq t_0.$$

Proof. We apply the induction principle to show that
$$S(t): \quad v(t) < w(t)$$
holds for all $t \in [t_0, \infty)$.

I. Clearly, $S(t_0)$ holds due to our assumption $v(t_0) < w(t_0)$.
II. Let $t \geq t_0$ be right-scattered and suppose that $S(t)$ is true. Then
$$v^\Delta(t) \leq g(t, v(t)) \leq g(t, w(t)) \leq w^\Delta(t)$$
since $v(t) < w(t)$ implies $g(t, v(t)) \leq g(t, w(t))$. Therefore
$$\begin{aligned} v(\sigma(t)) &= v(t) + \mu(t)v^\Delta(t) \\ &< w(t) + \mu(t)w^\Delta(t) \\ &= w(\sigma(t)) \end{aligned}$$
so that statement $S(\sigma(t))$ follows.
III. Let $t \geq t_0$ be right-dense and suppose that $S(t)$ is true. Since $v(t) < w(t)$, we have by continuity that there is a neighborhood U of t such that $v(r) < w(r)$ for all $r \in U$ and hence $S(r)$ holds for all $r \in U \cap (t, \infty)$.
IV. Let $t \geq t_0$ be left-dense and suppose that $S(r)$ is true for all $r \in [t_0, t)$. Then
$$v(r) < w(r) \quad \text{for all} \quad t_0 \leq r < t$$
and by continuity $v(t) \leq w(t)$. But as above
$$(w - v)^\Delta \geq 0 \quad \text{on} \quad [t_0, t].$$
It follows that $w - v$ is nondecreasing on $[t_0, t]$. Hence
$$(w - v)(t) \geq (w - v)(t_0) > 0$$
so that $S(t)$ holds.

Hence by the induction principle (Theorem 1.7) the proof is complete. □

Exercise 6.10. Discuss the possibility of replacing the condition $v(t_0) < w(t_0)$ in Theorem 6.9 by $v(t_0) \leq w(t_0)$ and concluding that $v(t) \leq w(t)$ for $t \geq t_0$.

Theorem 6.11 (Bihari's Inequality). *Suppose* $g : \mathbb{R} \to \mathbb{R}$ *is nondecreasing and* $y : \mathbb{T} \to \mathbb{R}$ *is such that* $g \circ y$ *is rd-continuous. Let* $p \geq 0$ *be rd-continuous and* $f : \mathbb{T} \to \mathbb{R}$ *differentiable. Then*
$$y(t) \leq f(t) + \int_{t_0}^t p(\tau)g(y(\tau))\Delta\tau \quad \text{for all} \quad t \geq t_0$$

implies

$$y(t) < w(t) \quad \text{for all} \quad t \geq t_0,$$

where w solves the initial value problem

$$w^\Delta = f^\Delta(t) + p(t)g(w), \quad w(t_0) = w_0 > f(t_0).$$

Proof. Let

$$v(t) := f(t) + \int_{t_0}^t p(\tau)g(y(\tau))\Delta\tau \quad \text{for} \quad t \geq t_0.$$

Then $v^\Delta(t) = f^\Delta(t) + p(t)g(y(t))$ and $y(t) \leq v(t)$ so that

$$v^\Delta(t) \leq f^\Delta(t) + p(t)g(v(t)).$$

Since $v(t_0) = f(t_0) < w_0 = w(t_0)$, the comparison result from Theorem 6.9 yields

$$v(t) < w(t) \quad \text{for all} \quad t \geq t_0.$$

Hence, since $y(t) \leq v(t)$, we arrive at the desired conclusion. $\qquad\square$

Corollary 6.12 ([187, Theorem 4.2]). *Suppose $g : \mathbb{R} \to \mathbb{R}$ is nondecreasing and $y : \mathbb{T} \to \mathbb{R}$ is such that $g \circ y$ is rd-continuous. Let $p \geq 0$ be rd-continuous and $\alpha \in \mathbb{R}$. Then*

$$y(t) \leq \alpha + \int_{t_0}^t p(\tau)g(y(\tau))\Delta\tau \quad \text{for all} \quad t \geq t_0$$

implies

$$y(t) < w(t) \quad \text{for all} \quad t \geq t_0,$$

where w solves the initial value problem

$$w^\Delta = p(t)g(w), \quad w(t_0) = w_0 > \alpha.$$

Proof. This is Theorem 6.11 with $f(t) \equiv \alpha$. $\qquad\square$

6.2. Hölder's and Minkowski's Inequalities

The following version of *Hölder's* inequality on time scales appears in [88, Lemma 2.2 (iv)], and its proof is similar to the proof of the classical Hölder inequality as given, e.g., in [147, Theorem 188].

Theorem 6.13 (Hölder's Inequality). *Let $a, b \in \mathbb{T}$. For rd-continuous functions $f, g : [a, b] \to \mathbb{R}$ we have*

$$\int_a^b |f(t)g(t)|\Delta t \leq \left\{\int_a^b |f(t)|^p \Delta t\right\}^{\frac{1}{p}} \left\{\int_a^b |g(t)|^q \Delta t\right\}^{\frac{1}{q}},$$

where $p > 1$ and $q = p/(p-1)$.

Proof. For nonnegative real numbers α and β, the basic inequality (see Exercise 6.14)

$$(6.1) \qquad\qquad \alpha^{1/p}\beta^{1/q} \leq \frac{\alpha}{p} + \frac{\beta}{q}$$

holds. Now suppose, without loss of generality, that

$$\left\{\int_a^b |f(t)|^p \Delta t\right\}\left\{\int_a^b |g(t)|^q \Delta t\right\} \neq 0.$$

Apply (6.1) to

$$\alpha(t) = \frac{|f(t)|^p}{\int_a^b |f(\tau)|^p \Delta\tau} \quad \text{and} \quad \beta(t) = \frac{|g(t)|^q}{\int_a^b |g(\tau)|^q \Delta\tau}$$

and integrate the obtained inequality between a and b (this is possible since all occurring functions are rd-continuous) to get

$$\int_a^b \frac{|f(t)|}{\left\{\int_a^b |f(\tau)|^p \Delta\tau\right\}^{1/p}} \frac{|g(t)|}{\left\{\int_a^b |g(\tau)|^q \Delta\tau\right\}^{1/q}} \Delta t = \int_a^b \alpha^{1/p}(t)\beta^{1/q}(t)\Delta t$$

$$\leq \int_a^b \left\{\frac{\alpha(t)}{p} + \frac{\beta(t)}{q}\right\} \Delta t$$

$$= \int_a^b \left\{\frac{1}{p}\frac{|f(t)|^p}{\int_a^b |f(\tau)|^p \Delta\tau} + \frac{1}{q}\frac{|g(t)|^q}{\int_a^b |g(\tau)|^q \Delta\tau}\right\} \Delta t$$

$$= \frac{1}{p}\int_a^b \left\{\frac{|f(t)|^p}{\int_a^b |f(\tau)|^p \Delta\tau}\right\} \Delta t + \frac{1}{q}\int_a^b \left\{\frac{|g(t)|^q}{\int_a^b |g(\tau)|^q \Delta\tau}\right\} \Delta t$$

$$= \frac{1}{p} + \frac{1}{q}$$

$$= 1.$$

This directly gives Hölder's inequality.   □

Exercise 6.14. Prove inequality (6.1).

The special case $p = q = 2$ yields the *Cauchy–Schwarz* inequality.

Theorem 6.15 (Cauchy–Schwarz Inequality). *Let* $a, b \in \mathbb{T}$. *For rd-continuous* $f, g : [a, b] \to \mathbb{R}$ *we have*

$$\int_a^b |f(t)g(t)|\Delta t \leq \sqrt{\left\{\int_a^b |f(t)|^2\Delta t\right\}\left\{\int_a^b |g(t)|^2\Delta t\right\}}.$$

Next, we can use Hölder's inequality to deduce *Minkowski's* inequality.

Theorem 6.16 (Minkowski's Inequality). *Let* $a, b \in \mathbb{T}$ *and* $p > 1$. *For rd-continuous* $f, g : [a, b] \to \mathbb{R}$ *we have*

$$\left\{\int_a^b |(f + g)(t)|^p\Delta t\right\}^{1/p} \leq \left\{\int_a^b |f(t)|^p\Delta t\right\}^{1/p} + \left\{\int_a^b |g(t)|^p\Delta t\right\}^{1/p}.$$

Proof. We apply Hölder's inequality (Theorem 6.13) with $q = p/(p-1)$ to obtain

$$\int_a^b |(f+g)(t)|^p \Delta t = \int_a^b |(f+g)|^{p-1}|(f+g)(t)| \Delta t$$

$$\leq \int_a^b |f(t)||(f+g)(t)|^{p-1} \Delta t + \int_a^b |g(t)||f+g|^{p-1}(t) \Delta t$$

$$\leq \left\{ \int_a^b |f(t)|^p \Delta t \right\}^{1/p} \left\{ \int_a^b |(f+g)(t)|^{(p-1)q} \Delta t \right\}^{1/q}$$

$$+ \left\{ \int_a^b |g(t)|^p \Delta t \right\}^{1/p} \left\{ \int_a^b |(f+g)(t)|^{(p-1)q} \Delta t \right\}^{1/q}$$

$$= \left[\left\{ \int_a^b |f(t)|^p \Delta t \right\}^{1/p} + \left\{ \int_a^b |g(t)|^p \Delta t \right\}^{1/p} \right] \left[\int_a^b |(f+g)(t)|^p \Delta t \right]^{1/q}.$$

We divide both sides of the above inequality by $\left[\int_a^b |(f+g)(t)|^p \Delta t \right]^{1/q}$ to arrive at Minkowski's inequality. □

6.3. Jensen's Inequality

The proof of the following version of *Jensen's* inequality on time scales follows closely the proof of the classical Jensen inequality (see for example [**142**, Exercise 3.42]). If $\mathbb{T} = \mathbb{R}$, then our version is the same as the classical Jensen inequality. However, if $\mathbb{T} = \mathbb{Z}$, then our version is the same as the well-known arithmetic-mean geometric-mean inequality. The version presented below can be found in [**3**].

Theorem 6.17 (Jensen's Inequality). *Let $a, b \in \mathbb{T}$ and $c, d \in \mathbb{R}$. If $g : [a, b] \to (c, d)$ is rd-continuous and $F : (c, d) \to \mathbb{R}$ is continuous and convex, then*

$$F\left(\frac{\int_a^b g(t) \Delta t}{b-a} \right) \leq \frac{\int_a^b F(g(t)) \Delta t}{b-a}.$$

Proof. Let $x_0 \in (c, d)$. Then (e.g., by [**142**, Exercise 3.42c]) there exists $\beta \in \mathbb{R}$ with

(6.2) $\qquad F(x) - F(x_0) \geq \beta(x - x_0) \qquad$ for all $\quad x \in (c, d)$.

Since $g \in C_{rd}$,

$$x_0 = \frac{\int_a^b g(\tau) \Delta \tau}{b-a}$$

is well defined. By Theorem 1.60 (v), the function $F \circ g$ is also rd-continuous, and hence we may apply (6.2) with $x = g(t)$ and integrate from a to b to obtain

$$
\begin{aligned}
\int_a^b F(g(t))\Delta t - (b-a)F\left(\frac{\int_a^b g(\tau)\Delta\tau}{b-a}\right) &= \int_a^b F(g(t))\Delta t - (b-a)F(x_0) \\
&= \int_a^b [F(g(t)) - F(x_0)]\,\Delta t \\
&\geq \beta \int_a^b [g(t) - x_0]\,\Delta t \\
&= \beta \left[\int_a^b g(t)\Delta t - x_0(b-a)\right] \\
&= 0.
\end{aligned}
$$

This directly yields Jensen's inequality.   □

Exercise 6.18. Prove (6.2).

Example 6.19. Let $\mathbb{T} = \mathbb{R}$. Obviously, $F = -\log$ is convex and continuous on $(0, \infty)$, so we apply Theorem 6.17 with $a = 0$ and $b = 1$ to obtain

$$
\log \int_0^1 g(t)dt \geq \int_0^1 \log g(t)dt
$$

and hence

$$
\int_0^1 g(t)dt \geq \exp\left(\int_0^1 \log g(t)dt\right)
$$

whenever $g : [0, 1] \to (0, \infty)$ is continuous.

Example 6.20. Let $N \in \mathbb{T} = \mathbb{N}$. Again we apply Jensen's inequality (Theorem 6.17) with $a = 1$, $b = N + 1$, and $g : \{1, 2, \ldots, N + 1\} \to (0, \infty)$ to find

$$
\begin{aligned}
\log\left\{\frac{1}{N}\sum_{t=1}^N g(t)\right\} &= \log\left\{\frac{1}{N}\int_1^{N+1} g(t)\Delta t\right\} \\
&\geq \frac{1}{N}\int_1^{N+1} \log g(t)\Delta t \\
&= \frac{1}{N}\sum_{t=1}^N \log g(t) \\
&= \log\left\{\prod_{t=1}^N g(t)\right\}^{1/N}
\end{aligned}
$$

and hence

$$
\frac{1}{N}\sum_{t=1}^N g(t) \geq \left\{\prod_{t=1}^N g(t)\right\}^{1/N}.
$$

This is the well-known *arithmetic-mean geometric-mean* inequality.

Example 6.21. Let $\mathbb{T} = 2^{\mathbb{N}_0}$ and $N \in \mathbb{N}$. We apply Theorem 6.17 with $a = 1$, $b = 2^N$, and $g : \{2^k : 0 \leq k \leq N\} \to (0, \infty)$ to find

$$\log\left\{\frac{\sum_{k=0}^{N-1} 2^k g(2^k)}{2^N - 1}\right\} = \log\left\{\frac{\int_1^{2^N} g(t)\Delta t}{2^N - 1}\right\}$$

$$\geq \frac{\int_1^{2^N} \log g(t)\Delta t}{2^N - 1}$$

$$= \frac{\sum_{k=0}^{N-1} 2^k \log g(2^k)}{2^N - 1}$$

$$= \frac{\sum_{k=0}^{N-1} \log(g(2^k))^{2^k}}{2^N - 1}$$

$$= \frac{\log\left\{\prod_{k=0}^{N-1}(g(2^k))^{2^k}\right\}}{2^N - 1}$$

$$= \log\left\{\prod_{k=0}^{N-1}(g(2^k))^{2^k}\right\}^{1/(2^N-1)}$$

and therefore

(6.3) $$\frac{\sum_{k=0}^{N-1} 2^k g(2^k)}{2^N - 1} \geq \left\{\prod_{k=0}^{N-1}(g(2^k))^{2^k}\right\}^{1/(2^N-1)}.$$

Exercise 6.22. Prove the analogue of (6.3) if $\mathbb{T} = q^{\mathbb{N}_0}$ with $q > 1$.

6.4. Opial Inequalities

Opial inequalities and many of their generalizations have various applications in the theory of differential and difference equations. This is very nicely illustrated in the book [19] *Opial Inequalities with Applications in Differential and Difference Equations* by Agarwal and Pang, which currently is the only book devoted solely to continuous and discrete versions of Opial inequalities. In this section we present several Opial inequalities that are valid on time scales. The results below are contained in [87]. Throughout we assume

$$0 \in \mathbb{T}$$

and let $h \in \mathbb{T}$ with $h > 0$.

Theorem 6.23 (Opial's Inequality). *For differentiable $x : [0, h] \to \mathbb{R}$ with $x(0) = 0$ we have*

$$\int_0^h \left|(x + x^\sigma)x^\Delta\right|(t)\Delta t \leq h \int_0^h \left|x^\Delta\right|^2(t)\Delta t$$

with equality when $x(t) = ct$ for some constant c.

Proof. Consider

$$y(t) = \int_0^t \left|x^\Delta(s)\right|\Delta s.$$

Then we have $y^\Delta = |x^\Delta|$ and $|x| \le y$, because

$$
\begin{aligned}
|x(t)| &= |x(t) - x(0)| \\
&= \left| \int_0^t x^\Delta(s) \Delta s \right| \\
&\le \int_0^t \left| x^\Delta(s) \right| \Delta s \\
&= y(t)
\end{aligned}
$$

so that

$$
\begin{aligned}
\int_0^h \left| (x + x^\sigma) x^\Delta \right| (t) \Delta t &\le \int_0^h \left[(|x| + |x^\sigma|) \left| x^\Delta \right| \right] (t) \Delta t \\
&\le \int_0^h \left[(y + y^\sigma) \left| x^\Delta \right| \right] (t) \Delta t \\
&= \int_0^h \left[(y + y^\sigma) y^\Delta \right] (t) \Delta t \\
&= \int_0^h (y^2)^\Delta(t) \Delta t \\
&= y^2(h) - y^2(0) \\
&= y^2(h) \\
&= \left\{ \int_0^h \left| x^\Delta(t) \right| \Delta t \right\}^2 \\
&\le h \int_0^h \left| x^\Delta \right|^2 (t) \Delta t,
\end{aligned}
$$

where we have used formula (1.1) and the Cauchy–Schwarz inequality (Theorem 6.15 with $f = x$ and $g \equiv 1$).

Now, let $\tilde{x}(t) = ct$ for some $c \in \mathbb{R}$. Then $\tilde{x}^\Delta(t) \equiv c$, and it is easy to check that the equation

$$
(6.4) \qquad \int_0^h \left| (\tilde{x} + \tilde{x}^\sigma) \tilde{x}^\Delta \right| (t) \Delta t = h \int_0^h \left| \tilde{x}^\Delta \right|^2 (t) \Delta t
$$

holds.   □

Exercise 6.24. Check that $\tilde{x}$ as given in Theorem 6.23 satisfies (6.4).

Exercise 6.25. Use Theorem 6.23 to derive Opial's inequality for $\mathbb{T} = \mathbb{R}$, $\mathbb{T} = \mathbb{Z}$, $\mathbb{T} = h\mathbb{Z}$, and $\mathbb{T} = q^{\mathbb{N}_0}$.

We now proceed to give an application of Theorem 6.23.

Example 6.26. Let y be a solution of the initial value problem

$$
(6.5) \qquad y^\Delta = 1 - t + \frac{y^2}{t}, \quad 0 < t \le 1, \quad y(0) = 0.
$$

Then $y \le \tilde{y}$ on $[0, 1]$, where $\tilde{y}(t) = t$ solves (6.5).

Proof. Clearly $\tilde{y}$ as defined above solves (6.5). We let y be a solution of (6.5) and consider R defined by

$$R(t) = 1 - t + \int_0^t \left|y^\Delta\right|^2 (s)\Delta s.$$

Let $t \in (0, 1]$. Then

$$
\begin{aligned}
\left|y^\Delta(t)\right| &= \left|1 - t + \frac{y^2(t)}{t}\right| \\
&\leq |1 - t| + \frac{1}{t}\left|y^2(t)\right| \\
&= 1 - t + \frac{1}{t}\left|\int_0^t (y^2)^\Delta(s)\Delta s\right| \\
&\leq 1 - t + \frac{1}{t}\int_0^t \left|(y^2)^\Delta(s)\right|\Delta s \\
&= 1 - t + \frac{1}{t}\int_0^t \left|(y + y^\sigma)y^\Delta\right|(s)\Delta s \\
&\leq 1 - t + \frac{1}{t}t\int_0^t \left|y^\Delta\right|^2 (s)\Delta s \\
&= R(t),
\end{aligned}
$$

where we haved used (1.1) and Opial's inequality (Theorem 6.23 with $x = y$ and $h = t$). Hence

$$R^\Delta(t) = -1 + \left|y^\Delta(t)\right|^2 \leq R^2(t) - 1 \quad \text{and} \quad R(0) = 1.$$

Now note that $R \geq \left|y^\Delta\right| \geq 0$ and hence $1 + R \in \mathcal{R}^+$ by Exercise 2.46. Thus, by Theorem 2.48 (i), $w = e_{1+R}(\cdot, 0) > 0$. Hence, because of $(R - 1)^\Delta = R^\Delta \leq R^2 - 1$, we have

$$
\begin{aligned}
0 &\geq \frac{(R - 1)^\Delta - (R^2 - 1)}{w^\sigma} \\
&= \frac{(R - 1)^\Delta w - (R^2 - 1)w}{ww^\sigma} \\
&= \frac{(R - 1)^\Delta w - (R - 1)(1 + R)w}{ww^\sigma} \\
&= \frac{(R - 1)^\Delta w - (R - 1)w^\Delta}{ww^\sigma} \\
&= \left(\frac{R - 1}{w}\right)^\Delta
\end{aligned}
$$

(we used the quotient rule from Theorem 1.20 (v)) so that

$$
\begin{aligned}
\left(\frac{R - 1}{w}\right)(t) &= \left(\frac{R - 1}{w}\right)(0) + \int_0^t \left(\frac{R - 1}{w}\right)^\Delta (s)\Delta s \\
&= \int_0^t \left(\frac{R - 1}{w}\right)^\Delta (s)\Delta s \\
&\leq 0
\end{aligned}
$$

and hence $R(t) \le 1$. Therefore

$$y^\Delta(t) \le |y^\Delta(t)| \le R(t) \le 1$$

and $y(t) = \int_0^t y^\Delta(s)\Delta s \le \int_0^t \Delta s = t = \tilde{y}(t)$. This holds for all $t \in (0,1]$, and since $y(0) = \tilde{y}(0) = 0$, we have $y \le \tilde{y}$ on $[0,1]$. $\square$

We next offer a generalization of Theorem 6.23, where $x(0)$ does not need to be equal to 0. Below we use the function $\text{dist}(x, \mathbb{T})$ defined by

$$\text{dist}(x, \mathbb{T}) = \inf\{|x - t| : t \in \mathbb{T}\} \quad \text{for} \quad x \in \mathbb{R}.$$

Theorem 6.27. *Let $x : [0, h] \to \mathbb{R}$ be differentiable. Then*

$$\int_0^h |(x + x^\sigma)x^\Delta|(t)\Delta t \le \alpha \int_0^h |x^\Delta(t)|^2 \Delta t + 2\beta \int_0^h |x^\Delta(t)|\Delta t,$$

where

(6.6) $$\alpha \in \mathbb{T} \quad \text{with} \quad \left|\frac{h}{2} - \alpha\right| = \text{dist}\left(\frac{h}{2}, \mathbb{T}\right)$$

and $\beta = \max\{|x(0)|, |x(h)|\}$.

Proof. We consider

$$y(t) = \int_0^t |x^\Delta(s)|\Delta s \quad \text{and} \quad z(t) = \int_t^h |x^\Delta(s)|\Delta s.$$

Then $y^\Delta = |x^\Delta|$, $z^\Delta = -|x^\Delta|$,

$$
\begin{aligned}
|x(t)| &\le |x(t) - x(0)| + |x(0)| \\
&= \left|\int_0^t x^\Delta(s)\Delta s\right| + |x(0)| \\
&\le \int_0^t |x^\Delta(s)|\Delta s + |x(0)| \\
&= y(t) + |x(0)|,
\end{aligned}
$$

and similarly $|x(t)| \le z(t) + |x(h)|$. Let $u \in [0, h] \cap \mathbb{T}$. Then

$$
\begin{aligned}
\int_0^u |(x + x^\sigma)x^\Delta|(t)\Delta t &\le \int_0^u [y(t) + |x(0)| + y^\sigma(t) + |x(0)|]y^\Delta(t)\Delta t \\
&= \int_0^u [(y + y^\sigma)y^\Delta](t)\Delta t + 2|x(0)| \int_0^u y^\Delta(t)\Delta t \\
&= y^2(u) + 2|x(0)|y(u) \\
&\le u \int_0^u |x^\Delta(t)|^2 \Delta t + 2|x(0)| \int_0^u |x^\Delta(t)|\Delta t,
\end{aligned}
$$

where we have used (1.1) and Theorem 6.15. Similarly, one shows

$$
\begin{aligned}
\int_u^h |(x + x^\sigma)x^\Delta|(t)\Delta t &\le z^2(u) + 2|x(h)|z(u) \\
&\le (h - u) \int_u^h |x^\Delta(t)|^2 \Delta t + 2|x(h)| \int_u^h |x^\Delta(t)|\Delta t.
\end{aligned}
$$

By putting $\nu(u) = \max\{u, h - u\}$ and adding the above two inequalities, we find

$$\int_0^h |(x + x^\sigma)x^\Delta|\,(t)\Delta t \le \nu(u) \int_0^h |x^\Delta(t)|^2 \Delta t + 2\beta \int_0^h |x^\Delta(t)|\,\Delta t.$$

This is true for any $u \in [0, h] \cap \mathbb{T}$, so it is also true if (note that $\mathbb{T}$ is closed)

$$\nu(u) \quad \text{is replaced by} \quad \min_{u \in [0,h] \cap \mathbb{T}} \nu(u).$$

However, this last quantity is equal to α. $\qquad\square$

Theorem 6.28. *Let $x : [0, h] \to \mathbb{R}$ be differentiable with $x(0) = x(h) = 0$. Then*

$$\int_0^h |(x + x^\sigma)x^\Delta|\,(t)\Delta t \le \alpha \int_0^h |x^\Delta(t)|^2 \Delta t,$$

where α is given in (6.6).

Proof. This follows easily from Theorem 6.27 since in this case we have $\beta = 0$. $\qquad\square$

Now we offer some of the various generalizations of the inequalities presented above. The continuous and/or discrete versions of those inequalities may be found in [19]. We have not included all of those results, but most of them may be proved using techniques similar to the ones presented in this section.

Theorem 6.29 (see [19, Theorem 2.5.1]). *Let p, q be positive and continuous on $[0, h]$, $\int_0^h \Delta t/p(t) < \infty$, and q nonincreasing. For differentiable $x : [0, h] \to \mathbb{R}$ with $x(0) = 0$ we have*

$$\int_0^h [q^\sigma |(x + x^\sigma)x^\Delta|]\,(t)\Delta t \le \left\{ \int_0^h \frac{\Delta t}{p(t)} \right\} \left\{ \int_0^h p(t)q(t)\,|x^\Delta(t)|^2 \Delta t \right\}.$$

Proof. We consider

$$y(t) = \int_0^t \sqrt{q^\sigma(s)}\,|x^\Delta(s)|\,\Delta s.$$

Then $y^\Delta = \sqrt{q^\sigma}\,|x^\Delta|$ and (note that $0 \le s < t$ implies $\sigma(s) \le t$, thus $q(\sigma(s)) \ge q(t)$; apply Theorem 1.77 (viii))

$$|x(t)| \le \int_0^t |x^\Delta(s)|\,\Delta s \le \int_0^t \sqrt{\frac{q^\sigma(s)}{q(t)}}\,|x^\Delta(s)|\,\Delta s = \frac{y(t)}{\sqrt{q(t)}}$$

so that (observe that $t \le \sigma(t)$ implies $q(t) \ge q(\sigma(t))$)

$$\begin{aligned}
\int_0^h [q^\sigma |(x + x^\sigma)x^\Delta|]\,(t)\Delta t &\le \int_0^h q^\sigma(t) \left(\frac{y(t)}{\sqrt{q(t)}} + \frac{y^\sigma(t)}{\sqrt{q^\sigma(t)}} \right) \frac{y^\Delta(t)}{\sqrt{q^\sigma(t)}}\,\Delta t \\
&\le \int_0^h (y(t) + y^\sigma(t))y^\Delta(t)\,\Delta t \\
&= y^2(h) \\
&= \left\{ \int_0^h \frac{1}{\sqrt{p(s)}}\sqrt{p(s)q^\sigma(s)}\,|x^\Delta(s)|\,\Delta s \right\}^2 \\
&\le \left\{ \int_0^h \frac{\Delta s}{p(s)} \right\} \left\{ \int_0^h p(s)q^\sigma(s)\,|x^\Delta(s)|^2 \Delta s \right\}.
\end{aligned}$$

Again we have used (1.1) and Theorem 6.15.   □

The following result involves higher order derivatives.

Theorem 6.30 (see [19, Chapter 3]). *Suppose* $l, n \in \mathbb{N}$. *For* n *times differentiable* $x : [0, h] \to \mathbb{R}$ *with* $x(0) = x^\Delta(0) = \cdots = x^{\Delta^{n-1}}(0) = 0$ *we have*

$$\int_0^h \left| \left\{ \sum_{k=0}^l x^k (x^\sigma)^{l-k} \right\} x^{\Delta^n} \right| (t) \Delta t \leq h^{nl} \int_0^h \left| x^{\Delta^n}(t) \right|^{l+1} \Delta t.$$

Proof. We consider

$$y(t) = \int_0^t \int_0^{\tau_{n-1}} \cdots \int_0^{\tau_2} \left\{ \int_0^{\tau_1} \left| x^{\Delta^n}(s) \right| \Delta s \right\} \Delta \tau_1 \Delta \tau_2 \cdots \Delta \tau_{n-1}.$$

Hence we have

$$y^\Delta(t) = \int_0^t \int_0^{\tau_{n-2}} \cdots \int_0^{\tau_2} \left\{ \int_0^{\tau_1} \left| x^{\Delta^n}(s) \right| \Delta s \right\} \Delta \tau_1 \Delta \tau_2 \cdots \Delta \tau_{n-2},$$

$$\cdots, \quad y^{\Delta^{n-1}}(t) = \int_0^t \left| x^{\Delta^n}(s) \right| \Delta s, \quad y^{\Delta^n}(t) = \left| x^{\Delta^n}(t) \right|,$$

and for $0 \leq t \leq h$

$$
\begin{aligned}
|x(t)| &\leq \int_0^t \left| x^\Delta(t_1) \right| \Delta t_1 \\
&\leq \int_0^t \int_0^{t_1} \left| x^{\Delta\Delta}(t_2) \right| \Delta t_2 \Delta t_1 \leq \cdots \leq y(t) \\
&= \int_0^t y^\Delta(s) \Delta s \leq \int_0^t y^\Delta(t) \Delta s \\
&\leq \int_0^h y^\Delta(t) \Delta s \\
&= h y^\Delta(t) \\
&\leq h^2 y^{\Delta\Delta}(t) \leq \cdots \leq h^{n-1} y^{\Delta^{n-1}}(t) \\
&= h^{n-1} f(t),
\end{aligned}
$$

where we put $f = y^{\Delta^{n-1}}$. Therefore

$$
\begin{aligned}
\int_0^h \left| \left\{ \sum_{k=0}^l x^k (x^\sigma)^{l-k} \right\} x^{\Delta^n} \right| (t) \Delta t &\leq \int_0^h \left\{ \sum_{k=0}^l |x|^k |x^\sigma|^{l-k} \left| x^{\Delta^n} \right| \right\} (t) \Delta t \\
&\leq \int_0^h \left\{ \sum_{k=0}^l (h^{n-1} f)^k (h^{n-1} f^\sigma)^{l-k} \left| x^{\Delta^n} \right| \right\} (t) \Delta t \\
&= h^{(n-1)l} \int_0^h \left\{ \sum_{k=0}^l f^k (f^\sigma)^{l-k} f^\Delta \right\} (t) \Delta t \\
&= h^{(n-1)l} \int_0^h (f^{l+1})^\Delta (t) \Delta t \\
&= h^{(n-1)l} f^{l+1}(h)
\end{aligned}
$$

$$
= h^{(n-1)l} \left\{ \int_0^h \left| x^{\Delta^n}(t) \right| \Delta t \right\}^{l+1}
$$

$$
\leq h^{(n-1)l} h^l \int_0^h \left| x^{\Delta^n}(t) \right|^{l+1} \Delta t
$$

$$
= h^{nl} \int_0^h \left| x^{\Delta^n}(t) \right|^{l+1} \Delta t,
$$

where we have used Theorem 1.77, Exercise 1.23, and Hölder's inequality (Theorem 6.13 with $p = (l+1)/l$ and $q = l+1$). $\qquad\square$

The following two results are easy corollaries of Theorem 6.30.

Theorem 6.31 (see [19, Theorem 3.2.1]). *Suppose $n \in \mathbb{N}$. For n times differentiable $x : [0, h] \to \mathbb{R}$ with $x(0) = x^\Delta(0) = \cdots = x^{\Delta^{n-1}}(0) = 0$ we have*

$$
\int_0^h \left| (x + x^\sigma) x^{\Delta^n} \right|(t) \Delta t \leq h^n \int_0^h \left| x^{\Delta^n}(t) \right|^2 \Delta t.
$$

Proof. This is Theorem 6.30 with $l = 1$. $\qquad\square$

Theorem 6.32 (see [19, Theorem 2.3.1]). *Let $l \in \mathbb{N}$. If $x : [0, h] \to \mathbb{R}$ is differentiable with $x(0) = 0$, then we have*

$$
\int_0^h \left| \left\{ \sum_{k=0}^l x^k (x^\sigma)^{l-k} \right\} x^\Delta \right|(t) \Delta t \leq h^l \int_0^h \left| x^\Delta(t) \right|^{l+1} \Delta t.
$$

Proof. This is Theorem 6.30 with $n = 1$. $\qquad\square$

We conclude this section with a *Wirtinger* type inequality due to Roman Hilscher; see [169].

Theorem 6.33 (Wirtinger Type Inequality). *Let M be positive and strictly monotone such that $M \in \mathrm{C}^1_{\mathrm{rd}}$. Then we have*

$$
\int_a^b \left| M^\Delta(t) \right| (y^\sigma(t))^2 \Delta t \leq \Psi \int_a^b \frac{M(t) M^\sigma(t)}{\left| M^\Delta(t) \right|} (y^\Delta(t))^2 \Delta t
$$

for any $y \in \mathrm{C}^1_{\mathrm{rd}}$ with $y(a) = y(b) = 0$, where

$$
\Psi = \left\{ \sqrt{\sup_{t \in [a,b]} \frac{M(t)}{M^\sigma(t)}} + \sqrt{\sup_{t \in [a,b]} \frac{\mu(t) \left| M^\Delta(t) \right|}{M^\sigma(t)} + \sup_{t \in [a,b]} \frac{M(t)}{M^\sigma(t)}} \right\}^2 .
$$

Proof. For convenience we delete the argument (t) in this proof. Let

$$
A = \int_a^b \left| M^\Delta \right| (y^\sigma)^2 \Delta t, \quad B = \int_a^b \frac{M M^\sigma}{\left| M^\Delta \right|} (y^\Delta)^2 \Delta t,
$$

$$
\alpha = \left(\sup_{[a,b]} \frac{M}{M^\sigma} \right)^{\frac{1}{2}}, \quad \beta = \sup_{[a,b]} \frac{\mu \left| M^\Delta \right|}{M^\sigma}.
$$

Without loss of generality we assume that M^Δ is of positive sign. Then we apply the Cauchy–Schwarz inequality (Theorem 6.15) and (1.1) (use also $y(a) = y(b) = 0$ for the third equal sign) to estimate

$$
\begin{aligned}
A &= \int_a^b M^\Delta (y^\sigma)^2 \Delta t \\
&= \int_a^b \left[(My^2)^\Delta - My^\Delta (y + y^\sigma) \right] \Delta t \\
&= -\int_a^b My^\Delta (y + y^\sigma) \Delta t \\
&\leq \int_a^b M \left| y^\Delta \right| |y + y^\sigma| \Delta t \\
&= \int_a^b M \left| y^\Delta \right| |2y^\sigma - \mu y^\Delta| \Delta t \\
&\leq 2 \int_a^b M \left| y^\Delta \right| |y^\sigma| \Delta t + \int_a^b \mu M (y^\Delta)^2 \Delta t \\
&= 2 \int_a^b \sqrt{\frac{MM^\sigma}{|M^\Delta|}} |y^\Delta| \sqrt{\frac{M}{M^\sigma}} |M^\Delta| |y^\sigma| \Delta t + \int_a^b \frac{\mu M^\Delta}{M^\sigma} \frac{MM^\sigma}{|M^\Delta|} (y^\Delta)^2 \Delta t \\
&\leq 2 \left\{ \int_a^b \frac{MM^\sigma}{|M^\Delta|} (y^\Delta)^2 \Delta t \right\}^{\frac{1}{2}} \left\{ \int_a^b \frac{M}{M^\sigma} |M^\Delta| (y^\sigma)^2 \Delta t \right\}^{\frac{1}{2}} + \beta B \\
&\leq 2\alpha \sqrt{AB} + \beta B.
\end{aligned}
$$

Therefore, by denoting $C = \sqrt{A/B}$, we find that $C^2 - 2\alpha C - \beta \leq 0$, and solving for $C \geq 0$ we obtain

$$
\frac{A}{B} = C^2 \leq (\alpha + \sqrt{\alpha^2 + \beta})^2 = \Psi
$$

so that the proof is complete. $\qquad\square$

Example 6.34. Let $a > 0$ and

$$
\Psi = \left\{ \left(\sup_{t \in [a,b]} \frac{\sigma(t)}{t} \right)^{\frac{1}{2}} + \left[\left(\sup_{t \in [a,b]} \frac{\mu(t)}{t} \right) + \left(\sup_{t \in [a,b]} \frac{\sigma(t)}{t} \right) \right]^{\frac{1}{2}} \right\}^2.
$$

Then

$$(6.7) \qquad \int_a^b (y^\Delta(t))^2 \Delta t \geq \frac{1}{\Psi} \int_a^b \frac{(y^\sigma(t))^2}{t\sigma(t)} \Delta t.$$

Proof. To show this we remark that $M(t) = \frac{1}{t}$ satisfies the assumptions of Theorem 6.33, and

$$
M^\Delta(t) = -\frac{1}{t\sigma(t)}
$$

holds by Example 1.25. Therefore

$$
\frac{M(t)}{M^\sigma(t)} = \frac{\sigma(t)}{t}, \quad \frac{\mu(t)|M^\Delta(t)|}{M^\sigma(t)} = \frac{\mu(t)}{t}, \quad \text{and} \quad \frac{M(t)M^\sigma(t)}{|M^\Delta(t)|} = 1,
$$

and (6.7) follows from Theorem 6.33. $\qquad\square$

As an application of Theorem 6.33 and Example 6.34 we now state a sufficient criterion for nonoscillation of a certain second order dynamic equation, see [**169**, Theorem 3].

Theorem 6.35. *Let $N \in \mathbb{T}$ and define*

$$\Psi_N = \left\{ \left(\sup_{t \geq N, t \in \mathbb{T}} \frac{\sigma(t)}{t} \right)^{\frac{1}{2}} + \left[\left(\sup_{t \geq N, t \in \mathbb{T}} \frac{\mu(t)}{t} \right) + \left(\sup_{t \geq N, t \in \mathbb{T}} \frac{\sigma(t)}{t} \right) \right]^{\frac{1}{2}} \right\}^2 .$$

If

$$0 < \limsup_{N \to \infty} \Psi_N = \Psi < \infty,$$

then the equation

$$y^{\Delta\Delta} + \frac{\lambda}{t\sigma(t)} y^{\sigma} = 0$$

is nonoscillatory for all $\lambda < \frac{1}{\Psi}$.

6.5. Lyapunov Inequalities

Lyapunov inequalities have proved to be useful tools in oscillation theory, disconjugacy, eigenvalue problems, and numerous other applications in the theory of differential and difference equations. A nice summary of continuous and discrete Lyapunov inequalities and their applications can be found in the survey paper [**93**] by Chen. In this section we present several versions of Lyapunov inequalities on time scales. The results below are contained in [**75**]. Throughout we assume $a, b \in \mathbb{T}$ with $a < b$.

We let $\mathbb{T}$ be any time scale, $q : \mathbb{T} \to \mathbb{R}$ be rd-continuous with $q(t) > 0$ for all $t \in \mathbb{T}$, and consider the Sturm–Liouville dynamic equation (4.1), i.e.,

$$(6.8) \qquad\qquad x^{\Delta\Delta} + q(t)x^{\sigma} = 0$$

together with the quadratic functional

$$\mathcal{F}(x) = \int_a^b \left\{ (x^{\Delta})^2 - q(x^{\sigma})^2 \right\} (t)\Delta t.$$

To prove a Lyapunov inequality for (6.8) we need two auxiliary results as follows.

Lemma 6.36. *If x solves (6.8) and if $\mathcal{F}(y)$ is defined, then*

$$\mathcal{F}(y) - \mathcal{F}(x) = \mathcal{F}(y - x) + 2(y - x)(b)x^{\Delta}(b) - 2(y - x)(a)x^{\Delta}(a).$$

Proof. Under the above assumptions we find

$$\mathcal{F}(y) - \mathcal{F}(x) - \mathcal{F}(y-x) = \int_a^b \left\{ (y^\Delta)^2 - q(y^\sigma)^2 - (x^\Delta)^2 + q(x^\sigma)^2 \right.$$
$$\left. -(y^\Delta - x^\Delta)^2 + q(y^\sigma - x^\sigma)^2 \right\} (t)\Delta t$$
$$= \int_a^b \left\{ (y^\Delta)^2 - q(y^\sigma)^2 - (x^\Delta)^2 + q(x^\sigma)^2 - (y^\Delta)^2 + 2y^\Delta x^\Delta - (x^\Delta)^2 \right.$$
$$\left. + q(y^\sigma)^2 - 2qy^\sigma x^\sigma + q(x^\sigma)^2 \right\} (t)\Delta t$$
$$= 2\int_a^b \left\{ y^\Delta x^\Delta - qy^\sigma x^\sigma + q(x^\sigma)^2 - (x^\Delta)^2 \right\} (t)\Delta t$$
$$= 2\int_a^b \left\{ y^\Delta x^\Delta + y^\sigma x^{\Delta\Delta} - x^\sigma x^{\Delta\Delta} - (x^\Delta)^2 \right\} (t)\Delta t$$
$$= 2\int_a^b \left\{ yx^\Delta - xx^\Delta \right\}^\Delta \Delta t$$
$$= 2\int_a^b \left\{ (y-x)x^\Delta \right\}^\Delta \Delta t$$
$$= 2(y(b)-x(b))x^\Delta(b) - 2(y(a)-x(a))x^\Delta(a),$$

where we have used the product rule from Theorem 1.20 (iii).   □

Lemma 6.37. *If $\mathcal{F}(y)$ is defined, then for any $r,s \in \mathbb{T}$ with $a \le r < s \le b$,*

$$\int_r^s (y^\Delta(t))^2 \Delta t \ge \frac{(y(s)-y(r))^2}{s-r}.$$

Proof. Under the above assumptions we define

$$x(t) = \frac{y(s)-y(r)}{s-r}t + \frac{sy(r)-ry(s)}{s-r}.$$

We then have

$$x(r) = y(r), \quad x(s) = y(s), \quad x^\Delta(t) = \frac{y(s)-y(r)}{s-r}, \quad \text{and} \quad x^{\Delta\Delta}(t) = 0.$$

Hence x solves the special Sturm–Liouville equation (6.8) where $q = 0$, and therefore we may apply Lemma 6.36 to $\mathcal{F}_0$ defined by

$$\mathcal{F}_0(x) = \int_r^s (x^\Delta)^2(t)\Delta t$$

to find

$$\mathcal{F}_0(y) = \mathcal{F}_0(x) + \mathcal{F}_0(y-x) + 2(y-x)(s)x^\Delta(s) - 2(y-x)(r)x^\Delta(r)$$
$$= \mathcal{F}_0(x) + \mathcal{F}_0(y-x)$$
$$\ge \mathcal{F}_0(x)$$
$$= \int_r^s \left\{ \frac{y(s)-y(r)}{s-r} \right\}^2 \Delta t$$
$$= \frac{(y(s)-y(r))^2}{s-r},$$

and this proves our claim.   □

Using Lemma 6.37, we now can prove one of the main results of this section, a Lyapunov inequality for Sturm–Liouville dynamic equations of the form (6.8).

Theorem 6.38 (Lyapunov's Inequality). *Let $q : \mathbb{T} \to (0, \infty)$ be positive-valued and rd-continuous. If the Sturm–Liouville dynamic equation (6.8) has a nontrivial solution x with $x(a) = x(b) = 0$, then the Lyapunov inequality*

$$(6.9) \qquad \int_a^b q(t)\Delta t \geq \frac{b-a}{f(d)},$$

holds, where $f : \mathbb{T} \to \mathbb{R}$ is defined by $f(t) = (t-a)(b-t)$, and where $d \in \mathbb{T}$ is such that

$$\left| \frac{a+b}{2} - d \right| = \text{dist}\left(\frac{a+b}{2}, \mathbb{T} \right).$$

Proof. Suppose x is a nontrivial solution of (6.8) with $x(a) = x(b) = 0$. But then we have from Lemma 6.36 (with $y = 0$) that

$$\mathcal{F}(x) = \int_a^b \left\{ (x^\Delta)^2 - q(x^\sigma)^2 \right\}(t)\Delta t = 0.$$

Since x is nontrivial, we have that M defined by (note that x is continuous by Theorem 1.16 (i))

$$(6.10) \qquad M = \max\left\{ x^2(t) : t \in [a, b] \cap \mathbb{T} \right\}$$

is positive. We now let $c \in [a, b] \cap \mathbb{T}$ be such that $x^2(c) = M$. Applying the above as well as Lemma 6.37 twice (once with $r = a$ and $s = c$ and a second time with $r = c$ and $s = b$) we find

$$
\begin{aligned}
M \int_a^b q(t)\Delta t \;&\geq\; \int_a^b \left\{ q(x^\sigma)^2 \right\}(t)\Delta t \\
&=\; \int_a^b (x^\Delta)^2(t)\Delta t \\
&=\; \int_a^c (x^\Delta)^2(t)\Delta t + \int_c^b (x^\Delta)^2(t)\Delta t \\
&\geq\; \frac{(x(c) - x(a))^2}{c - a} + \frac{(x(b) - x(c))^2}{b - c} \\
&=\; x^2(c) \left\{ \frac{1}{c - a} + \frac{1}{b - c} \right\} \\
&=\; M \frac{b - a}{f(c)} \\
&\geq\; M \frac{b - a}{f(d)},
\end{aligned}
$$

where the last inequality holds because of $f(d) = \max\{ f(t) : t \in [a, b] \}$. Hence, dividing by $M > 0$ yields the desired inequality. $\qquad\square$

Example 6.39. Here we briefly discuss the two important cases $\mathbb{T} = \mathbb{R}$ and $\mathbb{T} = \mathbb{Z}$. We use the notation from the proof of Theorem 6.38.

(i) If $\mathbb{T} = \mathbb{R}$, then

$$\min\left\{\left|\frac{a+b}{2} - s\right| : s \in [a,b]\right\} = 0 \quad \text{so that} \quad d = \frac{a+b}{2}.$$

Hence $f(d) = \frac{(b-a)^2}{4}$ and the Lyapunov inequality from Theorem 6.38 becomes

$$\int_a^b q(t)dt \geq \frac{4}{b-a}.$$

(ii) If $\mathbb{T} = \mathbb{Z}$, then we consider two cases. First, if $a+b$ is even, then

$$\min\left\{\left|\frac{a+b}{2} - s\right| : s \in [a,b]\right\} = 0 \quad \text{so that} \quad d = \frac{a+b}{2}.$$

Hence $f(d) = \frac{(b-a)^2}{4}$ and the Lyapunov inequality reads

$$\sum_{t=a}^{b-1} q(t) \geq \frac{4}{b-a}.$$

If $a+b$ is odd, then

$$\min\left\{\left|\frac{a+b}{2} - s\right| : s \in [a,b]\right\} = \frac{1}{2} \quad \text{so that} \quad d = \frac{a+b-1}{2}.$$

This time we have $f(d) = \frac{(b-a)^2-1}{4}$ and the Lyapunov inequality becomes

$$\sum_{t=a}^{b-1} q(t) \geq \frac{4}{b-a}\left\{\frac{1}{1-\frac{1}{(b-a)^2}}\right\}.$$

As an application of Theorem 6.38 we now prove a sufficient criterion for disconjugacy of (6.8). We use the notation of Theorem 6.38.

Theorem 6.40 (Sufficient Condition for Disconjugacy of (6.8)). *If q satisfies*

$$(6.11) \qquad \int_a^b q(t)\Delta t < \frac{b-a}{f(d)},$$

then (6.8) is disconjugate on $[a,b]$.

Proof. Suppose that (6.11) holds. For the sake of contradiction we assume (6.8) is not disconjugate. Then, by Theorem 4.57, there exists a nontrivial x with $x(a) = x(b) = 0$ such that $\mathcal{F}(x) \leq 0$. Using this x, we define M by (6.10) to find

$$\begin{aligned}
M\int_a^b q(t)\Delta t &\geq \int_a^b \left\{q(x^\sigma)^2\right\}(t)\Delta t \\
&\geq \int_a^b (x^\Delta)^2(t)\Delta t \\
&\geq \frac{M(b-a)}{f(d)},
\end{aligned}$$

where the last inequality follows precisely as in the proof of Theorem 6.38. Hence, after dividing by $M > 0$, we arrive at

$$\int_a^b q(t)\Delta t \geq \frac{b-a}{f(d)}$$

which contradicts (6.11) and hence completes the proof. $\square$

Remark 6.41. Note that in both conditions (6.9) and (6.11) we could replace $(b-a)/f(d)$ by $4/(b-a)$, and Theorems 6.38 and 6.40 would remain true. This is because for $a \le c \le b$ we have

$$\frac{1}{c-a} + \frac{1}{b-c} = \frac{(a+b-2c)^2}{(b-a)(c-a)(b-c)} + \frac{4}{b-a} \ge \frac{4}{b-a}.$$

For the remainder of this section we consider the *linear Hamiltonian dynamic system*

$$(6.12) \qquad x^\Delta = A(t)x^\sigma + B(t)u, \quad u^\Delta = -C(t)x^\sigma - A^*(t)u,$$

where $A, B, C \in C_{\mathrm{rd}}$ are $n \times n$-matrix-valued functions on $\mathbb{T}$ such that $I - \mu(t)A(t)$ is invertible and $B(t)$ and $C(t)$ are positive semidefinite for all $t \in \mathbb{T}$. A pair (x, u) is called *admissible* if it satisfies the first equation in (6.12). The corresponding *quadratic functional* is given by

$$\mathcal{F}(x,u) = \int_a^b \left\{ u^*Bu - (x^\sigma)^*Cx^\sigma \right\}(t)\Delta t.$$

As before we start with the following auxiliary lemma.

Lemma 6.42. *If (x,u) solves (6.12) and if (y,v) is admissible, then*

$$\mathcal{F}(y,v) - \mathcal{F}(x,u) = \mathcal{F}(y-x, v-u) + 2\,\mathrm{Re}\left[(y-x)^*(b)u(b) - (y-x)^*(a)u(a) \right].$$

Proof. Under the above assumptions we calculate

$$\mathcal{F}(y,v) - \mathcal{F}(x,u) - \mathcal{F}(y-x, v-u) = \int_a^b \{ v^*Bv - (y^\sigma)^*Cy^\sigma - u^*Bu$$

$$+ (x^\sigma)^*Cx^\sigma - [(v-u)^*B(v-u) - (y^\sigma - x^\sigma)^*C(y^\sigma - x^\sigma)] \}(t)\Delta t$$

$$= \int_a^b \{ -2u^*Bu + v^*Bu + u^*Bv + 2(x^\sigma)^*Cx^\sigma - (y^\sigma)^*Cx^\sigma - (x^\sigma)^*Cy^\sigma \}(t)\Delta t$$

$$= \int_a^b \{ -2u^*Bu + 2\,\mathrm{Re}[u^*Bv] + 2(x^\sigma)^*Cx^\sigma - 2\,\mathrm{Re}[(y^\sigma)^*Cx^\sigma] \}(t)\Delta t$$

$$= 2\,\mathrm{Re} \int_a^b \{ u^*(Bv - Bu) + [(x^\sigma)^* - (y^\sigma)^*]Cx^\sigma \}(t)\Delta t$$

$$= 2\,\mathrm{Re} \int_a^b \{ u^*(y^\Delta - Ay^\sigma - x^\Delta + Ax^\sigma) + [(x^\sigma)^* - (y^\sigma)^*][-u^\Delta - A^*u] \}(t)\Delta t$$

$$= 2\,\mathrm{Re} \int_a^b \{ u^*(y^\Delta - x^\Delta) + (y^\sigma - x^\sigma)^*u^\Delta + 2i\,\mathrm{Im}[u^*Ax^\sigma + (y^\sigma)^*A^*u] \}(t)\Delta t$$

$$= 2\,\mathrm{Re} \int_a^b \{ u^*(y^\Delta - x^\Delta) + (y^\sigma - x^\sigma)^*u^\Delta \}(t)\Delta t$$

$$= 2\,\mathrm{Re} \int_a^b \{ u^*(y^\Delta - x^\Delta) + (u^\Delta)^*(y^\sigma - x^\sigma) \}(t)\Delta t$$

$$= 2\,\mathrm{Re} \int_a^b \{ [u^*(y - x)]^\Delta \}(t)\Delta t$$

$$= 2\,\mathrm{Re} \{ u^*(b)[y(b) - x(b)] - u^*(a)[y(a) - x(a)] \}$$

$$= 2\,\mathrm{Re} \{ [y-x]^*(b)u(b) - [y-x]^*(a)u(a) \},$$

which is the conclusion we sought. □

We denote by $W(\cdot, r)$ the unique solution of the initial value problem

$$W^{\Delta} = -A^*(t)W, \quad W(r) = I,$$

where $r \in [a, b]$ is given, i.e., $W(t, r) = e_{-A^*}(t, r)$. Note that W exists due to our assumption on the invertibility of $I - \mu A$. Also note that $W(t, r) \equiv I$ provided $A(t) \equiv 0$. We define F by

$$F(s, r) := \int_r^s W^*(t, r)B(t)W(t, r)\Delta t.$$

Lemma 6.43. *Let W and F be defined as above. If (y, v) is admissible and if $r, s \in \mathbb{T}$ with $a \le r < s \le b$ such that $F(s, r)$ is invertible, then*

$$\int_r^s (v^*Bv)(t)\Delta t \ge [W^*(s, r)y(s) - y(r)]^* F^{-1}(s, r)[W^*(s, r)y(s) - y(r)].$$

Proof. Under the above assumptions we define

$$x(t) = W^{*^{-1}}(t, r)\left\{y(r) + F(t, r)F^{-1}(s, r)[W^*(s, r)y(s) - y(r)]\right\}$$

and

$$u(t) = W(t, r)F^{-1}(s, r)[W^*(s, r)y(s) - y(r)].$$

Then we have

$$x(r) = y(r), \quad x(s) = y(s), \quad u^{\Delta}(t) = -A^*(t)u(t),$$

and

$$
\begin{aligned}
x^{\Delta}(t) &= -W^{*^{-1}}(\sigma(t), r)(W^{\Delta}(t, r))^*x(t) + W^{*^{-1}}(\sigma(t), r)W^*(t, r)B(t)u(t) \\
&= W^{*^{-1}}(\sigma(t), r)W^*(t, r)A(t)x(t) + W^{*^{-1}}(\sigma(t), r)W^*(t, r)B(t)u(t) \\
&= [W(t, r)W^{-1}(\sigma(t), r)]^*[A(t)x(t) + B(t)u(t)].
\end{aligned}
$$

But

$$
\begin{aligned}
W(t, r)W^{-1}(\sigma(t), r) &= [W(\sigma(t), r) - \mu(t)W^{\Delta}(t, r)]W^{-1}(\sigma(t), r) \\
&= I + \mu(t)A^*(t)W(t, r)W^{-1}(\sigma(t), r)
\end{aligned}
$$

and therefore $[I - \mu(t)A^*(t)]W(t, r)W^{-1}(\sigma(t), r) = I$ so that

$$[I - \mu(t)A(t)]x^{\Delta}(t) = A(t)x(t) + B(t)u(t)$$

and hence

$$
\begin{aligned}
x^{\Delta}(t) &= A(t)x(t) + \mu(t)A(t)x^{\Delta}(t) + B(t)u(t) \\
&= A(t)x^{\sigma}(t) + B(t)u(t).
\end{aligned}
$$

Thus (x, u) solves the special Hamiltonian system (6.12) where $C = 0$, and we may apply Lemma 6.42 to $\mathcal{F}_0$ defined by

$$\mathcal{F}_0(x, u) = \int_r^s (u^*Bu)(t)\Delta t$$

to obtain

$$
\begin{aligned}
\mathcal{F}_0(y,v) &= \mathcal{F}_0(x,u) + \mathcal{F}_0(y-x,v-u) \\
&\quad +2\operatorname{Re}\{u^*(s)[y(s)-x(s)] - u^*(r)[y(r)-x(r)]\} \\
&= \mathcal{F}_0(x,u) + \mathcal{F}_0(y-x,v-u) \\
&\geq \mathcal{F}_0(x,u) \\
&= \int_r^s (u^*Bu)(t)\Delta t \\
&= [W^*(s,r)y(s) - y(r)]^* F^{-1}(r,s)[W^*(s,r)y(s) - y(r)]
\end{aligned}
$$

which shows our claim. $\square$

Remark 6.44. The assumption in Lemma 6.43 that $F(s,r)$ is invertible if $r < s$ can be dropped in case B is positive definite rather than positive semidefinite.

As before, we now may use Lemma 6.43 to derive a Lyapunov inequality for Hamiltonian systems (6.12).

Theorem 6.45 (Lyapunov's Inequality). *Assume (6.12) has a solution (x,u) such that x is nontrivial and satisfies $x(a) = x(b) = 0$. Let W and F be as defined above and suppose that $F(b,c)$ and $F(c,a)$ are invertible, where $\|x(c)\| = \max_{t\in[a,b]}\|x(t)\|$. Let λ be the largest eigenvalue of*

$$
F = \int_a^b W^*(t,c)B(t)W(t,c)\Delta t,
$$

and let $\nu(t)$ be the largest eigenvalue of $C(t)$. Then the Lyapunov inequality

$$
\int_a^b \nu(t)\Delta t \geq \frac{4}{\lambda}
$$

holds.

Proof. Suppose we are given a solution (x,u) of (6.12) such that $x(a) = x(b) = 0$. Lemma 6.42 then yields (using $y = v = 0$) that

$$
\mathcal{F}(x,u) = \int_a^b \{u^*Bu - (x^\sigma)^*Cx^\sigma\}(t)\Delta t = 0.
$$

So we apply Lemma 6.43 twice (once with $r = a$ and $s = c$ and a second time with $r = c$ and $s = b$) to obtain

$$
\begin{aligned}
\int_a^b [(x^\sigma)^*Cx^\sigma](t)\Delta t &= \int_a^b (u^*Bu)(t)\Delta t \\
&= \int_a^c (u^*Bu)(t)\Delta t + \int_c^b (u^*Bu)(t)\Delta t \\
&\geq x^*(c)W(c,a)F^{-1}(c,a)W^*(c,a)x(c) + x^*(c)F^{-1}(b,c)x(c) \\
&= x^*(c)[F^{-1}(b,c) - F^{-1}(a,c)]x(c) \\
&\geq 4x^*(c)F^{-1}x(c).
\end{aligned}
$$

Here we have used the relation $W(t,r)W(r,s) = W(t,s)$ from Theorem 5.21 (v) as well as the inequality $M^{-1} + N^{-1} \geq 4(M+N)^{-1}$ (see [**105**, Lemma 11, page 63] or

[**227**]). Now, by applying the Rayleigh–Ritz theorem [**176**, page 176] we conclude

$$
\begin{aligned}
\int_a^b \nu(t)\Delta t &\geq \int_a^b \nu(t)\frac{\|x^\sigma(t)\|^2}{\|x(c)\|^2}\Delta t \\
&= \frac{1}{\|x(c)\|^2}\int_a^b \nu(t)(x^\sigma(t))^* x^\sigma(t)\Delta t \\
&\geq \frac{1}{\|x(c)\|^2}\int_a^b (x^\sigma(t))^* C(t)x^\sigma(t)\Delta t \\
&\geq \frac{1}{\|x(c)\|^2}4x^*(c)F^{-1}x(c) \\
&\geq 4\min_{x\neq 0}\frac{x^* F^{-1} x}{x^* x} \\
&= \frac{4}{\lambda},
\end{aligned}
$$

and this finishes the proof.   □

Remark 6.46. If $A \equiv 0$, then $W \equiv I$ and $F = \int_a^b B(t)\Delta t$. If, in addition $B \equiv 1$, then $F = b - a$. Note how the Lyapunov inequality $\int_a^b \nu(t)\Delta t \geq \frac{4}{\lambda}$ reduces to $\int_a^b p(t)\Delta t \geq \frac{4}{b-a}$ for the scalar case as discussed earlier in this section.

It is possible to provide a slightly better bound than the one given in Theorem 6.45, similarly as in Theorem 6.38, but we shall not do so here. Instead we now give a disconjugacy criterion for systems (6.12) whose proof is similar to that of Theorem 6.40.

Theorem 6.47 (Sufficient Condition for Disconjugacy of (6.12)). *Using the notation from Theorem 6.45, if*

$$
\int_a^b \nu(t)\Delta t < \frac{4}{\lambda},
$$

then (6.12) *is disconjugate on* $[a, b]$.

We conclude this section with a result concerning the so-called *right-focal* boundary conditions, i.e., $x(a) = u(b) = 0$.

Theorem 6.48. *Assume* (6.12) *has a solution* (x, u) *such that x is nontrivial and $x(a) = u(b) = 0$. With the notation as in Theorem 6.45, the Lyapunov inequality*

$$
\int_a^b \nu(t)\Delta t \geq \frac{1}{\lambda}
$$

holds.

Proof. Suppose (x, u) is a solution of (6.12) such that $x(a) = u(b) = 0$ with $a < b$. Choose the point c in $(a, b]$ where $\|x(t)\|$ is maximal. Apply Lemma 6.42 with $y = v = 0$ to see that $\mathcal{F}(x, u) = 0$. Therefore,

$$
\int_a^b \left[(x^\sigma)^* C x^\sigma\right](t)\Delta t = \int_a^b (u^* B u)(t)\Delta t \geq \int_a^c (u^* B u)(t)\Delta t.
$$

Using Lemma 6.43 with $r = a$ and $s = c$, we get

$$
\begin{aligned}
\int_a^c (u^* Bu)(t)\Delta t &\geq [W^*(c,a)x(c) - x(a)]^* F^{-1}(c,a) [W^*(c,a)x(c) - x(a)] \\
&= x^*(c)W(c,a)F^{-1}(c,a)W^*(c,a)x(c) \\
&= -x^*(c)F^{-1}(a,c)x(c) \\
&= x^*(c) \left(\int_a^c W^*(t,c)B(t)W(t,c)\Delta t \right)^{-1} x(c) \\
&\geq x^*(c) \left(\int_a^b W^*(t,c)B(t)W(t,c)\Delta t \right)^{-1} x(c) \\
&= x^*(c)F^{-1}x(c).
\end{aligned}
$$

Hence,

$$
\int_a^b [(x^\sigma)^* C x^\sigma](t)\Delta t \geq x^*(c)F^{-1}x(c),
$$

and the same arguments as in the proof of Theorem 6.45 lead us to our final conclusion. □

6.6. Upper and Lower Solutions

This section follows closely a paper due to Elvan Akın [32]. Throughout we assume that f is continuous on $[a,b] \times \mathbb{R}$. We are concerned with the second order equation

(6.13) $$x^{\Delta\Delta} = f(t, x^\sigma)$$

and the BVP

(6.14) $$x^{\Delta\Delta} = f(t, x^\sigma) \text{ on } [a,b], \quad x(a) = A, \quad x(\sigma^2(b)) = B.$$

Similar results for difference equations can be found in Kelley and Peterson [191, Chapter 9]. We state without proof the famous Schauder–Tychonov fixed point theorem (see, e.g., [152, Theorem XII.0.2]).

Theorem 6.49 (Schauder–Tychonov Fixed Point Theorem). *Let* X *be a Banach space. Assume that* K *is a closed, bounded, convex subset of* X. *If* $T : K \to K$ *is compact, then* T *has a fixed point in* K.

We now use the Schauder–Tychonov fixed point theorem to prove an existence theorem for the BVP (6.14).

Theorem 6.50. *Let*

$$
D := \max_{t \in [a, \sigma^2(b)]} \int_a^{\sigma(b)} |G(t,s)|\Delta s,
$$

where $G(t,s)$ *is Green's function for the BVP*

$$
x^{\Delta\Delta} = 0, \quad x(a) = x(\sigma^2(b)) = 0.
$$

Assume that f *is continuous on* $[a,b] \times \mathbb{R}$. *If* $M > \max\{|A|, |B|\}$ *and* $D \leq \frac{M}{Q}$, *where* $Q > 0$ *satisfies*

$$
Q \geq \max\{|f(t,x)| : \ t \in [a,b], \ |x| \leq 2M\},
$$

then the BVP (6.14) has a solution.

Proof. Define C to be the Banach space of all continuous functions on $[a, \sigma^2(b)]$ equipped with the norm $\| \cdot \|$ defined by

$$\|x\| := \max_{t \in [a, \sigma^2(b)]} |x(t)| \quad \text{for all} \quad x \in C.$$

Let

$$K := \{x \in C : \|x\| \leq 2M\}.$$

It can be shown (see Exercise 6.51) that K is a closed, bounded, and convex subset of C. Define $T : K \to C$ by

$$Tx(t) := z(t) + \int_a^{\sigma(b)} G(t, s) f(s, x^\sigma(s)) \Delta s$$

for $t \in [a, \sigma^2(b)]$, where z is the solution of the BVP

$$z^{\Delta\Delta} = 0, \quad z(a) = A, \quad z(\sigma^2(b)) = B.$$

It can be shown that $T : K \to C$ is continuous. Let $x \in K$ and consider

$$
\begin{aligned}
|Tx(t)| &= \left| z(t) + \int_a^{\sigma(b)} G(t, s) f(s, x^\sigma(s)) \Delta s \right| \\
&\leq |z(t)| + \int_a^{\sigma(b)} |G(t, s)| |f(s, x^\sigma(s))| \Delta s \\
&\leq M + Q \int_a^{\sigma(b)} |G(t, s)| \Delta s \\
&\leq M + QD \\
&\leq M + Q \frac{M}{Q} \\
&= 2M
\end{aligned}
$$

for all $t \in [a, \sigma^2(b)]$. But this implies that $\|Tx\| \leq 2M$. Hence $T : K \to K$. Using the Arzelà–Ascoli theorem (see, e.g., [**142**, Theorem 4.44]) it can be shown that $T : K \to K$ is a compact operator. Hence T has a fixed point x in K by the Schauder–Tychonov theorem (Theorem 6.49). This implies that x is a solution of the BVP (6.14). $\qquad\square$

Note that, if $\mathbb{T} = \mathbb{R}$ or $\mathbb{T} = \mathbb{Z}$, then (see Exercise 4.79) $D \leq \frac{(\sigma^2(b)-a)^2}{8}$.

Exercise 6.51. Show that the set K in Theorem 6.50 is a closed, bounded, and convex subset of C. Show also that $T : K \to K$ is continuous and compact.

Corollary 6.52. *If f is continuous and bounded on $[a, b] \times \mathbb{R}$, then the BVP (6.14) has a solution.*

Proof. Choose $P > \sup\{|f(t, x)| : a \leq t \leq b, x \in \mathbb{R}\}$. Pick M large enough so that

$$D < \frac{M}{P} \quad \text{and} \quad |A| \leq M, \quad |B| \leq M.$$

Then there is a number $Q > 0$ such that

$$P \geq Q, \quad \text{where} \quad Q \geq \max\{|f(t, x)| : t \in [a, b], |x| \leq 2M\}.$$

Hence

$$D < \frac{M}{P} \leq \frac{M}{Q},$$

and thus (6.14) has a solution by Theorem 6.50.   □

Definition 6.53. We say that $\alpha \in C_{rd}^2$ is a *lower solution* of (6.13) on $[a, \sigma^2(b)]$ provided

$$\alpha^{\Delta\Delta}(t) \geq f(t, \alpha^\sigma(t)) \quad \text{for all} \quad t \in [a, b].$$

Similarly, $\beta \in C_{rd}^2$ is called an *upper solution* of (6.13) on $[a, \sigma^2(b)]$ provided

$$\beta^{\Delta\Delta}(t) \leq f(t, \beta^\sigma(t)) \quad \text{for all} \quad t \in [a, b].$$

Theorem 6.54. *Let f be continuous on $[a, b] \times \mathbb{R}$. Assume that there exist a lower solution α and an upper solution β of (6.13) with*

$$\alpha(a) \leq A \leq \beta(a) \quad and \quad \alpha(\sigma^2(b)) \leq B \leq \beta(\sigma^2(b))$$

such that

$$\alpha(t) \leq \beta(t) \quad for\ all \quad t \in [a, \sigma^2(b)].$$

Then the BVP (6.14) has a solution x with

$$\alpha(t) \leq x(t) \leq \beta(t) \quad for\ all \quad t \in [a, \sigma^2(b)].$$

Proof. Define the modification of f with respect to α and β for each fixed $t \in [a, b]$ by

$$F(t, x) = \begin{cases} f(t, \beta^\sigma(t)) + \frac{x - \beta^\sigma(t)}{1 + |x|} & \text{if} \quad x \geq \beta^\sigma(t) \\ f(t, x) & \text{if} \quad \alpha^\sigma(t) \leq x \leq \beta^\sigma(t) \\ f(t, \alpha^\sigma(t)) + \frac{x - \alpha^\sigma(t)}{1 + |x|} & \text{if} \quad x \leq \alpha^\sigma(t). \end{cases}$$

Note that F is continuous and bounded on $[a, b] \times \mathbb{R}$ and $F(t, x) = f(t, x)$ whenever $\alpha^\sigma(t) \leq x \leq \beta^\sigma(t)$ for $t \in [a, b]$. By Corollary 6.52, the BVP

$$x^{\Delta\Delta} = F(t, x^\sigma), \quad x(a) = A, \quad x(\sigma^2(b)) = B$$

has a solution x. To complete the proof it suffices to show that

$$\alpha(t) \leq x(t) \leq \beta(t) \quad \text{for all} \quad t \in [a, \sigma^2(b)].$$

We now show $x(t) \leq \beta(t)$ for $t \in [a, \sigma^2(b)]$. Assume not, then if $z := x - \beta$, it follows that z has a positive maximum in $(a, \sigma^2(b))$. Choose $c \in (a, \sigma^2(b))$ so that $z(c) = \max\{z(t) : t \in [a, \sigma^2(b)]\} > 0$ and $z(t) < z(c)$ for $t \in (c, \sigma^2(b))$.

There are four cases to consider: c left-dense and right-scattered, isolated, left-scattered and right-dense, and dense. We show that the first case is impossible and in the other cases we show that

$$z^\Delta(c) \leq 0 \quad \text{and} \quad z^{\Delta\Delta}(\rho(c)) \leq 0.$$

Case 1: $\rho(c) = c < \sigma(c)$. Clearly $z^\Delta(c) < 0$. Since $\lim_{t \to c^-} z^\Delta(t) = z^\Delta(c) < 0$, there exists $\delta > 0$ such that $z^\Delta(t) < 0$ on $(c - \delta, c]$. But then z is strictly decreasing on $(c - \delta, c]$, which contradicts the way c was chosen.

Case 2: $\rho(c) < c < \sigma(c)$. It is easy to check that $z^\Delta(c) < 0$ and $z^{\Delta\Delta}(\rho(c)) < 0$.

Case 3: $\rho(c) < c = \sigma(c)$. If $z^\Delta(c) > 0$, then $\lim_{t \to c^+} z^\Delta(t) = z^\Delta(c) > 0$. This implies that there exists $\delta > 0$ such that $z^\Delta(t) > 0$ on $[c, c + \delta)$, but then z is strictly increasing on $[c, c + \delta)$ which contradicts the way c was chosen. Therefore $z^\Delta(c) \leq 0$. Since $\rho(c)$ is right-scattered,

$$z^{\Delta\Delta}(\rho(c)) = \frac{z^\Delta(c) - z^\Delta(\rho(c))}{c - \rho(c)} \leq 0.$$

Case 4: $\rho(c) = c = \sigma(c)$. We show $z^\Delta(c) = 0$ and $z^{\Delta\Delta}(\rho(c)) \le 0$: Using the same proof as in Case 3 we find $z^\Delta(c) \le 0$. If $z^\Delta(c) < 0$, then $\lim_{t \to c} z^\Delta(t) = z^\Delta(c) < 0$. This implies that there exists $\delta > 0$ such that $z^\Delta(t) < 0$ on $(c - \delta, c]$. Hence z is strictly decreasing on $(c - \delta, c]$. But this contradicts the way c was chosen. Hence $z^\Delta(c) = 0$. If $z^{\Delta\Delta}(\rho(c)) > 0$, then (note that c is dense) there exists $\delta > 0$ such that $z^{\Delta\Delta}(t) > 0$ on $(c - \delta, c + \delta)$. Hence $z^\Delta(t)$ is strictly increasing on $(c - \delta, c + \delta)$. But $z^\Delta(c) = 0$ and hence $z^\Delta(t) > 0$ on $(c, c + \delta)$. This implies that z is strictly increasing on $[c, c + \delta)$. But this contradicts the way c was chosen. Therefore $z^{\Delta\Delta}(\rho(c)) \le 0$.

Hence

$$x(c) > \beta(c), \quad x^\Delta(c) \le \beta^\Delta(c), \quad x^{\Delta\Delta}(\rho(c)) \le \beta^{\Delta\Delta}(\rho(c)).$$

But

$$
\begin{aligned}
x^{\Delta\Delta}(\rho(c)) &= F(\rho(c), x^\sigma(\rho(c))) \\
&= f(\rho(c), \beta^\sigma(\rho(c))) + \frac{x^\sigma(\rho(c)) - \beta^\sigma(\rho(c))}{1 + |x^\sigma(\rho(c))|} \\
&= f(\rho(c), \beta^\sigma(\rho(c))) + \frac{x(c) - \beta(c)}{1 + |x(c)|} \\
&> f(\rho(c), \beta^\sigma(\rho(c))) \\
&\ge \beta^{\Delta\Delta}(\rho(c))
\end{aligned}
$$

since $\sigma(\rho(c)) = c$ (recall that we proved in Case 1 that $\rho(c) = c < \sigma(c)$ was not possible), $x(c) > \beta(c)$ and β is an upper solution of (6.13) on $[a, \sigma^2(b)]$. Hence $x^{\Delta\Delta}(\rho(c)) > \beta^{\Delta\Delta}(\rho(c))$. But this contradicts the fact that $x^{\Delta\Delta}(\rho(c)) \le \beta^{\Delta\Delta}(\rho(c))$. Therefore $x(t) \le \beta(t)$ for $t \in [a, \sigma^2(b)]$.

Similarly (Exercise 6.55), one can show that $\alpha(t) \le x(t)$ for $t \in [a, \sigma^2(b)]$. Therefore

$$\alpha(t) \le x(t) \le \beta(t) \quad \text{for all} \quad t \in [a, \sigma^2(b)],$$

and so x solves the BVP (6.14). $\qquad\square$

Exercise 6.55. Show (in the proof of Theorem 6.54) $\alpha(t) \le x(t)$ for $t \in [a, \sigma^2(b)]$.

Example 6.56. Consider the BVP

$$x^{\Delta\Delta} = -\cos x^\sigma, \quad x(0) = x(\sigma^2(b)) = 0.$$

First note that $\alpha \equiv 0$ is a lower solution on $[0, \sigma^2(b)]$ since

$$\alpha^{\Delta\Delta}(t) = 0 > -\cos 0 = -1.$$

Next, let $\beta(t) = \int_0^t (c - s)\Delta s$, where $c = \frac{1}{\sigma^2(b)} \int_0^{\sigma^2(b)} \tau \Delta\tau$. Then

$$\beta^{\Delta\Delta}(t) = -1 \le -\cos \beta^\sigma(t),$$

so β is an upper solution on $[0, \sigma^2(b)]$. Note that we have $\alpha(0) = \beta(0) = 0$ and $\alpha(\sigma^2(b)) = \beta(\sigma^2(b)) = 0$. Since β is a solution of the BVP

$$\beta^{\Delta\Delta} = -1, \quad \beta(0) = \beta(\sigma^2(b)) = 0,$$

it follows that $\beta(t) \ge 0$ on $[0, \sigma^2(b)]$. Therefore, by Theorem 6.54, we can conclude that there is a solution x with

$$0 \le x(t) \le \int_0^t (c - s)\Delta s \quad \text{for all} \quad t \in [0, \sigma^2(b)].$$

Theorem 6.57. *Let f be continuous on $[a,b] \times \mathbb{R}$. Assume that for each fixed $t \in [a,b]$, $f(t,x)$ is nondecreasing in $x \in \mathbb{R}$. Then the BVP (6.14) has a solution. If, for each $t \in [a,b]$, $f(t,x)$ is strictly increasing in $x \in \mathbb{R}$, then the BVP (6.14) has a unique solution.*

Proof. Choose $M \geq \max\{|f(t,0)| : t \in [a,b]\}$. Let u be the solution of the BVP

$$u^{\Delta\Delta} = M, \quad u(a) = u(\sigma^2(b)) = 0.$$

This implies that $u(t) \leq 0$ on $[a, \sigma^2(b)]$. Pick $K \geq \max\{|A|, |B|\}$. Set

$$\alpha(t) = u(t) - K \quad \text{for all} \quad t \in [a, \sigma^2(b)].$$

Then

$$\alpha^{\Delta\Delta}(t) = u^{\Delta\Delta}(t) = M \geq f(t,0) \geq f(t, \alpha^\sigma(t))$$

on $[a,b]$ since for each fixed $t \in [a,b]$, $f(t,x)$ is nondecreasing in $x \in \mathbb{R}$. Therefore α is a lower solution of (6.13) on $[a, \sigma^2(b)]$.

Next, let v be the solution of the BVP

$$v^{\Delta\Delta} = -M, \quad v(a) = v(\sigma^2(b)) = 0.$$

This implies that $v(t) \geq 0$ on $[a, \sigma^2(b)]$. Then set

$$\beta(t) = v(t) + K \quad \text{for all} \quad t \in [a, \sigma^2(b)].$$

It follows that

$$\beta^{\Delta\Delta}(t) = v^{\Delta\Delta}(t) = -M \leq f(t,0) \leq f(t, \beta^\sigma(t))$$

on $[a,b]$ since $f(t,x)$ is nondecreasing in x. Therefore β is an upper solution of (6.13) on $[a, \sigma^2(b)]$. Note that $\alpha(a) \leq A \leq \beta(a)$, $\alpha(\sigma^2(b)) \leq B \leq \beta(\sigma^2(b))$ and $\alpha(t) \leq \beta(t)$ on $[a, \sigma^2(b)]$. Therefore there exists a solution x of the BVP (6.14) with

$$\alpha(t) \leq x(t) \leq \beta(t) \quad \text{for all} \quad t \in [a, \sigma^2(b)]$$

by Theorem 6.54. Now assume that for each $t \in [a,b]$, $f(t,x)$ is strictly increasing in $x \in \mathbb{R}$. Assume x and y are distinct solutions of the BVP (6.14). Without loss of generality, assume $x(t) > y(t)$ at some points in $(a, \sigma^2(b))$. This implies that $z(t) := x(t) - y(t)$ has a positive maximum in $(a, \sigma^2(b))$. Hence there exists $c \in (a, \sigma^2(b))$ so that $z(c) = \max\{z(t) : t \in [a, \sigma^2(b)]\} > 0$ and $z(t) < z(c)$ for $t \in (c, \sigma^2(b)]$. After considering four cases as we did in the proof of Theorem 6.54 we get a similar contradiction. Therefore the BVP (6.14) has a unique solution. $\square$

The following example is a simple application of Theorem 6.57.

Example 6.58. If $c, d, e \in C_{rd}$ with $c(t) \geq 0$, $d(t) \geq 0$ on $[a,b]$, then the BVP

$$x^{\Delta\Delta} = c(t)x^\sigma + d(t)(x^\sigma)^3 + e(t), \quad x(a) = A, \quad x(\sigma^2(b)) = B$$

has a solution. If, in addition, $c(t) + d(t) > 0$ on $[a,b]$, then the above BVP has a unique solution.

The next theorem is a generalization of the uniqueness of solutions of initial value problems for (6.13).

Theorem 6.59. *Let f be continuous on $[a, b] \times \mathbb{R}$. Assume that solutions of initial value problems for (6.13) are unique. Let α and β be lower and upper solutions of (6.13) on $[a, \sigma^2(b)]$ such that $\alpha(t) \leq \beta(t)$ on $[a, \sigma^2(b)]$. If there exists $t_0 \in [a, \sigma(b)]$ such that*

$$\alpha(t_0) = \beta(t_0) \quad and \quad \alpha^\Delta(t_0) = \beta^\Delta(t_0),$$

then $\alpha(t) \equiv \beta(t)$ on $[a, \sigma^2(b)]$.

Proof. Assume $\alpha(t) \not\equiv \beta(t)$ on $[a, \sigma^2(b)]$. First consider the case where $t_0 < \sigma^2(b)$ and $\alpha(t) < \beta(t)$ for at least one point in $(t_0, \sigma^2(b)]$. Pick

$$t_1 = \max\{t : \ \alpha(s) = \beta(s), \ t_0 \leq s \leq t\} < \sigma^2(b).$$

We have two cases to consider. Case 1: $\sigma(t_1) = t_1$. In this case there exists $t_2 \in \mathbb{T}$ with $t_1 < t_2$ such that $\alpha(t) < \beta(t)$ on $(t_1, t_2]$. By Theorem 6.54, the BVP

$$x^{\Delta\Delta} = f(t, x^\sigma), \quad x(t_1) = \beta(t_1), \quad x(t_2) = \beta(t_2)$$

has a solution x_1 satisfying $\alpha(t) \leq x_1(t) \leq \beta(t)$ on $[t_1, t_2]$. Similarly, the BVP

$$x^{\Delta\Delta} = f(t, x^\sigma), \quad x(t_1) = \alpha(t_1), \quad x(t_2) = \alpha(t_2)$$

has a solution x_2 satisfying $\alpha(t) \leq x_2(t) \leq \beta(t)$ on $[t_1, t_2]$. Since $\alpha(t) \leq x_i(t) \leq \beta(t)$, $i = 1, 2$, $t \in [t_1, t_2]$, $x_1^\Delta(t_1) = x_2^\Delta(t_1)$. Since solutions of initial value problems are unique, $x_1(t) \equiv x_2(t)$. But this contradicts the fact that $x_1(t_2) \neq x_2(t_2)$.

Case 2: $\sigma(t_1) > t_1$. First we need to show that $t_1 > t_0$. Assume not, then $t_1 = t_0$. By our assumptions, $\alpha(\sigma(t_0)) = \beta(\sigma(t_0))$. But this contradicts the way t_1 was chosen and hence $t_1 > t_0$. There are two subcases. Subcase 1: $\rho(t_1) < t_1$. Since $\alpha(\rho(t_1)) = \beta(\rho(t_1))$, $\alpha(t_1) = \beta(t_1)$ and $\alpha(\sigma(t_1)) < \beta(\sigma(t_1))$,

$$\beta^{\Delta\Delta}(\rho(t_1)) > \alpha^{\Delta\Delta}(\rho(t_1)).$$

But

$$\begin{aligned}
\alpha^{\Delta\Delta}(\rho(t_1)) &\geq f(\rho(t_1), \alpha^\sigma(\rho(t_1))) \\
&= f(\rho(t_1), \alpha(t_1)) \\
&= f(\rho(t_1), \beta(t_1)) \\
&= f(\rho(t_1), \beta^\sigma(\rho(t_1))) \\
&\geq \beta^{\Delta\Delta}(\rho(t_1))
\end{aligned}$$

since α and β are lower and upper solutions of (6.13) on $[a, \sigma^2(b)]$. This is a contradiction.

Subcase 2: $\rho(t_1) = t_1$. By continuity

$$\beta^\Delta(t_1) = \lim_{t \to t_1^-} \beta^\Delta(t) = \lim_{t \to t_1^-} \alpha^\Delta(t) = \alpha^\Delta(t_1).$$

This implies that $\beta(\sigma(t_1)) = \alpha(\sigma(t_1))$ and we get a contradiction to the way t_1 was chosen. Therefore $\alpha(t) \equiv \beta(t)$ on $[t_0, \sigma^2(b)]$.

Next consider the other case where $a < t_0$ and $\alpha(t) < \beta(t)$ for at least one point in $[a, t_0)$. This time pick

$$t_1 = \min\{t : \ \alpha(s) = \beta(s), \ t \leq s \leq t_0\} > a.$$

We have two cases. Case 1: $\rho(t_1) = t_1$. In this case there exists $t_2 \in \mathbb{T}$ with $t_2 < t_1$ such that $\alpha(t) < \beta(t)$ on $[t_2, t_1)$. By Theorem 6.54, the BVP

$$x^{\Delta\Delta} = f(t, x^\sigma), \quad x(t_1) = \beta(t_1), \quad x(t_2) = \beta(t_2)$$

has a solution x_1 satisfying $\alpha(t) \le x_1(t) \le \beta(t)$ on $[t_2, t_1]$. Similarly, the BVP

$$x^{\Delta\Delta} = f(t, x^\sigma), \quad x(t_1) = \alpha(t_1), \quad x(t_2) = \alpha(t_2)$$

has a solution x_2 satisfying $\alpha(t) \le x_2(t) \le \beta(t)$ on $[t_2, t_1]$. Since $\alpha(t) \le x_i(t) \le \beta(t)$ for $i = 1, 2$, $t \in [t_2, t_1]$, $x_1^\Delta(t_1) = x_2^\Delta(t_1)$. Since solutions of initial value problems are unique, $x_1(t) \equiv x_2(t)$ on $[t_2, t_1]$. But this contradicts the fact that $x_1(t_2) \ne x_2(t_2)$.

Case 2: $\rho(t_1) < t_1$. In this case we have $\alpha(t_1) = \beta(t_1)$, $\alpha^\Delta(t_1) = \beta^\Delta(t_1)$, and $\alpha(\rho(t_1)) < \beta(\rho(t_1))$. Hence

$$\beta^{\Delta\Delta}(\rho(t_1)) > \alpha^{\Delta\Delta}(\rho(t_1)).$$

But

$$
\begin{aligned}
\alpha^{\Delta\Delta}(\rho(t_1)) &\ge f(\rho(t_1), \alpha^\sigma(\rho(t_1))) \\
&= f(\rho(t_1), \beta^\sigma(\rho(t_1))) \\
&\ge \beta^{\Delta\Delta}(\rho(t_1)),
\end{aligned}
$$

and so this is a contradiction. Therefore $\alpha(t) \equiv \beta(t)$ on $[a, t_0]$. Hence $\alpha(t) \equiv \beta(t)$ on $[a, \sigma^2(b)]$. $\qquad\square$

Exercise 6.60. By Theorem 6.59, show that if the linear equation $x^{\Delta\Delta} + p(t)x^\sigma = 0$ has a positive upper solution on $[a, \sigma^2(b)]$, then it is disconjugate on $[a, \sigma^2(b)]$.

In the next theorem we prove an existence-uniqueness theorem for solutions of the BVP (6.14), where we assume that $f(t, x)$ satisfies a one-sided Lipschitz condition.

Theorem 6.61. *Let f be continuous on $[a, b] \times \mathbb{R}$. Assume that solutions of initial value problems for (6.13) are unique, continuous with respect to initial conditions, and exist on $[a, \sigma^2(b)]$. Moreover, suppose there exists $k \in C_{rd}([a, b])$ with*

$$f(t, x) - f(t, y) \ge k(t)(x - y) \quad for \quad x \ge y, \ t \in [a, b].$$

If $x^{\Delta\Delta} = k(t)x^\sigma$ is disconjugate on $[a, \sigma^2(b)]$, then the BVP (6.14) has a unique solution.

Proof. Let $x(t, m)$ be the solution of the IVP

$$x^{\Delta\Delta} = f(t, x^\sigma), \quad x(a) = A, \quad x^\Delta(a) = m.$$

Define $S := \{x(\sigma^2(b), m) : m \in \mathbb{R}\}$. By continuity of solutions with respect to initial conditions, S is a connected set. We want to show that S is neither bounded above nor below. Fix $m_1 > m_2$ and let

$$w(t) := x(t, m_1) - x(t, m_2).$$

Note that $w(a) = 0$ and $w^\Delta(a) = m_1 - m_2 > 0$.

We now show that $w(t) > 0$ on $(a, \sigma^2(b)]$. Let

$$t_1 = \max\{t \in [a, \sigma^2(b)] : w(s) \ge 0 \text{ for } s \in [a, t]\}.$$

Then $t_1 \ge a$ and

(6.15) $$w(\sigma(t_1)) \le 0.$$

Now

$$w^{\Delta\Delta}(t) \;=\; x^{\Delta\Delta}(t, m_1) - x^{\Delta\Delta}(t, m_2)$$
$$=\; f(t, x^\sigma(t, m_1)) - f(t, x^\sigma(t, m_2))$$
$$\geq\; k(t)\left[x^\sigma(t, m_1) - x^\sigma(t, m_2)\right]$$

on $[a, t_1)$. Define an operator L by

$$Lx(t) := x^{\Delta\Delta}(t) - k(t)x^\sigma(t) \quad \text{for} \quad t \in [a, b].$$

Put $h(t) = Lw(t)$, and so $h(t) \geq 0$ on $[a, t_1)$. Then w is the solution of the IVP

$$Lx(t) = h(t), \quad x(a) = 0, \quad x^\Delta(a) = m_1 - m_2.$$

Let v be the solution of the IVP

$$Lx(t) = 0, \quad x(a) = 0, \quad x^\Delta(a) = m_1 - m_2.$$

It is clear that $w - v$ is the solution of the IVP

$$Lx(t) = h(t), \quad x(a) = x^\Delta(a) = 0.$$

By the variation of constants formula,

$$(w - v)(t) = \int_a^t k(t, s)h(s)\Delta s,$$

where $k(t, s)$ is the Cauchy function for $Lx = 0$. In particular

$$(w - v)(\sigma(t_1)) \;=\; \int_a^{\sigma(t_1)} k(\sigma(t_1), s)h(s)\Delta s$$
$$=\; \int_a^{t_1} k(\sigma(t_1), s)h(s)\Delta s + \int_{t_1}^{\sigma(t_1)} k(\sigma(t_1), s)h(s)\Delta s$$
$$=\; \int_a^{t_1} k(\sigma(t_1), s)h(s)\Delta s + k(\sigma(t_1), t_1)h(t_1)\mu(t_1)$$
$$=\; \int_a^{t_1} k(\sigma(t_1), s)h(s)\Delta s$$
$$\geq\; 0,$$

where we have used $h(s) \geq 0$ on $[a, t_1)$, the value of $\int_a^{t_1} k(\sigma(t_1), s)h(s)\Delta s$ does not depend on the value of the integrand at t_1, and $k(t, s) \geq 0$ for $t \geq \sigma(s)$ by disconjugacy of $Lx = 0$ on $[a, \sigma^2(b)]$. This implies that $w(\sigma(t_1)) \geq v(\sigma(t_1)) > 0$ using the fact that $Lx = 0$ is disconjugate on $[a, \sigma^2(b)]$. But this contradicts (6.15).

Hence $w(t) > 0$ on $(a, \sigma^2(b)]$. In particular

$$w(\sigma^2(b)) \geq v(\sigma^2(b)) > 0.$$

Fix m_2 and let $m_1 \to \infty$. This implies that

$$\lim_{m_1 \to \infty} x(\sigma^2(b), m_1) = \infty.$$

Therefore S is not bounded above. Fix m_1 and let $m_2 \to -\infty$. This implies that

$$\lim_{m_2 \to -\infty} x(\sigma^2(b), m_2) = -\infty.$$

Therefore S is not bounded below. Hence $S = \mathbb{R}$ and so $B \in S$. This implies that there exists $m_0 \in \mathbb{R}$ such that $x(\sigma^2(b), m_0) = B$. Hence the BVP (6.14) has a solution. Uniqueness follows immediately from the fact that $m_1 > m_2$ implies $x(\sigma^2(b), m_1) > x(\sigma^2(b), m_2)$. $\qquad\square$

6.7. Notes and References

Most of the results in this chapter can be found in the survey paper by Ravi Agarwal, Martin Bohner, and Allan Peterson [3] and in Elvan Akın's PhD thesis [33] (University of Nebraska, Lincoln, 2000); see also Akın [32].

The basic form of Gronwall's inequality is due to Hilger [160] and is also, along with several other consequences, contained in Keller [190]. Bihari type inequalities are discussed in [187]. The proof of Hölder's inequality follows closely the classical strategy from Hardy, Littlewood, and Polya [147] and is first mentioned for the time scales setting in Martin Bohner and Donald Lutz [88]. Minkowski's inequality and the Cauchy–Schwarz inequality are, as usual, easy consequences of Hölder's inequality. Jensen's inequality has been discussed with Elvan Akın and Billûr Kaymakçalan and can be found in the survey paper [3].

The section on Opial's inequality follows Bohner and Kaymakçalan [87]. For the original results we refer to Opial [221] (differential equations) and Lasota [207] (difference equations). Agarwal and Pang present a survey of continuous and discrete results in [19]. A Wirtinger type inequality due to Roman Hilscher [169] is given at the end of Section 6.4.

In Section 6.5, we included many of the results given in Martin Bohner, Stephen Clark, and Jerry Ridenhour [75]. The original work (for differential equations) can be found in Lyapunov [215]. Further results are due to Stephen Clark and Don Hinton (see [101] for differential equations results and [102] for difference equations results). A survey on continuous and discrete Lyapunov inequalities is presented in Chen [93]. Supplementing results from [75], we also refer to Ferhan Atıcı, Gusein Guseinov, and Billûr Kaymakçalan [44] and to Guseinov and Kaymakçalan [145].

CHAPTER 7

Linear Symplectic Dynamic Systems

7.1. Symplectic Systems and Special Cases

In this chapter we investigate so-called symplectic systems of dynamic equations, which have a variety of important equations as their special cases, e.g., linear Hamiltonian dynamic systems or Sturm–Liouville dynamic equations of higher (even) order. Many of the results in this chapter can be found in Došlý and Hilscher [**121**], and are extensions of results by Bohner and Došlý [**76**]. Throughout this chapter we denote by $\mathcal{J}$ the $2n \times 2n$-matrix

$$\mathcal{J} = \begin{pmatrix} 0 & I \\ -I & 0 \end{pmatrix}.$$

We start by recalling the concepts of symplectic and Hamiltonian matrices.

Definition 7.1. A $2n \times 2n$-matrix $\mathcal{A}$ is called

(i) *symplectic* if $\mathcal{A}^* \mathcal{J} \mathcal{A} = \mathcal{J}$;
(ii) *Hamiltonian* if $\mathcal{A}^* \mathcal{J} + \mathcal{J} \mathcal{A} = 0$.

Exercise 7.2. Show that a real 2×2-matrix $\mathcal{A}$ is symplectic iff $det A = 1$.

Note that $\mathcal{A}$ is Hamiltonian iff $\mathcal{A}^* \mathcal{J}$ is Hermitian. This is because of $\mathcal{J}^* = -\mathcal{J}$.

Lemma 7.3. *Let A, B, C, and D be $n \times n$-matrices. Then the $2n \times 2n$-matrix* $\begin{pmatrix} A & B \\ C & D \end{pmatrix}$ *is*

(i) *symplectic iff A^*C and B^*D are Hermitian and $A^*D - C^*B = I$;*
(ii) *Hamiltonian iff B and C are Hermitian and $D = -A^*$.*

Proof. We first calculate

$$\begin{pmatrix} A & B \\ C & D \end{pmatrix}^* \mathcal{J} \begin{pmatrix} A & B \\ C & D \end{pmatrix} = \begin{pmatrix} A^* & C^* \\ B^* & D^* \end{pmatrix} \begin{pmatrix} 0 & I \\ -I & 0 \end{pmatrix} \begin{pmatrix} A & B \\ C & D \end{pmatrix}$$

$$= \begin{pmatrix} -C^* & A^* \\ -D^* & B^* \end{pmatrix} \begin{pmatrix} A & B \\ C & D \end{pmatrix}$$

$$= \begin{pmatrix} A^*C - C^*A & A^*D - C^*B \\ B^*C - D^*A & B^*D - D^*B \end{pmatrix}$$

$$= \begin{pmatrix} A^*C - C^*A & A^*D - C^*B \\ -(A^*D - C^*B)^* & B^*D - D^*B \end{pmatrix},$$

and this is equal to the matrix $\mathcal{J} = \begin{pmatrix} 0 & I \\ -I & 0 \end{pmatrix}$ iff A^*C and B^*D are both Hermitian

and $A^*D - C^*B = I$. Next, we have

$$\begin{pmatrix} A & B \\ C & D \end{pmatrix}^* \mathcal{J} + \mathcal{J} \begin{pmatrix} A & B \\ C & D \end{pmatrix} = \begin{pmatrix} A^* & C^* \\ B^* & D^* \end{pmatrix} \begin{pmatrix} 0 & I \\ -I & 0 \end{pmatrix} + \begin{pmatrix} 0 & I \\ -I & 0 \end{pmatrix} \begin{pmatrix} A & B \\ C & D \end{pmatrix}$$

$$= \begin{pmatrix} -C^* & A^* \\ -D^* & B^* \end{pmatrix} + \begin{pmatrix} C & D \\ -A & -B \end{pmatrix}$$

$$= \begin{pmatrix} C - C^* & A^* + D \\ -D^* - A & B^* - B \end{pmatrix}$$

$$= \begin{pmatrix} C - C^* & A^* + D \\ -(A^* + D)^* & B^* - B \end{pmatrix},$$

and this is equal to the zero-matrix if and only if B and C are both Hermitian and $D = -A^*$.   □

Exercise 7.4. Find constants a, b, c, d so that the 4×4-matrix

$$\mathcal{A} = \begin{pmatrix} 2 & 0 & 2 & 0 \\ 0 & 1 & 0 & 2 \\ 3 & 0 & a & b \\ 0 & 3 & c & d \end{pmatrix}$$

is symplectic.

Lemma 7.5. *Let A, B, C, and D be $n \times n$-matrices. Then $\begin{pmatrix} A & B \\ C & D \end{pmatrix}$ is symplectic iff it is invertible with*

$$\begin{pmatrix} A & B \\ C & D \end{pmatrix}^{-1} = \begin{pmatrix} D^* & -B^* \\ -C^* & A^* \end{pmatrix}.$$

Proof. The calculation

$$\begin{pmatrix} D^* & -B^* \\ -C^* & A^* \end{pmatrix} \begin{pmatrix} A & B \\ C & D \end{pmatrix} = \begin{pmatrix} D^*A - B^*C & D^*B - B^*D \\ A^*C - C^*A & A^*D - C^*B \end{pmatrix}$$

together with Lemma 7.3 (i) proves the claim.   □

Remark 7.6. Because of Lemma 7.5, we also have

$$\begin{aligned}
I &= \begin{pmatrix} A & B \\ C & D \end{pmatrix} \begin{pmatrix} D^* & -B^* \\ -C^* & A^* \end{pmatrix} \\
&= \begin{pmatrix} AD^* - BC^* & BA^* - AB^* \\ CD^* - DC^* & DA^* - CB^* \end{pmatrix} \\
&= \begin{pmatrix} AD^* - BC^* & BA^* - AB^* \\ CD^* - DC^* & (AD^* - BC^*)^* \end{pmatrix}
\end{aligned}$$

and therefore condition (i) in Lemma 7.3 is also equivalent to

$$AB^* \text{ and } CD^* \text{ are Hermitian and } AD^* - BC^* = I.$$

Definition 7.7. A $2n \times 2n$-matrix-valued function $S \in C_{rd}$ is called *symplectic with respect to* $\mathbb{T}$ provided

$$(7.1) \qquad S^*(t)\mathcal{J} + \mathcal{J}S(t) + \mu(t)S^*(t)\mathcal{J}S(t) = 0 \quad \text{for all} \quad t \in \mathbb{T}.$$

The system

$$(7.2) \qquad z^{\Delta} = S(t)z, \quad \text{where } S \text{ satisfies } (7.1)$$

is then called a *linear symplectic dynamic system*.

Example 7.8. Here we consider the cases $\mathbb{T} = \mathbb{R}$ and $\mathbb{T} = \mathbb{Z}$.

(i) S is symplectic with respect to $\mathbb{R}$ if and only if it is *Hamiltonian*, i.e., iff

$$S^*(t)\mathcal{J} + \mathcal{J}S(t) = 0 \quad \text{holds for all} \quad t \in \mathbb{R}.$$

(ii) S is symplectic with respect to $\mathbb{Z}$ if and only if $I + S$ is *symplectic*, i.e., iff

$$(I + S(t))^* \mathcal{J}(I + S(t)) = \mathcal{J} \quad \text{holds for all} \quad t \in \mathbb{Z}.$$

This can be seen from the identity

$$(I + S)^* \mathcal{J}(I + S) = S^* \mathcal{J} + \mathcal{J}S + S^* \mathcal{J}S + \mathcal{J}.$$

Lemma 7.9. *If S is symplectic with respect to $\mathbb{T}$, then $I + \mu S$ is symplectic.*

Proof. The identity

$$\begin{aligned}(I + \mu S)^* \mathcal{J}(I + \mu S) &= \mathcal{J} + \mu S^* \mathcal{J} + \mu \mathcal{J}S + \mu^2 S^* \mathcal{J}S \\ &= \mathcal{J} + \mu [S^* \mathcal{J} + \mathcal{J}S + \mu S^* \mathcal{J}S]\end{aligned}$$

immediately establishes the lemma. □

Exercise 7.10. Show that the converse of Lemma 7.9 is not true in general.

Theorem 7.11. *If S is symplectic with respect to $\mathbb{T}$, then it is regressive (with respect to $\mathbb{T}$).*

Proof. Assume that S is symplectic with respect to $\mathbb{T}$. By Lemma 7.9, $I + \mu S$ is symplectic (on $\mathbb{T}$), and by Lemma 7.5, $I + \mu S$ is invertible (on $\mathbb{T}$). By the definition of regressivity, Definition 5.5, we conclude that S is regressive with respect to $\mathbb{T}$. □

An immediate conclusion of Theorem 7.11 and of Theorem 5.8 is the following.

Corollary 7.12. *Suppose $S \in C_{rd}$ is a $2n \times 2n$-matrix-valued function which is symplectic with respect to $\mathbb{T}$. Let $t_0 \in \mathbb{T}$ and $z_0 \in \mathbb{R}^{2n}$. Then the initial value problem*

$$z^\Delta = S(t)z, \quad z(t_0) = z_0$$

has a unique solution $z : \mathbb{T} \to \mathbb{R}^{2n}$.

Theorem 7.13. *Let A, B, C, and D be $n \times n$-matrix-valued functions on $\mathbb{T}$. Then*

$$\begin{pmatrix} A & B \\ C & D \end{pmatrix} \text{ is symplectic with respect to } \mathbb{T} \text{ if and only if}$$

(7.3)
$$\begin{cases} B^*(I + \mu D) = (I + \mu D)^* B, \\ C^*(I + \mu A) = (I + \mu A)^* C, \\ (I + \mu A)^* D = \mu C^* B - A^*, \end{cases}$$

i.e., $B^*(I + \mu D)$, $C^*(I + \mu A)$ are Hermitian, and $(I + \mu A)^* D = \mu C^* B - A^*$.

Proof. Assume that $\mathcal{S} = \begin{pmatrix} A & B \\ C & D \end{pmatrix}$ is symplectic with respect to $\mathbb{T}$. Using the

calculations from the proof of Lemma 7.3, we find

$$
\mathcal{S}^*\mathcal{J} + \mathcal{J}\mathcal{S} + \mu\mathcal{S}^*\mathcal{J}\mathcal{S} = \begin{pmatrix} A^* & C^* \\ B^* & D^* \end{pmatrix}\begin{pmatrix} 0 & I \\ -I & 0 \end{pmatrix} + \begin{pmatrix} 0 & I \\ -I & 0 \end{pmatrix}\begin{pmatrix} A & B \\ C & D \end{pmatrix}
$$

$$
+\mu\begin{pmatrix} A^* & C^* \\ B^* & D^* \end{pmatrix}\begin{pmatrix} 0 & I \\ -I & 0 \end{pmatrix}\begin{pmatrix} A & B \\ C & D \end{pmatrix}
$$

$$
= \begin{pmatrix} C - C^* & D + A^* \\ -(D + A^*)^* & B^* - B \end{pmatrix} + \mu\begin{pmatrix} A^*C - C^*A & A^*D - C^*B \\ -(A^*D - C^*B)^* & B^*D - D^*B \end{pmatrix}
$$

$$
= \begin{pmatrix} (I + \mu A)^*C - C^*(I + \mu A) & (I + \mu A)^*D + A^* - \mu C^*B \\ -[(I + \mu A)^*D + A^* - \mu C^*B]^* & B^*(I + \mu D) - (I + \mu D)^*B \end{pmatrix},
$$

and this is the zero-matrix iff $(I + \mu A)^*C$ and $(I + \mu D)^*B$ are both Hermitian and $(I + \mu A)^*D = \mu C^*B - A^*$. $\qquad \square$

Exercise 7.14. Show that z solves

$$
z^\sigma = (I + \mu(t)\mathcal{S}(t))z
$$

provided z is a solution of the symplectic system (7.2).

Symplectic systems are a very general class of objects. They cover as special cases so-called linear Hamiltonian systems. Those are defined next. We also will use the $2n \times 2n$-matrix $\mathcal{M}$ defined by

$$
\mathcal{M} = \begin{pmatrix} 0 & 0 \\ I & 0 \end{pmatrix}.
$$

Definition 7.15. A $2n \times 2n$-matrix-valued function $\mathcal{H}$ is called *Hamiltonian with respect to* $\mathbb{T}$ provided

(7.4) $\mathcal{H}(t)$ is Hamiltonian and $I - \mu(t)\mathcal{H}(t)\mathcal{M}^*\mathcal{M}$ is invertible for all $t \in \mathbb{T}$.

The system

(7.5) $z^\Delta = \mathcal{H}(t)[\mathcal{M}^*\mathcal{M}z^\sigma + \mathcal{M}\mathcal{M}^*z]$,   where $\mathcal{H}$ satisfies (7.4)

is then called a *linear Hamiltonian dynamic system*.

Remark 7.16. By Example 7.8 (i), we have that $\mathcal{H} = \begin{pmatrix} \mathcal{A} & \mathcal{B} \\ \mathcal{C} & -\mathcal{A}^* \end{pmatrix}$ for some $n \times n$-matrix-valued functions $\mathcal{A}, \mathcal{B}, \mathcal{C}$ such that $\mathcal{B}$ and $\mathcal{C}$ are Hermitian. Also,

$$\mathcal{M}^*\mathcal{M} = \begin{pmatrix} I & 0 \\ 0 & 0 \end{pmatrix} \quad \text{and} \quad \mathcal{M}\mathcal{M}^* = \begin{pmatrix} 0 & 0 \\ 0 & I \end{pmatrix}.$$

Hence, by putting $z = \begin{pmatrix} x \\ u \end{pmatrix}$, we can rewrite the system (7.5) as

$$x^\Delta = \mathcal{A}(t)x^\sigma + \mathcal{B}(t)u, \quad u^\Delta = \mathcal{C}(t)x^\sigma - \mathcal{A}^*(t)u.$$

Next, the calculation

$$I - \mu\mathcal{H}\mathcal{M}^*\mathcal{M} = I - \mu \begin{pmatrix} \mathcal{A} & \mathcal{B} \\ \mathcal{C} & -\mathcal{A}^* \end{pmatrix} \begin{pmatrix} I & 0 \\ 0 & 0 \end{pmatrix} = \begin{pmatrix} I - \mu\mathcal{A} & 0 \\ -\mu\mathcal{C} & I \end{pmatrix}$$

shows that $I - \mu\mathcal{H}\mathcal{M}^*\mathcal{M}$ is invertible iff $I - \mu\mathcal{A}$ is so. Therefore, condition (7.4) may be rewritten as

$$\mathcal{H} = \begin{pmatrix} \mathcal{A} & \mathcal{B} \\ \mathcal{C} & -\mathcal{A}^* \end{pmatrix}, \quad I - \mu\mathcal{A} \text{ is invertible and } \mathcal{B}, \mathcal{C} \text{ are Hermitian.}$$

Note also that, in case $\mathbb{T} = \mathbb{R}$, there is no difference between a symplectic system (7.2) and a Hamiltonian system (7.5), while these objects are usually not the same, e.g., for $\mathbb{T} = \mathbb{Z}$. Their exact relationship in general is described next.

Theorem 7.17. *Every Hamiltonian system (7.5) is also a symplectic system (7.2), with*

(7.6) $$\mathcal{S} = (I - \mu\mathcal{H}\mathcal{M}^*\mathcal{M})^{-1}\mathcal{H}.$$

Furthermore, if a symplectic system (7.2) is such that $I + \mu\mathcal{M}^*\mathcal{M}\mathcal{S}$ is invertible, then (and only then) it is also a Hamiltonian system (7.5), with

$$\mathcal{H} = \mathcal{S}(I + \mu\mathcal{M}^*\mathcal{M}\mathcal{S})^{-1}.$$

Proof. We assume first that z solves (7.5). Substitute z^σ by $z + \mu z^\Delta$ and observe $\mathcal{M}^*\mathcal{M} + \mathcal{M}\mathcal{M}^* = I$ to obtain $z^\Delta = \mathcal{H}(z + \mu\mathcal{M}^*\mathcal{M}z^\Delta)$ and hence $(I - \mu\mathcal{H}\mathcal{M}^*\mathcal{M})z^\Delta = \mathcal{H}z$ so that (7.6) (observe (7.4)) follows. It remains to verify that $\mathcal{S}$ defined by (7.6) is symplectic with respect to $\mathbb{T}$, i.e., satisfies condition (7.1). The remainder of this proof is Exercise 7.18. $\qquad\square$

Exercise 7.18. Finish the proof of Theorem 7.17.

We conclude this section by presenting three classes of important examples of symplectic systems: Sturm–Liouville dynamic equations, self-adjoint vector dynamic equations, and linear Hamiltonian dynamic systems.

Example 7.19. Recall the second order Sturm–Liouville equation (4.1) which has been studied in Chapter 4, namely

$$(p(t)y^\Delta)^\Delta + q(t)y^\sigma = 0.$$

We will now rewrite equation (4.1) as a symplectic system of the form (7.2). Introduce a "state variable" x and a "control variable" u by

$$x = y \quad \text{and} \quad u = py^\Delta.$$

Then (4.1) is equivalent to the system of dynamic equations

$$x^\Delta = \frac{1}{p(t)}u, \quad u^\Delta = -q(t)x^\sigma,$$

i.e., with $z = \begin{pmatrix} x \\ u \end{pmatrix}$, to

$$
\begin{aligned}
z^\Delta &= \begin{pmatrix} 0 & \frac{1}{p(t)} \\ -q(t) & 0 \end{pmatrix} \begin{pmatrix} x^\sigma \\ u \end{pmatrix} \\
&= \begin{pmatrix} 0 & \frac{1}{p(t)} \\ -q(t) & 0 \end{pmatrix} \left[\begin{pmatrix} 1 & 0 \\ 0 & 0 \end{pmatrix} z^\sigma + \begin{pmatrix} 0 & 0 \\ 0 & 1 \end{pmatrix} z \right] \\
&= \mathcal{H}(t) \left[\mathcal{M}^* \mathcal{M} z^\sigma + \mathcal{M} \mathcal{M}^* z \right]
\end{aligned}
$$

with

$$\mathcal{H} = \begin{pmatrix} 0 & \frac{1}{p} \\ -q & 0 \end{pmatrix}.$$

Note that

$$\mathcal{H}^* \mathcal{J} + \mathcal{J} \mathcal{H} = \begin{pmatrix} 0 & -q \\ \frac{1}{p} & 0 \end{pmatrix} \begin{pmatrix} 0 & 1 \\ -1 & 0 \end{pmatrix} + \begin{pmatrix} 0 & 1 \\ -1 & 0 \end{pmatrix} \begin{pmatrix} 0 & \frac{1}{p} \\ -q & 0 \end{pmatrix} = 0.$$

Hence $\mathcal{H}$ is Hamiltonian (and this could have been seen also by applying Lemma 7.3 (ii)). We next have

$$
\begin{aligned}
I - \mu \mathcal{H} \mathcal{M}^* \mathcal{M} \ &= \
\begin{pmatrix} 1 & 0 \\ 0 & 1 \end{pmatrix}
- \mu
\begin{pmatrix} 0 & \frac{1}{p} \\ -q & 0 \end{pmatrix}
\begin{pmatrix} 1 & 0 \\ 0 & 0 \end{pmatrix} \\[2mm]
&= \
\begin{pmatrix} 1 & 0 \\ 0 & 1 \end{pmatrix}
- \mu
\begin{pmatrix} 0 & 0 \\ -q & 0 \end{pmatrix} \\[2mm]
&= \
\begin{pmatrix} 1 & 0 \\ \mu q & 1 \end{pmatrix},
\end{aligned}
$$

which is invertible (and this could have been seen also by applying Remark 7.16). Therefore $\mathcal{H}$ is Hamiltonian with respect to $\mathbb{T}$. The system

$$
z^{\Delta} = \mathcal{H}(t) \left[\mathcal{M}^* \mathcal{M} z^{\sigma} + \mathcal{M} \mathcal{M}^* z \right]
$$

hence is a Hamiltonian system. By Theorem 7.17, it is also a symplectic system of the form (7.2), where $\mathcal{S}$ is given in formula (7.6) as

$$
\begin{aligned}
\mathcal{S} \ &= \ (I - \mu \mathcal{H} \mathcal{M}^* \mathcal{M})^{-1} \mathcal{H} \\[2mm]
&= \
\begin{pmatrix} 1 & 0 \\ \mu q & 1 \end{pmatrix}^{-1}
\begin{pmatrix} 0 & \frac{1}{p} \\ -q & 0 \end{pmatrix} \\[2mm]
&= \
\begin{pmatrix} 1 & 0 \\ -\mu q & 1 \end{pmatrix}
\begin{pmatrix} 0 & \frac{1}{p} \\ -q & 0 \end{pmatrix} \\[2mm]
&= \
\begin{pmatrix} 0 & \frac{1}{p} \\ -q & -\frac{\mu q}{p} \end{pmatrix}.
\end{aligned}
$$

Example 7.20. Similar to the previous example we may rewrite the self-adjoint vector equation (5.17) from Section 5.3, namely

$$
(P(t) Y^{\Delta})^{\Delta} + Q(t) Y^{\sigma} = 0,
$$

as a symplectic system. As in the previous example we introduce

$$
X = Y \quad \text{and} \quad U = P Y^{\Delta},
$$

and then (5.17) is equivalent to

$$
Z^{\Delta} = \mathcal{H}(t) \left[\mathcal{M}^* \mathcal{M} Z^{\sigma} + \mathcal{M} \mathcal{M}^* Z \right]
$$

with $Z = \begin{pmatrix} X \\ U \end{pmatrix}$ and

$$\mathcal{H} = \begin{pmatrix} 0 & P^{-1} \\ -Q & 0 \end{pmatrix}.$$

By Lemma 7.3 (ii), $\mathcal{H}$ is Hamiltonian and

$$
\begin{aligned}
I - \mu \mathcal{H} \mathcal{M}^* \mathcal{M} &= \begin{pmatrix} I & 0 \\ 0 & I \end{pmatrix} - \mu \begin{pmatrix} 0 & P^{-1} \\ -Q & 0 \end{pmatrix} \begin{pmatrix} I & 0 \\ 0 & 0 \end{pmatrix} \\
&= \begin{pmatrix} I & 0 \\ 0 & I \end{pmatrix} - \mu \begin{pmatrix} 0 & 0 \\ -Q & 0 \end{pmatrix} \\
&= \begin{pmatrix} I & 0 \\ \mu Q & I \end{pmatrix}
\end{aligned}
$$

is invertible, so that (7.4) is satisfied and $\mathcal{H}$ is Hamiltonian with respect to $\mathbb{T}$. By Theorem 7.17, the system can be rewritten as a symplectic system (7.2) with

$$
\begin{aligned}
\mathcal{S} &= (I - \mu \mathcal{H} \mathcal{M}^* \mathcal{M})^{-1} \mathcal{H} \\
&= \begin{pmatrix} I & 0 \\ \mu Q & I \end{pmatrix}^{-1} \begin{pmatrix} 0 & P^{-1} \\ -Q & 0 \end{pmatrix} \\
&= \begin{pmatrix} I & 0 \\ -\mu Q & I \end{pmatrix} \begin{pmatrix} 0 & P^{-1} \\ -Q & 0 \end{pmatrix} \\
&= \begin{pmatrix} 0 & P^{-1} \\ -Q & -\mu Q P^{-1} \end{pmatrix}.
\end{aligned}
$$

Example 7.21. Assume a Hamiltonian system

$$z^\Delta = \mathcal{H}(t) \left[\mathcal{M}^* \mathcal{M} z^\sigma + \mathcal{M} \mathcal{M}^* z \right]$$

is given with

$$\mathcal{H} = \begin{pmatrix} A & B \\ C & -A^* \end{pmatrix}$$

such that B and C are Hermitian and $I - \mu A$ is invertible (see Remark 7.16). We put

$$\tilde{A} = (I - \mu A)^{-1}.$$

Then, by Theorem 7.17, this system can be rewritten as a symplectic system of the form (7.2), where $\mathcal{S}$ is given in formula (7.6) as

$$
\begin{aligned}
\mathcal{S} &= (I - \mu \mathcal{H} \mathcal{M}^* \mathcal{M})^{-1} \mathcal{H} \\[2mm]
&= \left[\begin{pmatrix} I & 0 \\ 0 & I \end{pmatrix} - \mu \begin{pmatrix} A & B \\ C & -A^* \end{pmatrix} \begin{pmatrix} I & 0 \\ 0 & 0 \end{pmatrix} \right]^{-1} \mathcal{H} \\[2mm]
&= \begin{pmatrix} I - \mu A & 0 \\ -\mu C & I \end{pmatrix}^{-1} \mathcal{H} \\[2mm]
&= \begin{pmatrix} \tilde{A} & 0 \\ \mu C \tilde{A} & I \end{pmatrix} \begin{pmatrix} A & B \\ C & -A^* \end{pmatrix} \\[2mm]
&= \begin{pmatrix} \tilde{A} A & \tilde{A} B \\ \mu C \tilde{A} A + C & \mu C \tilde{A} B - A^* \end{pmatrix} \\[2mm]
&= \begin{pmatrix} \tilde{A} A & \tilde{A} B \\ C \tilde{A} & \mu C \tilde{A} B - A^* \end{pmatrix},
\end{aligned}
$$

where the last equal sign follows from

$$(7.7) \qquad I + \mu \tilde{A} A = \tilde{A}(I - \mu A) + \mu \tilde{A} A = \tilde{A}(I - \mu A + \mu A) = \tilde{A}.$$

It is not obvious at all that $\mathcal{S}$ is symplectic with respect to $\mathbb{T}$ although Theorem 7.17 guarantees this fact. It also can be seen directly using Theorem 7.13 as follows: First,

$$(I + \mu \tilde{A} A)^* C \tilde{A} = \tilde{A}^* C \tilde{A}$$

and

$$
\begin{aligned}
\left(I + \mu [\mu C \tilde{A} B - A^*] \right)^* \tilde{A} B &= [I - \mu A + \mu^2 B \tilde{A}^* C] \tilde{A} B \\
&= B + \mu^2 B \tilde{A}^* C \tilde{A} B
\end{aligned}
$$

are both Hermitian, and

$$
\begin{aligned}
(I + \mu \tilde{A} A)^* [\mu C \tilde{A} B - A^*] &= \tilde{A}^* [\mu C \tilde{A} B - A^*] \\
&= \mu (C \tilde{A})^* \tilde{A} B - (A \tilde{A})^* \\
&= \mu (C \tilde{A})^* \tilde{A} B - (\tilde{A} A)^*,
\end{aligned}
$$

where we have used formula (7.7) and the fact that A and $\tilde{A}$ commute, which can be seen from

$$A(I - \mu A) = (I - \mu A) A.$$

Exercise 7.22. Show that the higher order Sturm–Liouville difference equation

$$(7.8) \qquad \sum_{\nu=0}^{n} \Delta^{\nu}(r_k^{(\nu)} \Delta^{\nu} y_{k+n-\nu}) = 0 \quad \text{with} \quad r_k^{(n)} \neq 0$$

can be written in the form of (7.5), where

$$A = (a_{ij})_{1 \leq i,j \leq n} \quad \text{with} \quad a_{ij} = \begin{cases} 1 & \text{if} \quad j = i+1, \ 1 \leq i \leq n-1 \\ 0 & \text{otherwise,} \end{cases}$$

$$B_k = \text{diag}\left\{0, \ldots, 0, \frac{1}{r_k^{(n)}}\right\}, \quad C_k = \text{diag}\left\{r_k^{(0)}, \ldots, r_k^{(n-1)}\right\},$$

using the substitution

$$x_k = \begin{pmatrix} y_{k+n-1} \\ \Delta y_{k+n-2} \\ \vdots \\ \Delta^{n-1} y_k \end{pmatrix}, \quad u_k = \begin{pmatrix} \sum_{\nu=1}^{n}(-1)^{\nu-1}\Delta^{\nu-1}\left(r_k^{(\nu)}\Delta^{\nu} y_{k+n-\nu}\right) \\ \vdots \\ -\Delta\left(r_k^{(n)}\Delta^n y_k\right) + r_k^{(n-1)}\Delta^{n-1} y_{k+1} \\ r_k^{(n)}\Delta^n y_k \end{pmatrix}.$$

Write down the dynamic version of equation (7.8) and explain how it can be written as a system of the form (7.5).

7.2. Conjoined Bases

Theorem 7.23 (Wronskian Identity). *Let f and g both be solutions of (7.2). Then f^*Jg is a constant.*

Proof. We use Exercise 7.14 and the product formula to obtain

$$\begin{aligned} (f^*Jg)^{\Delta} &= (f^*)^{\Delta}Jg^{\sigma} + f^*Jg^{\Delta} \\ &= f^*S^*J(I + \mu S)g + f^*JSg \\ &= f^*\{S^*J + JS + \mu S^*JS\}g \\ &= 0. \end{aligned}$$

Hence, f^*Jg is constant according to Corollary 1.68 (ii).   □

Remark 7.24. Let $Z = \begin{pmatrix} X \\ U \end{pmatrix}$ and $\tilde{Z} = \begin{pmatrix} \tilde{X} \\ \tilde{U} \end{pmatrix}$ be solutions of (7.2), where X, U, $\tilde{X}$, $\tilde{U}$ are $n \times n$-matrix-valued. The Wronskian identity says that

$$
\begin{aligned}
Z^* J \tilde{Z} &= \begin{pmatrix} X \\ U \end{pmatrix}^* \begin{pmatrix} 0 & I \\ -I & 0 \end{pmatrix} \begin{pmatrix} \tilde{X} \\ \tilde{U} \end{pmatrix} \\[2mm]
&= \begin{pmatrix} X \\ U \end{pmatrix}^* \begin{pmatrix} \tilde{U} \\ -\tilde{X} \end{pmatrix} \\[2mm]
&= X^* \tilde{U} - U^* \tilde{X}
\end{aligned}
$$

is a constant $n \times n$-matrix-valued function on $\mathbb{T}$.

Definition 7.25. We say that a $2n \times n$-matrix-valued solution Z of (7.2) is a *conjoined solution* if $Z^* J Z \equiv 0$. A *conjoined basis* of (7.2) is a conjoined solution Z satisfying rank$Z \equiv n$. Two conjoined bases Z and $\tilde{Z}$ are called *normalized* provided $Z^* J \tilde{Z} \equiv I$.

Remark 7.26. If we put $Z = \begin{pmatrix} X \\ U \end{pmatrix}$ with $n \times n$-matrix-valued functions X and U, then

$$Z^* J Z = X^* U - U^* X$$

according to Remark 7.24. Hence Z is a conjoined solution of (7.2) if $\begin{pmatrix} X \\ U \end{pmatrix}$ solves (7.2) such that

$$X^* U \quad \text{is Hermitian on} \quad \mathbb{T}.$$

Also, two conjoined solutions $\begin{pmatrix} X \\ U \end{pmatrix}$ and $\begin{pmatrix} \tilde{X} \\ \tilde{U} \end{pmatrix}$ of (7.2) are normalized conjoined bases of (7.2) if

$$X^* \tilde{U} - U^* \tilde{X} \equiv I \quad \text{on} \quad \mathbb{T}.$$

Lemma 7.27. *Two solutions Z and $\tilde{Z}$ of (7.2) are normalized if and only if the $2n \times 2n$-matrix-valued function $\mathcal{Z} = (Z \ \tilde{Z})$ is symplectic.*

Proof. For $\mathcal{Z} = (Z \ \tilde{Z})$ we have

$$
\mathcal{Z}^* \mathcal{J} \mathcal{Z} - \mathcal{J} = \begin{pmatrix} Z^* \\ \tilde{Z}^* \end{pmatrix} \mathcal{J} (Z \ \tilde{Z}) - \mathcal{J}
$$

$$
= \begin{pmatrix} Z^* \mathcal{J} Z & Z^* \mathcal{J} \tilde{Z} - I \\ \tilde{Z}^* \mathcal{J} Z + I & \tilde{Z}^* \mathcal{J} \tilde{Z} \end{pmatrix}
$$

$$
= \begin{pmatrix} Z^* \mathcal{J} Z & Z^* \mathcal{J} \tilde{Z} - I \\ -(Z^* \mathcal{J} \tilde{Z} - I)^* & \tilde{Z}^* \mathcal{J} \tilde{Z} \end{pmatrix},
$$

and this is zero iff $Z^* \mathcal{J} Z = \tilde{Z}^* \mathcal{J} \tilde{Z} = 0$ and $Z^* \mathcal{J} \tilde{Z} = I$. $\qquad\square$

Remark 7.28. Suppose $\begin{pmatrix} X \\ U \end{pmatrix}$ and $\begin{pmatrix} \tilde{X} \\ \tilde{U} \end{pmatrix}$ are normalized conjoined bases of (7.2). By Lemma 7.27 and Lemma 7.5,

$$
\begin{pmatrix} X & \tilde{X} \\ U & \tilde{U} \end{pmatrix}^{-1} = \begin{pmatrix} \tilde{U}^* & -\tilde{X}^* \\ -U^* & X^* \end{pmatrix}.
$$

Hence we find that all of the following formulas hold:

$$
(7.9) \qquad \begin{cases} X^* U = U^* X, \ \tilde{X}^* \tilde{U} = \tilde{U}^* \tilde{X}, \\ X \tilde{X}^* = \tilde{X} X^*, \ U \tilde{U}^* = \tilde{U} U^*, \\ X^* \tilde{U} - U^* \tilde{X} = X \tilde{U}^* - \tilde{X} U^* = I. \end{cases}
$$

More formulas can be derived using Exercise 7.14: By denoting $\mathcal{S} = \begin{pmatrix} A & B \\ C & D \end{pmatrix}$, we have

$$
\begin{pmatrix} X & \tilde{X} \\ U & \tilde{U} \end{pmatrix}^\sigma = \begin{pmatrix} I + \mu A & \mu B \\ \mu C & I + \mu D \end{pmatrix} \begin{pmatrix} X & \tilde{X} \\ U & \tilde{U} \end{pmatrix}
$$

and hence because of Lemma 7.5

$$
\begin{pmatrix} X & \tilde{X} \\ U & \tilde{U} \end{pmatrix}^\sigma \begin{pmatrix} \tilde{U}^* & -\tilde{X}^* \\ -U^* & X^* \end{pmatrix} = \begin{pmatrix} I + \mu A & \mu B \\ \mu C & I + \mu D \end{pmatrix},
$$

which splits into the following four formulas:

$$(7.10) \qquad \begin{cases} X^\sigma \tilde{U}^* - \tilde{X}^\sigma U^* = I + \mu A, \\ \tilde{X}^\sigma X^* - X^\sigma \tilde{X}^* = \mu B, \\ U^\sigma \tilde{U}^* - \tilde{U}^\sigma U^* = \mu C, \\ \tilde{U}^\sigma X^* - U^\sigma \tilde{X}^* = I + \mu D. \end{cases}$$

Lemma 7.29. *Let Z be a conjoined basis of (7.2). Then there exists another conjoined basis $\tilde{Z}$ of (7.2) such that Z and $\tilde{Z}$ are normalized.*

Proof. Let $t_0 \in \mathbb{T}$ and observe that $Z^*(t_0)Z(t_0)$ is invertible due to the condition $\operatorname{rank} Z(t_0) = n$. By Corollary 7.12, there exists a unique $2n \times 2n$-matrix-valued solution $\tilde{Z}$ of (7.2) satisfying

$$\tilde{Z}(t_0) = \mathcal{J}^* Z(t_0)[Z^*(t_0)Z(t_0)]^{-1}.$$

By Theorem 7.23, $\tilde{Z}^* \mathcal{J} \tilde{Z}$ is a constant matrix, and this constant matrix is given by

$$\begin{aligned} \tilde{Z}^*(t_0)\mathcal{J}\tilde{Z}(t_0) &= [Z^*(t_0)Z(t_0)]^{-1} Z^*(t_0) \mathcal{J} \mathcal{J} \mathcal{J}^* Z(t_0)[Z^*(t_0)Z(t_0)]^{-1} \\ &= [Z^*(t_0)Z(t_0)]^{-1} Z^*(t_0) \mathcal{J} Z(t_0)[Z^*(t_0)Z(t_0)]^{-1} \\ &= 0 \end{aligned}$$

since $Z^*(t)\mathcal{J}Z(t) \equiv 0$. Hence $\tilde{Z}^*(t)\mathcal{J}\tilde{Z}(t) \equiv 0$ so that $\tilde{Z}$ is a conjoined solution of (7.2). Next, again by Theorem 7.23, $Z^* \mathcal{J} \tilde{Z}$ is a constant matrix, and this constant matrix is given by

$$\begin{aligned} Z^*(t_0)\mathcal{J}\tilde{Z}(t_0) &= Z^*(t_0) \mathcal{J} \mathcal{J}^* Z(t_0)[Z^*(t_0)Z(t_0)]^{-1} \\ &= Z^*(t_0) Z(t_0)[Z^*(t_0)Z(t_0)]^{-1} \\ &= I. \end{aligned}$$

Hence $Z^*(t)\mathcal{J}\tilde{Z}(t) \equiv I$ and in particular $\operatorname{rank}\tilde{Z} \equiv n$. Therefore $\tilde{Z}$ is a conjoined basis of (7.2), and Z and $\tilde{Z}$ are normalized. $\qquad \square$

We now consider the Riccati equation

$$(7.11) \quad RQ = 0 \text{ with } RQ := Q^\Delta - \begin{pmatrix} I \\ Q \end{pmatrix}^* \mathcal{J} \mathcal{S} \begin{pmatrix} I \\ Q \end{pmatrix} \left\{ I + \mu \begin{pmatrix} I \\ 0 \end{pmatrix}^* \mathcal{S} \begin{pmatrix} I \\ Q \end{pmatrix} \right\}^{-1}.$$

Hence, if Q solves (7.11), then necessarily

$$(7.12) \qquad \begin{pmatrix} I \\ 0 \end{pmatrix}^* \mathcal{S} \begin{pmatrix} I \\ Q \end{pmatrix} \in \mathcal{R}.$$

Note that if we put $\mathcal{S} = \begin{pmatrix} A & B \\ C & D \end{pmatrix}$, then

$$(7.13) \qquad RQ = Q^\Delta + Q^\sigma(A + BQ) - (C + DQ) \quad \text{with} \quad A + BQ \in \mathcal{R}.$$

Exercise 7.30. Show that the Riccati equation (5.21) from Chapter 5 is a special case of the Riccati equation (7.11).

Exercise 7.31. Show that Q solves (7.11) iff (7.12) holds with

$$Q^{\Delta} = \begin{pmatrix} I \\ Q^{\sigma} \end{pmatrix}^* \mathcal{J} S \begin{pmatrix} I \\ Q \end{pmatrix}.$$

Theorem 7.32. *Let $\mathcal{I} \subset \mathbb{T}$ be an interval. Then (7.2) has a conjoined basis $\begin{pmatrix} X \\ U \end{pmatrix}$ with X invertible on $\mathcal{I}$ if and only if there exists a Hermitian solution Q of* (7.11) *on $\mathcal{I}^{\kappa}$.*

Proof. First, assume that $\begin{pmatrix} X \\ U \end{pmatrix}$ is a conjoined basis of (7.2) such that X is invertible on $\mathcal{I}$. Put $Q = UX^{-1}$. Then Q is Hermitian and

$$\begin{aligned} Q^{\Delta} &= U^{\Delta}X^{-1} - U^{\sigma}(X^{\sigma})^{-1}X^{\Delta}X^{-1} \\ &= (U^{\Delta} - Q^{\sigma}X^{\Delta})X^{-1} \\ &= [CX + DU - Q^{\sigma}(AX + BU)]X^{-1} \\ &= C + DQ - Q^{\sigma}(A + BQ), \end{aligned}$$

and hence Q solves (7.11) because of (7.13).

Next, suppose that Q is a Hermitian solution of (7.11). Then $A + BQ \in \mathcal{R}$. Define, for some $t_0 \in \mathcal{I}$,

$$X = e_{A+BQ}(\cdot, t_0) \quad \text{and} \quad U = QX.$$

Then X is invertible by Theorem 5.21 (iii), UX^{-1} is Hermitian,

$$X^{\Delta} = (A + BQ)X = AX + BU,$$

and

$$\begin{aligned} U^{\Delta} &= Q^{\Delta}X + Q^{\sigma}X^{\Delta} \\ &= [C + DQ - Q^{\sigma}(A + BQ)]X + Q^{\sigma}(AX + BU) \\ &= CX + DQX - Q^{\sigma}AX - Q^{\sigma}BQX + Q^{\sigma}(AX + BU) \\ &= CX + DU, \end{aligned}$$

and hence $\begin{pmatrix} X \\ U \end{pmatrix}$ is indeed a conjoined basis of (7.2) such that X is invertible. $\square$

One result that follows from Remark 7.28 and which is needed for the proof of Theorem 7.43 is the following.

Lemma 7.33. *If* $\begin{pmatrix} X \\ U \end{pmatrix}$ *is a conjoined basis of* (7.2) *with*

$$\operatorname{Ker} X^{\sigma}(t) \subset \operatorname{Ker} X(t)$$

for some $t \in \mathbb{T}$*, then*

$$\operatorname{Ker}(X^{\sigma})^{*}(t) \subset \operatorname{Ker}(\mu(t)B^{*}(t)).$$

Proof. According to Lemma 7.29 there exists another conjoined basis $\begin{pmatrix} \tilde{X} \\ \tilde{U} \end{pmatrix}$ of

(7.2) such that $\begin{pmatrix} X \\ U \end{pmatrix}$ and $\begin{pmatrix} \tilde{X} \\ \tilde{U} \end{pmatrix}$ are normalized. Now, let $\alpha \in \operatorname{Ker}(X^{\sigma})^{*}$ (at the

argument t, which we drop in this proof) so that $(X^{\sigma})^{*}\alpha = 0$. Then, by (7.9) of Remark 7.28, we have

$$X^{\sigma}(\tilde{X}^{\sigma})^{*}\alpha = \tilde{X}^{\sigma}(X^{\sigma})^{*}\alpha = 0.$$

Hence

$$(\tilde{X}^{\sigma})^{*}\alpha \in \operatorname{Ker} X^{\sigma} \subset \operatorname{Ker} X$$

so that $X(\tilde{X}^{\sigma})^{*}\alpha = 0$. Therefore by (7.10) of Remark 7.28

$$
\begin{aligned}
\mu B^{*}\alpha &= \mu(\tilde{X}^{\sigma}X^{*} - X^{\sigma}\tilde{X}^{*})^{*}\alpha \\
&= \mu X(\tilde{X}^{\sigma})^{*}\alpha - \mu\tilde{X}(X^{\sigma})^{*}\alpha \\
&= 0.
\end{aligned}
$$

Therefore $\alpha \in \operatorname{Ker}(\mu B^{*})$. Altogether, $\operatorname{Ker}(X^{\sigma})^{*} \subset \operatorname{Ker}(\mu B^{*})$. $\qquad\square$

Definition 7.34. Let $t_{0} \in \mathbb{T}$. Suppose $\mathcal{Z} = (Z \ \tilde{Z})$ is the $2n \times 2n$-matrix-valued solution of

$$\mathcal{Z}^{\Delta} = \mathcal{S}(t)\mathcal{Z}, \quad \mathcal{Z}(t_{0}) = \mathcal{J}^{*}.$$

Then Z is called the *principal solution* of (7.2) (at t_{0}) while $\tilde{Z}$ is referred to as being the *associated solution* of (7.2) (at t_{0}). Together they are called the *special normalized conjoined bases* of (7.2) (at t_{0}).

Remark 7.35. Of course the special normalized conjoined bases Z and $\tilde{Z}$ are normalized conjoined bases since

$$\mathcal{Z}^{*}\mathcal{J}\mathcal{Z} \equiv \mathcal{Z}^{*}(t_{0})\mathcal{J}\mathcal{Z}(t_{0}) = \mathcal{J}\mathcal{J}\mathcal{J}^{*} = \mathcal{J}$$

implies that $\mathcal{Z}$ is symplectic (apply Lemma 7.27).

In order to introduce the concept of disconjugacy for systems (7.2), we shall need to recall the definition and basic properties of Moore–Penrose inverses. The existence and uniqueness of the matrix introduced in the next definition is easy to show (see, e.g. [56]).

Definition 7.36. Let M be an arbitrary matrix. The unique matrix N satisfying simultaneously

(i) $MNM = M$ and $NMN = N$;

(ii) MN and NM are both symmetric

is denoted by $M^\dagger$ and called the *Moore–Penrose inverse* of M.

For basic properties of the Moore–Penrose inverse we refer, e.g., to [56] or [30, Chapter 9]. However, here we prove the following useful lemma.

Lemma 7.37. *For two matrices V and W (with appropriate sizes) we have*

$$\operatorname{Ker} V \subset \operatorname{Ker} W \quad \text{iff} \quad W = WV^\dagger V \quad \text{iff} \quad W^\dagger = V^\dagger V W^\dagger.$$

Proof. First, suppose $W = WV^\dagger V$. Let $x \in \operatorname{Ker} V$ so that $Vx = 0$. But then

$$Wx = WV^\dagger Vx = 0$$

and hence $x \in \operatorname{Ker} W$. Therefore $\operatorname{Ker} V \subset \operatorname{Ker} W$. Next, assume $\operatorname{Ker} V \subset \operatorname{Ker} W$ holds. But then we have

$$\operatorname{Im} W^* = (\operatorname{Ker} W)^\perp \subset (\operatorname{Ker} V)^\perp = \operatorname{Im} V^*$$

(where $M^\perp$ denotes the orthogonal complement of a set M). Let c be arbitrary such that W^*c is defined. Hence

$$W^*c \in \operatorname{Im} W^* \subset \operatorname{Im} V^*$$

so there exists d such that $V^*d = W^*c$. Since the entire space is the direct sum of $\operatorname{Im} V$ and $(\operatorname{Im} V)^\perp = \operatorname{Ker} V^*$, we know that there exists d_1 and $d_2 \in \operatorname{Ker} V^*$ with $d = Vd_1 + d_2$. Now

$$
\begin{aligned}
W^*c &= V^*d = V^*Vd_1 + V^*d_2 = V^*Vd_1 \\
&= V^*VV^\dagger Vd_1 = V^*(VV^\dagger)^*Vd_1 = V^*(V^\dagger)^*V^*Vd_1 \\
&= V^*(V^\dagger)^*W^*c = (WV^\dagger V)^*c.
\end{aligned}
$$

Therefore $W = WV^\dagger V$. □

Exercise 7.38. Prove the other equivalence from Lemma 7.37.

Definition 7.39. A $2n \times n$-matrix-valued solution $\begin{pmatrix} X \\ U \end{pmatrix}$ of (7.2) is said to have

no focal points in $(a, b]$ if

$$
\begin{cases}
X(t) \text{ is invertible in all dense points } t \in (a, b], \\
\operatorname{Ker} X^\sigma(t) \subset \operatorname{Ker} X(t),\ X(t)(X^\sigma(t))^\dagger B(t) \geq 0 \text{ on } (a, b]^\kappa
\end{cases}
$$

holds. If the principal solution of (7.2) at a does not have any focal points in $(a, b]$, then system (7.2) is called *disconjugate* (on $[a, b]$).

For z with $\mathcal{M}Sz = \mathcal{M}z^\Delta$ we define a quadratic functional by

$$\mathcal{F}(z) = \int_a^b \{z^*(S^*\mathcal{M} + \mathcal{M}S + \mu S^*\mathcal{M}S)z\}(t)\Delta t.$$

Theorem 7.40. *If z solves (7.2), then*

$$\mathcal{F}(z) = z^*(b)\mathcal{M}z(b) - z^*(a)\mathcal{M}z(a).$$

Proof. Suppose z solves (7.2). Then we have

$$
\begin{aligned}
\mathcal{F}(z) &= \int_a^b \left\{ z^* S^* \mathcal{M} z + z^* \mathcal{M} z^\Delta + \mu z^* S^* \mathcal{M} z^\Delta \right\}(t) \Delta t \\
&= \int_a^b \left\{ z^* S^* \mathcal{M}(z + \mu z^\Delta) + z^* \mathcal{M} z^\Delta \right\}(t) \Delta t \\
&= \int_a^b \left\{ (z^*)^\Delta \mathcal{M} z^\sigma + z^* \mathcal{M} z^\Delta \right\}(t) \Delta t \\
&= \int_a^b (z^* \mathcal{M} z)^\Delta (t) \Delta t \\
&= z^*(b) \mathcal{M} z(b) - z^*(a) \mathcal{M} z(a),
\end{aligned}
$$

and this proves our claim.   □

Now we will relate positive definiteness of the quadratic functional $\mathcal{F}$ to disconjugacy of (7.2) and to solvability of the Riccati equation (7.11). Before stating this *Reid Roundabout Theorem*, which we shall not prove (for a proof see Roman Hilscher [172]), we need the following two definitions.

Definition 7.41. The quadratic functional $\mathcal{F}$ is called *positive definite* provided $\mathcal{F}(z) > 0$ for all admissible vectors z with $\mathcal{M} z(a) = \mathcal{M} z(b) = 0$ and $\mathcal{M} z \not\equiv 0$. Here, z is called *admissible* if $\mathcal{M} z \in C^1_{\mathrm{prd}}$, $\mathcal{M}^* z \in C_{\mathrm{prd}}$, and $\mathcal{M} z^\Delta = \mathcal{M} S z$.

Definition 7.42. The system (7.2) is called *dense normal* on the interval $[a, s]$ for a dense point s if the only solution z of (7.2) for which $\mathcal{M} z \equiv 0$ on $[a, s]$ is the zero solution $z \equiv 0$.

Theorem 7.43 (Reid Roundabout Theorem). *Suppose* (7.2) *is dense normal on any interval* $[a, s]$ *with* $s \leq b$ *and* s *dense. Then the following statements are equivalent:*

(i) $\mathcal{F}$ *is positive definite;*

(ii) *system* (7.2) *is disconjugate on* $[a, b]$*;*

(iii) *there exists a conjoined basis* $\begin{pmatrix} X \\ U \end{pmatrix}$ *of* (7.2) *with no focal points on* $(a, b]$ *and such that* X *is invertible on* $[a, b]$*;*

(iv) *there exists a symmetric solution* Q *of* (7.11) *on* $[a, b]$ *with*

$$
[I + \mu(A + BQ)]^{-1} B \geq 0 \quad \text{on} \quad (a, b]^\kappa.
$$

Proof. See Roman Hischer [172, Theorem 1].   □

7.3. Transformation Theory and Trigonometric Systems

In this section we show that symplectic systems are transformed into other symplectic systems when using the transformation

$$(7.14) \qquad\qquad z = \mathcal{R}(t) \bar{z} \quad \text{with symplectic} \quad \mathcal{R} \in C^1_{\mathrm{rd}}.$$

We also introduce trigonometric (or self-reciprocal) systems and show that every symplectic system may be transformed into a trigonometric system using a transformation of the form (7.14). Furthermore, this transformation preserves "oscillatory behavior", e.g., if the original system is disconjugate on a certain interval, then so is the transformed system.

Theorem 7.44. *Let $\mathcal{R} \in C^1_{rd}$ be a $2n \times 2n$-matrix-valued symplectic function. The transformation* (7.14) *transforms* (7.2) *into another symplectic system*

$$(7.15) \qquad\qquad \bar{z}^\Delta = \bar{\mathcal{S}}(t)\bar{z},$$

where $\bar{\mathcal{S}} = (\mathcal{R}^\sigma)^{-1}(\mathcal{S}\mathcal{R} - \mathcal{R}^\Delta)$.

Proof. First of all, since $\mathcal{R}$ is symplectic, it is invertible by Lemma 7.5. Using the formulas from Theorem 5.3, we find

$$
\begin{aligned}
\bar{z}^\Delta &= (\mathcal{R}^{-1}z)^\Delta \\
&= (\mathcal{R}^\sigma)^{-1}z^\Delta - (\mathcal{R}^\sigma)^{-1}\mathcal{R}^\Delta\mathcal{R}^{-1}z \\
&= (\mathcal{R}^\sigma)^{-1}\mathcal{S}z - (\mathcal{R}^\sigma)^{-1}\mathcal{R}^\Delta\mathcal{R}^{-1}z \\
&= (\mathcal{R}^\sigma)^{-1}\mathcal{S}\mathcal{R}\bar{z} - (\mathcal{R}^\sigma)^{-1}\mathcal{R}^\Delta\bar{z} \\
&= (\mathcal{R}^\sigma)^{-1}(\mathcal{S}\mathcal{R} - \mathcal{R}^\Delta)\bar{z} \\
&= \bar{\mathcal{S}}\bar{z},
\end{aligned}
$$

so $\bar{z}$ indeed solves (7.15). It remains to verify that (7.15) is a symplectic system, too. Our assumptions imply that $\bar{\mathcal{S}} \in C_{rd}$. Since $\mathcal{R}$ is symplectic, the formula

$$\mathcal{R}^* \mathcal{J} \mathcal{R} = \mathcal{J}$$

holds. Now the calculation

$$
\begin{aligned}
\bar{\mathcal{S}}^* \mathcal{J} + \mathcal{J}\bar{\mathcal{S}} + \mu\bar{\mathcal{S}}^* \mathcal{J}\bar{\mathcal{S}} &= (\mathcal{R}^\sigma\bar{\mathcal{S}})^* \mathcal{J}\mathcal{R}^\sigma + (\mathcal{R}^\sigma)^* \mathcal{J}\mathcal{R}^\sigma\bar{\mathcal{S}} + \mu(\mathcal{R}^\sigma\bar{\mathcal{S}})^* \mathcal{J}\mathcal{R}^\sigma\bar{\mathcal{S}} \\
&= (\mathcal{S}\mathcal{R} - \mathcal{R}^\Delta)^* \mathcal{J}\mathcal{R}^\sigma + (\mathcal{R}^\sigma)^* \mathcal{J}(\mathcal{S}\mathcal{R} - \mathcal{R}^\Delta) + \mu(\mathcal{S}\mathcal{R} - \mathcal{R}^\Delta)^* \mathcal{J}(\mathcal{S}\mathcal{R} - \mathcal{R}^\Delta) \\
&= \mathcal{R}^*\mathcal{S}^* \mathcal{J}\mathcal{R}^\sigma - (\mathcal{R}^\Delta)^* \mathcal{J}\mathcal{R}^\sigma + (\mathcal{R}^\sigma)^* \mathcal{J}\mathcal{S}\mathcal{R} - (\mathcal{R}^\sigma)^* \mathcal{J}\mathcal{R}^\Delta \\
&\qquad + \mu\mathcal{R}^*\mathcal{S}^* \mathcal{J}\mathcal{S}\mathcal{R} - \mu(\mathcal{R}^\Delta)^* \mathcal{J}\mathcal{S}\mathcal{R} - \mu\mathcal{R}^*\mathcal{S}^* \mathcal{J}\mathcal{R}^\Delta + \mu(\mathcal{R}^\Delta)^* \mathcal{J}\mathcal{R}^\Delta \\
&= \mathcal{R}^*\mathcal{S}^* \mathcal{J}(\mathcal{R}^\sigma - \mu\mathcal{R}^\Delta) - (\mathcal{R}^\Delta)^* \mathcal{J}\mathcal{R}^\sigma + (\mathcal{R}^\sigma - \mu\mathcal{R}^\Delta)^* \mathcal{J}\mathcal{S}\mathcal{R} \\
&\qquad - (\mathcal{R}^\sigma - \mu\mathcal{R}^\Delta)^* \mathcal{J}\mathcal{R}^\Delta + \mu\mathcal{R}^*\mathcal{S}^* \mathcal{J}\mathcal{S}\mathcal{R} \\
&= \mathcal{R}^*\mathcal{S}^* \mathcal{J}\mathcal{R} - (\mathcal{R}^\Delta)^* \mathcal{J}\mathcal{R}^\sigma + \mathcal{R}^* \mathcal{J}\mathcal{S}\mathcal{R} - \mathcal{R}^* \mathcal{J}\mathcal{R}^\Delta + \mu\mathcal{R}^*\mathcal{S}^* \mathcal{J}\mathcal{S}\mathcal{R} \\
&= \mathcal{R}^*\mathcal{S}^* \mathcal{J}\mathcal{R} - (\mathcal{R}^* \mathcal{J}\mathcal{R})^\Delta + \mathcal{R}^* \mathcal{J}\mathcal{S}\mathcal{R} + \mu\mathcal{R}^*\mathcal{S}^* \mathcal{J}\mathcal{S}\mathcal{R} \\
&= \mathcal{R}^*\mathcal{S}^* \mathcal{J}\mathcal{R} + \mathcal{R}^* \mathcal{J}\mathcal{S}\mathcal{R} + \mu\mathcal{R}^*\mathcal{S}^* \mathcal{J}\mathcal{S}\mathcal{R} \\
&= \mathcal{R}^*(\mathcal{S}^* \mathcal{J} + \mathcal{J}\mathcal{S} + \mu\mathcal{S}^* \mathcal{J}\mathcal{S})\mathcal{R} \\
&= 0
\end{aligned}
$$

shows that $\bar{\mathcal{S}}$ is symplectic with respect to $\mathbb{T}$. Hence (7.15) really is a symplectic dynamic system. $\qquad\square$

Remark 7.45. If the matrix-valued functions $\mathcal{S}$, $\bar{\mathcal{S}}$, and $\mathcal{R}$ from Theorem 7.44 are partitioned as

$$
\mathcal{S} = \begin{pmatrix} A & B \\ C & D \end{pmatrix}, \; \bar{\mathcal{S}} = \begin{pmatrix} \bar{A} & \bar{B} \\ \bar{C} & \bar{D} \end{pmatrix}, \quad \text{and} \quad \mathcal{R} = \begin{pmatrix} H & M \\ K & N \end{pmatrix},
$$

using $n \times n$-matrix-valued entries, then we have by Lemma 7.5

$$
\begin{pmatrix} \bar{A} & \bar{B} \\ \bar{C} & \bar{D} \end{pmatrix} = \bar{S} = (\mathcal{R}^\sigma)^{-1}(\mathcal{S}\mathcal{R} - \mathcal{R}^\Delta)
$$

$$
= \begin{pmatrix} N^* & -M^* \\ -K^* & H^* \end{pmatrix}^\sigma \left[\begin{pmatrix} A & B \\ C & D \end{pmatrix} \begin{pmatrix} H & M \\ K & N \end{pmatrix} - \begin{pmatrix} H^\Delta & M^\Delta \\ K^\Delta & N^\Delta \end{pmatrix} \right]
$$

$$
= \begin{pmatrix} N^* & -M^* \\ -K^* & H^* \end{pmatrix}^\sigma \begin{pmatrix} AH + BK - H^\Delta & AM + BN - M^\Delta \\ CH + DK - K^\Delta & CM + DN - N^\Delta \end{pmatrix},
$$

i.e., the following formulas hold:

$$
(7.16) \quad \begin{cases} \bar{A} = (M^\sigma)^*(K^\Delta - CH - DK) - (N^\sigma)^*(H^\Delta - AH - BK), \\ \bar{B} = (M^\sigma)^*(N^\Delta - CM - DN) - (N^\sigma)^*(M^\Delta - AM - BN), \\ \bar{C} = (K^\sigma)^*(H^\Delta - AH - BK) - (H^\sigma)^*(K^\Delta - CH - DK), \\ \bar{D} = (K^\sigma)^*(M^\Delta - AM - BN) - (H^\sigma)^*(N^\Delta - CM - DN). \end{cases}
$$

Of special interest are transformations (7.14) using matrices $\mathcal{R}$ of the form

$$
(7.17) \quad \mathcal{R} = \begin{pmatrix} H & 0 \\ K & H^{*-1} \end{pmatrix} \in C_{rd}^1 \quad \text{with } \mathcal{R} \text{ symplectic,}
$$

i.e., when $M = 0$ and $N = H^{*-1}$ in Remark 7.45. This is because such transformations preserve disconjugacy of the original system (7.2), as is shown below. Concerning transformations of the form (7.17), we have the following two results.

Corollary 7.46. *Using the notation from Remark 7.45, if*

$$
\mathcal{S} = \begin{pmatrix} A & B \\ C & D \end{pmatrix}, \quad \bar{\mathcal{S}} = \begin{pmatrix} \bar{A} & \bar{B} \\ \bar{C} & \bar{D} \end{pmatrix}, \quad \text{and} \quad \mathcal{R} = \begin{pmatrix} H & 0 \\ K & H^{*-1} \end{pmatrix},
$$

then the following formulas hold:

$$
(7.18) \quad \begin{cases} \bar{A} = (H^\sigma)^{-1}(AH - BK - H^\Delta), \\ \bar{B} = (H^\sigma)^{-1}BH^{*-1}, \\ \bar{C} = (K^\sigma)^*(H^\Delta - AH - BK) - (H^\sigma)^*(K^\Delta - CH - DK), \\ \bar{D} = (H^\Delta + D^*H^\sigma - B^*K^\sigma)^*H^{*-1}. \end{cases}
$$

Proof. This follows from Remark 7.45. The first three formulas of (7.18) follow easily from (7.16), while the last one holds because of

$$
\begin{aligned} \bar{D} &= -(K^\sigma)^*BH^{*-1} - (H^\sigma)^*((H^{*-1})^\Delta - DH^{*-1}) \\ &= (D^*H^\sigma - B^*K^\sigma)^*H^{*-1} - ((H^{-1})^\Delta H^\sigma)^* \\ &= (H^\Delta + D^*H^\sigma - B^*K^\sigma)^*H^{*-1} \end{aligned}
$$

since

$$-(H^{-1})^{\Delta} H^{\sigma} = H^{-1} H^{\Delta} (H^{\sigma})^{-1} H^{\sigma} = H^{-1} H^{\Delta}$$

by Theorem 5.3 (iv).   □

Theorem 7.47. *Let $\mathcal{R}$ be a transformation matrix as in (7.17), and suppose Z is a conjoined basis of (7.2) without focal points in $(a, b]$. Then $\bar{Z} = \mathcal{R}^{-1} Z$ is a conjoined basis of the (via (7.14)) transformed system (7.15) without focal points in $(a, b]$, either.*

Proof. As usual we denote $Z = \begin{pmatrix} X \\ U \end{pmatrix}$ and $\bar{Z} = \begin{pmatrix} \bar{X} \\ \bar{U} \end{pmatrix}$. We assume that Z has no focal points in $(a, b]$ (see Definition 7.39). Then by Lemma 7.5

$$\begin{pmatrix} \bar{X} \\ \bar{U} \end{pmatrix} = \bar{Z} = \mathcal{R}^{-1} Z = \begin{pmatrix} H^{-1} & 0 \\ -K^* & H^* \end{pmatrix} \begin{pmatrix} X \\ U \end{pmatrix} = \begin{pmatrix} H^{-1} X \\ H^* U - K^* X \end{pmatrix}.$$

Therefore $\bar{X} = H^{-1} X$. First, suppose t is dense. Then $X(t)$ is invertible and so is $\bar{X}(t) + H^{-1}(t) X(t)$. Next, let $t \in (a, b]^{\kappa}$. Suppose $c \in \operatorname{Ker} \bar{X}^{\sigma}(t)$. Then

$$X^{\sigma}(t) c = H^{\sigma}(t) \bar{X}^{\sigma}(t) c = 0$$

so that

$$c \in \operatorname{Ker} X^{\sigma}(t) \subset \operatorname{Ker} X(t)$$

and hence $c \in \operatorname{Ker} \bar{X}(t)$. Altogether, $\operatorname{Ker} \bar{X}^{\sigma}(t) \subset \operatorname{Ker} \bar{X}(t)$. To show that

$$\bar{X}(t) (\bar{X}^{\sigma}(t))^{\dagger} \bar{B}(t) \geq 0,$$

we consider two cases. First, if $\mu(t) = 0$, then t is dense and both $X(t)$ and $\bar{X}(t)$ are invertible. Therefore

$$\begin{aligned}
\bar{X} (\bar{X}^{\sigma})^{\dagger} \bar{B} &= \bar{B} \\
&= (H^{\sigma})^{-1} B H^{*-1} \\
&= H^{-1} B H^{*-1} \\
&= H^{-1} X (X^{\sigma})^{\dagger} B H^{*-1} \\
&\geq 0
\end{aligned}$$

holds at t. And, if $\mu(t) > 0$, then because of Lemma 7.33 and Lemma 7.37

$$\begin{aligned}
\bar{X} (\bar{X}^{\sigma})^{\dagger} \bar{B} &= \frac{1}{\mu} H^{-1} X ((H^{\sigma})^{-1} X^{\sigma})^{\dagger} (H^{\sigma})^{-1} \mu B H^{*-1} \\
&= \frac{1}{\mu} H^{-1} X (X^{\sigma})^{\dagger} X^{\sigma} ((H^{\sigma})^{-1} X^{\sigma})^{\dagger} (H^{\sigma})^{-1} X^{\sigma} (X^{\sigma})^{\dagger} \mu B H^{*-1} \\
&= H^{-1} X (X^{\sigma})^{\dagger} H^{\sigma} ((H^{\sigma})^{-1} X^{\sigma}) ((H^{\sigma})^{-1} X^{\sigma})^{\dagger} ((H^{\sigma})^{-1} X^{\sigma}) (X^{\sigma})^{\dagger} B H^{*-1} \\
&= H^{-1} X (X^{\sigma})^{\dagger} H^{\sigma} ((H^{\sigma})^{-1} X^{\sigma}) (X^{\sigma})^{\dagger} B H^{*-1} \\
&= H^{-1} X (X^{\sigma})^{\dagger} X^{\sigma} (X^{\sigma})^{\dagger} B H^{*-1} \\
&= H^{-1} X (X^{\sigma})^{\dagger} B H^{*-1} \\
&\geq 0
\end{aligned}$$

at t. This shows that $\bar{X}(t) (\bar{X}^{\sigma}(t))^{\dagger} \bar{B}(t) \geq 0$ holds in any case.   □

Corollary 7.46 may also be used to derive a reduction of order formula for systems (7.2).

Theorem 7.48 (Reduction of Order). *Suppose* $Z = \begin{pmatrix} X \\ U \end{pmatrix}$ *is a conjoined basis of*

(7.2) *such that* X *is invertible on* $[a, b]$. *Then* $\tilde{Z} = \begin{pmatrix} \tilde{X} \\ \tilde{U} \end{pmatrix}$ *given by*

(7.19) $\begin{cases} \tilde{X}(t) = X(t) \int_a^t \left((X^\sigma)^{-1} B X^{*-1} \right)(\tau) \Delta\tau, \\ \tilde{U}(t) = U(t) \int_a^t \left((X^\sigma)^{-1} B X^{*-1} \right)(\tau) \Delta\tau + X^{*-1}(t) \end{cases}$

is also a conjoined basis of (7.2), *and* Z *and* $\tilde{Z}$ *are normalized.*

Proof. We apply Corollary 7.46 with

$$\mathcal{R} = \begin{pmatrix} X & 0 \\ U & X^{*-1} \end{pmatrix}.$$

This transformation matrix is symplectic since Z is a conjoined basis:

$$\begin{pmatrix} X & 0 \\ U & X^{*-1} \end{pmatrix}^* \begin{pmatrix} 0 & I \\ -I & 0 \end{pmatrix} \begin{pmatrix} X & 0 \\ U & X^{*-1} \end{pmatrix} = \begin{pmatrix} X^* & U^* \\ 0 & X^{-1} \end{pmatrix} \begin{pmatrix} U & X^{*-1} \\ -X & 0 \end{pmatrix}$$

$$= \begin{pmatrix} 0 & I \\ -I & 0 \end{pmatrix}.$$

It transforms the system $z^\Delta = \mathcal{S}(t)z$ into $\bar{z}^\Delta = \bar{\mathcal{S}}\bar{z}$, where

$$\mathcal{S} = \begin{pmatrix} A & B \\ C & D \end{pmatrix} \quad \text{and} \quad \bar{\mathcal{S}} = \begin{pmatrix} \bar{A} & \bar{B} \\ \bar{C} & \bar{D} \end{pmatrix}$$

with (apply Corollary 7.46)

$$\begin{aligned}
\bar{A} &= -(X^\sigma)^{-1}(X^\Delta - AX - BU) = 0, \\
\bar{B} &= (X^\sigma)^{-1} B X^{*-1}, \\
\bar{C} &= (U^\sigma)^*(X^\Delta - AX - BU) - (X^\sigma)^*(U^\Delta - CX - DU) = 0, \\
\bar{D} &= (X^\Delta + D^* X^\sigma - B^* U^\sigma)^* X^{*-1} \\
&= \{[A + D^* + \mu(D^*A - B^*C)]X + [B - B^* + \mu(D^*B - B^*D)]U\} X^{*-1} \\
&= 0
\end{aligned}$$

because of (7.3) of Theorem 7.13. Hence the transformed system reads

$$\bar{z}^\Delta = \begin{pmatrix} 0 & (X^\sigma)^{-1}BX^{*-1} \\ 0 & 0 \end{pmatrix} \bar{z},$$

i.e.,

$$\bar{X}^\Delta = (X^\sigma)^{-1}BX^{*-1}\bar{U}, \quad \bar{U}^\Delta \equiv 0,$$

which has the principal solution

$$\bar{U} \equiv I, \quad \bar{X}(t) = \int_a^t \left((X^\sigma)^{-1}BX^{*-1} \right)(\tau)\Delta\tau.$$

Therefore, a solution of $z^\Delta = \mathcal{S}(t)z$ is given by

$$\begin{pmatrix} \tilde{X} \\ \tilde{U} \end{pmatrix} = \tilde{Z} = \mathcal{R}\bar{Z} = \begin{pmatrix} X & 0 \\ U & X^{*-1} \end{pmatrix} \begin{pmatrix} \bar{X} \\ \bar{U} \end{pmatrix} = \begin{pmatrix} X\bar{X} \\ U\bar{X} + X^{*-1}\bar{U} \end{pmatrix},$$

i.e., $\tilde{X}$ and $\tilde{U}$ are as in (7.19). The rest of the statement can be verified easily. $\square$

Remark 7.49. Given the notation from Theorem 7.48, we have that $\begin{pmatrix} X \\ U \end{pmatrix}$ and $\begin{pmatrix} \tilde{X} \\ \tilde{U} \end{pmatrix}$ are normalized conjoined bases of (7.2). Therefore, the solution of the initial value problem

$$Z^\Delta = \mathcal{S}(t)Z, \quad Z(a) = \begin{pmatrix} \hat{X}_a \\ \hat{U}_a \end{pmatrix}$$

is given by

$$(7.20) \qquad \begin{pmatrix} \hat{X} \\ \hat{U} \end{pmatrix} = \begin{pmatrix} X & \tilde{X} \\ U & \tilde{U} \end{pmatrix} \begin{pmatrix} V \\ W \end{pmatrix}, \qquad \text{i.e.,} \quad \hat{X} = XV + \tilde{X}W, \hat{U} = UV + \tilde{U}W$$

with

$$V = X^{-1}(a)\hat{X}_a, \quad W = X^*(a)[\hat{U}_a - U(a)V].$$

The solution given by (7.20) is a conjoined basis of (7.2) iff V^*W is Hermitian and

$$\text{rank}\begin{pmatrix} V \\ W \end{pmatrix} \equiv n.$$

We now turn our attention to so-called self-reciprocal (or trigonometric) systems.

Definition 7.50. The system (7.15) that results from (7.2) upon the transformation (7.14) with $\mathcal{R} \equiv \mathcal{J}$ is called the *reciprocal* of (7.2). If the reciprocal of (7.2) is the same system as (7.2), then (7.2) is called *self-reciprocal* or *trigonometric*.

Lemma 7.51. *The system (7.2) is self-reciprocal iff*

$$(7.21) \qquad \mathcal{J}^* S \mathcal{J} = S$$

holds. This is equivalent to S being of the form

$$(7.22) \qquad S = \begin{pmatrix} P & Q \\ -Q & P \end{pmatrix}$$

with

$$(7.23) \qquad Q^* - Q + \mu(Q^* P - P^* Q) = P^* + P + \mu(Q^* Q + P^* P) = 0.$$

Moreover, P and Q satisfy (7.23) iff $M = P + iQ$ satisfies

$$M^* + M + \mu M^* M = 0.$$

Proof. For $\mathcal{R} = \mathcal{J}$ we have

$$\bar{S} = (\mathcal{R}^\sigma)^{-1}(S\mathcal{R} - \mathcal{R}^\Delta) = \mathcal{J}^{-1} S \mathcal{J} = \mathcal{J}^* S \mathcal{J},$$

and this is equal to S if and only if (7.21) holds. Now suppose (7.21) holds, and put $S = \begin{pmatrix} A & B \\ C & D \end{pmatrix}$. Then

$$
\begin{aligned}
0 &= \mathcal{J}^* S \mathcal{J} - S = \begin{pmatrix} 0 & -I \\ I & 0 \end{pmatrix} \begin{pmatrix} A & B \\ C & D \end{pmatrix} \begin{pmatrix} 0 & I \\ -I & 0 \end{pmatrix} - \begin{pmatrix} A & B \\ C & D \end{pmatrix} \\
&= \begin{pmatrix} D & -C \\ -B & A \end{pmatrix} - \begin{pmatrix} A & B \\ C & D \end{pmatrix} = \begin{pmatrix} D-A & -C-B \\ -C-B & A-D \end{pmatrix},
\end{aligned}
$$

i.e., $D = A$ and $C = -B$, and hence S is of the form (7.22). Applying (7.3) from Theorem 7.13 yields (7.23). Of course it is easily verified that any S of the form (7.23) satisfies (7.21), provided (7.23) holds. Finally, put $M = P + iQ$ with matrices P and Q. Then

$$
\begin{aligned}
M^* + M + \mu M^* M &= (P+iQ)^* + P + iQ + \mu(P+iQ)^*(P+iQ) \\
&= P^* - iQ^* + P + iQ + \mu(P^* - iQ^*)(P+iQ) \\
&= P^* + P + \mu(P^* P + Q^* Q) + i[Q - Q^* + \mu(P^* Q - Q^* P)],
\end{aligned}
$$

and this is zero iff (7.23) holds.   □

Theorem 7.52. *Given a symplectic system (7.2), there exists a transformation matrix $\mathcal{R}$ of the form (7.17) which transforms (7.2) into a self-reciprocal system.*

Proof. Let $\begin{pmatrix} X \\ U \end{pmatrix}$ and $\begin{pmatrix} \tilde{X} \\ \tilde{U} \end{pmatrix}$ be normalized conjoined bases of (7.2), H any matrix

with $HH^* = XX^* + \tilde{X}\tilde{X}^*$, $K = (UX^* + \tilde{U}\tilde{X}^*)H^{*-1}$, and $\mathcal{R} = \begin{pmatrix} H & 0 \\ K & H^{*-1} \end{pmatrix}$,

$V = H^{-1}(X + i\tilde{X})$. Then we have

$$
\begin{aligned}
H^*K - K^*H &= H^{-1}(HH^*KH^* - HK^*HH^*)H^{*-1} \\
&= H^{-1}\left[(XX^* + \tilde{X}\tilde{X}^*)(UX^* + \tilde{U}\tilde{X}^*) - (XU^* + \tilde{X}\tilde{U}^*)(XX^* + \tilde{X}\tilde{X}^*)\right]H^{*-1} \\
&= H^{-1}\left[X(X^*U - U^*X)X^* + \tilde{X}(\tilde{X}^*\tilde{U} - \tilde{U}^*\tilde{X})\tilde{X}^* \right.\\
&\qquad \left. + X(X^*\tilde{U} - U^*\tilde{X})\tilde{X}^* + \tilde{X}(\tilde{X}^*U - \tilde{U}^*X)X^*\right]H^{*-1} \\
&= H^{-1}(X\tilde{X}^* - \tilde{X}X^*)H^{*-1} \\
&= 0,
\end{aligned}
$$

where we have used (7.9) from Remark 7.28 for the last two equal signs. Next,

$$
\begin{aligned}
V^\Delta &= (H^{-1})^\Delta(X + i\tilde{X}) + (H^\sigma)^{-1}(X + i\tilde{X})^\Delta \\
&= -(H^\sigma)^{-1}H^\Delta H^{-1}(X + i\tilde{X}) + (H^\sigma)^{-1}(AX + BU + i(A\tilde{X} + B\tilde{U})) \\
&= (H^\sigma)^{-1}\left\{-H^\Delta + [(AX + BU) \right.\\
&\qquad \left. i(A\tilde{X} + B\tilde{U})](X^* - i\tilde{X}^*)(X^* - i\tilde{X}^*)^{-1}(X + i\tilde{X})^{-1}H\right\}V \\
&= (H^\sigma)^{-1}\left\{-H\delta + [(A(XX^* + \tilde{X}\tilde{X}^*) + B(UX^* + \tilde{U}\tilde{X}^*)]H^{*-1} \right.\\
&\qquad \left. + [iA(-X\tilde{X}^* + \tilde{X}X^*) + iB(-U\tilde{X}^* + UX^*)]H^{*-1}\right\}V \\
&= (H^\sigma)^{-1}\left\{-H^\Delta H^* + A(XX^* + \tilde{X}\tilde{X}^*) + B(UX^* + \tilde{U}\tilde{X}^*) + iB\right\}H^{*-1}V \\
&= \left\{(H^\sigma)^{-1}(-H^\Delta + AH + BK) + i(H^\sigma)^{-1}BH^{*-1}\right\}V = (P + iQ)V,
\end{aligned}
$$

with

$$
P := (H^\sigma)^{-1}(-H^\Delta + AH + BK) \quad \text{and} \quad Q := (H^\sigma)^{-1}BH^{*-1}.
$$

Denote $S := H^{-1}X$ and $C := -H^{-1}\tilde{X}$. Then one can directly verify that (S, C) and $(-C, S)$ form normalized conjoined bases of the self-reciprocal system

$$
S^\Delta = P(t)S + Q(t)C, \quad C^\Delta = -Q(t)S + P(t)C.
$$

Moreover, by (7.9) and (7.10) we have

$$
U = KS + H^{*-1}C \quad \text{and} \quad \tilde{U} = -KC + H^{*-1}S.
$$

Since the matrix V is unitary, i.e., $V^*V = I$, we have $(V^*V)^\Delta = 0$, and hence, with $M = P + iQ$,

$$
\begin{aligned}
0 &= (V^*V)^\Delta \\
&= (V^*)^\Delta V^\sigma + V^*V^\Delta \\
&= V^*M^*(V + \mu MV) + V^*MV \\
&= V^*(M^* + M + \mu M^*M)V.
\end{aligned}
$$

Thus $M^* + M + \mu M^*M = 0$, which means that the transformation (7.14) with $\mathcal{R}$ as in (7.17) really transforms (7.2) into a trigonometric system for the new variables $s = H^{-1}x$, $c = -K^*x + H^*u$.   □

7.4. Notes and References

The terminology of this chapter, in particular "symplectic" systems and "Hamiltonian" systems, has been discussed in detail with Calvin Ahlbrandt and Stefan Hilger, and we could as well have chosen other terminology for the main objects presented in this chapter. The discrete case features a major difference between the two systems under consideration, while they are the same objects in the continuous case.

Hamiltonian differential systems are studied in William Reid [238] and Werner Kratz [193] (see also [54, 101, 114, 196, 197]) while results on Hamiltonian difference systems are summarized in Ahlbrandt and Peterson [30]. These systems have been introduced by Erbe and Yan in [137]; see also [138, 139, 140]. Further work on discrete Hamiltonian systems can be found in [1, 23, 36, 38, 61, 62, 64, 66, 68, 71, 72, 77, 82, 83, 84, 117, 120, 224]. Higher order Sturm–Liouville difference equations are treated in [65, 70, 118]. Some matrix theory is needed in this chapter, and we make use of the classical references [143, 176, 177, 244]. In particular, we want to refer to Ben-Israel and Greville [56] for an extensive study of generalized inverse matrices.

Martin Bohner and Ondřej Došlý initiated the study on discrete symplectic systems in [76], and further results are contained in [69]. Trigonometric symplectic systems are introduced in Bohner and Došlý [78] and studied in Douglas Anderson [37, 38, 80].

Results on Hamiltonian dynamic equations are contained in [2, 24, 81, 86, 166, 167, 168]. Ondřej Došlý and Roman Hilscher initiated the study of symplectic dynamic systems on time scales in [121], and it was pursued by Calvin Ahlbrandt, Martin Bohner, and Jerry Ridenhour [24] and Roman Hilscher [172]. In particular, Theorem 7.43 is Theorem 1 in [172].

CHAPTER 8

Extensions

In this last chapter we present some possible forms of extensions of the theory of time scales. One such extension is the model of a measure chain as developed by Stefan Hilger in [160], which will be discussed in the first section of this chapter. We also present so-called alpha derivatives as introduced by Ahlbrandt, Bohner, and Ridenhour in [24]. Many results presented in this book can be derived for these more general models, and research in these areas continues.

8.1. Measure Chains

Let $\mathbb{T}$ be some set. According to Hilger [159, 160], a triple $(\mathbb{T}, \leq, \nu)$ is called a *(strong) measure chain* provided it satisfies the following three axioms.

Axiom 8.1. The relation "$\leq$" satisfies (for all $r, s, t \in \mathbb{T}$):

 (i) $t \leq t$ (reflexive);
 (ii) if $r \leq s$ and $s \leq t$, then $r \leq t$ (transitive);
 (iii) if $r \leq s$ and $s \leq r$, then $r = s$ (antisymmetric);
 (iv) either $r \leq s$ or $s \leq r$ (total).

Axiom 8.2. Any nonvoid subset of $\mathbb{T}$ which is bounded above has a least upper bound (i.e., the chain $(\mathbb{T}, \leq)$ is *conditionally complete*).

Axiom 8.3. The mapping $\nu : \mathbb{T} \times \mathbb{T} \to \mathbb{R}$ has the following properties (for all $r, s, t \in \mathbb{T}$):

 (i) $\nu(r, s) + \nu(s, t) = \nu(r, t)$ (cocycle property);
 (ii) if $r > s$, then $\nu(r, s) > 0$ (strong isotony);
 (iii) ν is continuous (continuity).

Now the *jump operators* σ and ρ are defined as before by

$$\sigma(t) = \inf\{s \in \mathbb{T} : s > t\} \quad \text{and} \quad \rho(t) = \sup\{s \in \mathbb{T} : s < t\}$$

(where $\sigma(t) = t$ if $t = \max \mathbb{T}$ and $\rho(t) = t$ if $t = \min \mathbb{T}$), while the *graininess* is defined as

$$\mu(t) = \nu(\sigma(t), t).$$

The notions of left-dense, left-scattered, right-dense, right-scattered, and $\mathbb{T}^\kappa$ are also defined as before.

Definition 8.4. Let $\mathbb{X}$ be some Banach space with norm $|\cdot|$ and let $f : \mathbb{T} \to \mathbb{X}$ be a mapping. We say that f is differentiable at $t \in \mathbb{T}$ if there exists $f^\Delta(t) \in \mathbb{X}$ such that for each $\varepsilon > 0$ there exists a neighborhood U of t with

$$|f(\sigma(t)) - f(s) - f^\Delta(t)\nu(\sigma(t), s)| \leq \varepsilon |\nu(\sigma(t), s)| \quad \text{for all} \quad s \in U.$$

In this case $f^\Delta(t)$ is said to be the derivative of f at t.

Theorem 8.5. *Let $f : \mathbb{T} \to \mathbb{X}$ be a mapping and let $t \in \mathbb{T}$.*

(i) *If $t \in \mathbb{T}^\kappa$, then f has at most one derivative at t.*
(ii) *If $t \notin \mathbb{T}^\kappa$, then f has each $x \in \mathbb{X}$ as a derivative at t.*
(iii) *If f is differentiable at t, then it is continuous at t.*
(iv) *The mapping $\nu(\cdot, t)$ is differentiable with derivative*

$$\nu^\Delta(\cdot, t) = 1.$$

(v) *If f is continuous at t and if t is right-scattered, then f is differentiable at t with derivative*

$$f^\Delta(t) = \frac{f(\sigma(t)) - f(t)}{\mu(t)}.$$

Proof. We only prove part (iv) as the rest of this proof is identical to Exercise 1.11, Exercise 1.12, and the proof of Theorem 1.16. Now define f by $f = \nu(\cdot, t)$. Let $r \in \mathbb{T}$ and let $\varepsilon > 0$ be given. Observe that the cocycle property of Axiom 8.3 (i) says that

$$\nu(\sigma(r), s) + \nu(s, t) = \nu(\sigma(r), t) \quad \text{for all} \quad s \in \mathbb{T}.$$

Therefore we find

$$
\begin{aligned}
|f(\sigma(r)) - f(s) - 1 \cdot \nu(\sigma(r), s)| &= |\nu(\sigma(r), t) - \nu(s, t) - \nu(\sigma(r), s)| \\
&= 0 \\
&\leq \varepsilon |\nu(\sigma(r), s)|.
\end{aligned}
$$

This shows that f is differentiable at r with $f^\Delta(r) = 1$. □

Exercise 8.6. Finish the proof of Theorem 8.5.

Example 8.7. Consider the measure chain $(\mathbb{T}, \leq, \nu)$, where $\mathbb{T}$ is a closed subset of the reals, "$\leq$" is the usual order relation between real numbers, and

$$\nu(r, s) = r - s \quad \text{for all} \quad r, s \in \mathbb{T}.$$

Then $\mathbb{T}$ is a time scale as discussed in this book.

Example 8.8 (Pospíšil, [**233**, Example 1]). Let $\{a_k\}_{k \in \mathbb{N}_0}$ be an increasing sequence of real numbers and let $\{d_k\}_{k \in \mathbb{N}}$ be a sequence of positive real numbers. Let

$$\mathbb{T} = \{(k, \tau) \in \mathbb{N}_0 \times \mathbb{R} : a_k \leq \tau \leq a_{k+1}\}.$$

As "$\leq$" we use the lexicographic ordering on $\mathbb{T}$, i.e.,

$$(k, \xi) \leq (l, \eta) \quad \text{iff} \quad k < l \text{ or } k = l, \xi \leq \eta.$$

Finally, we define $\nu : \mathbb{T} \times \mathbb{T} \to \mathbb{R}$ by

$$\nu((k, \xi), (l, \eta)) = \xi - \eta + \sum_{i=l+1}^{k} d_i$$

(we use the convention $\sum_{i=n}^{m} \alpha_i = -\sum_{i=m+1}^{n-1} \alpha_i$ for $m < n - 1$ and $\sum_{i=n}^{n-1} \alpha_i = 0$). Then $(\mathbb{T}, \leq, \nu)$ is a measure chain. The forward jump operator σ is given by

$$\sigma((k, \tau)) = \begin{cases} (k, \tau) & \text{if} \quad \tau \neq a_{k+1} \\ (k+1, \tau) & \text{if} \quad \tau = a_{k+1}, \end{cases}$$

while the graininess μ is given by

$$\mu((k,\tau)) = \begin{cases} 0 & \text{if} \quad \tau \neq a_{k+1} \\ d_{k+1} & \text{if} \quad \tau = a_{k+1}. \end{cases}$$

This measure chain can serve to model a process whose continuous evolution is usually interrupted by an event (impulse, catastrophe, etc.) of "size" d_i at time instants a_i, $i \in \mathbb{N}$. In such a case, (k,τ) denotes a time instant with "absolute time distance" τ from beginning a_0 after k occurrences of the interrupting events.

Theorem 8.9. *Let $f, g : \mathbb{T} \to \mathbb{X}$ be differentiable at $t \in \mathbb{T}^\kappa$. Then:*

(i) *The mapping $\alpha f + \beta g$ (if α and β are constants) is differentiable with*

$$(\alpha f + \beta g)^\Delta(t) = \alpha f^\Delta(t) + \beta g^\Delta(t).$$

(ii) *The mapping $f \cdot g$ (if "·" is bilinear and continuous) is differentiable with*

$$(f \cdot g)^\Delta(t) = f(\sigma(t)) \cdot g^\Delta(t) + f^\Delta(t)g(t).$$

(iii) *The mapping $f \cdot g^{-1}$ (if g is algebraically invertible) is differentiable with*

$$(f \cdot g^{-1})^\Delta(t) = \left[f^\Delta(t) - (f \cdot g^{-1})(t) \cdot g^\Delta(t)\right] \cdot g^{-1}(\sigma(t)).$$

Proof. The proof is identical to the proof of Theorem 1.20. □

Exercise 8.10. Prove Theorem 8.9. Define the notions from Chapter 1, e.g., pre-differentiability, pre-antiderivative, integral, and antiderivative for the more general setting of measure chains. Also prove the mean value theorem (precisely as it is stated in Theorem 1.67) for this more general setting of measure chains. Finally, prove Corollary 8.11, following closely the proof of Corollary 1.68.

Corollary 8.11. *Let $f : \mathbb{T} \to \mathbb{X}$ be pre-differentiable with D and let U be some compact interval of $\mathbb{T}$ with endpoints r and s. Then the estimate*

$$|f(s) - f(r)| \leq \left\{ \sup_{t \in U^\kappa \cap D} |f^\Delta(t)| \right\} |\nu(s,r)|$$

holds.

Theorem 8.12. *Suppose $f_n : \mathbb{T} \to \mathbb{X}$ is pre-differentiable with D for each $n \in \mathbb{N}$. Assume that for each $t \in \mathbb{T}^\kappa$ there exists a compact interval neighborhood $U(t)$ such that the sequence*

$$\{f_n^\Delta\}_{n \in \mathbb{N}} \quad \text{converges uniformly on} \quad U(t) \cap D.$$

(i) *If $\{f_n\}_{n \in \mathbb{N}}$ converges at some $t_0 \in U(t)$ for some $t \in \mathbb{T}^\kappa$, then it converges uniformly on $U(t)$.*

(ii) *If $\{f_n\}_{n \in \mathbb{N}}$ converges at some $t_0 \in \mathbb{T}$, then it converges uniformly on $U(t)$ for all $t \in \mathbb{T}^\kappa$.*

(iii) *The limit mapping $f = \lim_{n \to \infty} f_n$ is pre-differentiable with D and we have*

$$f^\Delta(t) = \lim_{n \to \infty} f_n(t) \quad \text{for all} \quad t \in D.$$

Proof. Part (i). Let $t \in \mathbf{T}^\kappa$. Then there exists $N \in \mathbf{N}$ such that

$$\sup_{s \in U(t) \cap D} |(f_n - f_m)^\Delta(s)| \quad \text{is finite for all} \quad m, n \geq N.$$

Suppose $\{f_n(t_0)\}_{n \in \mathbf{N}}$ converges for some $t_0 \in U(t)$. Let $m, n \geq N$ and $r \in U(t)$. Then

$$
\begin{aligned}
|f_n(r) - f_m(r)| &= |f_n(r) - f_m(r) - [f_n(t_0) - f_m(t_0)] + f_n(t_0) - f_m(t_0)| \\
&\leq |f_n(t_0) - f_m(t_0)| + |(f_n - f_m)(r) - (f_n - f_m)(t_0)| \\
&\leq |f_n(t_0) - f_m(t_0)| + \left\{ \sup_{s \in U(t) \cap D} |(f_n - f_m)^\Delta(s)| \right\} |\nu(t_0, r)|
\end{aligned}
$$

by Corollary 8.11. Hence $\{f_n\}_{n \in \mathbf{N}}$ converges uniformly on $U(t)$. Thus $\{f_n\}_{n \in \mathbf{N}}$ is a locally uniformly convergent sequence.

Part (ii). Assume now that $\{f_n(t_0)\}_{n \in \mathbf{N}}$ converges for some $t_0 \in \mathbf{T}$. We use the induction principle to show that the statement

$$S(t): \quad \{f_n(t)\}_{n \in \mathbf{N}} \text{ converges}$$

holds for each $t \in [t_0, \infty)$.

I. Clearly, $S(t_0)$ holds by the assumption.
II. Now suppose that t is right-scattered and that $S(t)$ holds. Then

$$f_n(\sigma(t)) = f_n(t) + \mu(t) f_n^\Delta(t)$$

converges by the assumption. This implies that $S(\sigma(t))$ holds as well.
III. Next, suppose that t is right-dense and that $S(t)$ holds. Then $\{f_n\}_{n \in \mathbf{N}}$ converges (even uniformly) on $U(t)$ by part (i) and so $S(r)$ holds for all $r \in U(t) \cap (t, \infty)$.
IV. Finally, suppose that t is left-dense and that $S(r)$ holds for all $t_0 \leq r < t$. Since $U(t) \cap [t_0, t) \neq \emptyset$, we find again by part (i) that $\{f_n\}_{n \in \mathbf{N}}$ converges (even uniformly) on $U(t)$; in particular $S(t)$ is true.

These observations show that $S(t)$ is indeed true for all $t \in [t_0, \infty)$. We use the dual version of the induction principle for the negative direction (see Remark 1.8) to show that $S(t)$ is also true for all $t \in (-\infty, t_0]$. While the first part of this has already been shown, the third and the fourth part follow again by (i). Only for the second part one has to observe that for left-scattered t we have

$$f_n(\rho(t)) = f_n(t) - \mu(\rho(t)) f_n^\Delta(\rho(t)).$$

Part (iii). Let $t \in D$. Without loss of generality we can assume that $\sigma(t) \in U(t)$. Let $\varepsilon > 0$. By the proof of (i), there exists $N \in \mathbf{N}$ such that

$$|(f_n - f_m)(r) - (f_n - f_m)(\sigma(t))| \leq \left\{ \sup_{s \in U(t) \cap D} |(f_n - f_m)^\Delta(s)| \right\} |\nu(\sigma(t), r)|$$

for all $r \in U(t)$ and all $m, n \geq N$. Since $\{f_n^\Delta\}_{n \in \mathbf{N}}$ converges uniformly on $U(t) \cap D$, there exists $\tilde{N} \geq N$ such that

$$\sup_{s \in U(t) \cap D} |(f_n - f_m)^\Delta(s)| \leq \frac{\varepsilon}{3} \quad \text{for all} \quad m, n \geq \tilde{N}.$$

Hence,

$$|(f_n - f_m)(r) - (f_n - f_m)(\sigma(t))| \leq \frac{\varepsilon}{3}|\nu(\sigma(t), r)|$$

for all $r \in U(t)$ and all $m, n \geq \tilde{N}$ so that, by letting $m \to \infty$,

$$|(f_n - f)(r) - (f_n - f)(\sigma(t))| \leq \frac{\varepsilon}{3}|\nu(\sigma(t), r)|$$

for all $r \in U(t)$ and all $n \geq \tilde{N}$. Let

$$g = \lim_{n \to \infty} f_n^\Delta.$$

Then there exists $M \geq \tilde{N}$ such that

$$\left|f_M^\Delta(\sigma(t)) - g(\sigma(t))\right| \leq \frac{\varepsilon}{3},$$

and since f_M is differentiable at t, there also exists a neighborhood W of t with

$$\left|f_M(\sigma(t)) - f_M(r) - f_M^\Delta(t)\nu(\sigma(t), r)\right| \leq \frac{\varepsilon}{3}|\nu(\sigma(t), r)| \quad \text{for all} \quad r \in W.$$

Altogether we have now for all $r \in U(t) \cap W$

$$
\begin{aligned}
|f(\sigma(t)) - f(r) - g(t)\nu(\sigma(t), r)| &\leq |(f_M - f)(\sigma(t)) - (f_M - f)(r)| \\
&\quad + |[f_M^\Delta(t) - g(t)]\nu(\sigma(t), r)| + |f_M(\sigma(t)) - f_M(r) - f_M^\Delta(t)\nu(\sigma(t), r)| \\
&\leq \frac{\varepsilon}{3}|\nu(\sigma(t), r)| + \frac{\varepsilon}{3}|\nu(\sigma(t), r)| + \frac{\varepsilon}{3}|\nu(\sigma(t), r)| \\
&= \varepsilon|\nu(\sigma(t), r)|
\end{aligned}
$$

so that indeed f is differentiable at t with $f^\Delta(t) = g(t)$. $\qquad \square$

Theorem 8.13. *Let $t_0 \in \mathbb{T}$, $x_0 \in \mathbb{X}$, and a regulated map $f : \mathbb{T}^\kappa \to \mathbb{X}$ be given. Then there exists exactly one pre-differentiable function F satisfying*

$$F^\Delta(t) = f(t) \quad \text{for all} \quad t \in D, \quad F(t_0) = x_0.$$

Proof. Let $n \in \mathbb{N}$. We now use the induction principle (Theorem 1.7) to show that

$$S(t): \quad \begin{cases} \text{There exists a pre-differentiable } (F_{nt}, D_{nt}), \ F_{nt} : [t_0, t] \to \mathbb{X} \text{ with} \\ F_{nt}(t_0) = x_0 \quad \text{and} \quad |F_{nt}^\Delta(s) - f(s)| \leq \frac{1}{n} \text{ for } s \in D_{nt} \end{cases}$$

for all $t \in [t_0, \infty)$.

I. Let

$$D_{nt_0} = \emptyset \quad \text{and} \quad F_{nt_0}(t_0) = x_0.$$

Then statement $S(t_0)$ follows.

II. Suppose t is right-scattered and $S(t)$ is true. Define

$$D_{n\sigma(t)} = D_{nt} \cup \{t\}$$

and $F_{n\sigma(t)}$ on $[t_0, \sigma(t)]$ by

$$F_{n\sigma(t)}(s) = \begin{cases} F_{nt}(s) & \text{if} \quad s \in D_{nt} \\ F_{nt}(t) + \mu(t)f(t) & \text{if} \quad s = \sigma(t). \end{cases}$$

Then $F_{n\sigma(t)}(t_0) = F_{nt}(t_0) = x_0$,

$$|F_{n\sigma(t)}^\Delta(s) - f(s)| = |F_{nt}^\Delta(s) - f(s)| \leq \frac{1}{n} \quad \text{for} \quad s \in D_{nt},$$

and

$$\left|F^{\Delta}_{n\sigma(t)}(t) - f(t)\right| = \left|\frac{F_{n\sigma(t)}(\sigma(t)) - F_{n\sigma(t)}(t)}{\mu(t)} - f(t)\right|$$

$$= \left|\frac{F_{nt}(t) + \mu(t)f(t) - F_{n\sigma(t)}(t)}{\mu(t)} - f(t)\right|$$

$$= \left|\frac{\mu(t)f(t)}{\mu(t)} - f(t)\right|$$

$$= 0$$

$$\leq \frac{1}{n},$$

and therefore statement $S(\sigma(t))$ is valid.

III. Suppose t is right-dense and $S(t)$ is true. Since t is right-dense and f is regulated,

$$f(t^+) = \lim_{s \to t, s > t} f(s) \quad \text{exists.}$$

Hence there is a neighborhood U of t with

(8.1) $|f(s) - f(t^+)| \leq \frac{1}{n} \quad \text{for all} \quad s \in U \cap (t, \infty).$

Let $r \in U \cap (t, \infty)$. Define

$$D_{nr} = [D_{nt} \setminus \{t\}] \cup (t, r]^{\kappa}$$

and F_{nr} on $[t_0, r]$ by

$$F_{nr}(s) = \begin{cases} F_{nt}(s) & \text{if} \quad s \in [t_0, t] \\ F_{nt}(t) + f(t^+)\nu(s, t) & \text{if} \quad s \in (t, r]. \end{cases}$$

Then F_{nr} is continuous at t and hence on $[t_0, r]$. Also the function F_{nr} is differentiable on $(t, r]^{\kappa}$ with

$$F^{\Delta}_{nr}(s) \equiv f(t^+) \quad \text{for all} \quad s \in (t, r].$$

Hence F_{nr} is pre-differentiable on $[t_0, t)$. Since t is right-dense, we have that F_{nr} is pre-differentiable with D_{nr}. Because of $S(t)$ and (8.1), we also have

$$|F^{\Delta}_{nr}(s) - f(s)| \leq \frac{1}{n} \quad \text{for all} \quad s \in D_{nr}.$$

Therefore the statement $S(r)$ is true for all $r \in U \cap (t, \infty)$.

IV. Now suppose that t is left-dense and that the statements $S(r)$ are true for all $r < t$. Since f is regulated,

$$f(t^-) = \lim_{s \to t, s < t} f(s) \quad \text{exists.}$$

Hence there exists a neighborhood U of t with

(8.2) $|f(s) - f(t^-)| \leq \frac{1}{n} \quad \text{for all} \quad s \in U \cap (-\infty, t).$

Fix some $r \in U \cap (-\infty, t)$ and define

$$D_{nt} = \begin{cases} D_{nr} \cup (r, t) & \text{if} \quad r \text{ is right-dense} \\ D_{nr} \cup [r, t) & \text{if} \quad r \text{ is right-scattered} \end{cases}$$

and F_{nt} on $[t_0, t]$ by

$$F_{nt}(s) = \begin{cases} F_{nr}(s) & \text{if} \quad s \in [t_0, r] \\ F_{nr}(t) + f(t^-)\nu(s, t) & \text{if} \quad s \in (r, t]. \end{cases}$$

Again, F_{nt} is continuous at r and hence on $[t_0, t]$. The function F_{nt} might not be differentiable at r if r is right-dense, but since

$$F_{nt}^\Delta(s) \equiv f(t^-) \quad \text{for all} \quad s \in (r, t],$$

F_{nt} is indeed pre-differentiable with D_{nt}, and $S(r)$ together with (8.2) imply

$$\left| F_{nt}^\Delta(s) - f(s) \right| \le \frac{1}{n} \quad \text{for all} \quad s \in D_{nt}.$$

Hence statement $S(t)$ holds.

By the induction principle, $S(t)$ is true for all $t \ge t_0$, $t \in \mathbb{T}$. Similarly, we can show $S(t)$ is valid for $t \le t_0$. Hence F_n is pre-differentiable with D_n, $F_n(t_0) = x_0$, and

$$\left| F_n^\Delta(t) - f(t) \right| \le \frac{1}{n} \quad \text{for all} \quad t \in D_n.$$

Now let

$$F = \lim_{n \to \infty} F_n \quad \text{and} \quad D = \bigcap_{n \in \mathbb{N}} D_n.$$

Then $F(t_0) = x_0$, F is pre-differentiable on D, and

$$F^\Delta(t) = \lim_{n \to \infty} F_n^\Delta(t) = f(t) \quad \text{for all} \quad t \in D$$

by Theorem 8.12. $\qquad\square$

8.2. Nonlinear Theory

Most of the dynamical equations considered in this book, with a few exceptions, are linear. For an extensive collection of results on nonlinear problems, we refer to the book by Kaymakçalan, Lakshmikantham, and Sivasundaram [185]. In this section, however, we wish to present the basic existence and uniqueness results for equations of the form

$$y^\Delta = f(t, y).$$

These results are not needed in this book, except for the fundamental existence result of the matrix exponential in Theorem 5.8, which claims that the IVP

$$y^\Delta = A(t)y + f(t), \quad y(t_0) = y_0$$

has a unique solution, provided $f : \mathbb{T} \to \mathbb{R}^n$ is rd-continuous and $A \in \mathcal{R}$ is an $n \times n$-matrix-valued function on $\mathbb{T}$. Of course, Theorem 2.35 is the scalar case corresponding to Theorem 5.8, but it has been proved separately in Chapter 2 without actually referring to Theorem 5.8.

Definition 8.14. Let $\mathbb{T}$ be a time scale and $\mathbf{X}$ be a Banach space. A function $f : \mathbb{T} \times \mathbf{X} \to \mathbf{X}$ is called

(i) *rd-continuous*, if g defined by $g(t) = f(t, x(t))$ is rd-continuous for any continuous function $x : \mathbb{T} \to \mathbf{X}$;

(ii) *regressive* at $t \in \mathbb{T}^\kappa$, if the mapping

$$\text{id} + \mu(t)f(t, \cdot) : \mathbf{X} \to \mathbf{X} \quad \text{is invertible}$$

(where id is the identity function), and f is called *regressive* on $\mathbb{T}^\kappa$, if f is regressive at each $t \in \mathbb{T}^\kappa$;

(iii) *bounded* on a set $S \subset \mathbb{T} \times \mathbf{X}$, if there exists a constant $M > 0$ such that

$$|f(t, x)| \leq M \quad \text{for all} \quad (t, x) \in S;$$

(iv) *Lipschitz continuous* on a set $S \subset \mathbb{T} \times \mathbf{X}$, if there exists a constant $L > 0$ such that

$$|f(t, x_1) - f(t, x_2)| \leq L|x_1 - x_2| \quad \text{for all} \quad (t, x_1), (t, x_2) \in S.$$

Exercise 8.15. Show that every Lipschitz continuous function $f : \mathbb{T} \times \mathbf{X} \to \mathbf{X}$ is regressive on $\mathbb{T}^\kappa$, provided the *Lipschitz constant L* satisfies

$$L\mu(t) < 1 \quad \text{for all} \quad t \in \mathbb{T}^\kappa.$$

Theorem 8.16 (Local Existence and Uniqueness). *Let $\mathbb{T}$ be a time scale, $\mathbf{X}$ a Banach space, $t_0 \in \mathbb{T}$, $x_0 \in \mathbf{X}$, $a > 0$ with $\inf \mathbb{T} \leq t_0 - a$ and $\sup \mathbb{T} \geq t_0 + a$, and put*

$$I_a = (t_0 - a, t_0 + a) \quad \text{and} \quad U_b = \{x \in \mathbf{X} : |x - x_0| < b\}.$$

Suppose that $f : I_a \times U_b \to \mathbf{X}$ is rd-continuous, bounded (with bound $M > 0$), and Lipschitz continuous (with constant $L > 0$). Then the IVP

$$(8.3) \qquad\qquad x^\Delta = f(t, x), \quad x(t_0) = x_0$$

has exactly one solution on $[t_0 - \alpha, t_0 + \alpha]$, where

$$\alpha = \min\left\{a, \frac{b}{M}, \frac{1 - \varepsilon}{L}\right\} \quad \text{for some} \quad \varepsilon > 0.$$

If t_0 is right-scattered and $\alpha < \mu(t_0)$, then the unique solution exists on the interval $[t_0 - \alpha, \sigma(t_0)]$.

Proof. Let $C(A, B)$ denote the space of all continuous functions $f : A \to B$. We put $C = C(I_\alpha, U_b)$. If $\lambda \in C$, then the function g defined by $g(t) = f(t, \lambda(t))$ is rd-continuous according to Definition 8.14 (i), and hence g has an antiderivative according to Theorem 1.74. Hence the operator $F : C \to C(I_\alpha, \mathbf{X})$ defined by

$$F(\lambda)(t) = x_0 + \int_{t_0}^t f(\tau, \lambda(\tau))\Delta\tau, \quad \lambda \in C$$

is well defined. We now show that F satisfies the assumption of *Banach's fixed point theorem*. First, let $\lambda \in C$. Then

$$|\lambda(t) - x_0| < b \quad \text{for all} \quad t \in I_\alpha$$

and hence

$$|F(\lambda)(t) - x_0| = \left|\int_{t_0}^t f(\tau, \lambda(\tau))\Delta\tau\right| \leq |t - t_0|M < \alpha M \leq b$$

for all $t \in I_\alpha$. Thus $F(\lambda) : I_\alpha \to U_b$. In addition, F is differentiable and hence continuous by Theorem 1.16 (i), and hence

$$F(C) \subset C$$

so that F maps C into itself. Next, let $\lambda_1, \lambda_2 \in C$. Then

$$
\begin{aligned}
|F(\lambda_1)(t) - F(\lambda_2)(t)| &= \left| \int_{t_0}^{t} (f(\tau, \lambda_1(\tau)) - f(\tau, \lambda_2(\tau))) \, \Delta\tau \right| \\
&\leq |t - t_0| L \|\lambda_1 - \lambda_2\| \\
&< \alpha L \|\lambda_1 - \lambda_2\| \\
&\leq (1 - \varepsilon)\|\lambda_1 - \lambda_2\|
\end{aligned}
$$

so that

$$
\|F(\lambda_1) - F(\lambda_2)\| \leq (1 - \varepsilon)\|\lambda_1 - \lambda_2\|
$$

(where $\|\cdot\|$ is the corresponding supremum norm). So F is a contraction mapping. By Banach's fixed point theorem, the equation

$$
\lambda = F(\lambda)
$$

has exactly one solution $\tilde{\lambda} \in C$, and

$$
\tilde{\lambda}^{\Delta}(t) = f(t, \tilde{\lambda}(t)), \quad \tilde{\lambda}(t_0) = x_0
$$

so that $\tilde{\lambda}$ is the unique solution of (8.3). If t_0 is right-scattered, then

$$
\tilde{\lambda}(\sigma(t_0)) = \tilde{\lambda}(t_0) + \mu(t_0)\tilde{\lambda}^{\Delta}(t_0) = x_0 + \mu(t_0)f(t_0, x_0)
$$

is uniquely determined. □

"Going forward" in time is easy to do under the conditions of Theorem 8.16, but for "going backward" in time, Lipschitz continuity is not sufficient as can be seen from the following example. Instead, regressivity helps.

Example 8.17. Consider the time scale $\mathbb{T} = \mathbb{Z}$ and the initial value problem

$$
x^{\Delta} = -x, \quad x(0) = 1.
$$

Here, $f(t, x) = -x$ is Lipschitz continuous with $L = 1$, as is seen from

$$
|f(t, x_1) - f(t, x_2)| = |x_1 - x_2|.
$$

We have

$$
x(k) = 0 \quad \text{for all} \quad k \in \mathbb{N}.
$$

However, the solution x does not exist at $-k$ for $k \in \mathbb{N}$. Note also that f is not regressive on $\mathbb{T}$ because

$$
x + \mu(t)f(t, x) = x - x = 0.
$$

Similarly, a solution could exist for all times but may not be unique, if regressivity is not satisfied.

Theorem 8.18. *Assume in addition to the assumptions from Theorem 8.16 that f is regressive. If t_0 is left-scattered and $\alpha > \mu(\rho(t_0))$, then the solution of (8.3) exists on $[\rho(t_0), t_0 + \alpha]$.*

Proof. The following needs to be satisfied:

$$
x_0 = \tilde{\lambda}(t_0) = \tilde{\lambda}(\rho(t_0)) + \mu(\rho(t_0))f(\rho(t_0), \tilde{\lambda}(\rho(t_0))) = (\mathrm{id} + \mu f(\rho(t_0), \cdot)) \, \tilde{\lambda}(\rho(t_0)).
$$

Because of regressivity,

$$
\tilde{\lambda}(\rho(t_0)) = (\mathrm{id} + \mu f(\rho(t_0), \cdot))^{-1} x_0
$$

is uniquely determined. □

Definition 8.19. We say that the IVP (8.3) has a *maximal solution* $\tilde{x} : I_{\max} \to X$ with *maximal interval of existence* $I_{\max}$ provided the following holds: If $J \subset \mathbb{T}$ is an interval and $x : J \to X$ is a solution of (8.3), then

$$J \subset I_{\max} \quad \text{and} \quad \tilde{x}(t) = x(t) \text{ for all } t \in J.$$

Now we can prove the following global existence-uniqueness result.

Theorem 8.20 (Global Existence and Uniqueness). *Let* $f : \mathbb{T} \times X \to X$ *be rd-continuous and regressive. Suppose that for each* $(t, x) \in \mathbb{T} \times X$ *there exists a neighborhood* $I_a \times U_b$ *as in the statement of Theorem 8.16 such that* f *is bounded on* $I_a \times U_b$ *and such that the Lipschitz condition*

$$|f(t, x_1) - f(t, x_2)| \le L(t, x)|x_1 - x_2| \quad \text{for all} \quad (t, x_1), (t, x_2) \in I_a \times U_b$$

holds, where $L(t, x) > 0$. *Then the IVP (8.3) has exactly one maximal solution* $\tilde{\lambda} : I_{\max} \to X$, *and the maximal interval of existence* $I_{\max} = I_{\max}(t_0, x_0)$ *is open.*

Proof. Because of Theorem 8.16 and Theorem 8.18, (8.3) is uniquely solvable on an interval $[t_0^-, t_0^+]$ with $t_0^-, t_0^+ \in \mathbb{T}$ and $t_0^- < t_0 < t_0^+$. We extend this local solution in both directions until it is not possible to extend it further, and hence obtain a solution $\tilde{\lambda} : I \to X$. Now suppose λ is another solution of (8.3). Then

$$\lambda(t) = \tilde{\lambda}(t) \quad \text{for all} \quad t \in [t_0^-, t_0^+]$$

because of Theorem 8.16. We want to show that $\lambda = \tilde{\lambda}$ on all of $I_{\max}$. Consider the set

$$M = \{t \ge t_0 : \lambda(t) \ne \tilde{\lambda}(t)\}.$$

For the sake of achieving a contradiction, assume that $M \ne \emptyset$. Since M is also bounded below by t_0^+,

$$t^* := \inf M \in \mathbb{T} \quad \text{exists.}$$

If $t \in [t_0, t^*)$, then $t \notin M$, so $\lambda(t) = \tilde{\lambda}(t)$. Assume t^* is left-dense. Then

$$\lambda(t^*) = \lim_{\substack{t \to t^* \\ t \in [t_0, t^*)}} \lambda(t) = \lim_{\substack{t \to t^* \\ t \in [t_0, t^*)}} \tilde{\lambda}(t) = \tilde{\lambda}(t^*)$$

so that $t^* \notin M$. If t^* is left-scattered, then $\rho(t^*) \notin M$, but $\rho(t^*) \ge t_0$ since $t^* > t_0$ (note that $t_0^+ \notin M$), and therefore $\lambda(\rho(t^*)) = \tilde{\lambda}(\rho(t^*))$. Hence

$$
\begin{aligned}
\lambda(t^*) &= \lambda(\rho(t^*)) + \mu(\rho(t^*))\lambda^\Delta(\rho(t^*)) \\
&= \lambda(\rho(t^*)) + \mu(\rho(t^*))f(\rho(t^*), \lambda(\rho(t^*))) \\
&= \tilde{\lambda}(\rho(t^*)) + \mu(\rho(t^*))f(\rho(t^*), \tilde{\lambda}(\rho(t^*))) \\
&= \tilde{\lambda}(\rho(t^*)) + \mu(\rho(t^*))\tilde{\lambda}^\Delta(\rho(t^*)) \\
&= \tilde{\lambda}(t^*)
\end{aligned}
$$

so that again $t^* \notin M$. Hence $t^* \notin M$ in either case. Put $x^* = \lambda(t^*) = \tilde{\lambda}(t^*)$ and consider the IVP

$$(8.4) \qquad\qquad x^\Delta = f(t, x), \quad x(t^*) = x^*.$$

By Theorem 8.16 (and Theorem 8.18), (8.4) has a unique solution on $[t^{*-}, t^{*+}]$, where $t^{*-}, t^{*+} \in \mathbb{T}$ with $t^{*-} < t^* < t^{*+}$. Since both λ and $\tilde{\lambda}$ solve (8.4), it follows that

$$\lambda(t) = \tilde{\lambda}(t) \quad \text{for all} \quad t \in [t^*, t^{*+}],$$

but this contradicts the definition of t^*. Hence $M = \emptyset$. Similarly one can show that

$$N = \{t \leq t_0 : \lambda(t) \neq \tilde{\lambda}(t)\}$$

is empty, too, but this part will be left to the reader in Exercise 8.21. $\qquad\square$

Exercise 8.21. Assume that the set N in the proof of Theorem 8.20 is not empty. Then define $\tilde{t} = \sup N$, derive (similarly as in the proof of Theorem 8.20) a contradiction, and conclude that $N = \emptyset$.

Now Theorems 8.16, 8.18, and 8.20 imply the following.

Corollary 8.22. *Suppose the assumptions of Theorem 8.20 are satisfied, and let a, b, and M be as in Theorem 8.16. Then (8.3) has a unique solution on $\left[t_0 - \frac{b}{M}, t_0 + \frac{b}{M}\right]$.*

Exercise 8.23. Prove Corollary 8.22.

Theorem 8.24. *Suppose the assumptions of Theorem 8.20 are satisfied, and assume that there are positive and continuous functions p and q with*

$$|f(t,x)| \leq p(t)|x| + q(t) \quad \text{for all} \quad (t,x) \in \mathbb{T} \times \mathbf{X}.$$

Then each solution of $x^\Delta = f(t,x)$ exists on all of $\mathbb{T}$.

Proof. Let $t_0 \in \mathbb{T}$ be given. We only show that a solution exists on $[t_0, \infty)$, but the existence of this solution on $(-\infty, t_0]$ follows similarly (see Exercise 8.25). For the sake of achieving a contradiction, we assume that a solution λ exists only on $[t_0, s)$ with $\sup \mathbb{T} \geq s$. If s is left-scattered, then $\rho(s) \in [t_0, s)$. But then the solution λ exists also at s, namely by putting

$$\lambda(s) = \lambda(\rho(s)) + \mu(\rho(s))f(\rho(s), \lambda(\rho(s))),$$

a contradiction. If s is left-dense, then there exists $r \in [t_0, s)$ with

$$s - r =: \delta < \frac{1}{\|p\|}, \quad \text{where} \quad \|p\| = \max_{t_0 \leq t \leq s} |p(t)|.$$

Now put

$$b = \frac{\|p\||\lambda(r)| + \|q\|}{1 - \delta\|p\|}\delta > 0, \quad \text{where} \quad \|q\| = \max_{t_0 \leq t \leq s} |q(t)|.$$

If $|x - \lambda(r)| \leq b$ and $t_0 \leq t \leq s$, then

$$
\begin{aligned}
|f(t,x)| &\leq p(t)|x| + q(t) \\
&\leq \|p\|\|x\| + \|q\| \\
&\leq \|p\|(|\lambda(r)| + |x - \lambda(r)|) + \|q\| \\
&\leq \|p\|(|\lambda(r)| + b) + \|q\| \\
&= \|p\| \left(|\lambda(r)| + \frac{\|p\||\lambda(r)| + \|q\|}{1 - \delta\|p\|}\delta \right) + \|q\| \\
&= \|p\|\frac{|\lambda(r)| - \delta|\lambda(r)|\|p\| + \|p\||\lambda(r)|\delta + \|q\|\delta}{1 - \delta\|p\|} + \|q\| \\
&= \|p\|\frac{|\lambda(r)| + \|q\|\delta}{1 - \delta\|p\|} + \|q\| \\
&= \frac{\|p\||\lambda(r)| + \|p\|\|q\|\delta + \|q\| - \delta\|p\|\|q\|}{1 - \delta\|p\|} \\
&= \frac{\|p\||\lambda(r)| + \|q\|}{1 - \delta\|p\|} \\
&= \frac{b}{\delta}
\end{aligned}
$$

so that by Corollary 8.22 the solution exists on

$$
\left[r, r + \frac{b}{b/\delta} \right] = [r, r + \delta] = [r, s],
$$

a contradiction.   □

Exercise 8.25. Show that the solution from the proof of Theorem 8.24 exists also on all of $(-\infty, t_0]$.

8.3. Alpha Derivatives

According to Ahlbrandt, Bohner, and Ridenhour [24], a *(generalized) time scale* is a pair $(\mathbb{T}, \alpha)$ satisfying

(i) $\mathbb{T} \subset \mathbb{R}$ is nonempty such that every Cauchy sequence in $\mathbb{T}$ converges to a point in $\mathbb{T}$ with the possible exception of Cauchy sequences which converge to a finite infimum or finite supremum of $\mathbb{T}$;

(ii) α maps $\mathbb{T}$ into $\mathbb{T}$.

We think of α as a *(generalized) jump function*. If $\alpha = \sigma$, then this is consistent with our previous definition. The case $\alpha = \rho$ is discussed in the next section.

Let us define the *interior* of $\mathbb{T}$ relative to α to be the set

$$
\mathbb{T}^\kappa = \{ t \in \mathbb{T} : \text{ either } \alpha(t) \neq t \text{ or } \alpha(t) = t \text{ and } t \text{ is not isolated} \}.
$$

Definition 8.26 (The Alpha Derivative). Let $(\mathbb{T}, \alpha)$ be a time scale. A function $g : \mathbb{T} \to \mathbb{R}$ is said to be alpha differentiable at $t \in \mathbb{T}^\kappa$ if

(i) g is defined in a neighborhood $\mathcal{U}$ of t;

(ii) g is defined at $\alpha(t)$;

(iii) there exists a unique real number $g_\alpha(t)$, called the alpha derivative of g at t, such that for each $\varepsilon > 0$, there exists a neighborhood $\mathcal{N}$ of t with $\mathcal{N} \subseteq \mathcal{U}$ and

$$(8.5) \qquad |g(\alpha(t)) - g(s) - (\alpha(t) - s)g_\alpha(t)| \le \varepsilon|\alpha(t) - s| \qquad \text{for every} \quad s \in \mathcal{N}.$$

We also use the notation $\frac{dg(t)}{d\alpha(t)}$ for $g_\alpha(t)$. This agrees with standard notation for derivatives if $\mathbb{T} = \mathbb{R}$ and $\alpha(t) = t$. If $\alpha = \sigma$, then the sigma derivative is the usual delta derivative. If $\alpha = \rho$, Hilger's left jump function, then the rho derivative is the left-hand difference quotient at left-scattered points.

Example 8.27. Let

$$\mathbb{T} = [0,1] \cup \{-1, 2, 3\}.$$

The point $t = 3$ is not in the interior of $\mathbb{T}$ relative to σ whereas -1 is not in the interior of $\mathbb{T}$ relative to ρ.

Exercise 8.28. Discuss the set $\mathbb{T} = \mathbb{R}$ with two jump functions $\alpha(t) = t$ and $\beta(t) = t + 1$ and how it gives rise to *differential-difference operators* $(f_\alpha)_\beta$.

Remark 8.29. If t is an isolated point of $\mathbb{T}$ with $\alpha(t) = t$, then the singleton set $\mathcal{N} = \{t\}$ is a neighborhood of t and for any real L and any $\varepsilon > 0$, the condition

$$|g(\alpha(t)) - g(s) - (\alpha(t) - s)L| = \varepsilon|\alpha(t) - s| = 0$$

holds for every $s \in \mathcal{N}$, i.e., for $s = t$. Since we demand uniqueness of $g_\alpha(t)$, we conclude that no function can be alpha differentiable at a point $t \in \mathbb{T} \setminus \mathbb{T}^\kappa$.

Lemma 8.30. *If $t \in \mathbb{T}^\kappa$, g is defined in a neighborhood $\mathcal{U}$ of t, $g(\alpha(t))$ is defined, and there exist numbers L_1 and L_2 such that for each $\varepsilon > 0$, there exist neighborhoods $\mathcal{N}_1$ and $\mathcal{N}_2$ of t with*

$$|g(\alpha(t)) - g(s) - (\alpha(t) - s)L_i| \le \varepsilon|\alpha(t) - s|$$

for every $s \in \mathcal{N}_i$, then $L_1 = L_2$ and $g_\alpha(t)$ exists.

Proof. Assume $L_1 \ne L_2$ and set $\varepsilon = |L_1 - L_2|/4$. Then for $s \in \mathcal{N}_1 \cap \mathcal{N}_2$ we have the inequalities

$$-\varepsilon|\alpha(t) - s| \le g(\alpha(t)) - g(s) - (\alpha(t) - s)L_1 \le \varepsilon|\alpha(t) - s|$$

and

$$-\varepsilon|\alpha(t) - s| \le -g(\alpha(t)) + g(s) + (\alpha(t) - s)L_2 \le \varepsilon|\alpha(t) - s|.$$

Add these inequalities and use $2\varepsilon = |L_1 - L_2|/2$ for the conclusion that

$$|\alpha(t) - s| \cdot |L_1 - L_2| \le (1/2)|L_1 - L_2| \cdot |\alpha(t) - s|.$$

If $\alpha(t) \ne t$, choose $s = t$ for a contradiction. If $\alpha(t) = t$ and t is not isolated, then there exists a point $s \in \mathcal{N}_1 \cap \mathcal{N}_2$ other than t which also gives a contradiction. Thus if $t \in \mathbb{T}^\kappa$ and one such L exists, then $g_\alpha(t)$ exists. $\square$

Theorem 8.31. *If f is alpha differentiable at t, then it is continuous at t.*

Proof. Let $\varepsilon > 0$. Define

$$\varepsilon^* = \varepsilon[1 + |f_\alpha(t)| + 2|\alpha(t) - t|]^{-1},$$

where without loss of generality $\varepsilon^* \in (0,1)$. Since f is α-differentiable at t, there exists a neighborhood $\mathcal{N}$ of t (without loss of generality the diameter of $\mathcal{N}$ is $\leq \varepsilon^*$) such that

$$|f(\alpha(t)) - f(s) - (\alpha(t) - s)f_\alpha(t)| \leq \varepsilon^*|\alpha(t) - s| \quad \text{for all} \quad s \in \mathcal{N}.$$

Let $s \in \mathcal{N}$. Then

$$
\begin{aligned}
|f(t) - f(s)| &= |\{f(\alpha(t)) - f(s) - (\alpha(t) - s)f_\alpha(t)\} \\
&\quad - \{f(\alpha(t)) - f(t) - (\alpha(t) - t)f_\alpha(t)\} + (t - s)f_\alpha(t)| \\
&\leq \varepsilon^*|\alpha(t) - s| + \varepsilon^*|\alpha(t) - t| + |t - s||f_\alpha(t)| \\
&\leq \varepsilon^*[|\alpha(t) - t| + |t - s| + |\alpha(t) - t| + |f_\alpha(t)|] \\
&< \varepsilon^*[1 + |f_\alpha(t)| + 2|\alpha(t) - t|] \\
&= \varepsilon.
\end{aligned}
$$

Hence f is continuous at t. $\qquad\square$

Since the choice of $\mathbb{T} = \mathbb{R}$ and $\alpha(t) = t$ gives the usual calculus, we know that there exist examples of functions which are continuous at a point but not alpha differentiable there.

Theorem 8.32. *If $(\mathbb{T}, \alpha)$ is a time scale and $t \in \mathbb{T}^\kappa$ has $\alpha(t) \neq t$, f is continuous at t, and $f(\alpha(t))$ is defined, then f is alpha differentiable at t with*

$$(8.6) \qquad f_\alpha(t) = \frac{df(t)}{d\alpha(t)} = \frac{f(\alpha(t)) - f(t)}{\alpha(t) - t}.$$

Proof. Note that if $f_\alpha(t)$ exists, then this form of $f_\alpha(t)$ follows from equation (8.5) (with $g = f$) since every neighborhood of t must contain $s = t$ and $t \in \mathbb{T}^\kappa$ implies that the derivative is unique. To see that f is alpha differentiable at t, let $\varepsilon > 0$ be given. Since $\alpha(t) - t \neq 0$, the function g defined for $s \in \mathbb{T}$, $s \neq \alpha(t)$, by

$$g(s) = \frac{f(\alpha(t)) - f(s)}{\alpha(t) - s}$$

is continuous at t. Thus there exists a neighborhood $\mathcal{N}$ of t such that for $f_\alpha(t)$ given by (8.6) and $s \in \mathcal{N}$,

$$|f(\alpha(t)) - f(s) - (\alpha(t) - s)f_\alpha(t)| = |g(s) - g(t)| \cdot |\alpha(t) - s| \leq \varepsilon|\alpha(t) - s|.$$

Hence f is alpha differentiable at t. $\qquad\square$

This suggests the useful formula

$$(8.7) \qquad f^\alpha(t) = f(t) + h(t)f_\alpha(t) \quad \text{with } \textit{step-size} \quad h(t) := \alpha(t) - t,$$

which holds if $f_\alpha(t)$ exists. As in the notation f^σ, we write $f^\alpha = (f \circ \alpha)$.

Theorem 8.33 (Product Rules). *If $f, g : \mathbb{T} \to \mathbb{R}$ are alpha differentiable at $t \in \mathbb{T}^\kappa$, then fg is alpha differentiable at t and the formulas*

$$(fg)_\alpha = f_\alpha g^\alpha + f g_\alpha = f_\alpha g + f^\alpha g_\alpha$$

hold at t.

Proof. Let $\varepsilon > 0$ be given. Define
$$\varepsilon^* = \varepsilon[1 + |f(t)| + |g^\alpha(t)| + |g_\alpha(t)|]^{-1},$$
where without loss of generality $\varepsilon^* \in (0,1)$. Hence there exist neighborhoods $\mathcal{N}_1$, $\mathcal{N}_2$, and $\mathcal{N}_3$ of t with
$$|f(\alpha(t)) - f(s) - (\alpha(t) - s)f_\alpha(t)| \le \varepsilon^*|\alpha(t) - s| \quad \text{for all} \quad s \in \mathcal{N}_1,$$
$$|g(\alpha(t)) - g(s) - (\alpha(t) - s)g_\alpha(t)| \le \varepsilon^*|\alpha(t) - s| \quad \text{for all} \quad s \in \mathcal{N}_2,$$
and (from Theorem 8.31)
$$|f(t) - f(s)| \le \varepsilon^* \quad \text{for all} \quad s \in \mathcal{N}_3.$$
Put $\mathcal{N} = \mathcal{N}_1 \cap \mathcal{N}_2 \cap \mathcal{N}_3$ and let $s \in \mathcal{N}$. Then
$$\begin{aligned}
&|(fg)(\alpha(t)) - (fg)(s) - (\alpha(t) - s)[f_\alpha(t)g(\alpha(t)) + f(t)g_\alpha(t)]| \\
&= |[f(\alpha(t)) - f(s) - (\alpha(t) - s)f_\alpha(t)]g(\alpha(t)) \\
&\quad + [g(\alpha(t)) - g(s) - (\alpha(t) - s)g_\alpha(t)]f(t) \\
&\quad + [g(\alpha(t)) - g(s) - (\alpha(t) - s)g_\alpha(t)][f(s) - f(t)] \\
&\quad + (\alpha(t) - s)g_\alpha(t)[f(s) - f(t)]| \\
&\le \varepsilon^*|\alpha(t) - s| \cdot |g(\alpha(t))| + \varepsilon^*|\alpha(t) - s||f(t)| \\
&\quad + \varepsilon^*\varepsilon^*|\alpha(t) - s| + \varepsilon^*|\alpha(t) - s||g_\alpha(t)| \\
&= \varepsilon^*|\alpha(t) - s| \cdot [|g(\alpha(t))| + |f(t)| + \varepsilon^* + |g_\alpha(t)|] \\
&\le \varepsilon^*|\alpha(t) - s| \cdot [1 + |f(t)| + |g(\alpha(t))| + |g_\alpha(t)|] \\
&= \varepsilon|\alpha(t) - s|.
\end{aligned}$$

By Lemma 8.30, existence of one derivative at a point $t \in \mathbb{T}^\kappa$ implies uniqueness. Thus $(fg)_\alpha = f_\alpha g^\alpha + fg_\alpha$. The second formula follows from the first since products commute. $\square$

The product formulas in the matrix case are valid if we also require that the matrix product of the functions is defined. The definition of the alpha derivative of a matrix follows from the choice of 2-norm in the definition of alpha derivatives. Then use transposes twice: $(FG)_\alpha = [(G^T F^T)_\alpha]^T$ to derive the second formula from the first. Since $\|A^T\|_2 = \|A\|_2$ [244, Theorem 2.10, p. 180], the choice of 2-norm assures that the operations of transpose and alpha differentiation commute.

Exercise 8.34. State and prove a quotient rule for alpha derivatives.

Theorem 8.35 (Chain Rule). *Let $(\mathbb{T}, \alpha)$ and $(\tilde{\mathbb{T}}, \tilde{\alpha})$ be time scales related by a function $g : \mathbb{T} \to \tilde{\mathbb{T}}$. Let $w : \tilde{\mathbb{T}} \to \mathbb{R}$ and let $z = w \circ g$. Suppose that t is a point of $\mathbb{T}^\kappa$ such that g has the property $g(\alpha(t)) = \tilde{\alpha}(g(t))$. If $g_\alpha(t)$ and $w_{\tilde{\alpha}}(g(t))$ exist, then $z_\alpha(t)$ exists and satisfies the chain rule*
$$z_\alpha = (w \circ g)_\alpha = (w_{\tilde{\alpha}} \circ g)g_\alpha \quad at \quad t.$$

Proof. Let $\varepsilon > 0$ be given. Define
$$\varepsilon^* = \varepsilon[1 + |g_\alpha(t)| + |w_{\tilde{\alpha}}(g(t))|]^{-1},$$
where $\varepsilon^* \in (0,1)$ without loss of generality. According to the assumptions, there exist neighborhoods $\mathcal{N}$ of t and $\mathcal{V}$ of $g(t)$ on which, respectively,
$$|g(\alpha(t)) - g(s) - (\alpha(t) - s)g_\alpha(t)| \le \varepsilon^*|\alpha(t) - s|, \quad s \in \mathcal{N}$$

and

$$|w(\tilde{\alpha}(g(t))) - w(r) - (\tilde{\alpha}(g(t)) - r)w_{\tilde{\alpha}}(g(t))| \leq \varepsilon^*|\tilde{\alpha}(g(t)) - r|, \quad r \in \mathcal{V}.$$

Since g is alpha differentiable at $t \in \mathbb{T}^\kappa$, it is continuous at t by Theorem 8.31, and there exists a neighborhood $\mathcal{U}$ of t such that $s \in \mathcal{U}$ implies $g(s) \in \mathcal{V}$.

Put $\mathcal{N}_1 = \mathcal{N} \cap \mathcal{U}$ and let $s \in \mathcal{N}_1$. Then $s \in \mathcal{N}$, $g(s) \in \mathcal{V}$, and

$$
\begin{aligned}
&|w(g(\alpha(t))) - w(g(s)) - (\alpha(t) - s)[w_{\tilde{\alpha}}(g(t))g_\alpha(t)]| \\
&= \; |w(g(\alpha(t))) - w(g(s)) - (\tilde{\alpha}(g(t)) - g(s))w_{\tilde{\alpha}}(g(t)) \\
&\quad + [\tilde{\alpha}(g(t)) - g(s) - (\alpha(t) - s)g_\alpha(t)]w_{\tilde{\alpha}}(g(t))| \\
&\leq \; \varepsilon^*|\tilde{\alpha}(g(t)) - g(s)| + \varepsilon^*|\alpha(t) - s||w_{\tilde{\alpha}}(g(t))| \\
&\leq \; \varepsilon^* \{\, |\tilde{\alpha}(g(t)) - g(s) - (\alpha(t) - s)g_\alpha(t)| + |\alpha(t) - s||g_\alpha(t)| \\
&\quad + |\alpha(t) - s||w_{\tilde{\alpha}}(g(t))| \,\} \\
&\leq \; \varepsilon^* \{\, \varepsilon^*|\alpha(t) - s| + |\alpha(t) - s||g_\alpha(t)| + |\alpha(t) - s||w_{\tilde{\alpha}}(g(t))| \,\} \\
&= \; \varepsilon^*|\alpha(t) - s| \{\, \varepsilon^* + |g_\alpha(t)| + |w_{\tilde{\alpha}}(g(t))| \,\} \\
&\leq \; \varepsilon^* \{\, 1 + |g_\alpha(t)| + |w_{\tilde{\alpha}}(g(t))| \,\} |\alpha(t) - s| \\
&= \; \varepsilon|\alpha(t) - s|.
\end{aligned}
$$

This establishes our chain rule.   □

The hypothesis $g(\alpha(t)) = \tilde{\alpha}(g(t))$ holds in the usual chain rule of calculus because in that setting $\alpha(t) = t$ and $\tilde{\alpha}(x) = x$.

Example 8.36 (The Discrete Chain Rule). Suppose that $\mathbb{T} = \mathbb{N}$ and $\alpha = \sigma$, i.e., $\alpha(t) = t + 1$ for $t \in \mathbb{N}$. Let $g(t) := 1/t$ on $\mathbb{N}$, let $\tilde{\mathbb{T}} = \{1/t : t \in \mathbb{N}\}$, consider $w : \tilde{\mathbb{T}} \to \mathbb{R}$ defined by $w(x) = x^2$. Then $z(t) = w(g(t)) = 1/t^2$ and the induced jump function $\tilde{\alpha}$ on $\tilde{\mathbb{T}}$ such that $\tilde{\alpha}(x) = g(\alpha(t))$ for $x = g(t)$ is $\tilde{\alpha}(x) = x/(x+1)$. Thus

$$w_{\tilde{\alpha}}(x) = \frac{w(\tilde{\alpha}(x)) - w(x)}{\tilde{\alpha}(x) - x},$$

$x/(x+1) = 1/(t+1)$, and the chain rule reads

$$z_\alpha(t) = \frac{1}{(t+1)^2} - \frac{1}{t^2} = \frac{\frac{1}{(t+1)^2} - \frac{1}{t^2}}{\frac{1}{t+1} - \frac{1}{t}} \cdot \left[\frac{1}{t+1} - \frac{1}{t}\right] = w_{\tilde{\alpha}}\left(\frac{1}{t}\right) \cdot g_\alpha(t).$$

Theorem 8.37 (Differentiation of the Inverse Function). *Let $(\mathbb{T}, \alpha)$ be a time scale. Suppose that $g : \mathbb{T} \to \mathbb{R}$ is strictly monotone and $(\tilde{\mathbb{T}}, \tilde{\alpha})$ is the induced time scale defined by $\tilde{\mathbb{T}} := g(\mathbb{T})$ with jump function $\tilde{\alpha}(x) := g(\alpha(t))$ for $x = g(t)$. Let f be the inverse function of g. Then $g \circ \alpha = \tilde{\alpha} \circ g$ on $\mathbb{T}$. Assume that t is a point of $\mathbb{T}^\kappa$ such that $g_\alpha(t)$ exists and $f_{\tilde{\alpha}}$ exists at $x = g(t)$. Then $g_\alpha(t)$ is nonzero and*

$$(8.8) \qquad\qquad\qquad\qquad \frac{1}{g_\alpha(t)} = f_{\tilde{\alpha}}(x).$$

With these developments of time scales $(\mathbb{T}, \alpha)$ and their properties one could readily convert many of the results in this book to this more general setting by replacing delta derivatives by alpha derivatives. One gap in our development has been in generalizing the result that an rd-continuous function has a delta antiderivative. Perhaps some type of a class of alpha continuous functions would have the analogous property of having alpha antiderivatives.

Exercise 8.38. Develop some of the results given in this book for alpha derivatives in place of delta derivatives.

8.4. Nabla Derivatives

If we let $\alpha = \rho$ in the definition of the alpha derivative (see Section 8.3), then we get what Atıcı and Guseinov [**43**] call the nabla derivative which they denote by $f^{\nabla}(t)$. We give several properties of this nabla derivative here. We define

$$\nu(t) := t - \rho(t) \quad \text{and} \quad f^{\rho}(t) := f(\rho(t)).$$

Below we also need the set $\mathbb{T}_\kappa$ which is derived from the time scale $\mathbb{T}$ as follows: If $\mathbb{T}$ has a right-scattered minimum m, then $\mathbb{T}_\kappa = \mathbb{T} - \{m\}$. Otherwise, $\mathbb{T}_\kappa = \mathbb{T}$. Similar to the proof of Theorem 1.16 one can prove the following theorem.

Theorem 8.39. *Assume* $f : \mathbb{T} \to \mathbb{R}$ *is a function and let* $t \in \mathbb{T}_\kappa$. *Then we have the following:*

(i) *If* f *is nabla differentiable at* t, *then* f *is continuous at* t.

(ii) *If* f *is continuous at* t *and* t *is left-scattered, then* f *is nabla differentiable at* t *with*

$$f^{\nabla}(t) = \frac{f(t) - f(\rho(t))}{\nu(t)}.$$

(iii) *If* t *is left-dense, then* f *is nabla differentiable at* t *iff the limit*

$$\lim_{s \to t} \frac{f(t) - f(s)}{t - s}$$

exists as a finite number. In this case

$$f^{\nabla}(t) = \lim_{s \to t} \frac{f(t) - f(s)}{t - s}.$$

(iv) *If* f *is nabla differentiable at* t, *then*

$$f^{\rho}(t) = f(t) - \nu(t) f^{\nabla}(t).$$

Example 8.40. If $\mathbb{T} = \mathbb{R}$, then

$$f^{\nabla}(t) = f'(t),$$

and if $\mathbb{T} = \mathbb{Z}$, then

$$f^{\nabla}(t) = \nabla f(t) := f(t) - f(t-1),$$

where ∇ is the backward difference operator.

Similar to the proof of Theorem 1.20 one can prove the following theorem.

Theorem 8.41. *Assume* $f, g : \mathbb{T} \to \mathbb{R}$ *are nabla differentiable at* $t \in \mathbb{T}_\kappa$. *Then:*

(i) *The sum* $f + g : \mathbb{T} \to \mathbb{R}$ *is nabla differentiable at* t *with*

$$(f + g)^{\nabla}(t) = f^{\nabla}(t) + g^{\nabla}(t).$$

(ii) *For any constant* α, $\alpha f : \mathbb{T} \to \mathbb{R}$ *is nabla differentiable at* t *with*

$$(\alpha f)^{\nabla}(t) = \alpha f^{\nabla}(t).$$

(iii) *The product* $fg : \mathbb{T} \to \mathbb{R}$ *is nabla differentiable at* t, *and we get the product rules*

$$(fg)^{\nabla}(t) = f^{\nabla}(t)g(t) + f^{\rho}(t)g^{\nabla}(t) = f(t)g^{\nabla}(t) + f^{\nabla}(t)g^{\rho}(t).$$

(iv) *If $f(t)f^\rho(t) \neq 0$, then $\frac{1}{f}$ is nabla differentiable at t with*

$$\left(\frac{1}{f}\right)^\nabla (t) = -\frac{f^\nabla(t)}{f(t)f^\rho(t)}.$$

(v) *If $g(t)g^\rho(t) \neq 0$, then $\frac{f}{g}$ is nabla differentiable at t, and we get the quotient rule*

$$\left(\frac{f}{g}\right)^\nabla (t) = \frac{f^\nabla(t)g(t) - f(t)g^\nabla(t)}{g(t)g^\rho(t)}.$$

Definition 8.42. A function $F : \mathbb{T} \to \mathbb{R}$ is called a *nabla antiderivative* (or nabla antiderivative) of $f : \mathbb{T} \to \mathbb{R}$ provided $F^\nabla(t) = f(t)$ holds for all $t \in \mathbb{T}_\kappa$. We then define the *integral* of f by

$$\int_a^t f(\tau)\nabla\tau = F(t) - F(a) \quad \text{for all} \quad t \in \mathbb{T}.$$

In order to find a class of functions which possess a nabla antiderivative, we introduce now the set of ld-continuous functions.

Definition 8.43 (Ld-Continuity). Let $f : \mathbb{T} \to \mathbb{R}$ be a function. We say that f is ld-continuous if it is continuous at each left-dense point in $\mathbb{T}$ and $\lim_{s \to t^+} f(s)$ exists as a finite number for all right-dense points $t \in \mathbb{T}$.

Exercise 8.44. In each of the following determine if f is ld-continuous:

(i) $f = \rho$, the backward jump operator;
(ii) $f = g^\rho$, where $g : \mathbb{R} \to \mathbb{R}$ is continuous;
(iii) $f : \mathbb{R} \to \mathbb{R}$ is continuous and $\mathbb{T} = \mathbb{R}$;
(iv) $f : \mathbb{T} \to \mathbb{T}$ and every point in $\mathbb{T}$ is isolated.

Similar as in Theorem 1.74 one can prove the following theorem.

Theorem 8.45 (Existence of a nabla Antiderivative). *Every ld-continuous function has a nabla antiderivative.*

Similar to the proof of Theorem 1.75 one can prove the following theorem.

Theorem 8.46. *If $f : \mathbb{T} \to \mathbb{R}$ is ld-continuous and $t \in \mathbb{T}_\kappa$, then*

$$\int_{\rho(t)}^t f(\tau)\nabla\tau = f(t)\nu(t).$$

Similar to the proof of Theorem 1.77 one can prove the following theorem.

Theorem 8.47. *If $a, b, c \in \mathbb{T}$, $\alpha \in \mathbb{R}$, and $f, g : \mathbb{T} \to \mathbb{R}$ are ld-continuous; then*

(i) $\int_a^b [f(t) + g(t)]\nabla t = \int_a^b f(t)\nabla t + \int_a^b g(t)\nabla t$;
(ii) $\int_a^b \alpha f(t)\nabla t = \alpha \int_a^b f(t)\nabla t$;
(iii) $\int_a^b f(t)\nabla t = -\int_b^a f(t)\nabla t$;
(iv) $\int_a^b f(t)\nabla t = \int_a^c f(t)\nabla t + \int_c^b f(t)\nabla t$;
(v) $\int_a^b f(\rho(t))g^\nabla(t)\nabla t = (fg)(b) - (fg)(a) - \int_a^b f^\nabla(t)g(t)\nabla t$;
(vi) $\int_a^b f(t)g^\nabla(t)\nabla t = (fg)(b) - (fg)(a) - \int_a^b f^\nabla(t)g(\rho(t))\nabla t$;
(vii) $\int_a^a f(t)\nabla t = 0$.

Similar to the proof of Theorem 1.79 one can prove the following theorem.

Theorem 8.48. *Assume $a, b \in \mathbb{T}$ and $f : \mathbb{T} \to \mathbb{R}$ is ld-continuous.*

(i) *If $\mathbb{T} = \mathbb{R}$, then*

$$\int_a^b f(t)\nabla t = \int_a^b f(t)dt,$$

where the integral on the right is the Riemann integral from calculus.

(ii) *If $\mathbb{T}$ consists of only isolated points, then*

$$\int_a^b f(t)\nabla t = \begin{cases} \sum_{t \in (a,b]} f(t)\nu(t) & \text{if} \quad a < b \\ 0 & \text{if} \quad a = b \\ -\sum_{t \in (b,a]} f(t)\nu(t) & \text{if} \quad a > b. \end{cases}$$

(iii) *If $\mathbb{T} = h\mathbb{Z}$, where $h > 0$, then*

$$\int_a^b f(t)\nabla t = \begin{cases} \sum_{k=\frac{a+h}{h}}^{\frac{b}{h}} f(kh)h & \text{if} \quad a < b \\ 0 & \text{if} \quad a = b \\ -\sum_{k=\frac{b+h}{h}}^{\frac{a}{h}} f(kh)h & \text{if} \quad a > b. \end{cases}$$

(iv) *If $\mathbb{T} = \mathbb{Z}$, then*

$$\int_a^b f(t)\nabla t = \begin{cases} \sum_{t=a+1}^{b} f(t) & \text{if} \quad a < b \\ 0 & \text{if} \quad a = b \\ -\sum_{t=b+1}^{a} f(t) & \text{if} \quad a > b. \end{cases}$$

In the next theorem we give a relationship between the delta derivative and the nabla derivative.

Theorem 8.49.   (i) *Assume that $f : \mathbb{T} \to \mathbb{R}$ is delta differentiable on $\mathbb{T}^\kappa$. Then f is nabla differentiable at t and*

$$(8.9) \qquad\qquad f^\nabla(t) = f^\Delta(\rho(t))$$

for $t \in \mathbb{T}_\kappa$ such that $\sigma(\rho(t)) = t$. If, in addition, f^Δ is continuous on $\mathbb{T}^\kappa$, then f is nabla differentiable at t and (8.9) holds for any $t \in \mathbb{T}_\kappa$.

(ii) *Assume that $f : \mathbb{T} \to \mathbb{R}$ is nabla differentiable on $\mathbb{T}_\kappa$. Then f is delta differentiable at t and*

$$(8.10) \qquad\qquad f^\Delta(t) = f^\nabla(\sigma(t))$$

for $t \in \mathbb{T}^\kappa$ such that $\rho(\sigma(t)) = t$. If, in addition, f^∇ is continuous on $\mathbb{T}_\kappa$, then f is delta differentiable at t and (8.10) holds for any $t \in \mathbb{T}^\kappa$.

The following theorem also appears in Atıcı and Guseinov [43].

Theorem 8.50. *If f, f^Δ, f^∇ are continuous, then*

(i) $\left[\int_a^t f(t,s)\Delta s\right]^\Delta = \int_a^t f^\Delta(t,s)\Delta s + f(\sigma(t),t);$

(ii) $\left[\int_a^t f(t,s)\Delta s\right]^\nabla = \int_a^t f^\nabla(t,s)\Delta s + f(\rho(t),\rho(t));$

(iii) $\left[\int_a^t f(t,s)\nabla s\right]^\Delta = \int_a^t f^\Delta(t,s)\nabla s + f(\sigma(t),\sigma(t));$

(iv) $\left[\int_a^t f(t,s)\nabla s\right]^\nabla = \int_a^t f^\nabla(t,s)\nabla s + f(\rho(t),t).$

Atıcı and Guseinov [**43**] study the following form of a self-adjoint equation

(8.11) $[p(t)x^\Delta]^\nabla + q(t)x = 0$ for $t \in \mathbb{T}^*,$

where

$$\mathbb{T}^* := \mathbb{T}^\kappa \cap \mathbb{T}_\kappa.$$

Definition 8.51. We say that $k(t,s)$ is the Cauchy function for (8.11) provided that for each fixed $s \in \mathbb{T}^*$, $k(t,s)$ is the unique solution of the IVP

$$[p(t)x^\Delta]^\nabla + q(t)x = 0, \quad x(s) = 0, \quad x^\Delta(\rho(s)) = \frac{1}{p(\rho(s))}.$$

Example 8.52. Find the Cauchy function for the dynamic equation

$$[p(t)x^\Delta]^\nabla = 0.$$

Since for each fixed $s \in \mathbb{T}^*$, $k(t,s)$ is a solution of the dynamic equation (8.11), we have that

$$[p(t)k^\Delta(t,s)]^\nabla = 0.$$

It follows that

$$p(t)k^\Delta(t,s) = \alpha(s)$$

and consequently

$$k^\Delta(t,s) = \frac{\alpha(s)}{p(t)}.$$

Using $k^\Delta(\rho(s),s) = 1/p(\rho(s))$, we get $\alpha(s) = 1$ and hence $k^\Delta(t,s) = 1/p(t)$. Integrating from s to t we obtain the desired result

$$k(t,s) = \int_s^t \frac{1}{p(\tau)}\Delta\tau.$$

As a special case of this we get that $k(t,s) = t - s$ is the Cauchy function for the dynamic equation $(x^\Delta)^\nabla = 0$.

Theorem 8.53 (Variation of Constants Formula). *Let $a \in \mathbb{T}^*$ and assume that h is ld-continuous. Then the solution of the IVP*

$$[p(t)x^\Delta]^\nabla + q(t)x = h(t), \quad x(a) = x^\Delta(a) = 0$$

is given by

$$x(t) = \int_a^t k(t,s)h(s)\nabla s,$$

where $k(t,s)$ is the Cauchy function for (8.11).

Proof. Let

$$x(t) = \int_a^t k(t,s)h(s)\nabla s,$$

where $a \in \mathbb{T}^*$ and $k(t,s)$ is the Cauchy function for (8.11). Then $x(a) = 0$. Also using Theorem 8.50 we get

$$\begin{aligned} x^\Delta(t) &= \int_a^t k^\Delta(t,s)h(s)\nabla s + k(\sigma(t),\sigma(t))h(\sigma(t)) \\ &= \int_a^t k^\Delta(t,s)h(s)\nabla s. \end{aligned}$$

It follows that $x^\Delta(a) = 0$. Also

$$p(t)x^\Delta(t) = \int_a^t p(t)k^\Delta(t,s)h(s)\nabla s.$$

Taking the nabla derivative of both sides and using Theorem 8.50 we get

$$[p(t)x^\Delta(t)]^\nabla = \int_a^t [p(t)k^\Delta(t,s)]^\nabla h(s)\nabla s + p(\rho(t))k^\Delta(\rho(t),t)h(t)$$

$$= \int_a^t [p(t)k^\Delta(t,s)]^\nabla h(s)\nabla s + h(t).$$

It follows that

$$Lx(t) = \int_a^t Lx(t,s)h(s)\nabla s + h(t) = h(t),$$

and the proof is complete.   □

Example 8.54. Use the variation of constants formula to solve the IVP

$$(x^\Delta)^\nabla = t, \quad x(0) = x^\Delta(0) = 0$$

given that $\mathbb{T} = \mathbb{Z}$. By the variation of constants formula

$$\begin{aligned}
x(t) &= \int_a^t k(t,s)h(s)\nabla s \\
&= \int_0^t (t-s)s\nabla s \\
&= t\int_0^t s\nabla s - \int_0^t s^2\nabla s \\
&= (t-1)\int_0^t s^{(1)}\nabla s - \int_0^t s^{(2)}\nabla s \\
&= (t-1)\sum_{s=1}^t s^{(1)} - \sum_{s=1}^t s^{(2)} \\
&= (t-1)\left\{\frac{s^{(2)}}{2}\right\}_{s=1}^{s=t+1} - \left\{\frac{s^{(3)}}{3}\right\}_{s=1}^{s=t+1} \\
&= \frac{(t-1)(t+1)^{(2)}}{2} - \frac{(t+1)^{(3)}}{3} \\
&= \frac{(t+1)^{(3)}}{3!}.
\end{aligned}$$

Exercise 8.55. Solve the IVP in Example 8.54 by integrating both sides of the dynamic equation from 0 to t first with respect to the nabla integral and then with respect to the delta integral.

Exercise 8.56. Prove some of the results given in this book for the case of nabla derivatives.

8.5. Notes and References

In this last chapter we discussed several possibilities on how to extend the results presented in the previous chapters of this book to more general cases. The extension on measure chains is featured in the first section, and the main existence and uniqueness theorems are proved in Sections 8.1 and 8.2 and are used at various places throughout the book for the special time scales case. Many papers in the literature use the word "measure chains", however, do not deal with the measure chain case in its full generality, but with the corresponding special case of a "time scale". All of the results presented in the first section are due to Stefan Hilger, who derived them in his PhD thesis [159] and published them in [160]. Example 8.8 is taken from Zdenek Pospíšil [233]. Most of the results on dynamic equations contained in this book are for the linear case, except for Section 6.6 concerning lower and upper solutions and the existence and uniqueness theorems in Section 8.2. The book by Kaymakçalan, Lakshmikantham, and Sivasundaram [185] features many results for the nonlinear case, and further work can be found in [17, 98, 100, 107, 181, 184, 188, 199, 202, 211, 212]. For the proof of the global and local existence theorems in Section 8.2 we closely followed Stefan Keller [190].

In Section 8.3 we follow Calvin Ahlbrandt, Martin Bohner, and Jerry Ridenhour by introducing the concept of alpha derivatives from [24]. The theory presented in this book is a special case with $\alpha = \sigma$ (the Hilger delta derivative), and the special case with $\alpha = \rho$ (the Atıcı–Guseinov "nabla derivative") is discussed in Section 8.4 (see Ferhan Atıcı and Gusein Guseinov [43]). The basic rules for alpha derivatives are presented in Section 8.3: Product rule, quotient rule, chain rule, substitution rule, etc. are available, but one missing concept is the appropriate analogue of rd-continuity, as in the $\alpha = \sigma$ case (this concept is also not missing in the $\alpha = \rho$ case, where it is simply replaced by an analogous ld-continuity concept). This class of functions that have alpha antiderivatives leaves room for further research, and once found, can help to derive many of the results presented in this book also for the case of general alpha derivatives. This concept can also accommodate the case of differential-difference equations, as is pointed out in Exercise 8.28. A classic reference for differential-difference equations is Bellman and Cooke [55]. The study of impulsive equations (see, e.g., Bainov, Lakshmikantham, and Simeonov [203]), hybrid systems (see Lakshmikantham and Vatsala [201]), or half-linear equations (see Pavel Řehák's PhD thesis [249] (Masaryk University Brno, Czech Republic, 2000)) on time scales are topics of further research. Finally, partial dynamic equations have been introduced by Calvin Ahlbrandt and Christina Morian in [21]. Research in this area is very likely to continue.

Solutions to Selected Problems

Chapter 1

1.5: The forward jump operator $\sigma : \mathbb{T} \to \mathbb{T}$ is onto iff there are no points in $\mathbb{T}$ that are left-dense and right-scattered at the same time (note that this condition implies: If $\inf \mathbb{T} \in \mathbb{T}$, then $\inf \mathbb{T}$ is right-dense). Also, σ is one-to-one iff there are no points in $\mathbb{T}$ that are left-scattered and right-dense at the same time (note that this condition implies: If $\sup \mathbb{T} \in \mathbb{T}$, then $\sup \mathbb{T}$ is left-dense).

1.6: $\max \mathbb{T} - \min \mathbb{T}$.

1.14: $f^{\Delta}(t) = t + \sigma(t)$, $g^{\Delta}(t) = \frac{1}{\sqrt{t} + \sqrt{\sigma(t)}}$.

1.19: (ii) $f^{\Delta}(t) = \sqrt{t^2 + 1} + t$, $t \in \mathbb{N}_0^{\frac{1}{2}}$;

(iii) $f^{\Delta}(t) = 2t + \frac{1}{2}$, $t \in \mathbb{T} = \{\frac{n}{2} : n \in \mathbb{N}_0\}$;

(iv) $(t^3 + 1)^{\frac{2}{3}} + t(t^3 + 1)^{\frac{1}{3}} + t^2$, $t \in \mathbb{N}_0^{\frac{1}{3}}$.

1.23: $(f^{n+1})^{\Delta} = \{\sum_{k=0}^{n} f^k (f^{\sigma})^{n-k}\} f^{\Delta}$, $n \in \mathbb{N}$. Use mathematical induction to prove this.

1.29: (i) 0;

(ii) 0;

(iii) it possibly does not exist.

1.30: It is not $t\rho(t)$. We do not know about a "closed" formula for the solution, however we will get back to this in Section 1.6.

1.34: $\Delta^n(fg)(t) = \sum_{k=0}^{n} \binom{n}{k} \Delta^{n-k} f(t+k) \Delta^k g(t)$.

1.35: Equation (1.3) is not true in general, however it does hold for $\mathbb{T} = \mathbb{R}$ or $\mathbb{T} = \mathbb{Z}$ or more generally, whenever the graininess μ is constant.

1.36: (i) $f^{\sigma\Delta} = (1 + \mu^{\Delta}) f^{\Delta\sigma}$;

(ii) $f^{\sigma\sigma\Delta} = (1 + \mu^{\Delta}) f^{\sigma\Delta\sigma} = (1 + \mu^{\Delta})(1 + \mu^{\Delta^{\sigma}}) f^{\Delta\sigma\sigma}$;

(iii) $f^{\sigma^n \Delta} = \left[\prod_{i=0}^{n-1} \left(1 + \mu^{\Delta\sigma^i} \right) \right] f^{\Delta\sigma^n}$.

1.42: (i) $\sigma^{\Delta} \equiv q$;

(ii) $\mu^{\Delta} \equiv q - 1$.

1.43: $f^{\Delta^3} = \left[f^{\sigma^3} - (1 + q + q^2) f^{\sigma^2} + (q + q^2 + q^3) f^{\sigma} - q^3 f \right] / (q\mu)^3$.

1.50:   (i) $u(t) = \alpha \frac{1}{t(t+2)(t+1)^2}$, for $t \in \mathbb{N}$;

       (ii) $u(t) = \beta(2t+1)2^t$, for $t \in \mathbb{N}_0$.

1.64:   (i) The function f is regulated on $\mathbb{T}$, is rd-continuous on $\mathbb{T}$, and is pre-differentiable with $D = \mathbb{T}^\kappa$.

      (ii) The function f is not regulated on $\mathbb{T}$, is not rd-continuous on $\mathbb{T}$, and is not pre-differentiable.

     (iii) f is regulated and rd-continuous on $\mathbb{T}$.

1.81: If $\mathbb{T} = [0,1] \cup [2,3]$, then

$$\int_0^t s\Delta s = \begin{cases} \frac{t^2}{2} & \text{if} \quad 0 \leq t \leq 1 \\ \frac{t^2}{2} - \frac{1}{2} & \text{if} \quad 2 \leq t \leq 3. \end{cases}$$

1.83: $\int_1^\infty \frac{1}{t^2}\Delta t = q$.

1.84: $\int_a^\infty \frac{1}{t\sigma(t)}\Delta t = \frac{1}{a}$.

1.86: $(f \circ g)^\Delta(t) = 4t^3 + 6t^2 + 4t + 1 \neq 4t^3 + 2t^2 + 2t + 1 = f^\Delta(g(t))g^\Delta(t)$ for $t \neq 0$.

1.89: $c = \sqrt{6.5}$. Note that c is in the real interval $[2, \sigma(2)] = [2,3]$.

1.95: $(w \circ \nu)^\Delta(t) = w^{\tilde{\Delta}}(\nu(t))\nu^\Delta(t) = 8t^3 + 12t^2 + 8t + 2$.

1.100: $\frac{1}{6}t^{(3)}$.

1.101: $2^{t^3} - 1$.

1.106: $\frac{1}{\prod_{j=0}^n \sum_{k=0}^j q^k} t^{n+1}$.

1.107: $h_0(t,0) = 1$, $h_1(t,0) = t$,

$$h_2(t,0) = \begin{cases} \frac{t^2}{2} & \text{if} \quad 0 \leq t \leq 1 \\ \frac{t^2}{2} - 2 & \text{if} \quad 3 \leq t \leq 4, \end{cases}$$

and

$$h_3(t,0) = \begin{cases} \frac{t^3}{6} & \text{if} \quad 0 \leq t \leq 1 \\ \frac{t^3}{6} - 2t + \frac{8}{3} & \text{if} \quad 3 \leq t \leq 4. \end{cases}$$

Chapter 2

2.9: $z + w$.

2.29: $p + q$.

2.37: $\frac{e_p(t,r)}{1+\mu(s)p(s)}$.

2.40: $\frac{1}{3}e_{\frac{3}{t+2\mu(t)}}(t,t_0) - \frac{1}{3}$.

2.54: $e_1(t_n, 0) = \prod_{k=1}^{n-1}(1 + \alpha_k)$ for $t_n \in \mathbb{T}$.

2.56: $e_\alpha(t, t_0) = \prod_{s\in[t_0,t)}(1 + (q-1)\alpha s)$ if $t > t_0$.

2.60: If $\mathbb{T} = [1, \infty)$, $e_{\frac{\lambda}{t}}(t, 1) = t^\lambda$. If $\mathbb{T} = \mathbb{N}$, $e_{\frac{\lambda}{t}}(t, 1) = \frac{\Gamma(t+\lambda)}{\Gamma(t)\Gamma(\lambda+1)} = \frac{(t+\lambda-1)^{(\lambda)}}{\Gamma(\lambda)}$.

If $\mathbb{T} = \mathbb{N}$ and $\lambda \in \mathbb{N}$, then $e_{\frac{\lambda}{t}}(t, 1) = \frac{(t+\lambda-1)^{(\lambda)}}{\lambda!}$.

2.79:   (i) $y(t) = \frac{1}{4}e^{2t} - \frac{1}{2}t - \frac{1}{4}$;

(ii) $y(t) = t3^{t-1}$;

(iii) $y(t) = e_p(t, t_0) \int_{t_0}^t \frac{\Delta \tau}{1 + \mu(\tau)p(\tau)}$.

Chapter 3

3.15: $(1 + \mu\lambda_1)(1 + \mu\lambda_2) = 1 + \mu(\lambda_1 + \lambda_2 + \mu\lambda_1\lambda_2) = 1 + \mu(\mu\beta - \alpha)$.

3.19:   (i) $e_p(t, s)$;

(ii) $e_{-p}(t, s)$;

(iii) $\cosh_p(t, s)$;

(iv) $\sinh_p(t, s)$.

3.23: $\sinh_\alpha(t, 0) = \frac{(1+\alpha h)^{\frac{t}{h}} - (1-\alpha h)^{\frac{t}{h}}}{2}$ and $\cosh_\alpha(t, 0) = \frac{(1+\alpha h)^{\frac{t}{h}} + (1-\alpha h)^{\frac{t}{h}}}{2}$ when $\mathbb{T} = h\mathbb{Z}$. No.

3.27:   (i) $\cos_p(t, s) + i\sin_p(t, s) = e_{ip}(t, t_0)$;

(ii) $\cos_p(t, s) - i\sin_p(t, s) = e_{-ip}(t, t_0)$;

(iii) $\frac{\cos_p(s,t_0)\cos_p(t,t_0) + \sin_p(s,t_0)\sin_p(t,t_0)}{\cos_p^2(s,t_0) + \sin_p^2(s,t_0)} = \cos_p(t, s)$;

(iv) $\frac{\cos_p(s,t_0)\sin_p(t,t_0) - \sin_p(s,t_0)\cos_p(t,t_0)}{\cos_p^2(s,t_0) + \sin_p^2(s,t_0)} = \sin_p(t, s)$.

3.28: $\sin_\alpha(t, 0) = \sin(\alpha t)$ and $\cos_\alpha(t, 0) = \cos(\alpha t)$ when $\mathbb{T} = \mathbb{R}$, and $\sin_\alpha(t, 0) = \frac{(1+i\alpha)^t - (1-i\alpha)^t}{2i}$ and $\cos_\alpha(t, 0) = \frac{(1+i\alpha)^t + (1-i\alpha)^t}{2}$ when $\mathbb{T} = \mathbb{Z}$.

3.29: $e_1(t, 0) = (\sqrt{t})!2^{\sqrt{t}}$, $\sin_1 = \frac{e_i - e_{-i}}{2i}$ and $\cos_1 = \frac{e_i - e_{-i}}{2}$ (both at $(t, 0)$), where $e_i(t, 0) = \prod_{k=1}^{\sqrt{t}}[1 + (2k-1)i]$ and $e_{-i}(t) = \prod_{k=1}^{\sqrt{t}}[1 - (2k-1)i]$.

3.44:   (i) $p(t) \equiv 5$, $q(t) = t$;

(ii) $p(t) \equiv 5$, $q(t) = 1 + t$;

(iii) $p(t) = q(t) = t$;

(iv) $y(t) = c_1 e^{\frac{3}{2}t^2} + c_2 e^{-\frac{3}{2}t^2} \int_0^t e^{3\tau - \frac{3}{2}\tau^2} d\tau$, $t \in \mathbb{R}$;

(v) $y(t) = c_1 t! + c_2 \sum_{\tau=0}^{t-1} \frac{6^\tau}{(\tau+1)!}$, $t \in \mathbb{N}_0$.

3.48: $x(t) = \alpha e^{4t} + \beta e^{-4t}$ when $\mathbb{T} = \mathbb{R}$ and $x(t) = \alpha 5^t + \beta \left(\frac{1}{5}\right)^t$ when $\mathbb{T} = \mathbb{Z}$.

3.68: $t^\alpha \ln t$ when $\mathbb{T} = \mathbb{R}$ and $(t + \alpha - 1)^{(\alpha)} \sum_{\tau=1}^{t-1} \frac{1}{\tau+\alpha}$ when $\mathbb{T} = \mathbb{Z}$ and $\alpha \in \mathbb{N}$.

3.69: $t_0 e_{\frac{\alpha}{t}}(t, t_0) = t e_{\frac{\alpha-1}{\sigma(t)}}(t, t_0)$ for $t \in \mathbb{T}$.

3.72:   (i) $x(t) = c_1 2^t + c_2 4^t$ for $\mathbb{T} = [1, \infty)$, $x(t) = c_1(t+1)^{(2)} + c_2(t+3)^{(4)}$ for $\mathbb{T} = \mathbb{N}$;

(ii) $x(t) = c_1 2^t + c_2(-2)^t$ for $\mathbb{T} = [1, \infty)$, $x(t) = c_1(t+1)^{(2)} + c_2(t+3)^{(4)}$ for $\mathbb{T} = \mathbb{N}$;

(iii) $x(t) = c_1 2^t + c_2 4^t$ for $\mathbb{T} = [1, \infty)$, $x(t) = c_1(t+1)^{(2)} + c_2(t-4)^{(-3)}$ for $\mathbb{T} = \mathbb{N}$;

(iv) $x(t) = c_1 t^2 + c_2 t^2 \log t$ and $x(t) = c_1(t+1)^{(2)} + c_2(t+1)^{(2)} \sum_{\tau=1}^{t-1} \frac{1}{\tau+2}$
for $\mathbb{T} = [1,\infty)$ and for $\mathbb{T} = \mathbb{N}$, respectively;

(v) $x(t) = c_1 t^3 + c_2 t^3 \log t$ and $x(t) = c_1(t+2)^{(3)} + c_2(t+2)^{(3)} \sum_{\tau=1}^{t-1} \frac{1}{\tau+3}$
for $\mathbb{T} = [1,\infty)$ and for $\mathbb{T} = \mathbb{N}$, respectively.

3.76: (ii) $y(t) = c_1 e_1(t,t_0) + c_2 e_2(t,t_0) + e_2(t,t_0) \int_{t_0}^t \frac{1}{1+2\mu(\tau)} \Delta\tau$.

3.81: (ii) $y(t) = c_1 e_1(t,t_0) + c_2 e_2(t,t_0) + e_2(t,t_0) \int_{t_0}^t \frac{1}{1+2\mu(\tau)} \Delta\tau$;

(iii) $y(t) = c_1 e_1(t,t_0) + c_2 e_{-2}(t,t_0) - \frac{5}{4} - \frac{1}{2}t$;

(iv) $y(t) = c_1 \sin_1(t,t_0) + c_2 \cos_1(t,t_0) + \frac{1}{10} e_3(t,t_0)$.

3.88: $\mathcal{L}\{x^{\Delta^n}\}(z) = z^n \mathcal{L}\{x\}(z) - \sum_{j=0}^{n-1} z^j x^{\Delta^{n-j-1}}(0)$.

3.102: (i) $x(t) = e_1(t,0)$;

(ii) $x(t) = \frac{1}{3}\sinh_3(t,0)$;

(iv) $x(t) = \frac{3}{4}\sinh_4(t,0)$;

(v) $x(t) = e_3(t,0)\cos_{\frac{4}{1+3\mu}}(t,0) - \frac{1}{4}e_3(t,0)\sin_{\frac{4}{1+3\mu}}(t,0)$;

(vii) $x(t) = -1 + \frac{1}{2}e_1(t,0) + \frac{1}{2}\cos_1(t,0) + \frac{1}{2}\sin_1(t,0)$.

3.103: Let $g(t) = t$. For $\mathbb{T}_0 = \mathbb{R}$ we get the formula $\mathcal{L}\{gf\}(z) = -\frac{d}{dz}\mathcal{L}\{f\}(z)$.
For $\mathbb{T}_0 = \mathbb{Z}$ we get $\mathcal{Z}\{gf\}(z) = -z\frac{d}{dz}\mathcal{Z}\{f\}(z)$.

3.105: For $\mathbb{T}_0 := \overline{q^{\mathbb{Z}}}$ and $f(t) = t$, $1 * f = f * 1 = \frac{t^2}{1+q} = h_2(t,0)$.

3.109: (i) $x(t) = \frac{5}{2}e_{-1}(t,0) - \frac{1}{2}e_3(t,0)$;

(iii) $x(t) = -e_2(t,0) + 2e_4(t,0)$;

(iv) $x(t) = \frac{1}{2}\sinh_2(t,0)$.

3.115: (i) $x(t) = \frac{1}{5}\sin_2(t,0) - \frac{2}{15}\sin_3(t,0)$;

(iii) $x(t) = e_4(t,0) - e_1(t,0)$;

(iv) $x(t) = e_2(t,0)\int_0^t \frac{1}{1+2\mu(s)}\Delta s$;

(v) $x(t) = \frac{1}{5\alpha^3}\cos_5(t,0) + \frac{1}{\alpha}h_2(t,0) - \frac{1}{\alpha^3}$.

Chapter 4

4.3: $x^{\Delta\Delta}(t) + \left(\frac{q+1-qe^{(1-q)at}-e^{(q-1)qat}}{q(q-1)^2t^2}\right)x^\sigma(t) = 0$.

4.4: $x(t) = t + \beta$.

4.16: (i) $\alpha = -\frac{1}{2\mu}$, $p = e_\alpha$, $q = \frac{3}{2\mu^2}p$;

(ii) $(e^{-3t}x')' + 5e^{-3t}x = 0$;

(iii) $\Delta\left[\left(\frac{1}{2}^t\right)\Delta x(t)\right] + \left(\frac{1}{2}\right)^t x(t+1) = 0$;

(iv) $\left(e_{-9\mu/(1+9\mu^2)}(t,t_0)x^\Delta\right)^\Delta + \frac{9}{1+9\mu^2(t)}e_{-9\mu/(1+9\mu^2)}(t,t_0)x^\sigma = 0$.

4.18:   (i) $(e_2(t,t_0)x^\Delta)^\Delta + 3e_2(t,t_0)x^\sigma = 0;$

  (ii) $(e_2(t,t_0)x^\Delta)^\Delta + x^\sigma = 0;$

  (iii) $(e_2(t,t_0)e_3(t,t_0)x^\Delta)^\Delta + 3e_3(t,t_0)x^\sigma = 0;$

  (iv) $\left(\frac{e_3(t,t_0)}{e_2(t,t_0)}x^\Delta\right)^\Delta + e_3(t,t_0)x^\sigma = 0.$

4.25:   (i) $x(t) = \frac{1}{2}t^2$ if $t \in \mathbb{R}$, $x(t) = \frac{1}{2}t(t-1)$ if $t \in \mathbb{Z}$, $x(t) = \frac{1}{2}t(t-h)$ if $t \in h\mathbb{Z}$;

  (ii) $x(t) = t - \sin t.$

4.39:   (i) $z(t) = \left(\frac{1}{9}\right)^t \frac{2+3c+2ct}{1+ct}$ and $z(t) = \frac{3+2t}{t \cdot 9^t};$

  (ii) $z(t) = \left(\frac{c}{1+ct}\right)$ and $z(t) = \frac{1}{t}.$

4.40:   (i) $z(t) = \frac{-\sin t + c\cos t}{\cos t + c\sin t}$ and $z(t) = \cot t;$

  (ii) $z(t) = \frac{2+c+2c\ln t}{t^4 + ct^4 \ln t}$ and $z(t) = \frac{1+2\ln t}{t^4 \ln t};$

  (iii) $z(t) = \frac{2e^{4t} - 6ce^{-4t}}{1+ce^{-8t}}$ and $z(t) = -6e^{4t};$

  (iv) $z(t) = \frac{2e^{-5t} + 3ce^{-4t}}{1+ce^t}$ and $z(t) = 3e^{-5t};$

  (v) $z(t) = \frac{2+c+ct}{1+ct}e^{-4t}$ and $z(t) = \frac{1+t}{te^{4t}}.$

4.58: If $p > 0$, use Theorem 4.57. If $p \neq 0$, consider the solution $x(t) = (-1)^t$ of $(p(t)x^\Delta)^\Delta = 0$ with $p(t) = (-1)^t$ and $\mathbb{T} = \mathbb{Z}$.

4.63:   (i) $u(t) = e^{2t}$, $v(t) = te^{2t};$

  (ii) $u(t) = \sin t$, $v(t) = \cos t;$

  (iii) $u(t) = 2^t$, $v(t) = 4^t;$

  (iv) $u(t) = t^2$, $v(t) = t^3.$

4.66: $\int_1^\infty \frac{c}{q(q-1)t^2}\Delta t = \frac{c}{q-1}$ is finite (not infinite).

4.78:   (i) $x(t) = \frac{1}{2}t^2 - 25t;$

  (ii) $x(t) = \frac{3 \cdot 2^{10} - 3}{3^{10} - 1} + \frac{3^{11} - 3 \cdot 2^{10}}{3^{10} - 1}\left(\frac{1}{3}\right)^t - 3\left(\frac{2}{3}\right)^t;$

  (iv) $x(t) = \frac{1}{2}e^t + \frac{1}{2}e^{1-t} - \frac{e+1}{2}.$

4.91:   (ii) $u(t,s) = \frac{\cos(t-s) - \sin(t-s)}{2}$, $v(t,s) = \frac{\cos(t-s) + \sin(t-s)}{2}$, $x(t) \equiv 4.$

Chapters 5 – 8

5.7: $\det(I + \mu A) = 1 + \mu \operatorname{tr} A + \mu^2 \det A = 1 + \mu(\operatorname{tr} A + \mu \det A)$ if A is 2×2-matrix-valued.

5.11: This follows from $A(I + \mu A) = (I + \mu A)A.$

5.15:   (i) $A \ominus A = 0;$

  (ii) $\ominus(\ominus A) = A;$

  (iii) $A \ominus B \in \mathcal{R};$

(iv) $A \ominus B = (A - B)(I + \mu A)^{-1}$;

(v) $\ominus(A \ominus B) = B \ominus A$;

(vi) $\ominus(A \oplus B) = (\ominus A) \oplus (\ominus B)$.

5.20: $e_A(t, 1) = \prod_{s \in \mathbf{T} \cap (0, t)}(I + sA)$, order as: $(I + \rho(t)A) \cdots (I + 2A)(I + A)$.

5.34: (i) $x(t) = c_1 e_{-2}(t, t_0) \begin{pmatrix} 1 \\ 5 \end{pmatrix} + c_2 e_3(t, t_0) \begin{pmatrix} 1 \\ 1 \end{pmatrix}$;

(iii) $x(t) = c_1 \begin{pmatrix} \cos_3(t, t_0) - 3\sin_3(t, t_0) \\ 5\cos_3(t, t_0) \end{pmatrix} + c_2 \begin{pmatrix} \sin_3(t, t_0) + 3\cos_3(t, t_0) \\ 5\sin_3(t, t_0) \end{pmatrix}$.

5.37: For the first matrix, in general $x(t) = e_A(t, t_0)\xi$, where

$$e_A(t, t_0) = e_1(t, t_0)I + \int_{t_0}^t e_2(t, \sigma(s))e_1(s, t_0)\Delta s \begin{pmatrix} -1 & 1 \\ -2 & 2 \end{pmatrix}.$$

If $\mathbf{T} = \mathbb{R}$, then we have $x(t) = c_1\begin{pmatrix} 2e^t - e^{2t} \\ 2e^t - 2^{2t} \end{pmatrix} + c_2\begin{pmatrix} -e^t + e^{2t} \\ -e^t + 2e^{2t} \end{pmatrix}$. If $\mathbf{T} = \mathbb{Z}$, then $x(t) = c_1\begin{pmatrix} 3^t \\ 2 \cdot 3^t - 2^{t+1} \end{pmatrix} + c_2\begin{pmatrix} 2^t - 3^t \\ 3 \cdot 2^t - 2 \cdot 3^t \end{pmatrix}$.

5.75: $\Lambda_i = \frac{\Lambda - \lambda_i I}{1 + \mu \lambda_i}$, where $\Lambda = \text{diag}(\lambda_1, \lambda_2, \dots, \lambda_n)$.

5.93: (i) Not regressive.

(iv) Is regressive.

5.102: (i) $(\lambda_2 - \lambda_1)(\lambda_3 - \lambda_1)(\lambda_3 - \lambda_2)e_{\lambda_1 \oplus \lambda_2 \oplus \lambda_3}(t, t_0)$.

(ii) 1.

6.24: Both sides of (6.4) are equal to $c^2 h^2$.

6.25: $\int_0^h |xx'|(t)dt \leq (h/2)\int_0^h (x')^2(t)dt$ if $\mathbf{T} = \mathbb{R}$.

7.4: $a = \frac{7}{2}$, $b = c = 0$, $d = 7$.

8.55: $x(t) = \frac{(t+1)^{(3)}}{3!}$.

Bibliography

[1] R. Agarwal, C. Ahlbrandt, M. Bohner, and A. Peterson. Discrete linear Hamiltonian systems: A survey. *Dynam. Systems Appl.*, 8(3-4):307–333, 1999. Special Issue on "Discrete and Continuous Hamiltonian Systems", edited by R. P. Agarwal and M. Bohner.

[2] R. Agarwal, M. Bohner, D. O'Regan, and A. Peterson. Dynamic equations on time scales: A survey. *J. Comput. Appl. Math.*, 2001. Special Issue on "Dynamic Equations on Time Scales", edited by R. P. Agarwal, M. Bohner, and D. O'Regan. To appear.

[3] R. Agarwal, M. Bohner, and A. Peterson. Inequalities on time scales: A survey. 2001. Submitted.

[4] R. P. Agarwal. Difference calculus with applications to difference equations. In W. Walter, editor, *International Series of Numerical Mathematics*, volume 71, pp. 95–110, Birkhäuser, Basel, 1984. General Inequalities 4, Oberwolfach.

[5] R. P. Agarwal. *Difference Equations and Inequalities*. Marcel Dekker, Inc., New York, 1992.

[6] R. P. Agarwal. Continuous and discrete maximum principles and their applications to boundary value problems. In G. S. Ladde and M. Sambandham, editors, *Proceedings of Dynamic Systems and Applications*, volume 2, pp. 11–18, Atlanta, 1996. Dynamic Publishers. Second International Conference on Dynamic Systems and Applications.

[7] R. P. Agarwal. *Focal Boundary Value Problems for Differential and Difference Equations*, volume 436 of *MIA*. Kluwer Academic Publishers, Dordrecht, 1998.

[8] R. P. Agarwal and M. Bohner. Quadratic functionals for second order matrix equations on time scales. *Nonlinear Anal.*, 33:675–692, 1998.

[9] R. P. Agarwal and M. Bohner. Basic calculus on time scales and some of its applications. *Results Math.*, 35(1-2):3–22, 1999.

[10] R. P. Agarwal, M. Bohner, and D. O'Regan. Time scale boundary value problems on infinite intervals. *J. Comput. Appl. Math.*, 2001. Special Issue on "Dynamic Equations on Time Scales", edited by R. P. Agarwal, M. Bohner, and D. O'Regan. To appear.

[11] R. P. Agarwal, M. Bohner, and D. O'Regan. Time scale systems on infinite intervals. *Nonlinear Anal.*, 2001. To appear.

[12] R. P. Agarwal, M. Bohner, and P. J. Y. Wong. Eigenvalues and eigenfunctions of discrete conjugate boundary value problems. *Comput. Math. Appl.*, 38(3-4):159–183, 1999.

[13] R. P. Agarwal, M. Bohner, and P. J. Y. Wong. Positive solutions and eigenvalues of conjugate boundary value problems. *Proc. Edinburgh Math. Soc.*, 42:349–374, 1999.

[14] R. P. Agarwal, M. Bohner, and P. J. Y. Wong. Sturm–Liouville eigenvalue problems on time scales. *Appl. Math. Comput.*, 99:153–166, 1999.

[15] R. P. Agarwal, S. R. Grace, and D. O'Regan. *Oscillation Theory for Difference and Functional Differential Equations*. Kluwer Academic Publishers, Dordrecht, 2000.

[16] R. P. Agarwal and D. O'Regan. Triple solutions to boundary value problems on time scales. *Appl. Math. Lett.*, 13(4):7–11, 2000.

[17] R. P. Agarwal and D. O'Regan. Nonlinear boundary value problems on time scales. *Nonlinear Anal.*, 2001. To appear.

[18] R. P. Agarwal, D. O'Regan, and P. J. Y. Wong. *Positive Solutions of Differential, Difference and Integral Equations*. Kluwer Academic Publishers, Dordrecht, 1999.

[19] R. P. Agarwal and P. Y. H. Pang. *Opial Inequalities with Applications in Differential and Difference Equations*. Kluwer Academic Publishers, Dordrecht, 1995.

[20] R. P. Agarwal and P. J. Y. Wong. *Advanced Topics in Difference Equations*. Kluwer Academic Publishers, Dordrecht, 1997.

[21] C. Ahlbrandt and C. Morian. Partial differential equations on time scales. *J. Comput. Appl. Math.*, 2001. Special Issue on "Dynamic Equations on Time Scales", edited by R. P. Agarwal, M. Bohner, and D. O'Regan. To appear.

[22] C. D. Ahlbrandt. Continued fraction representations of maximal and minimal solutions of a discrete matrix Riccati equation. *SIAM J. Math. Anal.*, 24(6):1597–1621, 1993.

[23] C. D. Ahlbrandt. Equivalence of discrete Euler equations and discrete Hamiltonian systems. *J. Math. Anal. Appl.*, 180:498–517, 1993.

[24] C. D. Ahlbrandt, M. Bohner, and J. Ridenhour. Hamiltonian systems on time scales. *J. Math. Anal. Appl.*, 250:561–578, 2000.

[25] C. D. Ahlbrandt, M. Bohner, and T. Voepel. Variable change for Sturm–Liouville differential operators on time scales. 1999. In preparation.

[26] C. D. Ahlbrandt and M. Heifetz. Discrete Riccati equations of filtering and control. In S. Elaydi, J. Graef, G. Ladas, and A. Peterson, editors, *Conference Proceedings of the First International Conference on Difference Equations*, pp. 1–16, San Antonio, 1994. Gordon and Breach.

[27] C. D. Ahlbrandt, M. Heifetz, J. W. Hooker, and W. T. Patula. Asymptotics of discrete time Riccati equations, robust control, and discrete linear Hamiltonian systems. *Panamer. Math. J.*, 5:1–39, 1996.

[28] C. D. Ahlbrandt and J. W. Hooker. Riccati matrix difference equations and disconjugacy of discrete linear systems. *SIAM J. Math. Anal.*, 19(5):1183–1197, 1988.

[29] C. D. Ahlbrandt and A. Peterson. The (n,n)-disconjugacy of a $2n^{\text{th}}$ order linear difference equation. *Comput. Math. Appl.*, 28:1–9, 1994.

[30] C. D. Ahlbrandt and A. C. Peterson. *Discrete Hamiltonian Systems: Difference Equations, Continued Fractions, and Riccati Equations*, volume 16 of *Kluwer Texts in the Mathematical Sciences*. Kluwer Academic Publishers, Boston, 1996.

[31] C. D. Ahlbrandt and J. Ridenhour. Putzer algorithms for matrix exponentials, matrix powers, and matrix logarithms. 2000. In preparation.

[32] E. Akın. Boundary value problems for a differential equation on a measure chain. *Panamer. Math. J.*, 10(3):17–30, 2000.

[33] E. Akın. *Boundary value problems, oscillation theory, and the Cauchy functions for dynamic equations on a measure chain*. PhD thesis, University of Nebraska–Lincoln, 2000.

[34] E. Akın, M. Bohner, L. Erbe, and A. Peterson. Existence of bounded solutions for second order dynamic equations. 2001. In preparation.

[35] E. Akın, L. Erbe, B. Kaymakçalan, and A. Peterson. Oscillation results for a dynamic equation on a time scale. *J. Differ. Equations Appl.*, 2001. To appear.

[36] D. Anderson. *Discrete Hamiltonian systems*. PhD thesis, University of Nebraska–Lincoln, 1997.

[37] D. Anderson. Discrete trigonometric matrix functions. *Panamer. Math. J.*, 7(1):39–54, 1997.

[38] D. Anderson. Normalized prepared bases for symplectic matrix systems. *Dynam. Systems Appl.*, 8(3-4):335–344, 1999. Special Issue on "Discrete and Continuous Hamiltonian Systems", edited by R. P. Agarwal and M. Bohner.

[39] D. Anderson. Positivity of Green's function for an n-point right focal boundary value problem on measure chains. *Math. Comput. Modelling*, 31(6-7):29–50, 2000.

[40] D. Anderson. Eigenvalue intervals for a two-point boundary value problem on a measure chain. *J. Comput. Appl. Math.*, 2001. Special Issue on "Dynamic Equations on Time Scales", edited by R. P. Agarwal, M. Bohner, and D. O'Regan. To appear.

[41] D. Anderson and R. Avery. Existence of three positive solutions to a second-order boundary value problem on a measure chain. *J. Comput. Appl. Math.*, 2001. Special Issue on "Dynamic Equations on Time Scales", edited by R. P. Agarwal, M. Bohner, and D. O'Regan. To appear.

[42] D. Anderson and A. Peterson. Asymptotic properties of solutions of a 2nth-order differential equation on a time scale. *Math. Comput. Modelling*, 32(5-6):653–660, 2000. Boundary value problems and related topics.

[43] F. M. Atıcı and G. Sh. Guseinov. On Green's functions and positive solutions for boundary value problems on time scales. *J. Comput. Appl. Math.*, 2001. Special Issue on "Dynamic Equations on Time Scales", edited by R. P. Agarwal, M. Bohner, and D. O'Regan. To appear.

[44] F. M. Atıcı, G. Sh. Guseinov, and B. Kaymakçalan. On Lyapunov inequality in stability theory for Hill's equation on time scales. *J. Inequal. Appl.*, 5:603–620, 2000.

[45] F. M. Atıcı, G. Sh. Guseinov, and B. Kaymakçalan. Stability criteria for dynamic equations on time scales with periodic coefficients. In *Conference Proceedings of the Third International Conference on Dynamic Systems and Applications*, Atlanta, 2001. Dynamic publishers. To appear.

[46] F. V. Atkinson. *Discrete and Continuous Boundary Problems*. Academic Press, New York, 1964.

[47] B. Aulbach. *Continuous and Discrete Dynamics Near Manifolds of Equilibria*, volume 1058 of *Lecture Notes in Mathematics*. Springer-Verlag, Berlin, 1984.

[48] B. Aulbach. *Analysis auf Zeitmengen*. Universität Augsburg, Augsburg, 1990. Lecture notes.

[49] B. Aulbach and S. Hilger. A unified approach to continuous and discrete dynamics. In *Qualitative Theory of Differential Equations (Szeged, 1988)*, volume 53 of *Colloq. Math. Soc. János Bolyai*, pp. 37–56. North-Holland, Amsterdam, 1990.

[50] B. Aulbach and S. Hilger. Linear dynamic processes with inhomogeneous time scale. In *Nonlinear Dynamics and Quantum Dynamical Systems (Gaussig, 1990)*, volume 59 of *Math. Res.*, pp. 9–20. Akademie Verlag, Berlin, 1990.

[51] B. Aulbach and C. Pötzsche. Reducibility of linear dynamic equations on measure chains. *J. Comput. Appl. Math.*, 2001. Special Issue on "Dynamic Equations on Time Scales", edited by R. P. Agarwal, M. Bohner, and D. O'Regan. To appear.

[52] J. H. Barrett. A Prüfer transformation for matrix differential equations. *Proc. Amer. Math. Soc.*, 8:510–518, 1957.

[53] G. Baur. An oscillation theorem for the Sturm–Liouville problem with self-adjoint boundary conditions. *Math. Nachr.*, 138:189–194, 1988.

[54] G. Baur and W. Kratz. A general oscillation theorem for self-adjoint differential systems with applications to Sturm–Liouville eigenvalue problems and quadratic functionals. *Rend. Circ. Mat. Palermo*, 38(2):329–370, 1989.

[55] R. Bellman and K. L. Cooke. *Differential-Difference Equations*. Academic Press, New York, 1963.

[56] A. Ben-Israel and T. N. E. Greville. *Generalized Inverses: Theory and Applications*. John Wiley & Sons, Inc., New York, 1974.

[57] Z. Benzaid and D. A. Lutz. Asymptotic representation of solutions of perturbed systems of linear difference equations. *Stud. Appl. Math.*, 77:195–221, 1987.

[58] J. P. Bézivin. Sur les équations fonctionnelles aux q-différences. *Aequationes Math.*, 43:159–176, 1993.

[59] T. G. Bhaskar. Comparison theorem for a nonlinear boundary value problem on time scales. *J. Comput. Appl. Math.*, 2001. Special Issue on "Dynamic Equations on Time Scales", edited by R. P. Agarwal, M. Bohner, and D. O'Regan. To appear.

[60] G. D. Birkhoff. Existence and oscillation theorem for a certain boundary value problem. *Trans. Amer. Math. Soc.*, 10:259–270, 1909.

[61] M. Bohner. Controllability and disconjugacy for linear Hamiltonian difference systems. In S. Elaydi, J. Graef, G. Ladas, and A. Peterson, editors, *Conference Proceedings of the First International Conference on Difference Equations*, pp. 65–77, San Antonio, 1994. Gordon and Breach.

[62] M. Bohner. *Zur Positivität diskreter quadratischer Funktionale*. PhD thesis, Universität Ulm, 1995. English Edition: On positivity of discrete quadratic functionals.

[63] M. Bohner. An oscillation theorem for a Sturm–Liouville eigenvalue problem. *Math. Nachr.*, 182:67–72, 1996.

[64] M. Bohner. Linear Hamiltonian difference systems: disconjugacy and Jacobi-type conditions. *J. Math. Anal. Appl.*, 199(3):804–826, 1996.

[65] M. Bohner. On disconjugacy for Sturm–Liouville difference equations. *J. Differ. Equations Appl.*, 2(2):227–237, 1996.

[66] M. Bohner. Riccati matrix difference equations and linear Hamiltonian difference systems. *Dynam. Contin. Discrete Impuls. Systems*, 2(2):147–159, 1996.

[67] M. Bohner. Inhomogeneous discrete variational problems. In S. Elaydi, I. Győri, and G. Ladas, editors, *Conference Proceedings of the Second International Conference on Difference Equations (Veszprém, 1995)*, pp. 89–97, Amsterdam, 1997. Gordon and Breach.

[68] M. Bohner. Positive definiteness of discrete quadratic functionals. In C. Bandle, editor, *General Inequalities, 7 (Oberwolfach, 1995)*, volume 123 of *Internat. Ser. Numer. Math.*, pp. 55–60, Birkhäuser, Basel, 1997.

[69] M. Bohner. Symplectic systems and related discrete quadratic functionals. *Facta Univ. Ser. Math. Inform.*, 12:143–156, 1997.

[70] M. Bohner. Asymptotic behavior of discretized Sturm–Liouville eigenvalue problems. *J. Differ. Equations Appl.*, 3:289–295, 1998.

[71] M. Bohner. Discrete linear Hamiltonian eigenvalue problems. *Comput. Math. Appl.*, 36(10-12):179–192, 1998.

[72] M. Bohner. Discrete Sturmian Theory. *Math. Inequal. Appl.*, 1(3):375–383, 1998.

[73] M. Bohner. Calculus of variations on time scales. 2000. In preparation.

[74] M. Bohner and J. Castillo. Mimetic methods on measure chains. *Comput. Math. Appl.*, 2001. To appear.

[75] M. Bohner, S. Clark, and J. Ridenhour. Lyapunov inequalities on time scales. *J. Inequal. Appl.*, 2001. To appear.

[76] M. Bohner and O. Došlý. Disconjugacy and transformations for symplectic systems. *Rocky Mountain J. Math.*, 27(3):707–743, 1997.

[77] M. Bohner and O. Došlý. Positivity of block tridiagonal matrices. *SIAM J. Matrix Anal. Appl.*, 20(1):182–195, 1998.

[78] M. Bohner and O. Došlý. Trigonometric transformations of symplectic difference systems. *J. Differential Equations*, 163:113–129, 2000.

[79] M. Bohner and O. Došlý. The discrete Prüfer transformation. *Proc. Amer. Math. Soc.*, 2001. To appear.

[80] M. Bohner and O. Došlý. Trigonometric systems in oscillation theory of difference equations. In *Conference Proceedings of the Third International Conference on Dynamic Systems and Applications*, Atlanta, 2001. Dynamic publishers. To appear.

[81] M. Bohner, O. Došlý, and R. Hilscher. Linear Hamiltonian dynamic systems on time scales: Sturmian property of the principal solution. *Nonlinear Anal.*, 2001. To appear.

[82] M. Bohner, O. Došlý, and W. Kratz. Inequalities and asymptotics for Riccati matrix difference operators. *J. Math. Anal. Appl.*, 221(1):262–286, 1998.

[83] M. Bohner, O. Došlý, and W. Kratz. A Sturmian theorem for recessive solutions of linear Hamiltonian difference systems. *Appl. Math. Lett.*, 12:101–106, 1999.

[84] M. Bohner, O. Došlý, and W. Kratz. Discrete Reid roundabout theorems. *Dynam. Systems Appl.*, 8(3-4):345–352, 1999. Special Issue on "Discrete and Continuous Hamiltonian Systems", edited by R. P. Agarwal and M. Bohner.

[85] M. Bohner and P. W. Eloe. Higher order dynamic equations on measure chains: Wronskians, disconjugacy, and interpolating families of functions. *J. Math. Anal. Appl.*, 246:639–656, 2000.

[86] M. Bohner and R. Hilscher. An eigenvalue problem for linear Hamiltonian systems on time scales. 2000. In preparation.

[87] M. Bohner and B. Kaymakçalan. Opial inequalities on time scales. *Ann. Polon. Math.*, 2001. To appear.

[88] M. Bohner and D. A. Lutz. Asymptotic behavior of dynamic equations on time scales. *J. Differ. Equations Appl.*, 7(1):21–50, 2001.

[89] M. Bohner and A. Peterson. A survey of exponential functions on time scales. *Rev. Cubo Mat. Educ.*, 2001. To appear.

[90] M. Bohner and A. Peterson. First and second order linear dynamic equations on measure chains. *J. Differ. Equations Appl.*, 2001. To appear.

[91] M. Bohner and A. Peterson. Laplace transform and Z-transform: Unification and extension. 2001. Submitted.

[92] J. E. Castillo, J. M. Hyman, M. J. Shashkov, and S. Steinberg. High-order mimetic finite difference methods on nonuniform grids. *Houston J. Math.*, 1996. ICOSAHOM 95.

[93] S. Chen. Lyapunov inequalities for differential and difference equations. *Fasc. Math.*, 23:25–41, 1991.

[94] S. Chen. Disconjugacy, disfocality, and oscillation of second order difference equations. *J. Differential Equations*, 107:383–394, 1994.

[95] S. Chen and L. Erbe. Oscillation and nonoscillation for systems of self-adjoint second-order difference equations. *SIAM J. Math. Anal.*, 20(4):939–949, 1989.

[96] F. B. Christiansen and T. M. Fenchel. *Theories of Populations in Biological Communities*, volume 20 of *Lecture Notes in Ecological Studies*. Springer-Verlag, Berlin, 1977.

[97] C. J. Chyan, J. M. Davis, J. Henderson, and W. K. C. Yin. Eigenvalue comparisons for differential equations on a measure chain. *Electron. J. Differential Equations*, 35:1–7, 1998.

[98] C. J. Chyan and J. Henderson. Eigenvalue problems for nonlinear differential equations on a measure chain. *J. Math. Anal. Appl.*, 245:547–559, 2000.

[99] C. J. Chyan and J. Henderson. Twin solutions of boundary value problems for differential equations on measure chains. *J. Comput. Appl. Math.*, 2001. Special Issue on "Dynamic Equations on Time Scales", edited by R. P. Agarwal, M. Bohner, and D. O'Regan. To appear.

[100] C. J. Chyan, J. Henderson, and H. C. Lo. Positive solutions in an annulus for nonlinear differential equations on a measure chain. *Tamkang J. Math.*, 30(3):231–240, 1999.

[101] S. Clark and D. B. Hinton. A Lyapunov inequality for linear Hamiltonian systems. *Math. Inequal. Appl.*, 1(2):201–209, 1998.

[102] S. Clark and D. B. Hinton. Discrete Lyapunov inequalities. *Dynam. Systems Appl.*, 8(3-4):369–380, 1999. Special Issue on "Discrete and Continuous Hamiltonian Systems", edited by R. P. Agarwal and M. Bohner.

[103] E. A. Coddington and N. Levinson. *Theory of Ordinary Differential Equations*. McGraw-Hill Book Company, Inc., New York, 1955.

[104] W. A. Coppel. *Stability and Asymptotic Behavior of Differential Equations*. D. C. Heath and Company, Boston, 1965.

[105] W. A. Coppel. *Disconjugacy*, volume 220, *Lecture Notes in Mathematics*. Springer-Verlag, Berlin, 1971.

[106] J. M. Davis, J. Henderson, and K. R. Prasad. Upper and lower bounds for the solution of the general matrix Riccati differential equation on a time scale. *J. Comput. Appl. Math.*, 2001. Special Issue on "Dynamic Equations on Time Scales", edited by R. P. Agarwal, M. Bohner, and D. O'Regan. To appear.

[107] J. M. Davis, J. Henderson, K. R. Prasad, and W. Yin. Solvability of a nonlinear second order conjugate eigenvalue problem on a time scale. *Abstr. Appl. Anal.*, 2001. To appear.

[108] J. M. Davis, J. Henderson, and D. T. Reid. Right focal eigenvalue problems on a measure chain. *Math. Sci. Res. Hot-Line*, 3(4):23–32, 1999.

[109] G. Derfel, E. Romanenko, and A. Sharkovsky. Long-time properties of solutions of simplest nonlinear q-difference equations. *J. Differ. Equations Appl.*, 6(5):485–511, 2000.

[110] J. Dieudonné. *Foundations of Modern Analysis*. Prentice Hall, New Jersey, 1966.

[111] R. Donahue. The development of a transform method for use in solving difference equations. Honors thesis, University of Dayton, 1987.

[112] O. Došlý. Principal and nonprincipal solutions of symplectic dynamic systems on time scales. In *Proceedings of the 6th Colloquium on the Qualitative Theory of Differential Equations (Szeged, 1999)*, pp. No. 5, 14 pp. (electronic). Electron. J. Qual. Theory Differ. Equ., Szeged, 2000.

[113] O. Došlý. On some properties of trigonometric matrices. *Čas. Pěst. Mat.*, 112:188–196, 1987.

[114] O. Došlý. On transformations of self-adjoint linear differential systems and their reciprocals. *Ann. Polon. Math.*, 50:223–234, 1990.

[115] O. Došlý. Oscillation criteria and the discreteness of the spectrum of self-adjoint, even order, differential operators. *Proc. Roy. Soc. Edinburgh*, 119A:219–232, 1991.

[116] O. Došlý. Oscillation theory of self-adjoint equations and some its applications. *Tatra Mt. Math. Publ.*, 4:39–48, 1994.

[117] O. Došlý. Transformations of linear Hamiltonian difference systems and some of their applications. *J. Math. Anal. Appl.*, 191:250–265, 1995.

[118] O. Došlý. Factorization of disconjugate higher order Sturm–Liouville difference operators. *Comput. Math. Appl.*, 36(4):227–234, 1998.

[119] O. Došlý and S. Hilger. A necessary and sufficient condition for oscillation of the Sturm–Liouville dynamic equation on time scales. *J. Comput. Appl. Math.*, 2001. Special Issue on "Dynamic Equations on Time Scales", edited by R. P. Agarwal, M. Bohner, and D. O'Regan. To appear.

[120] O. Došlý and R. Hilscher. Linear Hamiltonian difference systems: Transformations, recessive solutions, generalized reciprocity. *Dynam. Systems Appl.*, 8(3-4):401–420, 1999. Special Issue on "Discrete and Continuous Hamiltonian Systems", edited by R. P. Agarwal and M. Bohner.

[121] O. Došlý and R. Hilscher. Disconjugacy, transformations and quadratic functionals for symplectic dynamic systems on time scales. *J. Differ. Equations Appl.*, 2001. To appear.

[122] Z. Drici. Practical stability of large-scale uncertain control systems on time scales. *J. Differ. Equations Appl.*, 2(2):139–159, 1996.

[123] Z. Drici and V. Lakshmikantham. Stability of conditionally invariant sets and controlled uncertain systems on time scales. *Mathematical Problems in Engineering*, 1:1–10, 1995.

[124] M. S. P. Eastham. *The Asymptotic Solution of Linear Differential Systems. Applications of the Levinson Theorem.* Oxford University Press, Oxford, 1989.

[125] S. Elaydi. *An Introduction to Difference Equations.* Undergraduate Texts in Mathematics. Springer-Verlag, New York, 1996.

[126] P. Eloe. The method of quasilinearization and dynamic equations on compact measure chains. *J. Comput. Appl. Math.*, 2001. Special Issue on "Dynamic Equations on Time Scales", edited by R. P. Agarwal, M. Bohner, and D. O'Regan. To appear.

[127] P. W. Eloe and J. Henderson. Analogues of Fekete and Decartes systems of solutions for difference equations. *J. Approx. Theory*, 59:38–52, 1989.

[128] L. Erbe and S. Hilger. Sturmian theory on measure chains. *Differential Equations Dynam. Systems*, 1(3):223–244, 1993.

[129] L. Erbe, R. Mathsen, and A. Peterson. Existence, multiplicity, and nonexistence of positive solutions to a differential equation on a measure chain. *J. Comput. Appl. Math.*, 113(1-2):365–380, 2000.

[130] L. Erbe, R. Mathsen, and A. Peterson. Factoring linear differential operators on measure chains. *J. Inequal. Appl.*, 2001. To appear.

[131] L. Erbe and A. Peterson. Green's functions and comparison theorems for differential equations on measure chains. *Dynam. Contin. Discrete Impuls. Systems*, 6(1):121–137, 1999.

[132] L. Erbe and A. Peterson. Eigenvalue conditions and positive solutions. *J. Differ. Equations Appl.*, 6(2):165–191, 2000. Special Issue on "Fixed point theory with applications in nonlinear analysis", edited by R. P. Agarwal and D. O.'Regan.

[133] L. Erbe and A. Peterson. Positive solutions for a nonlinear differential equation on a measure chain. *Math. Comput. Modelling*, 32(5-6):571–585, 2000. Boundary value problems and related topics.

[134] L. Erbe and A. Peterson. Averaging techniques for self-adjoint matrix equations on a measure chain. 2001. Submitted.

[135] L. Erbe and A. Peterson. Oscillation criteria for second order matrix dynamic equations on a time scale. *J. Comput. Appl. Math.*, 2001. Special Issue on "Dynamic Equations on Time Scales", edited by R. P. Agarwal, M. Bohner, and D. O'Regan. To appear.

[136] L. Erbe and A. Peterson. Riccati equations on a measure chain. *Dynam. Systems Appl.*, 2001. To appear.

[137] L. Erbe and P. Yan. Disconjugacy for linear Hamiltonian difference systems. *J. Math. Anal. Appl.*, 167:355–367, 1992.

[138] L. Erbe and P. Yan. Qualitative properties of Hamiltonian difference systems. *J. Math. Anal. Appl.*, 171:334–345, 1992.

[139] L. Erbe and P. Yan. Oscillation criteria for Hamiltonian matrix difference systems. *Proc. Amer. Math. Soc.*, 119(2):525–533, 1993.

[140] L. Erbe and P. Yan. On the discrete Riccati equation and its applications to discrete Hamiltonian systems. *Rocky Mountain J. Math.*, 25(1):167–178, 1995.

[141] L. Erbe and B. G. Zhang. Oscillation of second order linear difference equations. *Chinese J. Math.*, 16(4):239–252, 1988.

[142] G. Folland. *Real Analysis: Modern Techniques and Their Applications.* John Wiley & Sons, Inc., New York, second edition, 1999.

[143] F. R. Gantmacher. *The Theory of Matrices*, volume 1. Chelsea Publishing Company, New York, 1959. Originally published in Moscow 1954.

[144] I. M. Gelfand and S. V. Fomin. *Calculus of Variations.* Prentice Hall, Inc., Englewood Cliffs, 1963.

[145] G. Sh. Guseinov and B. Kaymakçalan. On a disconjugacy criterion for second order dynamic equations on time scales. *J. Comput. Appl. Math.*, 2001. Special Issue on "Dynamic Equations on Time Scales", edited by R. P. Agarwal, M. Bohner, and D. O'Regan. To appear.

[146] D. Hankerson. Right and left disconjugacy in difference equations. *Rocky Mountain J. Math.*, 20(4):987–995, 1990.

[147] G. H. Hardy, J. E. Littlewood, and G. Pólya. *Inequalities*. Cambridge University Press, Cambridge, 1959.

[148] B. Harris, R. J. Krueger, and W. F. Trench. Trench's canonical form for a disconjugate nth order linear difference equation. *Panamer. Math. J.*, 8:55–71, 1998.

[149] W. A. Harris and D. A. Lutz. On the asymptotic integration of linear differential systems. *J. Math. Anal. Appl.*, 48(1):1–16, 1974.

[150] W. A. Harris and D. A. Lutz. Asymptotic integration of adiabatic oscillators. *J. Math. Anal. Appl.*, 51(1):76–93, 1975.

[151] W. A. Harris and D. A. Lutz. A unified theory of asymptotic integration. *J. Math. Anal. Appl.*, 57(3):571–586, 1977.

[152] P. Hartman. *Ordinary Differential Equations*. John Wiley & Sons, Inc., New York, 1964.

[153] P. Hartman. Difference equations: disconjugacy, principal solutions, Green's functions, complete monotonicity. *Trans. Amer. Math. Soc.*, 246:1–30, 1978.

[154] J. Henderson. Countably many solutions for a boundary value problems on a measure chain. *J. Differ. Equations Appl.*, 6:363–367, 2000.

[155] J. Henderson. Multiple solutions for 2mth order Sturm–Liouville boundary value problems on a measure chain. *J. Differ. Equations Appl.*, 6:417–429, 2000.

[156] J. Henderson and K. R. Prasad. Comparison of eigenvalues for Lidstone boundary value problems on a measure chain. *Comput. Math. Appl.*, 38(11-12):55–62, 1999.

[157] J. Henderson and K. R. Prasad. Existence and uniqueness of solutions of three-point boundary value problems on time scales by solution matching. *Nonlinear Stud.*, 38(11-12):55–62, 1999.

[158] J. Henderson and W. K. C. Yin. Existence of solutions for third order boundary value problems on a time scale. 2001. Submitted.

[159] S. Hilger. *Ein Maßkettenkalkül mit Anwendung auf Zentrumsmannigfaltigkeiten*. PhD thesis, Universität Würzburg, 1988.

[160] S. Hilger. Analysis on measure chains – a unified approach to continuous and discrete calculus. *Results Math.*, 18:18–56, 1990.

[161] S. Hilger. Smoothness of invariant manifolds. *J. Funct. Anal.*, 106:95–129, 1992.

[162] S. Hilger. Generalized theorem of Hartman–Grobman on measure chains. *J. Austral. Math. Soc. Ser. A*, 60(2):157–191, 1996.

[163] S. Hilger. Differential and difference calculus – unified! *Nonlinear Anal.*, 30(5):2683–2694, 1997.

[164] S. Hilger. Special functions, Laplace and Fourier transform on measure chains. *Dynam. Systems Appl.*, 8(3-4):471–488, 1999. Special Issue on "Discrete and Continuous Hamiltonian Systems", edited by R. P. Agarwal and M. Bohner.

[165] S. Hilger. Matrix Lie theory and measure chains. *J. Comput. Appl. Math.*, 2001. Special Issue on "Dynamic Equations on Time Scales", edited by R. P. Agarwal, M. Bohner, and D. O'Regan. To appear.

[166] R. Hilscher. Linear Hamiltonian systems on time scales: Transformations. *Dynam. Systems Appl.*, 8(3-4):489–501, 1999. Special Issue on "Discrete and Continuous Hamiltonian Systems", edited by R. P. Agarwal and M. Bohner.

[167] R. Hilscher. Linear Hamiltonian systems on time scales: Positivity of quadratic functionals. *Math. Comput. Modelling*, 32(5-6):507–527, 2000. Boundary value problems and related topics.

[168] R. Hilscher. On disconjugacy for vector linear Hamiltonian systems on time scales. In *Communications in difference equations (Poznan, 1998)*, pp. 181–188. Gordon and Breach, Amsterdam, 2000.

[169] R. Hilscher. A time scales version of a Wirtinger type inequality and applications. *J. Comput. Appl. Math.*, 2001. Special Issue on "Dynamic Equations on Time Scales", edited by R. P. Agarwal, M. Bohner, and D. O'Regan. To appear.

[170] R. Hilscher. Inhomogeneous quadratic functionals on time scales. *J. Math. Anal. Appl.*, 253(2):473–481, 2001.

[171] R. Hilscher. Positivity of quadratic functionals on time scales: Necessity. *Math. Nachr.*, 2001. To appear.

[172] R. Hilscher. Reid roundabout theorem for symplectic dynamic systems on time scales. *Appl. Math. Optim.*, 2001. To appear.

[173] R. Hilscher and V. Zeidan. Discrete optimal control: the accessory problem and necessary optimality conditions. *J. Math. Anal. Appl.*, 243(2):429–452, 2000.

[174] J. W. Hooker, M. K. Kwong, and W. T. Patula. Oscillatory second order linear difference equations and Riccati equations. *SIAM J. Math. Anal.*, 18(1):54–63, 1987.

[175] J. W. Hooker and W. T. Patula. Riccati type transformations for second-order linear difference equations. *J. Math. Anal. Appl.*, 82:451–462, 1981.

[176] R. A. Horn and C. R. Johnson. *Matrix Analysis.* Cambridge University Press, Cambridge, 1991.

[177] R. A. Horn and C. R. Johnson. *Topics in Matrix Analysis.* Cambridge University Press, Cambridge, 1991.

[178] E. Jury. *Theory and Applications of the z-transform.* John Wiley & Sons, Inc., New York, 1964.

[179] E. R. Kaufmann. Smoothness of solutions of conjugate boundary value problems on a measure chain. *Electron. J. Differential Equations*, pp. No. 54, 10 pp. (electronic), 2000.

[180] B. Kaymakçalan. A survey of dynamic systems on measure chains. *Funct. Differ. Equ.*, 6(1-2):125–135, 1999.

[181] B. Kaymakçalan. Lyapunov stability theory for dynamic systems on time scales. *J. Appl. Math. Stochastic Anal.*, 5(3):275–281, 1992.

[182] B. Kaymakçalan. Existence and comparison results for dynamic systems on time scales. *J. Math. Anal. Appl.*, 172(1):243–255, 1993.

[183] B. Kaymakçalan. Monotone iterative method for dynamic systems on time scales. *Dynam. Systems Appl.*, 2:213–220, 1993.

[184] B. Kaymakçalan. Stability analysis in terms of two measures for dynamic systems on time scales. *J. Appl. Math. Stochastic Anal.*, 6(5):325–344, 1993.

[185] B. Kaymakçalan, V. Lakshmikantham, and S. Sivasundaram. *Dynamic Systems on Measure Chains*, volume 370 of *Mathematics and its Applications.* Kluwer Academic Publishers, Dordrecht, 1996.

[186] B. Kaymakçalan and S. Leela. A survey of dynamic systems on time scales. *Nonlinear Times Digest*, 1:37–60, 1994.

[187] B. Kaymakçalan, S. A. Özgün, and A. Zafer. Gronwall and Bihari type inequalities on time scales. In *Conference Proceedings of the Second International Conference on Difference Equations (Veszprém, 1995)*, pp. 481–490, Amsterdam, 1997. Gordon and Breach.

[188] B. Kaymakçalan, S. A. Özgün, and A. Zafer. Asymptotic behavior of higher-order nonlinear equations on time scales. *Comput. Math. Appl.*, 36(10-12):299–306, 1998. Advances in difference equations, II.

[189] B. Kaymakçalan and L. Rangarajan. Variation of Lyapunov's method for dynamic systems on time scales. *J. Math. Anal. Appl.*, 185(2):356–366, 1994.

[190] S. Keller. *Asymptotisches Verhalten invarianter Faserbündel bei Diskretisierung und Mittelwertbildung im Rahmen der Analysis auf Zeitskalen.* PhD thesis, Universität Augsburg, 1999.

[191] W. G. Kelley and A. C. Peterson. *Difference Equations: An Introduction with Applications.* Academic Press, San Diego, second edition, 2001.

[192] W. Kratz. An oscillation theorem for self-adjoint differential systems and an index result for corresponding Riccati matrix differential equations. *Math. Proc. Cambridge Philos. Soc.*, 118:351–361, 1995.

[193] W. Kratz. *Quadratic Functionals in Variational Analysis and Control Theory*, volume 6 of *Mathematical Topics.* Akademie Verlag, Berlin, 1995.

[194] W. Kratz. An inequality for finite differences via asymptotics for Riccati matrix difference equations. *J. Differ. Equations Appl.*, 4:229–246, 1998.

[195] W. Kratz. Sturm–Liouville difference equations and banded matrices. *Arch. Math. (Brno)*, 36, 2000.

[196] W. Kratz and A. Peyerimhoff. An elementary treatment of the theory of Sturmian eigenvalue problems. *Analysis*, 4:73–85, 1984.

[197] W. Kratz and A. Peyerimhoff. A treatment of Sturm–Liouville eigenvalue problems via Picone's identity. *Analysis*, 5:97–152, 1985.

[198] V. Lakshmikantham and B. Kaymakçalan. Monotone flows and fixed points for dynamic systems on time scales. *Comput. Math. Appl.*, 28(1-3):185–189, 1994.

[199] V. Lakshmikantham and S. Sivasundaram. Stability of moving invariant sets and uncertain dynamic systems on time scales. *Comput. Math. Appl.*, 36(10-12):339–346, 1998. Advances in difference equations, II.

[200] V. Lakshmikantham and D. Trigiante. *Theory of Difference Equations: Numerical Methods and Applications*. Academic Press, New York, 1988.

[201] V. Lakshmikantham and A. S. Vatsala. Hybrid systems on time scales. *J. Comput. Appl. Math.*, 2001. Special Issue on "Dynamic Equations on Time Scales", edited by R. P. Agarwal, M. Bohner, and D. O'Regan. To appear.

[202] V. Laksmikantham. Monotone flows and fixed points for dynamic systems on time scales in a Banach space. *Appl. Anal.*, 56(1-2):175–184, 1995.

[203] V. Laksmikantham, D. D. Bainov, and P. S. Simeonov. *Theory of Impulsive Differential Equations*. World Scientific Publishing Co., Singapore, 1989.

[204] V. Laksmikantham and S. Leela. *Differential and Integral Inequalities, Vol. I*. Academic Press, New York, 1969.

[205] V. Laksmikantham, N. Shahzad, and S. Sivasundaram. Nonlinear variation of parameters formula for dynamical systems on measure chains. *Dynam. Contin. Discrete Impuls. Systems*, 1(2):255–265, 1995.

[206] V. Lakshmikantham and S. Sivasundaram. Stability of moving invariant sets and uncertain dynamic systems on time scales. In *First International Conference on Nonlinear Problems in Aviation and Aerospace (Daytona Beach, FL, 1996)*, pp. 331–340. Embry-Riddle Aeronaut. Univ. Press, Daytona Beach, FL, 1996.

[207] A. Lasota. A discrete boundary value problem. *Ann. Polon. Math.*, 20:183–190, 1968.

[208] B. Lawrence. A variety of differentiability results for a multi-point boundary value problem. *J. Comput. Appl. Math.*, 2001. Special Issue on "Dynamic Equations on Time Scales", edited by R. P. Agarwal, M. Bohner, and D. O'Regan. To appear.

[209] S. Leela. Uncertain dynamic systems on time scales. *Mem. Differential Equations Math. Phys.*, 12:142–148, 1997. International Symposium on Differential Equations and Mathematical Physics (Tbilisi, 1997).

[210] S. Leela and S. Sivasundaram. Dynamic systems on time scales and superlinear convergence of iterative process. In *Inequalities and Applications*, pp. 431–436. World Scientific Publishing Co., River Edge, NJ, 1994.

[211] S. Leela and A. S. Vatsala. Dynamic systems on time scales and generalized quasilinearization. *Nonlinear Stud.*, 3(2):179–186, 1996.

[212] S. Leela and A. S. Vatsala. Generalized quasilinearization for dynamics systems on time scales and superlinear convergence. In G. S. Ladde and M. Sambandham, editors, *Proceedings of Dynamic Systems and Applications (Atlanta, GA, 1995)*, volume 2, pp. 333–339, Atlanta, 1996. Dynamic Publishers. Second International Conference on Dynamic Systems and Applications.

[213] A. Levin. Nonoscillation of solutions of the equation $x^{(n)} + p_1(t)x^{(n-1)} + \cdots + p_n x = 0$. *Uzbek Mat. Zh.*, 24(2):43–96, 1969.

[214] N. Levinson. The asymptotic nature of solutions of linear differential equations. *Duke Math. J.*, 15:111–126, 1948.

[215] A. M. Lyapunov. Problème général de la stabilité du mouvement. *Ann. Fac. Sci. Toulouse Math.*, 9:203–474, 1907.

[216] R. E. Mickens. *Difference Equations: Theory and Applications*. Van Nostrand Reinhold Co., New York, second edition, 1990.

[217] R. E. Mickens. *Nonstandard Finite Difference Models of Differential Equations*. World Scientific Publishing Co., River Edge, NJ, second edition, 1994.

[218] M. Morelli and A. Peterson. A third-order differential equation on a time scale. *Math. Comput. Modelling*, 32(5-6):565–570, 2000. Boundary value problems and related topics.

[219] K. N. Murty and G. V. S. R. Deekshitulu. Nonlinear Lyapunov type matrix system on time scales – existence uniqueness and sensitivity analysis. *Dynam. Systems Appl.*, 7(4):451–460, 1998.

[220] K. N. Murty and Y. S. Rao. Two-point boundary value problems in homogeneous time-scale linear dynamic process. *J. Math. Anal. Appl.*, 184(1):22–34, 1994.

[221] Z. Opial. Sur une inégalité. *Ann. Polon. Math.*, 8:29–32, 1960.

[222] T. Peil and A. Peterson. Criteria for C-disfocality of a self-adjoint vector difference equation. *J. Math. Anal. Appl.*, 179(2):512–524, 1993.

[223] A. Peterson. Boundary value problems for an nth order linear difference equation. *SIAM J. Math. Anal.*, 15(1):124–132, 1984.

[224] A. Peterson. C-disfocality for linear Hamiltonian difference systems. *J. Differential Equations*, 110(1):53–66, 1994.

[225] A. Peterson and J. Ridenhour. Atkinson's superlinear oscillation theorem for matrix difference equations. *SIAM J. Math. Anal.*, 22(3):774–784, 1991.

[226] A. Peterson and J. Ridenhour. Oscillation of second order linear matrix difference equations. *J. Differential Equations*, 89:69–88, 1991.

[227] A. Peterson and J. Ridenhour. A disconjugacy criterion of W.T. Reid for difference equations. *Proc. Amer. Math. Soc.*, 114(2):459–468, 1992.

[228] A. Peterson and J. Ridenhour. A disfocality criterion for an nth order difference equation. In S. Elaydi, J. Graef, G. Ladas, and A. Peterson, editors, *Conference Proceedings of the First International Conference on Difference Equations*, pp. 411–418, San Antonio, 1994. Gordon and Breach.

[229] A. Peterson and J. Ridenhour. The (2, 2)-disconjugacy of a fourth order difference equation. *J. Differ. Equations Appl.*, 1(1):87–93, 1995.

[230] M. Picone. Sulle autosoluzione e sulle formule di maggiorazione per gli integrali delle equazioni differenziali lineari ordinarie autoaggiunte. *Math. Z.*, 28:519–555, 1928.

[231] G. Polya. On the mean-value theorem corresponding to a given linear homogeneous differential equation. *Trans. Amer. Math. Soc.*, 24:312–324, 1922.

[232] J. Popenda. Oscillation and nonoscillation theorems for second-order difference equations. *J. Math. Anal. Appl.*, 123:34–38, 1987.

[233] Z. Pospíšil. Hyperbolic sine and cosine functions on measure chains. *Nonlinear Anal.*, 2001. To appear.

[234] C. Pötzsche. Chain rule and invariance principle on measure chains. *J. Comput. Appl. Math.*, 2001. Special Issue on "Dynamic Equations on Time Scales", edited by R. P. Agarwal, M. Bohner, and D. O'Regan. To appear.

[235] H. Prüfer. Neue Herleitung der Sturm–Liouvilleschen Reihenentwicklung stetiger Funktionen. *Math. Ann.*, 95:499–518, 1926.

[236] W. T. Reid. A Prüfer transformation for differential systems. *Pacific J. Math.*, 8:575–584, 1958.

[237] W. T. Reid. *Ordinary Differential Equations*. John Wiley & Sons, Inc., New York, 1971.

[238] W. T. Reid. *Sturmian Theory for Ordinary Differential Equations*. Springer-Verlag, New York, 1980.

[239] R. J. Sacker and G. R. Sell. A spectral theory for linear differential systems. *J. Differential Equations*, 27:320–358, 1978.

[240] J. Schinas. On the asymptotic equivalence of linear difference equations. *Riv. Math. Univ. Parma*, 4(3):111–118, 1977.

[241] S. Siegmund. A spectral notion for dynamic equations on time scales. *J. Comput. Appl. Math.*, 2001. Special Issue on "Dynamic Equations on Time Scales", edited by R. P. Agarwal, M. Bohner, and D. O'Regan. To appear.

[242] S. Sivasundaram. Method of mixed monotony for dynamic systems on time scales. *Nonlinear Differential Equations Appl.*, 1(1-2):93–102, 1995.

[243] S. Sivasundaram. Convex dependence of solutions of dynamic systems on time scales relative to initial data. *Stability Control Theory Methods Appl.*, 5:329–334, 1997. Advances in Nonlinear Dynamics.

[244] G. W. Stewart. *Introduction to Matrix Computations*. Academic Press, New York, 1973.

[245] C. Sturm. Sur les équations différentielles linéaires du second ordre. *J. Math. Pures Appl.*, 1:106–186, 1836.

[246] W. F. Trench. Canonical forms and principal systems for general disconjugate equations. *Trans. Amer. Math. Soc.*, 189:319–327, 1974.

[247] W. J. Trijtzinsky. Analytic theory of linear q-difference equations. *Acta Math.*, 61:1–38, 1933.

[248] A. S. Vatsala. Strict stability criteria for dynamic systems on time scales. *J. Differ. Equations Appl.*, 3(3-4):267–276, 1998.

[249] P. Řehák. *Half-linear discrete oscillation theory*. PhD thesis, Masaryk University Brno, 2000.

[250] W. Wasow. *Asymptotic Expansions for Ordinary Differential Equations*. John Wiley & Sons, Inc., New York, 1965.

[251] P. J. Y. Wong. Optimal Abel–Gontscharoff interpolation error bounds on measure chains. *J. Comput. Appl. Math.*, 2001. Special Issue on "Dynamic Equations on Time Scales", edited by R. P. Agarwal, M. Bohner, and D. O'Regan. To appear.

[252] B. G. Zhang. Oscillation and asymptotic behavior of second order difference equations. *J. Math. Anal. Appl.*, 173:58–68, 1993.

[253] C. Zhang. Sur la sommabilité des séries entières solutions d'équations aux q-différences, I. *C. R. Acad. Sci. Paris Sér. I Math.*, 327:349–352, 1998.

Index